SAS® System for Statistical Graphics

First Edition

SAS Institute Inc.
SAS Campus Drive
Cary, NC 27513

Editing:	Jennifer M. Ginn, John M. West
Composition:	Gail C. Freeman, Cynthia M. Hopkins, Pamela A. Troutman, David S. Tyree
Proofreading:	Patsy P. Blessis, Heather B. Dees, Josephine P. Pope, Susan E. Willard
Graphic Design:	Creative Services Department

The correct bibliographic citation for this manual is as follows: Michael Friendly, *SAS® System for Statistical Graphics, First Edition,* Cary, NC: SAS Institute Inc., 1991. 697 pp.

SAS® System for Statistical Graphics, First Edition

Contents

Reference Aids

Figures

Tables

Acknowledgments

The preparation of this book evolved from a short course on statistical graphics given through the Statistical Consulting Service of the Institute for Survey Research at York University. I am indebted to a number of friends and colleagues whose efforts have added to this book in many ways. Paul Herzberg provided thoughtful readings and suggestions throughout the development of the manuscript. John Fox suggested the use of the Duncan data, shared some of his unpublished work, and reviewed several chapters. Georges Monette, Dick Goranson, and Hervé Abdi also reviewed portions of the manuscript.

Grateful acknowledgment and appreciation are extended to several people whose ideas and suggestions influenced this book. Keith Muller, University of North Carolina, offered helpful comments on power analysis. Forrest Young, University of North Carolina, provided the notes from his course on exploratory multivariate data analysis and helpful discussion of multivariate techniques. I have a special debt to John Tukey, William Cleveland, and John Hartigan, whose contributions to statistical graphics have influenced me greatly.

I have also benefitted from the ideas and thoughtful advice, by electronic mail, of contributors to the *SAS-L Discussion Group*, including Kevin Broekhoven, Sandra Gardner, Barry Grau, Edward Lyons, Dan Jacobs, Nathan Mamias, Phillip Miller, Peter O'Neill, and Paul Thompson. I am grateful to Terry Allen for sharing the code behind the sunflower plot and for other graphical ideas.

Many people at York University and at SAS Institute have helped make this book a reality—too many to mention all individually. Special thanks go to Marshal Linfoot and Daniel Bloom for assistance with all manner of computing problems; to Dominic Nolan, Mike Daigle, and the Computer Operations staff at York for the care and patience with which they produced my hundreds of plots; to Warren Kuhfeld and Warren Sarle for detailed help with statistical and graphical issues; to Mike Kalt for help with many technical SAS/GRAPH software questions; and to technical reviewers Donna Fulenwider, Marje Martin, Eddie Routten, David Schlotzhauer, and Sandra Schlotzhauer for their many helpful comments. It was a pleasure to work with Mimi Saffer and the Publications staff on this project.

x

About the Author

Michael Friendly received his doctorate in psychology from Princeton University, where he specialized in psychometrics and cognitive psychology. He is associate professor and associate director of the Statistical Consulting Service at York University in Ontario, Canada. Dr. Friendly teaches graduate-level courses in multivariate data analysis and computer methods in the Department of Psychology, where he has been teaching the SAS System for over ten years. His research interests generally apply quantitative and computer methods to problems in cognitive psychology, including the cognitive aspects of extracting information from graphical displays. He is the author of *Advanced Logo: A Language for Learning* (1988).

Using This Book

Purpose

SAS System for Statistical Graphics, First Edition shows how to use the SAS System to produce meaningful, incisive displays of statistical data. The emphasis is on statistical graphics rather than presentation graphics. As such, the book focuses on

☐ data displays that reveal aspects of data not easily captured in numerical summaries or tabular displays

☐ diagnostic displays that help the viewer decide if assumptions of an analysis are met.

While the bulk of the examples use graphs produced with SAS/GRAPH software, the book also includes graphs produced with SAS/QC and SAS/IML software, as well as printer graphics produced with base SAS software. In some cases, the same type of display is produced twice, once with printer graphics and once with SAS/GRAPH software, to emphasize the similarities and differences between both the SAS statements required and the output produced. Often, the data displayed in the graphs is created by SAS/STAT procedures.

Although the SAS System and SAS/GRAPH software provide the basic tools for statistical analysis and graphical data display, many modern methods of statistical graphics (for example, diagnostic plots, enhanced scatterplots, and plots for multivariate data) are not provided directly in SAS procedures. Hence, the primary goals of this book are to explain what kinds of graphical displays are useful for different questions and how these displays can be produced with the SAS System. The book deals with the major types of data and statistical analyses encountered across a wide variety of disciplines. However, space considerations force the exclusion of a number of useful, but highly specialized methods. For example, there is no treatment of methods for time-series data.

The examples used in this book were developed and selected to illustrate graphical displays that are useful in data analysis, research, and teaching, and those examples describe only the features of the SAS System that are important to the application. For example, the GFONT procedure is used in this book, not for its own sake, but to construct specialized plotting symbols for a statistical purpose.

For the most part, this book is applicable to both Version 5 and Version 6 of the SAS System. Therefore, it does not contain features specific to either version, except where they make a substantial difference in accomplishing some task. Also, procedures from the SUGI Supplemental Library are not generally used, because the library may not be available at all sites and it is not available in the current release of Version 6. However, the IDPLOT procedure is discussed in this book because it is important for plotting observations with descriptive labels.

Audience

This book is written basically for two audiences. The primary audience consists of those people who have some basic to intermediate-level experience with the SAS System and with statistics. They want to learn how to use the SAS System to produce graphical displays that aid in the analysis of data. The secondary audience consists of people who have relatively more experience with statistics and with the SAS System. These users want to refine their knowledge and use of the SAS System for statistical graphics.

Neither audience is assumed to be familiar with the special characteristics of Annotate data sets or with the coordination of DATA steps and graphics procedure steps to produce custom graphics. The Annotate facilities are used extensively in this book to incorporate graphical features that are not provided by the procedures themselves. Each application in the book describes the most important SAS/GRAPH features, the Annotate facilities, and the programming considerations used in that application. Nevertheless, not all such features can be explained in a book of this scope.

Both audiences are also not assumed to be familiar with SAS macros. A number of examples in the book use SAS macros to package together the steps required for a particular display. The source code for all the macro programs appears in Appendix 1, "Macro Programs." These programs can be ordered in machine-readable form.

Note: The diskette containing these programs is available to purchasers of the book through the Book Sales Department. (There is a shipping and handling charge for the diskette.) You can contact Book Sales by telephone (919-677-8000) or by fax machine (919-677-8166). When ordering, please specify the diskette size and order number (either 5¼″, order #A56144 or 3½″, order #A56145).

Some of the macro programs may be complex for the audience, so the use of these macros is explained, and several of the macro programs are described in detail. However, many of the details of macro writing are left for the more advanced user to study. Less advanced users can read the usage sections in the text and view the macro programs somewhat like new SAS procedures.

Prerequisites

There are two types of knowledge you must have to understand and use this book: SAS software knowledge and statistical knowledge. The two following sections explain the prerequisites that are required in both of these areas.

SAS Processing Prerequisites

It is assumed that the reader has a basic knowledge of SAS processing, including DATA and PROC steps, using output data sets from one procedure as the input to another, and so on. It is also assumed that the reader has at least some familiarity with SAS/GRAPH software, access to the *SAS/GRAPH User's Guide, Version 5 Edition* or *SAS/GRAPH Software: Reference, Version 6, First Edition, Volume 1* and *Volume 2*, and knowledge of local procedures and conventions for producing SAS/GRAPH output at his or her site.

Statistical Prerequisites

It is assumed that the reader has a sound, basic knowledge of standard, classical inferential statistics, such as would be acquired in an intermediate- or graduate-level statistics course in most disciplines. Basic ideas related to scatterplots, correlation, regression, and analysis of variance are assumed without much explanation, though references to more complete statistical treatments are provided.

However, the book also uses some statistical techniques with which the reader may not be familiar, including methods of exploratory data analysis and techniques for multivariate analysis. Where used, these are briefly explained, but it is *not* intended that the treatment be a complete tutorial on the topic. Rather, the reader interested in more background on these techniques should pursue the references given.

How to Use This Book

This section provides an overview of the information contained in this book and how it is organized.

Organization

SAS System for Statistical Graphics is organized so that the reader can quickly find how to display data of a particular type in relation to a particular type of analysis, task, or question. The book is grouped into five major parts, according to the type of data to be analyzed. The progression is from graphical displays for univariate data to bivariate and regression data, experimental design, and multivariate data. Chapters within each part generally deal with particular methods of analysis, and the sections of a chapter address graphical displays for tasks or questions related to that type of data and analysis.

The parts and chapters of the book are as follows:

Part 1: Introduction

Part 1 provides an introduction to statistical graphics. Topics discussed include why you need to plot data, the various roles of graphics in data analysis, and strategies for what to plot and how to plot it.

Chapter 1, "Introduction to Statistical Graphics"

Part 2: Univariate Displays

Part 2 presents graphical methods for univariate data.

Chapter 2, "Graphical Methods for Univariate Data"

Chapter 3, "Plotting Theoretical and Empirical Distributions"

Part 3: Bivariate Displays

Part 3 describes exploratory methods and graphical techniques for bivariate data. It also describes techniques for plotting data and fitted values for linear, polynomial, and multiple predictor regression models.

Chapter 4, "Scatterplots"

Chapter 5, "Plotting Regression Data"

Part 4.: Graphical Comparisons and Experimental Design Data

Part 4 is concerned with exploratory and confirmatory graphical methods for comparing groups.

Chapter 6, "Comparing Groups"

Chapter 7, "Plotting ANOVA Data"

Part 5: Multivariate Data

Part 5 presents graphical techniques for multivariate data.

Chapter 8, "Displaying Multivariate Data"

Chapter 9, "Multivariate Statistical Methods"

Chapter 10, "Displaying Categorical Data"

Part 6: Appendices

Part 6 presents the three appendices that contain the macro programs, data sets, and color versions of some of the output used in this book.

Appendix 1, "Macro Programs"

Appendix 2, "Data Sets"

Appendix 3, "Color Output"

Conventions

This section covers the typographical conventions this book uses.

Typographical Conventions

This book uses several type styles. The following list summarizes style conventions:

roman	is the basic type style used for most text.
UPPERCASE ROMAN	is used for references in text to SAS language elements.

italic	is used to define terms in text, to show formulas, and to present axis labels and variable values. It is also used for elements from equations when they appear in text.
bold	is used in headings. It is also used to indicate matrices in text and equations.
bold italic	is used for vectors that appear both in equations and text.
`monospace`	is used to show examples of programming code. In most cases, this book uses lowercase type for SAS code. You can enter your own SAS code in lowercase, uppercase, or a mixture of the two. The SAS System changes your variable names to uppercase, but character variable values remain lowercase if you have entered them they way. Enter the case for titles, notes, and footnotes exactly as you want them to appear in your output.

Additional Documentation

The *Publications Catalog*, published twice a year, gives detailed information on the many publications available from SAS Institute. To obtain a free catalog, please send your request to the following address:

> SAS Institute Inc.
> Book Sales Department
> SAS Campus Drive
> Cary, NC 27513
> 919-677-8000

SAS Series in Statistical Applications

SAS System for Statistical Graphics is one in a series of statistical applications guides developed by SAS Institute. Each book in the SAS series in statistical applications covers a well-defined statistical topic by describing and illustrating relevant SAS procedures.

Other books currently available in the series include

☐ *SAS System for Elementary Statistical Analysis* (order #A5619) teaches you how to perform a variety of data analysis tasks and interpret your results. Written in tutorial style, the guide provides the essential information you need, without overwhelming you with extraneous details. This approach makes the book a ready guide for the business user and an excellent tool for teaching fundamental statistical concepts. Topics include comparing two or more groups, simple regression and basic diagnostics, as well as basic DATA steps.

☐ *SAS System for Forecasting Time Series, 1986 Edition* (order #A5629) describes how SAS/ETS software can be used to perform univariate and multivariate time-series analyses. Early chapters introduce linear regression and autoregression using simple models. Later chapters discuss the ARIMA model and its special applications, state space modeling, spectral analysis, and cross-spectral analysis. The SAS procedures ARIMA, SPECTRA, and STATESPACE are featured, with mention of other procedures such as AUTOREG, FORECAST, and X11.

☐ *SAS System for Linear Models, Third Edition* (order #A56140) presents an introduction to analyzing linear models with the SAS System. In addition to the most current programming conventions for both Version 5 and Version 6 of the SAS System, this book contains information about new features and capabilities of several SAS procedures. Most statistical analyses are based on linear models, and most analyses can be performed by three SAS procedures: ANOVA, GLM, and REG. This book is written to make it easier for you to apply these procedures to your data analysis problems.

☐ *SAS System for Regression, Second Edition* (order #A56141) describes SAS procedures for performing regression analyses. Simple regression (a single independent variable) and multiple variable models are discussed as well as polynomial models, log-linear models, nonlinear models, spline functions, and restricted linear models. Features of the AUTOREG, GPLOT, NLIN, PLOT, PRINCOMP, REG, and RSREG procedures are covered.

Part 1
Introduction

Chapter 1   **Introduction to Statistical Graphics**

Chapter **1** Introduction to Statistical Graphics

1.1 Introduction

> The preliminary examination of most data is facilitated by the use of diagrams. Diagrams prove nothing, but bring outstanding features readily to the eye; they are therefore no substitute for such critical tests as may be applied to the data, but are valuable in suggesting such tests, and in explaining the conclusions founded upon them. (Fisher 1925, p. 27)

Statistical graphics are designed for a purpose—usually to display quantitative information visually in a way that communicates some information to the viewer. Like good writing, producing an effective graphical display requires an understanding of purpose—what aspects of the data are to be communicated or emphasized. This chapter discusses different functions that graphical display of statistical information can serve and some psychological principles that should be understood in designing graphical displays that communicate effectively. The final sections give capsule reviews of other writings on standards for graphical presentation and describe some general strategies for choosing how and what to plot.

In order to illustrate these general ideas of graphical data analysis, this chapter uses graphs and data sets (most drawn from later chapters) that provide a visual overview of the book. This chapter does not explain how these graphs are drawn using the SAS System because that is the goal of the remainder of the book. But it does refer you to the sections where these graphs are explained. If you are anxious to get to the "how to do it," by all means skip ahead and come back to the "what to do and why" later.

1.2 Advantages of Plotting Data

Statistical analysis and statistical graphics are both tools for understanding numerical data, but they have complementary aims and functions. Statistical analysis is designed to capsulize numerical information with a few well-chosen summary values. Graphical methods can also summarize, but their main function is to reveal the data—to expose what we might not have been prepared to see.

When we compress a set of data into a few numbers, we hope this will aid our understanding. However, most statistical summaries assume that data are reasonably "weil-behaved." This section shows how we can be misled by messy data if we only look at the standard numerical summaries.

1.2.1 Means and Standard Deviations: Potentially Misleading Measures

Suppose we have data from several groups we want to compare, such as the four sets in Output 1.1. The first thing many people do is to calculate the sample mean, $\bar{x}$, and the sample standard deviation, s, for each group.

Output 1.1
Four Data Sets with the Same Means and Standard Deviations

I	SET1	SET2	SET3	SET4
1	40.50	41.64	35.00	44.50
2	41.50	58.36	37.00	45.00
3	42.50	42.29	42.00	45.50
4	43.50	57.71	53.90	46.00
5	44.50	42.93	53.00	46.50
6	45.50	57.07	50.60	47.00
7	46.50	43.57	50.50	47.50
8	47.50	56.43	53.80	48.00
9	48.50	44.21	52.50	48.50
10	49.50	55.79	53.60	49.00
11	50.50	44.86	50.40	49.50
12	51.50	55.14	52.20	50.00
13	52.50	45.50	52.70	50.50
14	53.50	54.50	52.40	51.00
15	54.50	46.14	52.70	51.50
16	55.50	53.86	51.40	52.00
17	56.50	46.79	53.80	52.50
18	57.50	53.21	52.90	53.00
19	58.50	47.43	56.81	72.71
20	59.50	52.57	42.79	49.79
MEAN	50.00	50.00	50.00	50.00
STD	5.92	5.92	5.92	5.92

The plot of these values in Output 1.2, showing $\bar{x} \pm s$ for each group, appears to indicate that all four sets are identical.*

Output 1.2   *Means and Standard Deviation Plot*

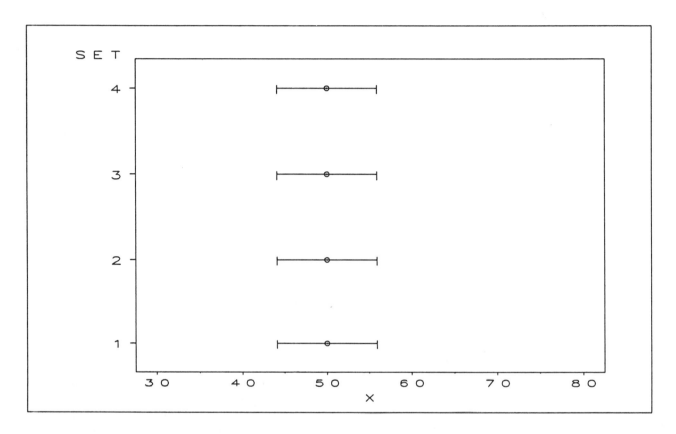

Means and standard deviations are good numerical summaries for "nice" data, but they can be misleading when the data contain outliers or separate clumps of observations.

One way to avoid being misled is to plot the raw data directly rather than numerical summaries of the data. A plot of the raw data values, shown in Output 1.3, indicates clearly that these four sets of numbers have very different distributions.

---

* The idea for this example comes from Cleveland (1985). In a sample of size n observations, $n-2$ can be chosen freely and the remaining 2 can be chosen to make the mean and standard deviation any desired values.

Output 1.3   *Raw Data Plot*

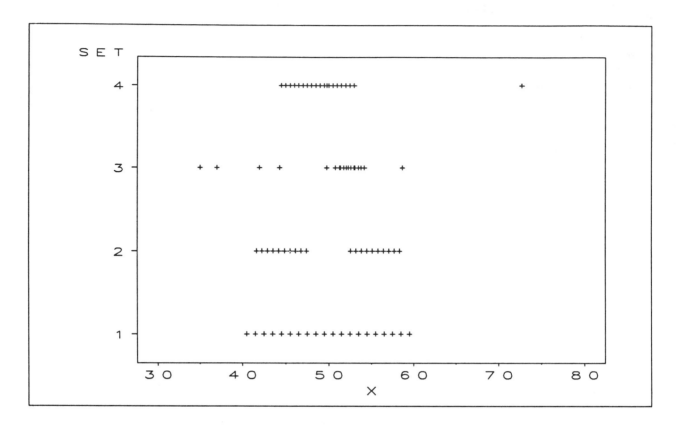

If the data values are distributed approximately symmetrically about the middle, as in Sets 1 and 2, then the sample mean, $\bar{x}=(\Sigma x_i)/n$, is a good representative measure of central tendency. But note that in Set 2, the value $\bar{x}=50$ falls between the two separate clusters of values and so is not really representative of any individual data value.

While the mean and standard deviation have many desirable properties for inferential statistics, they lack the property of *resistance to outliers*. Just a single deviant data value, as in Set 4, can have a substantial effect on both measures, and the further that value is from the rest, the greater is its impact. One simple way to deal with this, of course, is to plot—and look at—the raw data before calculating and plotting the standard numerical summaries.

Another way to avoid being misled is to use robust summary values that are resistant to outliers and other peculiarities. For the case at hand, the sample median and interquartile range are measures of location and spread that are not influenced by even a moderate number of deviant observations.* Output 1.4 shows the same data summarized by the median with a bar extending from the lower quartile to the upper quartile. These measures reflect only the observations in the middle of the distribution, and so are resistant to the influence of wild observations.

---

*   In particular, up to 25 percent of the observations in either tail can be made arbitrarily large without affecting the median and interquartile range.

Output 1.4   *Median and Interquartile Range Display*

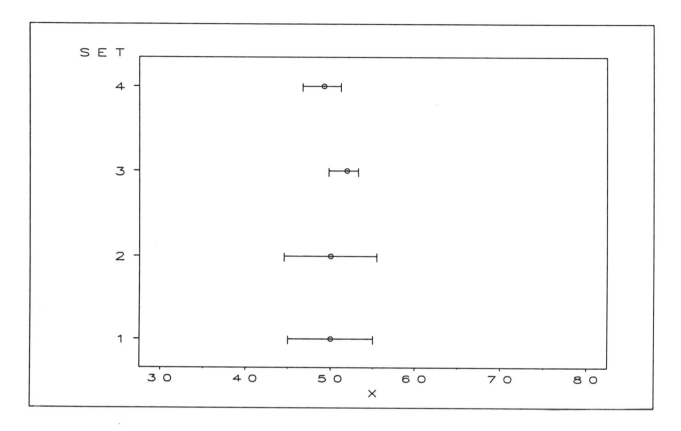

Output 1.4 shows the behavior of the middle 50 percent of the data and does justice to most of the raw data. For example, notice that Sets 3 and 4 have substantially less spread than Sets 1 and 2, and that the typical values in Set 3 are higher than those in the other sets of data.

The median and interquartile range display is not influenced by wild data values, but it does not show their presence either. If we want to see a summary of the data, but indicate potential outliers as well, the boxplot display shown in Output 1.5 provides a reasonable compromise between summary and exposure. The middle of the data, which is likely to be well-behaved, is summarized by the median and interquartile range. The rectangular box shows the interquartile range with the location of the median and mean marked by the line across the box and the star respectively. Observations in the extremes are plotted individually if they are outside a central interval defined in terms of the resistant summary values. (Boxplots are described in detail in Section 2.4, "Boxplots," and Section 6.3, "Comparative Boxplots.")

Output 1.5
Boxplot Display

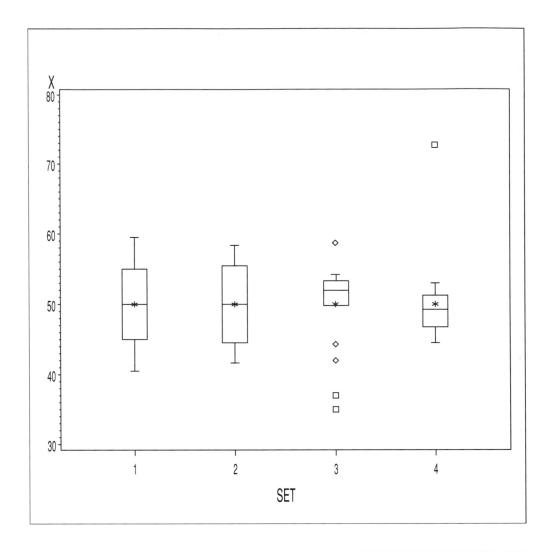

1.2.2 Anscombe's Data Set

With more complex data, the temptation to bypass visual exploration is often greater. Unfortunately, the possibilities for the data to depart from the assumptions of the standard statistical methods increase as the data become more complex. In regression analysis, for example, even with only one predictor variable, we often overlook the importance of plotting the data to determine whether the assumptions of the model are reasonable.

To demonstrate the consequences of a too hasty rush to regression, Anscombe (1973) constructed four sets of data that had identical statistics—means, standard deviations, correlations, and regression coefficients. These data are shown in Output 1.6.*

---

* From "Graphs in Statistical Analysis" by F. J. Anscombe, *The American Statistician*, Volume 27. Copyright © 1973 by the American Statistical Association. Reprinted by permission of the American Statistical Association.

Output 1.6
Anscombe's Data Set

```
        Set 1          Set 2          Set 3          Set 4
      ---------      ---------      ---------      ---------
       X    Y         X    Y         X    Y         X    Y
      ---- ----      ---- ----      ---- ----      ---- ----
       4   4.26       4   3.10       4   5.39      19  12.50
       5   5.68       5   4.74       5   5.73       8   6.89
       6   7.24       6   6.13       6   6.08       8   5.25
       7   4.82       7   7.26       7   6.42       8   7.91
       8   6.95       8   8.14       8   6.77       8   5.76
       9   8.81       9   8.77       9   7.11       8   8.84
      10   8.04      10   9.14      10   7.46       8   6.58
      11   8.33      11   9.26      11   7.81       8   8.47
      12  10.84      12   9.13      12   8.15       8   5.56
      13   7.58      13   8.74      13  12.74       8   7.71
      14   9.96      14   8.10      14   8.84       8   7.04
```

From the raw data in Output 1.6, you can see that the first three data sets share the same X values, but the Y values differ, and the last data set has all X values equal except one. Nevertheless, from just looking at standard regression output, you could be led to believe that all four sets are the same. For example, the standard summary statistics from a regression analysis for each set of data are shown in Table 1.1.

Table 1.1
Summary Statistics for Anscombe's Data

		Summary Statistics			
		Set 1	Set 2	Set 3	Set 4
Mean X		9.0	9.0	9.0	9.0
	Y	7.5	7.5	7.5	7.5
Std	X	3.317	3.317	3.317	3.317
	Y	2.030	2.030	2.030	2.030
r(X,Y)		.816	.816	.816	.816
Intercept		3.0	3.0	3.0	3.0
Slope		0.5	0.5	0.5	0.5

These four sets of data are plotted in Output 1.7. All four have the same linear regression line, but just a glance at these plots shows how widely the data sets differ. In fact, data set 1 is the only one for which the simple linear regression summary statistics really represent the data. In data set 2, the correlation between X and Y would be perfect if it were not for the one discrepant point. In data set 3, the relation is clearly nonlinear, but a linear regression analysis fits a line with the same slope and intercept as in the other data sets. In data set 4, there is no relation between X and Y, but the single data point at $X=19$ is enough to produce a correlation of 0.816.

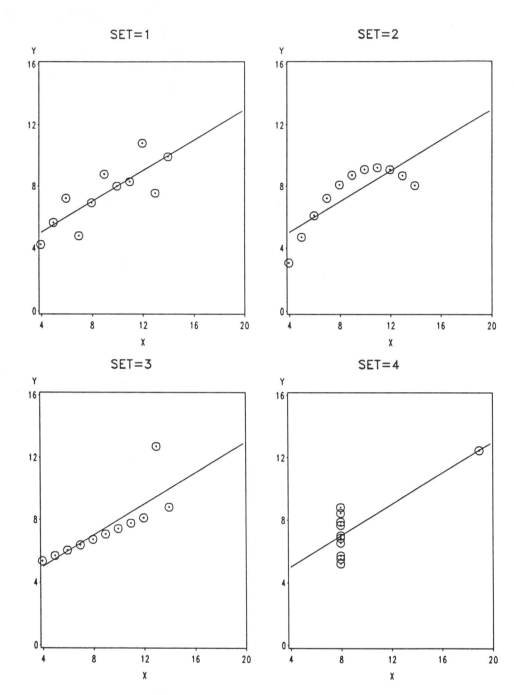

As in the previous example, classical statistical summaries are optimal when the data are well-behaved, but they are not resistant to the effects of unusual observations. As in a political caucus, the most extreme observations can have the most impact in defining the result.

This example, like the one in Table 1.1, is contrived. We would not expect to find such extreme collections as data set 4 in practice. Yet, cases such as data set 2 and 3 may be more common than we suspect, and the picture quickly reveals what is hidden in the numerical summary. Indeed, Tukey (1977) claims that "the greatest value of a picture is when it forces us to notice what we never expected to see" (p. vi).

In this case, too, there are other possibilities between plotting the raw data and looking only at the standard, least squares regression results. For example,

Section 4.4, "Interpolated Curves and Smoothings," describes several methods for smoothing scatterplots, including a robust method called lowess smoothing that is illustrated in Output 1.8.

Output 1.8
Anscombe's Data with Smooth Curves Fit by the Lowess Technique

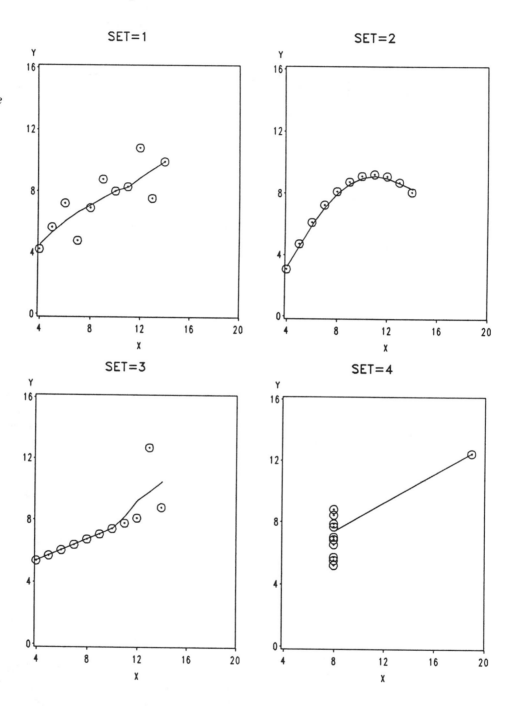

Lowess uses a form of robust, weighted regression where discrepant observations receive small or zero weights and does a reasonably good job of fitting the first three sets of data. (Data set 4, however, is so pathological that even lowess fits a line with a nonzero slope.) There are also many graphical methods for diagnosing the defects built into data sets 2—4 and for detecting unduly influential observations, some of which are presented in Section 5.4.4, "Leverage and Influential Observations."

The moral of these cautionary tales is not that classical statistical methods are ill-founded or fraught with danger. Rather, it is that visual exploration—plotting your data—must go hand in hand with numerical summary and statistical inference.

1.3 The Roles of Graphics in Data Analysis

Graphing data is a form of communication, much like writing. Like good writing, producing an effective graphical display requires an understanding of its purpose—what aspects of the data are to be communicated to the viewer. In writing we communicate most effectively when we know our audience and tailor the message appropriately. So, too, we may construct a graph in different ways to use ourselves, to present at a conference or meeting of our colleagues, or to publish in a research report. At least four different goals can be identified for graphical display:

□ reconnaissance

□ exploration

□ model building, testing, and confirmation

□ communication and persuasion.

The first three are most relevant to data analysis and serve to help us study and understand a body of data; graphs designed for these purposes are emphasized in this book.

Some of the graphical displays one might use for these four purposes are illustrated in this section with data on major league baseball players in the 1986 season. The data set BASEBALL is described and listed in Section A2.3, "The BASEBALL Data Set: Baseball Data." It contains 19 measures of hitting and fielding performance for 322 regular and substitute hitters in the 1986 year, their career performance statistics, and their salaries at the start of the 1987 season. The major goal of the analysis is to predict a player's salary from the performance data. The graphical methods used are described here only briefly. References are provided to those sections where the methods are discussed in detail.

1.3.1 Reconnaissance

Reconnaissance refers to a preliminary examination of the terrain we are about to explore, which is particularly relevant to large data sets. For small or medium-sized data sets, you can usually just plunge right in. But large data sets present special problems, and it is useful to understand what makes them difficult.

The difficulty in understanding data tends to grow in proportion to the size of the data set, some function of the number of observations and number of variables. Whereas a set of 30 observations can be readily analyzed by hand, a set of 300 observations requires the data analysis capabilities of a data analysis system such as the SAS System. However, larger data sets—3,000 or 30,000 observations—require sophisticated data management facilities of the SAS System as well. With large data sets, simple graphical displays such as ordinary (unenhanced) scatterplots may quickly lose effectiveness, as it becomes more difficult to spot interesting patterns or unusual values.

The divisions between small, medium, and large data sets are arbitrary, but numbers such as 30, 300, 3,000, and more suggest that the mental load tends to increase roughly as a logarithmic function of the number of observations.

However, problems of comprehension increase at a much faster rate in relation to the number of variables. Conceptually, we tend to divide statistical techniques on a one . . . two . . . many scale, corresponding to univariate, bivariate, and multivariate problems, and the organization of this book reflects this division.

Example: Baseball Data

By these criteria, the BASEBALL data set is of medium size in terms of observations, but it has a large number of variables. A good deal of preliminary work is reflected in the way the BASEBALL data set was set up as a SAS data set. This work included several steps:

naming

>   Descriptive names and labels were specified for the variables. Some variables in the BASEBALL data set are statistics for just 1986 (such as HITS); others are the player's career statistics for the same performance measure (HITSC).

creating value formats

>   Some of the classification variables in the data set, such as a player's TEAM, LEAGUE, and POSITION, are represented in coded form in the data set. The FORMAT procedure was used to create formats for printing and recoding these variables.

screening data

>   The UNIVARIATE procedure was used to screen each of the performance variables for anomalies—missing data or obvious errors or both. Salary data turned out to be missing for 59 players, and the distribution of salary appeared highly skewed.

Graphical methods can be particularly effective for dealing with such data sets because a well-designed display can portray far more quantitative information in a comprehensible form than can be shown by other means. Output 1.9 is an example of a technique for multivariate data called a scatterplot matrix. (A color version of this output appears as Output A3.1 in Appendix 3, "Color Output.") It shows 20 scatterplots of all pairs of the variables BATAVGC (batting average), RUNS, HITS, YEARS (in the major leagues), and SALARY, with players in the American and National leagues distinguished by the color and shape of the plotting symbol. Thus, the plot shows over 10,000 numbers (20 plots × 322 points × 2 coordinates) with a visual classification of each point.

Output 1.9 *Scatterplot Matrix for Baseball Data (American League: Black +; National League: Red ✕)*

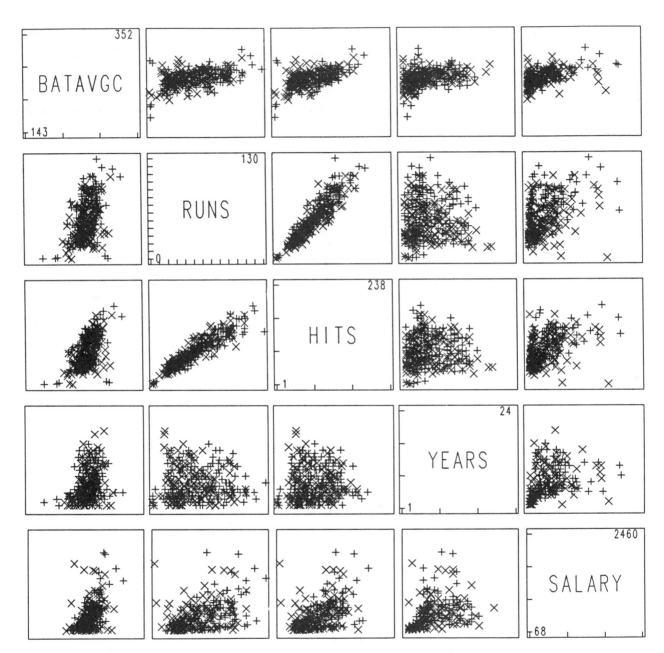

Note that to make the scatterplot matrix effective, many of the usual components of a graph have been omitted. There are no tick marks, value labels, or axis labels on the individual plots. Instead, the variable names and scale information (minimum and maximum value) have been placed in the diagonal cells of the matrix, where they identify the row and column variable. This reduction of some visual information, to help us see more than we could otherwise comprehend in a graphical display, is a simple example of a principle of visual thinning discussed in Section 1.6, "Strategies for Graphical Data Analysis."

Plotting all pairs of variables in a set can provide an effective overview of the relationships among them. In each row of Output 1.9, the same variable is plotted on the vertical axis; in each column, the same variable appears on the horizontal axis. In particular, the last column and the bottom row of the display show plots of salary against each of the first four variables. (The construction of the scatterplot matrix is discussed in Section 8.3, "Draftsman's Display and Scatterplot Matrix.") We can see that SALARY has a small positive relation to each of the other four variables; however, most of the points in the last column of the display are squeezed to the left, reflecting the skewed distribution of salary. There is also a very strong relation between the variables RUNS and HITS, but neither of these variables appears to be strongly related to years in the major leagues.

This reconnaissance has

□ led to the creation of a SAS data set with meaningful variable names and labels

□ provided an overview of the data

□ strongly suggested the need to transform salary.

1.3.2 Exploration

Exploratory graphs are used to

□ help detect patterns or unusual observations in the data

□ diagnose data properties or problems

□ suggest data transformations, substantive hypotheses, statistical analyses, and models.

These techniques generally do not require assumptions about the properties of the data and are designed to be resistant to the effects of unusual data values.

For detecting patterns in data, the boxplot shown in Output 1.5 is an example of an exploratory graphical method for one or more collections of univariate data. The lowess technique, which fits a smooth, robust curve to a scatterplot, is an exploratory method for bivariate data described in Section 4.4.2, "Lowess Smoothing." Glyph plots, star plots, and the scatterplot matrix are examples of exploratory methods for plotting more than two variables. These and other techniques for detecting patterns in multivariate data are discussed in Chapter 8, "Displaying Multivariate Data."

Diagnostic displays are designed to focus on a specific property of data, such as symmetry, equality of variance, or additivity (lack of interaction). These displays generally plot some measures calculated from the data rather than plotting the data directly. Often they are designed so the calculated quantities plot as a horizontal line when the data has that property. One example is given below for the baseball data.

Example: Baseball Data

For the baseball data, diagnostic plots can be used to help select a transformation of salary to make the distribution of salary more symmetric, or, in regression, to help equalize residual variance. Output 1.10 shows a *symmetry transformation plot* (discussed in Section 3.6, "Plots for Assessing Symmetry") for salary. The variables plotted are chosen so that a symmetric distribution would plot as a horizontal line with an intercept of 0 (the dashed line in Output 1.10).

Output 1.10
Symmetry
Transformation
Plot for Salary

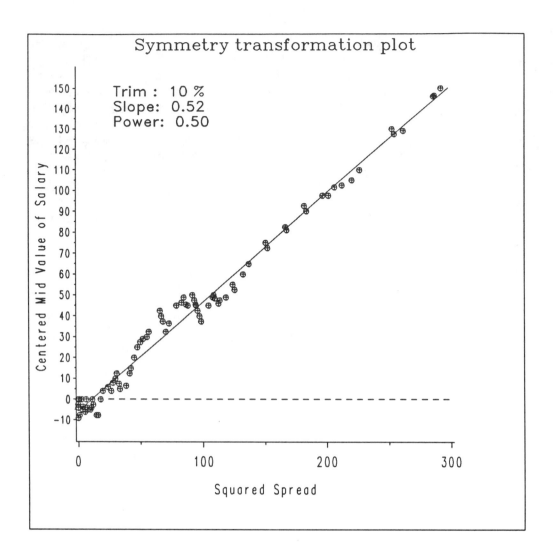

When the points plot as a line with a nonzero slope, b, then the transformation $y \rightarrow y^{1-b}$ is the suggested power transformation to make the distribution symmetric. The power .50 corresponds to $\sqrt{\text{SALARY}}$. (Diagnostic plots for equal variance, described in Section 6.4, "Diagnostic Plot for Equalizing Variability," suggest that log(SALARY) is actually a more useful transformation for this data.)

1.3.3 Model Building, Testing, and Confirmation

In linear models for analysis of variance and regression, model building is concerned with selecting the predictor variables to include in the final model and determining the form those variables should take. For the second task, plots of the response variable against the predictors can help determine if a linear relationship is suitable or suggest a transformation if the relationship appears nonlinear.

Including too many predictors in the model (over-fitting) typically leads to substantial shrinkage (the model will not validate well in future samples); omitting important predictors leads to bias. A statistic called C_P (described in Section 5.7, "Plots of C_P against P for Variable Selection") measures both badness of fit and bias in selecting the variables to be included in the model. Plotting C_P against the number of fitted parameters, P, helps assess the trade-off between goodness of fit and bias.

Statistical inference allows a limited sample of data to be used to draw precise conclusions—with a known risk of being wrong—about the population from which the data arose. These inferences depend, however, on the validity of the assumptions of the inference model.

Graphical methods help to verify whether the data meet these assumptions and can suggest corrective action if they do not. For example, most parametric models assume that residuals from the model are normally distributed. Normal probability plots, described in Section 3.5, "Quantile Plots," show how well this assumption is met. When the data clearly depart from normality, diagnostic plots such as Output 1.10 can be used to select a transformation that makes the data at least symmetric.

In some statistical methods, graphical techniques for detecting violations of the assumptions or other data problems are highly developed. For regression data, for example, the methods discussed in Chapter 5, "Plotting Regression Data," can help to avoid the problems illustrated in Anscombe's data.

Example: Baseball Data

Assuming we have decided to work with log(SALARY), plots of this variable against the predictors show whether these variables are linearly related to log(SALARY). Output 1.11 shows the relation between log(SALARY) and number of years in the major leagues, with a smooth curve fit by the lowess technique. The scatterplot has also been enhanced with a boxplot at the bottom of the figure to show the distribution of the YEARS variable (see Section 4.5, "Enhanced Scatterplots").

Output 1.11
Log(SALARY)
Plotted against
Years in the Major
Leagues

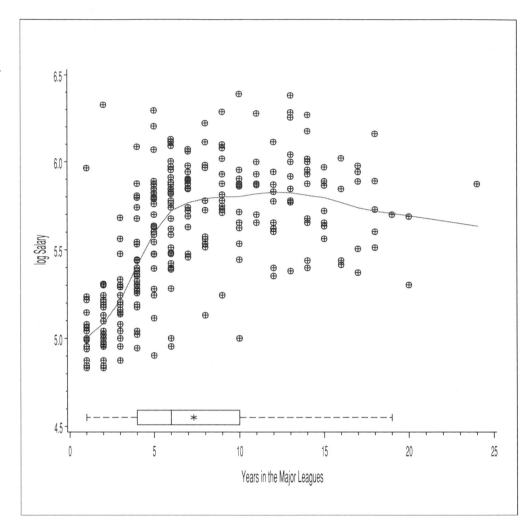

The lowess curve that shows log(SALARY) increases roughly linearly up to about six years (the median), then levels off or declines slightly. In building a regression model to predict salary, this information suggests redefining the YEARS variable as "early experience," which can be defined as min(YEARS, 6).

The C_P plot for model selection is illustrated in Output 1.12. (A color version of this output appears as Output A3.2 in Appendix 3.) The plot compares a measure of badness of fit and bias for all possible subsets of predictors. Each point in the plot shows the value of the C_P statistic for one model using a subset of the variables YEARS (up to six), RUNS, HITS, BATAVG (batting average), and fielding (measured by PUTOUTS plus ASSISTS) to predict log(SALARY). Each point is labeled with the first letter of the predictors in the model, so the point labeled YHR represents the model using YEARS, HITS, and RUNS.

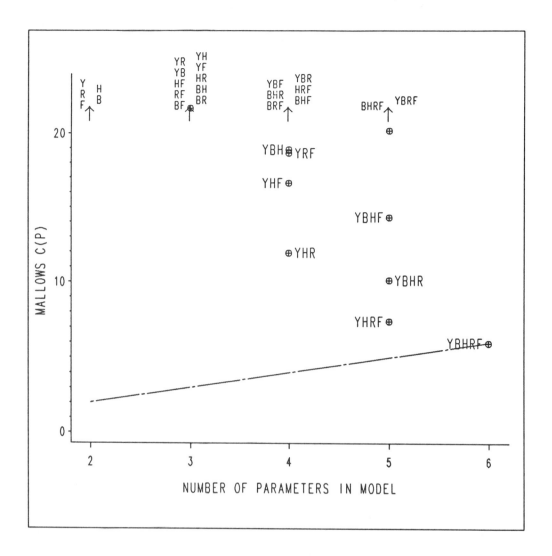

In this plot "good" models have values of C_P near the dashed line; "poor" models have large values of C_P and are shown off-scale, so they do not spoil the resolution of the smaller values. From this plot, it appears that only two possible selections have reasonable C_P value: the model YHRF uses YEARS, HITS, RUNS, and fielding; the model YBHRF uses all of the predictors. (By construction, the point for the full model always falls on the line.) While there are other criteria for selecting a regression model, a plot such as Output 1.12 helps to narrow the possibilities, sometimes greatly.

1.3.4 Communication and Persuasion

The final step in data analysis is usually to communicate what we have found to others. In this step, we can construct graphical displays of the data to serve different objectives. In some cases, the purpose may be to provide a visual summary or a detailed record as a substitute for a table. More often, however, the purpose is to attract the viewer's attention or to highlight the findings by presenting a conclusion in strong visual fashion.

As a result, presentation graphs tend to be simpler in form and content than graphs for analysis. Not incidentally, graphs designed to persuade or attract attention are also more highly decorated or embellished. This public purpose of graphical display is important: if a graph is not examined, it might as well not have been drawn (Lewandowski and Spence 1989). Although the primary focus of this book is on graphs for data analysis, it is hoped that the techniques described contribute to more effective communication graphics as well.

Example: Baseball Data

A variety of communicative graphs for the baseball data could be made for different purposes or audiences. One idea, shown in Output 1.13, is to summarize a series of linear regression models with a graph showing the increase in the multiple R^2 and the decrease in residual variance as additional predictors are added to the model.

The leftmost boxplot shows the distribution of residuals from a model in which the player's team is the only predictor. In the second model, effects for player's position are added as well. The last three models add successively years in the major leagues (up to six years), batting performance, and fielding performance. Values of R^2 are shown numerically at the top of the figure; residual standard deviation is shown visually by the size of each box. Players with large residuals are identified for the model including team, position, and years, and for the final model. (There is a longer story that could be told about this figure. It turns out that some of the large residuals in the middle boxplot correspond to errors in the original data set. The large residuals in the final model are special cases, too.)

Output 1.13
Summarizing a Series of Regression Models Predicting Log (SALARY)

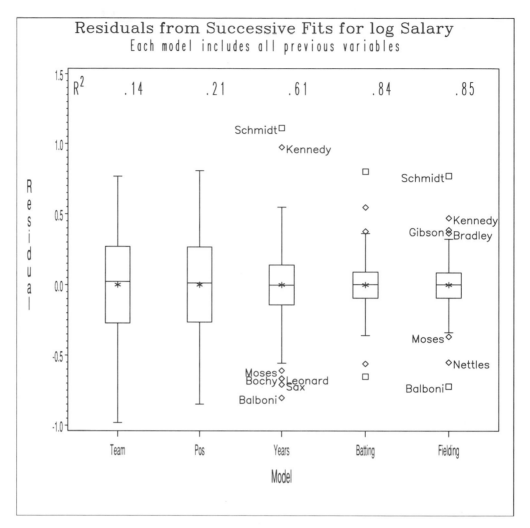

The purpose of Output 1.13 is to show, for an audience that understands boxplots, how the prediction of log(SALARY) improves as additional variables are added to the model. For other purposes or other audiences, this graph may be too complex or require too much explanation to make its point. A simpler display, which plots the R^2 value (labeled Explained Variance) against the last effect included in the model, is shown in Output 1.14.*

---

* This plot may be slightly misleading because MODEL is a discrete variable and the models are shown equally spaced. Other possibilities would be to show the R^2 values as a bar chart or to plot these values against the number of parameters fitted in each model.

Output 1.14
Simplified
Summary Display
for Baseball Salary
Data

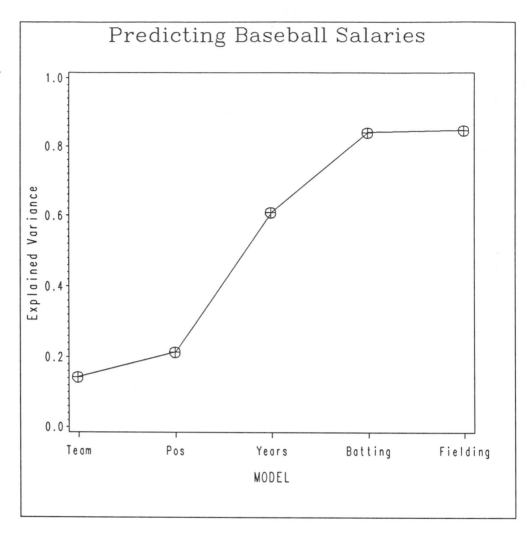

1.4 Psychological Principles of Data Display

Designing good graphics is an art, but one that ought to depend on science. In particular, graphical displays rely on our ability to see and understand patterns in pictures.

In constructing a graph, quantitative and qualitative information is encoded by visual features such as position, size, texture, symbols, and color. This translation is reversed when a person studies a graph—the representation of numerical magnitude and categorical grouping must be extracted from the visual display. This section briefly discusses current theory and research about our cognitive abilities to extract information from graphs and the principles of constructing graphs that enhance or detract from this.

1.4.1 Elementary Perceptual Tasks of Graphical Perception

One way to determine the psychological properties that make for effective graphical displays is to identify the elementary perceptual tasks that are used in decoding information from a graph and to study the relative accuracy with which people can perform these tasks. It should then be possible to understand existing graphical methods in terms of the elementary perceptual processes they depend on and to design new displays based on processes that are easier or more accurate.

This bottom-up approach has been the basis of important work by Cleveland and McGill (1984a, 1985, 1987). They identify the ten elementary visual codes used to portray quantitative information in graphs, shown in Figure 1.1. For instance, in viewing a pie chart, people make judgments of angles, while a divided bar chart requires judgments of length.*

---

* From "Graphical Perception: Theory, Experimentation and Application to the Development of Graphical Methods" by William S. Cleveland and Robert McGill, *Journal of the American Statistical Association*, Volume 79. Copyright © 1984 by the American Statistical Association. Reprinted by permission of the American Statistical Association.

Figure 1.1   *Elementary Codes for Graphical Comparison*

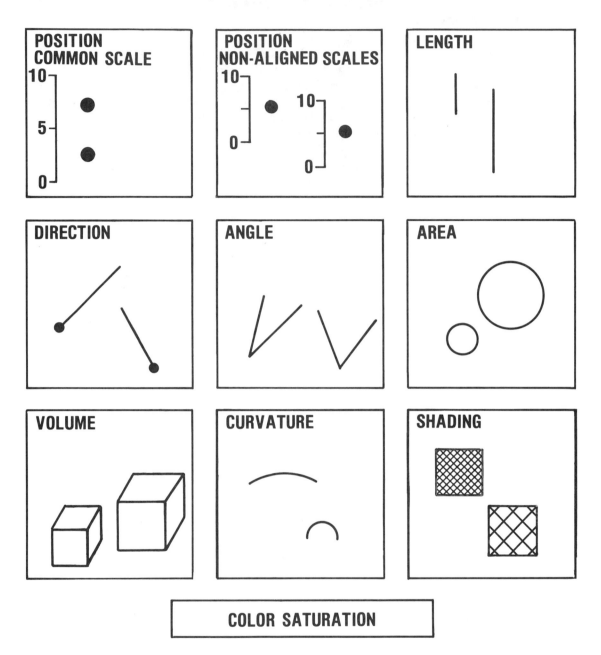

Their experiments typically ask people to compare two magnitudes represented by a particular code, such as position along a common scale, and judge what percent the smaller is of the larger.

Cleveland and McGill find, for example, that these comparative judgments based on position along a common scale are more accurate than judgments based on angle or area. A series of experiments that pit the various elementary codes against one another led Cleveland and McGill (1984a, 1986) to propose that the codes are ordered by accuracy, as shown in Table 1.2. Codes higher in the ordering are judged more accurately when the task is to compare two magnitudes.

Table 1.2
Order of Accuracy
of Elementary
Graphical Tasks

Rank	Graphical Codes
1	Position along a common scale
2	Position along nonaligned scales
3	Length, direction, angle
4	Area
5	Volume
6	Shading, color

Thus, the implication of this ordering is that graphs can be most accurately judged when they use elementary tasks as high in the ordering as possible. The dot chart (described in Section 2.5, "Dot Charts"), for example, is suggested by Cleveland (1984b) as a replacement for the divided bar chart. The dot chart uses position along a common scale to convey the magnitude of subgroups, while the divided bar chart relies on judgments of length, which are lower in the ordering.

Another example is the use of length and angle of lines to represent additional variables in a glyph plot (Section 8.2, "Glyph Plots") rather than the use of size and color of a circle in a bubble plot. The papers by Cleveland and McGill contain many other examples of how graphical methods can be improved by taking account of the accuracy of human perceptual judgments.

Nevertheless, it should be noted that the ordering in Table 1.2 applies to situations where one must mentally estimate the size of one magnitude relative to another. Other types of judgments may yield a different ordering. For example, when people judged what proportion one value is of a total, Simkin and Hastie (1987) reported that judgments based on angles (pie chart) were as accurate as those based on position (simple bar chart), but both angle and position judgments were more accurate than those based on length (divided bar chart).

1.4.2 Cognitive Model of Visual Information Processing

An alternate view looks at graph perception as a form of human information processing, studied by cognitive psychologists. Figure 1.2 depicts schematically the modal model of the contemporary theory of human information processing, which attempts to account for how people interpret and understand both visual and verbal information.

Figure 1.2   *Cognitive Model of Information Processing*

Stages:

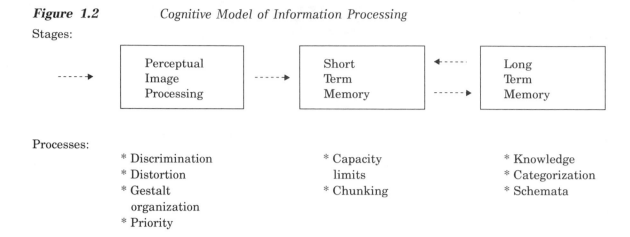

Processes:

* Discrimination	* Capacity	* Knowledge
* Distortion	limits	* Categorization
* Gestalt	* Chunking	* Schemata
organization		
* Priority		

The application of this model of information processing to graphical display was first described by Kosslyn (1985) in a review of five recent influential books on graphical display. This paper is also important reading for anyone interested in the psychological basis of display design.

Briefly, the model represented in Figure 1.2 proposes that human visual information processing can be divided into the three stages shown. Each stage is governed by processes that place constraints on the design of graphical display. Kosslyn (1985) describes in detail how each of the processes listed in Figure 1.2 can affect graphical perception and their implications for display design. The discussion below illustrates how the analysis of one of these processes, discrimination, can be applied to designing more effective graphs.

Discrimination refers to our ability to resolve separate graphic symbols. To be clearly seen, plotting symbols and labels must be a certain minimal size, allowing for possible photo-reduction in printing. To provide this, the programs in this book make liberal use of the H= options to control the height of plotting symbols, labels, and legends. To be clearly differentiated, plotting symbols, lines, and so on must differ by a minimal proportion. This means, for example, that if size of a plotting symbol is to convey quantitative information, a given difference in size will be harder to discriminate between large symbols than between smaller ones.

Discriminability is also an issue in plotting discrete data, where many values can fall at the same plot location. The sunflower plot described in Section 4.6, "Plotting Discrete Data," provides one solution to this problem.

More recently, Kosslyn (1989) extends this model as a way of analyzing the information in a graphical display to reveal whether it meets a set of acceptability principles; violations of these principles reveal the source of difficulties a viewer might have in using the display.

1.5 Graphical Standards for Data Display

Much has been written about the do's and don'ts of making graphs.* Nevertheless, "drawing graphs, like motor-car driving and love-making, is one of those activities which almost every researcher thinks he or she can do well without instruction" (Wainer and Thissen 1981). Unfortunately, naive graph-making can also lead to unsatisfactory results. Wainer (1984), in a tongue-in-cheek review entitled "How to Display Data Badly," illustrates how many presentation graphics drawn from government, corporate, and media reports fail to satisfy three precepts of good data display:

□ Show the data: have something to say.

□ Show the data accurately: use an appropriate visual metaphor to convey the order and magnitude of what is plotted.

□ Show the data clearly: do not obscure what is important.

But graphic blunders are not confined to the popular press. A detailed study by Cleveland (1984a) of the 377 graphs contained in 249 research papers in one volume of the journal *Science* found that 30 percent had one or more errors.

On the positive side, there are a number of good sources for learning how to do graphs effectively. Among the classics, Schmid and Schmid (1979) cover most of the basic design principles, details of scale construction, selection of line and pattern styles, and so forth for an enormous variety of graphs, charts, and mapping techniques. Though a great deal of their material is oriented to manual construction (such as how to shape a pencil point with sandpaper) and to simple presentation graphics (picture charts), their treatment is generally sound and highly readable. An updated version of the book (Schmid 1983) focuses somewhat more on statistical graphics and gives many examples of good and bad graphical practice.

For sheer creativity and scope, Bertin's *Semiology of Graphics* (1983) is unsurpassed. Bertin undertakes the systematic study of graphics as a sign system—the relation between a graphic sign and its meaning—and provides the most comprehensive treatment of the use of visual variables for data display ever written. While this is an important book for the serious student and researcher of graphics, it is apt be be somewhat overwhelming for the novice.

Among recent writers on graphical practice and theory, William Cleveland has been the most influential and productive. *The Elements of Graphing Data* (1985) contains the clearest presentation of principles of graph construction I have read. Cleveland provides for graphical communication what Strunk and White (1979) do for writing: simple rules, illustrated with graphs that break those rules and examples of how the same data can be presented more effectively. Some of these rules expand on the precept to show the data clearly:

□ Make the data stand out.

□ Avoid superfluous graphic elements.

---

* This section is not meant as a summary of the important precepts for displaying data effectively. To do that, it would be necessary to illustrate each idea concretely. Instead, the goal here is to refer you to other sources that treat this topic with the depth and clarity it deserves.

□ Use visually prominent graphical elements to show the data.

□ Do not clutter the data region.

□ Overlap plotting symbols only if they are visually distinguishable.

□ Superpose data sets only when they can be readily discriminated.

□ Preserve visual clarity under reduction and reproduction.

In an earlier book, *Graphical Methods for Data Analysis*, (Chambers et al. 1983), Cleveland and his colleagues at AT&T Bell Laboratories present an excellent, nontechnical description of many of the methods of graphical data analysis covered here. Because they do not have the task of explaining how these methods are implemented in software, their treatment of the statistical ideas and interpretation of data can be more extensive than is possible here.

Each of these books is my favorite in some category: Schmid for the beginner, Bertin for the expert, Cleveland for the best practical advice, and Chambers et al. for statistical method. There is an aesthetic to graphical display as well, and no one has captured this as well as Tufte's *The Visual Display of Quantitative Information* (1983). In addition to providing a sensible account of graphical standards and the visual principles of data display, Tufte's book is a model for the excellence in graphical design he wishes to promote, a joy to look at as well as to read.

1.6 Strategies for Graphical Data Analysis

The ideas and references cited in the previous section relate mainly to the syntax and semantics (meaning) of graphical communication: representing quantitative information accurately and clearly. Beyond this, it is useful to consider the pragmatic level of graphical communication: how well a given graph serves the viewer's needs (Kosslyn 1989).

One pragmatic consideration is to adapt the amount of information shown to the purpose of the display. As with writing, saying too much is as bad as saying too little. Another pragmatic principle is to tune the form and content of the display to what is most important for the viewer to see. These ideas give rise to some general strategies for graphic display that are illustrated in this book.

1.6.1 Match the Form of Display to What Should Be Seen Most Easily

Different forms of data display make it easy or difficult to perform different types of analyses:

□ determining the actual value of a variable

□ judging the difference between two values

□ assessing a trend

□ judging the difference between two trends (an interaction).

You should think, therefore, about exactly what questions should be most readily answered and select a display form that makes that information most accessible to inspection.

Consider the table below, which shows the mean value of a response in relation to two variables, called LEVEL and TYPE:

		LEVEL				
		1	2	3	4	5
TYPE						
	A	20.0	25.0	32.0	41.0	53.0
	B	10.0	22.5	35.0	47.5	60.0

These data are displayed in Output 1.15 as a bar chart in the left panel and as a connected line graph in the right panel. (A color version of this graph appears as Output A3.3 in Appendix 3.)

Output 1.15   *Matching Graphic Form to the Question*

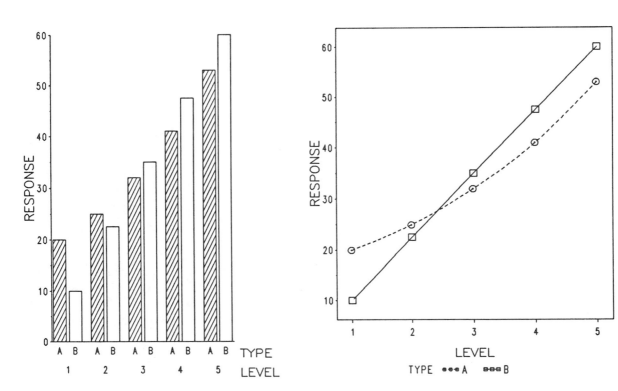

If you want to compare the response to TYPE A and B for particular levels and see which is greater or judge their ratio, this information is more readily extracted from the bar chart because the heights of adjacent bars are easier to compare than points on different curves.

However, it is far easier to see the trend across levels and to see that the variables TYPE and LEVEL interact differently (have different trends) in the line graph than in the bar chart. This is because the connected points on each curve in the line graph have simple descriptions (for example, "straight line" for TYPE B and "positively accelerated" for TYPE A) we can apprehend as one conceptual unit. On the other hand, if you want to extract the exact value of mean response for LEVEL 3, TYPE A, for example, this is done most easily from the table. This example suggests the following guidelines:

□ Use a table when individual actual values should be seen.

□ Use a grouped bar chart when ratios and differences of nearby values are to be seen.

□ Use a line graph when trends and interactions are to be seen.

1.6.2 Make Visual Comparisons Easy

Effective visual comparison depends strongly on how and what we plot. Our eyes are better at making some comparisons than others and can often be misled.

Use Visual Grouping to Connect Values to Be Compared

Once you have selected the basic form of the display, arrange things so that values to be compared most readily are grouped visually. In the grouped bar chart, it is easier to make comparisons among adjacent bars; in a line graph, points on the same curve are usually more easily compared than points on separate curves.

For example, Output 1.16 shows the bar chart from Output 1.15 with TYPE as the midpoint variable together with a bar chart of the same data with LEVEL as the midpoint variable. (A color version of this output appears as Output A3.4 in Appendix 3.) It is easier to compare the response for type A and B for a given level in the right panel of Output 1.16; it is easier to compare levels for a given type in the left panel.

Output 1.16   *Visual Grouping: Comparing Bars in a Grouped Bar Chart*

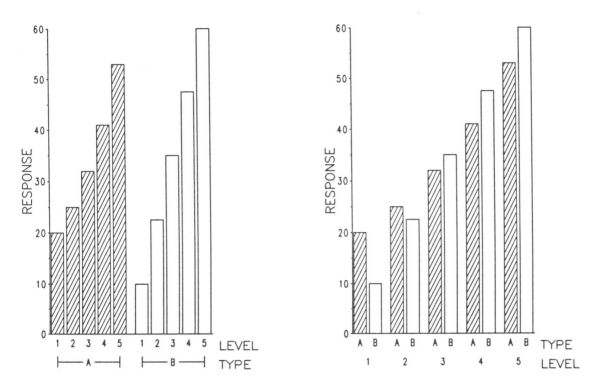

Similarly, Output 1.17 compares the original line graph of this data (a plot of RESPONSE against LEVEL with TYPE as a parameter) with a plot of RESPONSE against TYPE with LEVEL as a parameter. (A color version of this output appears as Output A3.5 in Appendix 3.) In a line graph, it is easier to compare connected points (across the horizontal variable) than points on separate curves (levels of the parameter variable). In the left panel, we see the trends across LEVEL directly; to compare the response across TYPE takes more effort. In the right panel, the comparisons across TYPE are more direct and it is hard to see the trends across LEVEL.

Output 1.17   *Visual Grouping: Comparing Connected Points in Line Graphs*

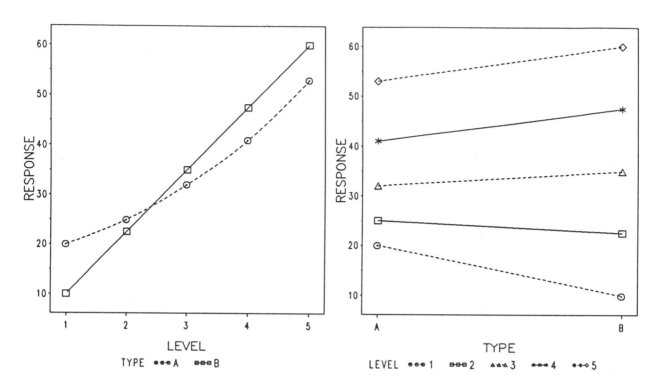

Plot the Difference between Curves Directly

If your goal is to show a difference between curves, you can construct a more effective display by plotting the difference itself. The left panel of Output 1.18 shows two curves (hyperbolas) of the form $y = a + b/x$. In this plot, it appears that the two curves are quite close together for small x values and diverge as x increases. In fact, the curves have a constant difference, Δ, which is plotted against x in the right panel.

Output 1.18   *Visual Comparisons: Seeing the Difference between Curves*

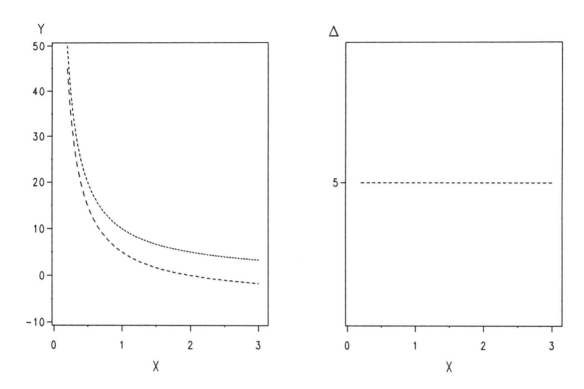

The left panel is simply no good for showing the difference between the curves because the eye-brain system tends to judge the minimum difference, which is perpendicular to the curves, rather than the difference in the vertical direction (Cleveland and McGill 1984a). Try to focus your attention on just the vertical distance between the curves in the left panel. Even with great mental effort, it is hard to overcome the impression that the curves get farther apart going from left to right. The right panel plots the actual difference between the curves, showing that the difference is constant.

Show an Expected Pattern as a Horizontal Line

Similarly, in data analysis many graphs have the goal of showing how actual results—the data—compare with expected results—the fit of some model. A graph showing a histogram of data values together with a smooth curve representing a theoretical distribution (Section 3.4, "Histogram Smoothing and Density Estimation") is one example. Another is the graph of the actual y value against the fitted value, $\hat{y}$, in a regression model.

In these cases, we will usually obtain the most effective graph by arranging things so that either

□ the expected pattern appears as a straight line, or

□ the expected pattern appears as a line whose slope is 0.

These two suggestions are based on the ideas that our eyes are better at perceiving straight lines than curves and better at detecting departure from a flat line than a tilted one (Tukey 1977).

Thus, Output 1.19 shows a histogram of the square root of the total runs scored by players in the baseball data, together with a fitted normal distribution.* The message conveyed by this figure is coarse: the smooth curve fits to some degree, but we cannot see the pattern of deviations between the actual and fitted values.

Output 1.19
Histogram with
Fitted Normal
Distribution

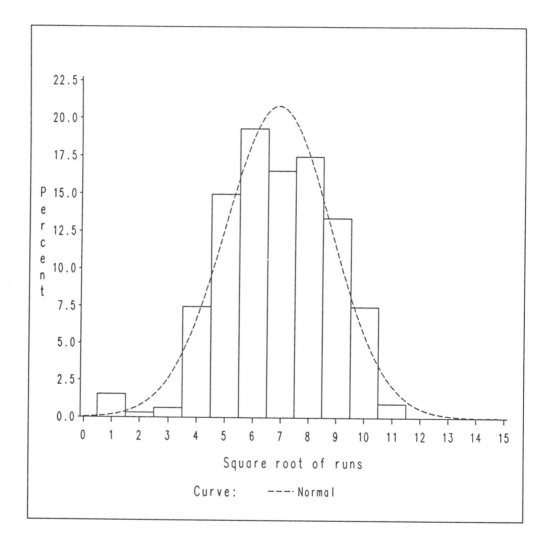

A more effective way to plot these data is the theoretical quantile-quantile plot (see Section 3.5), where quantiles of the data are plotted against the expected (normal) quantiles, so agreement between actual and predicted appears as points along the line $y=x$. The left panel of Output 1.20 shows this plot for the data of Output 1.19, and although the deviations are small it is clear that the points differ systematically from the line at both sides of the graph.

---

* Counted variables, such as number of runs or hits, are often approximately symmetrical on a square root scale.

Output 1.20   *Normal Quantile-Quantile Plots for SQRTRUNS*

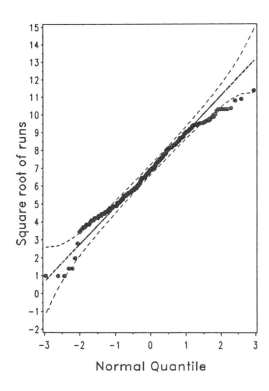
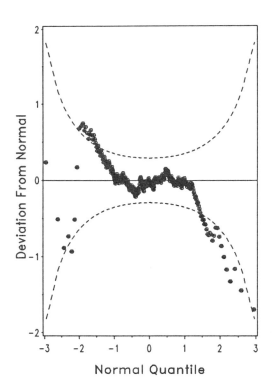

The right panel of Output 1.20 illustrates the second point. Here the ordinate is the difference between the actual points and the expected values, so a perfect fit corresponds to the flat line, $y=0$. Now it is even easier to see the departure of the points from the flat reference line at the left and right sides of the graph.

1.6.3 Remove the Obvious to Look beneath the Surface

Quite often the most notable—and therefore most obvious—features of a set of data have the greatest weight in determining the visual message of a graph. If the purpose of graphing is to see beyond the obvious, we must remove the outer layer and plot what lies beneath. Very often this means

□ transforming data or setting observations aside to remove an undue influence of extreme values

□ fitting a model and plotting the residuals from this model to see what is not explained.

The first point is illustrated in Output 1.21, which shows side-by-side boxplots for infant mortality rates of 101 countries classified by region in the world. Because of the few large values in Asia/Oceania, most of the data are squashed down to the bottom portion of the figure, making visual comparisons difficult.

Output 1.21
Boxplots for Infant
Mortality Rate, by
Region

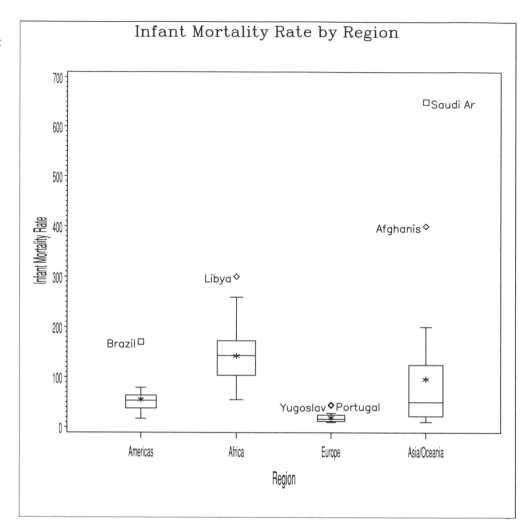

To make more useful plots of these data, try one of these things:

□ Temporarily set aside the few largest values to show the rest more clearly. (Or show them off-scale, as in Output 1.12.)

□ Transform infant mortality rate to pull the outlying values closer to the rest. (Taking logs does nicely; see Output 6.5.)

□ Plot the data for each region on a separate scale.

The second point is illustrated in Output 1.22. The raw data plot in panel (a) shows a nearly perfect linear relation. The plot of residuals from the line in panel (b) shows a systematic and interpretable pattern.

Output 1.22   *Plotting Residuals Shows What Remains to be Explained*

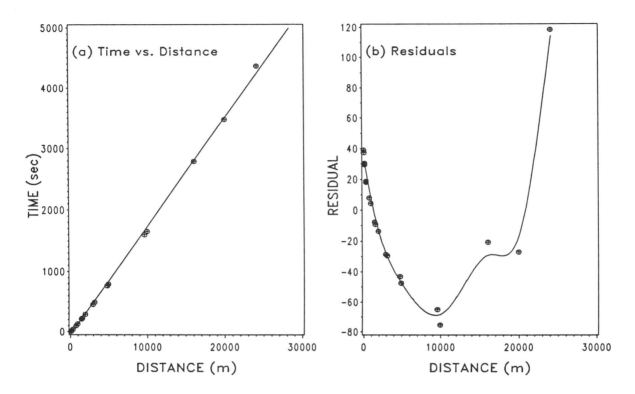

The panel on the left shows the world record times for men's track events in relation to the distance of the event, from the 1973 *Guinness Book of World Records* (McWhirter and McWhirter). Because the races range from the 100 meter race to the 15 mile race, most of the points appear in the bottom left corner.

This panel shows the obvious: the longer the race, the longer it takes to run. The approximately linear relation through the origin indicates that there is close to a proportional relationship between time and distance: doubling the distance increases the record time by a constant multiple. Moreover, the fit is exceptionally good: the linear relation accounts for 99.88% of the variance.

To go beyond this description, we can plot the discrepancies between actual running time and what is predicted by the linear relation. This residual, which we can think of as excess time, is plotted against distance in the right panel of Output 1.22. Now we see something that is news: among the sprints, actual times go from longer than predicted to shorter than predicted, most likely reflecting a "starting block" effect, the time required to reach full speed from a standstill. The rise in the residuals towards the right of this plot can be attributed to runners slowing down as the distance increases.

1.6.4 Enhance Weak Visual Patterns

If the goal is to show the pattern in the data, but the visual effect is weak, you can help the viewer by overlaying the plot with a smoothed version of the pattern.

For bivariate data, the simplest example of this technique is plotting a regression line or smooth curve on top of the data. Consider Output 1.23, which shows data from the 1969 U.S. draft lottery, used to determine the order in which eligible men would be drafted by assigning a priority value at random to each birth date. These data are discussed in Section 4.4.2, "Lowess Smoothing."

For the present, note that there does not appear to be any relation between priority value and month: the values appear to be randomly distributed as intended.

The same data are plotted in Output 1.24, this time with a smoothed curve drawn through the points. The smoothed curve decreases over the last few months of the year, indicating a defect in the method of random assignment. With the aid of the smoothed curve, we can now see evidence of a greater prevalence of high values in the early months and low values in the later months that was masked in Output 1.23. In other words, overlaying the plot with a smoothed curve reveals a tendency for the values to decrease over months of the year. Techniques for smoothing bivariate data are described in Section 4.4, "Interpolated Curves and Smoothings."

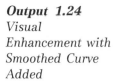

Output 1.24
Visual
Enhancement with
Smoothed Curve
Added

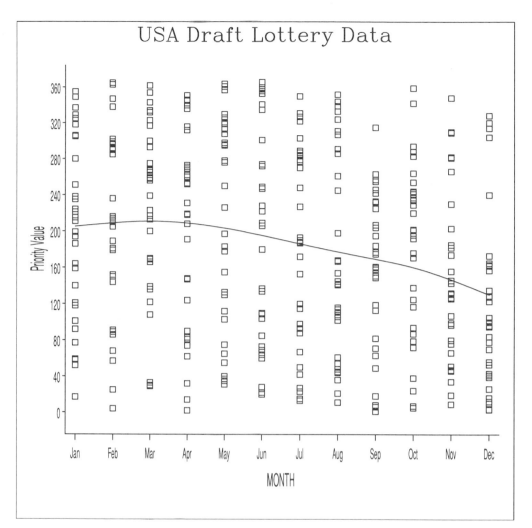

1.6.5 Reduce Extraneous Visual Elements

To deal with large amounts of data visually, it also helps to remove graphic elements that do not contribute to decoding graphic patterns. Tufte (1983) presents numerous examples of graphic components whose only purpose is decorative and measures the quality of a graph by its data-ink ratio, defined as the ratio of the amount of ink used to portray the data to the total amount of ink in the whole display. Unnecessary grid lines, redundant data-ink, and excessive use of patterns, for example, all lower the data-ink ratio.

A more general idea, which I call *visual thinning*, is to design graphs to allocate visual attention in a nonuniform way so as to focus attention on aspects of the data in relation to their worthiness for examination. The boxplot is an example of this principle applied to display of a single variable. For observations in the middle of the distribution, only the center and spread are portrayed collectively by the box; the whisker lines show the range of most of the data. Outside observations, which may need to be examined or set aside, are displayed individually, with more dramatic plotting symbols for points that are far outside.

Output 1.25 illustrates visual thinning applied to bivariate data. In this plot, the open squares represent the median draft priority value for each month; the smoothed curve in the middle traces the median value over months of the year. The stars and the two other curves plot the upper and lower quartiles of the distributions of priority values, but the individual data values are not plotted. (This figure is discussed more fully in Section 4.4.2.)

Output 1.25
Visual Thinning

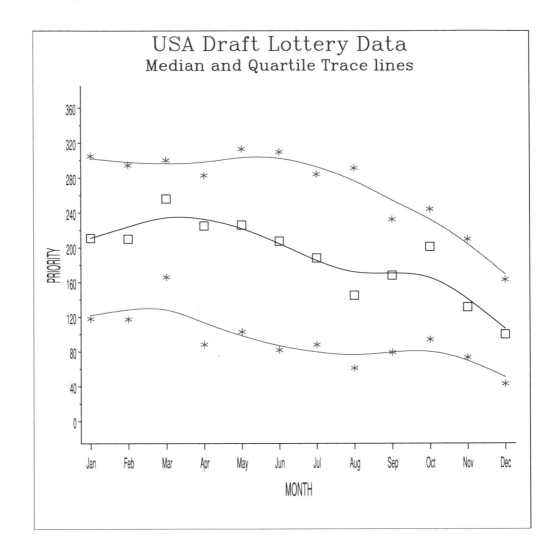

Now we can see something else: not only does the priority value decrease towards the end of the year, but the quartile curves tend to get closer together, indicating that the values become less variable as well. (To examine this apparent change in variability, we should also plot the difference between these quartile curves over time.) And the median values in March, August, and October tend to deviate somewhat from the pattern of gradual decline characteristic of the rest of the months.

1.6.6 Be Prepared to Iterate and Polish

Graphing data, like data analysis itself, is usually an iterative process. You should not expect one plot to show all there is to be seen. Just as one analysis can suggest further questions or hypotheses to test, one plot will often reveal some feature of the data that can be sharpened by plotting the data differently.

Sometimes the indications in one plot suggest plotting the same data, but in a different way. If the observations in a bivariate display are classified in several ways, for example, by age and sex, you may want to make one plot using age to determine the plotting symbol. If this plot shows clusters of points grouped according to age, another plot could be done using sex or both age and sex to determine the plotting symbol. The data ellipse, summarizing the relations for each group, is another useful technique. See Section 4.5.

For multivariate data, which are the most difficult to visualize, this may mean trying out several different display techniques such as those described in Chapter 8. For example, star plots and profiles are designed to show configurational properties of all the scores for each observation. These plots may be more useful for spotting outliers or for suggesting clustering. A scatterplot matrix or glyph plot for the same data may be more useful for showing the pattern of intercorrelations among the variables.

In other cases, an initial plot may suggest plotting new data or expressing the old data in a different way. For example, the plot of the track data in Output 1.22 is unsatisfactory in several respects:

□ Most of the points fall in the bottom left corner.

□ A linear relation between time and distance implies that all races are run at the same rate of speed, which is patently false.

Output 1.26 and Output 1.27 show several subsequent graphs that followed an examination of Output 1.22.

Output 1.26 displays an analysis of log(TIME) plotted against log(DISTANCE). These plots attempt to overcome the bunching of points near the origin in Output 1.22 by plotting time and distance on a log scale. The linear relation in logs has a slope of 1.13, and the regression results have a slightly better fit than the raw data ($R^2 = .9992$).* But the main improvement is that the distance values are more evenly spaced on a log scale than on the original scale. Moreover, the residuals from this fit in panel (b) now suggest a clearer separation between the short races (up to 220 yards) and the longer races.

---

* Another indication that the analysis of time and distance should be done with logs is that a linear relation between log(TIME) and log(DISTANCE) with a slope of 1.1 is also found for time-distance records in walking, swimming, speed skating, and horse racing (Harris 1972).

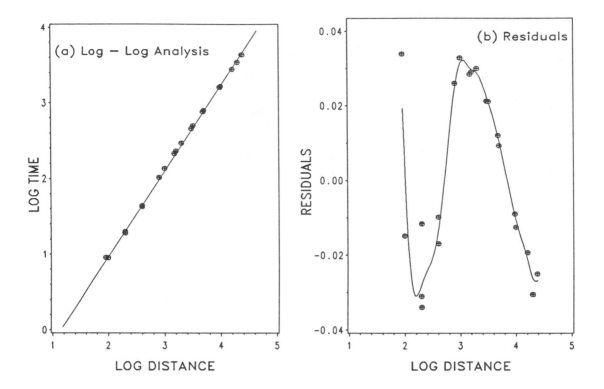

Output 1.27 shows plots designed to examine rate of speed directly. Here, average velocity=(distance/time) is plotted in relation to distance of the race. The first plot made, of velocity against (raw) distance, was unrevealing, because most of the points were clustered at the small distance values. So in the next iteration, velocity was plotted against log(DISTANCE), giving panel (a) of Output 1.27. The smooth curve suggests an increase in speed over the shortest races, followed by an approximately linear decrease in speed over middle distance races and a shallower decrease in speed over the longest distances. Panel (b) shows the residuals from a linear fit—the dashed line in panel (a)—to help us study this pattern in a more detailed way.

Output 1.27   *Iteration: Plotting a Different Measure*

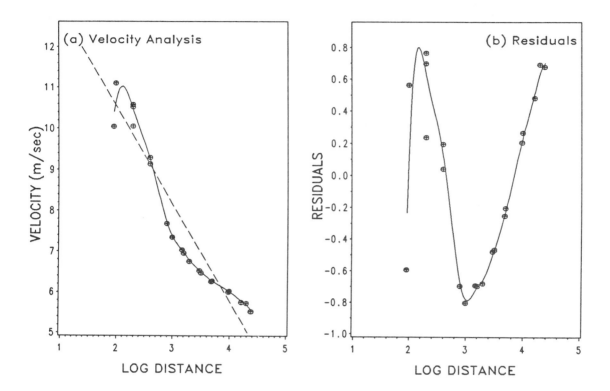

Once we have determined what and how to plot, there is another level of iteration: just as writing often goes through numerous revisions to produce a final draft, we should expect to draw a graph several times before it looks right. The early drafts are not shown here, but figures such as Outputs 1.26 and 1.27 might be drawn five or more times to get the final version.

Each of the figures shown here started with the simplest plot specifications. Choosing the plotting symbols, fonts, axis labels, and so forth to make the graphs more aesthetically pleasing took another pass or two. A final check that a graph would remain legible when reduced in size (carried out with a reducing photocopy machine) often led to additional polishing.

This might seem like a great deal of work. However, one of the virtues of SAS/GRAPH software is that even the simplest plot specifications, accepting all the defaults for symbols, labels, and fonts, produce a graph that can be looked at and used. You can decide if additional polishing is worthwhile or abandon that plot and try something different.

1.7 Data Sets

As in this chapter, the rest of the book uses a variety of data sets, with different characteristics, to illustrate methods of statistical graphics. A number of these data sets appear in several chapters where different features of the data or different variables are discussed. These data sets are described in Appendix 2, "Data Sets," which lists the data in the form of the DATA steps used to create them as SAS data sets.

When you analyze a given data set in several different SAS programs, it is wasteful to duplicate the data in each SAS file. Instead, all of the programs in this book assume that the data sets in Appendix 2 are stored as separate SAS files. The baseball data, for example, are contained in BASEBALL SAS.* Then, to analyze the data, you can use the %INCLUDE statement to have the DATA step and data inserted into the current program when it is executed by the SAS System. So, to analyze the data in BASEBALL SAS, start a new SAS program with the line**

```
%include BASEBALL;
```

Not only does this save space, but if you want to change the data set in any way (for example, to add a variable), there is only one copy of the data file to change. In addition, in the programs in the book, the statement

```
%include BASEBALL;
```

serves as a reference to the data set description in Appendix 2.

---

*   The form of filenames varies across operating systems. On VMS, MVS, and PC/DOS systems, for example, the file would be named BASEBALL.SAS. Consult the *SAS Companion* for your operating system.

**   On the MVS system, the %INCLUDE statement must be of the form
```
%include fileref(BASEBALL) ;
```
where *fileref* identifies a partitioned data set containing the baseball data as a member. On other operating systems, you may need to use a FILENAME statement to associate the name BASEBALL with the name of the external SAS file.

Part 2
Univariate Displays

Chapter 2 Graphical Methods for Univariate Data

2.1 Introduction

This chapter discusses methods for plotting the simplest type of data—a single set of data on one quantitative variable. This variable may be something directly observed or measured, or it may reflect the results of statistical analysis, such as residuals from a fitted model. For a first look at a set of numbers, we are usually interested in three properties:

☐ typical or central value

☐ spread or variability

☐ shape—symmetry or skewness.

Basic displays that show these properties, including histograms and stem and leaf plots, are described and illustrated in this chapter. The schematic plot or boxplot is also discussed as a means to focus attention selectively on these three important features of a set of data. The boxplot also highlights potential outliers—values that are widely separated from the rest of the data—and the behavior of the data in the tails. Another novel display, the dot chart, is useful when each data value has a name or label associated with it, and we want to display the names along with the data values.

2.2   Histograms and Bar Charts

Histograms and bar charts are often the first graphical techniques that come to mind for portraying the distribution of a set of scores. In the *histogram*, we partition the range of the data values into intervals of equal length, count the number of observations that fall into each interval, and plot these frequencies as the lengths of rectangles centered at the midpoint of each interval. The term *bar chart* is more general; the lengths of rectangles plotted in a bar chart can represent the percent of scores in the interval, cumulative frequency or percent, or even the sum or mean of another variable.

Bar charts of several types can be produced in the SAS System using the CHART procedure or the GCHART procedure in SAS/GRAPH software for high-resolution displays. The CAPABILITY procedure in SAS/QC software can also produce histograms, as well as other graphical displays of a frequency distribution. This section explains how to produce vertical bar charts of frequencies with PROC GCHART and PROC CAPABILITY. Bar charts of means are also described here. The use of PROC CAPABILITY to plot a histogram together with a fitted theoretical curve is discussed in Section 3.4, "Histogram Smoothing and Density Estimation."

2.2.1   Frequency Bar Charts

To produce a vertical frequency bar chart with PROC GCHART, use these statements:

```
PROC GCHART DATA=SASdataset;
   VBAR variable / options;
   other statements
```

where *variable* is the variable in the data set whose values are displayed on the midpoint axis (horizontal). By default, the response variable (vertical) in the VBAR chart is the frequency of scores for each value on the midpoint axis.

PROC CHART and PROC GCHART can also produce other kinds of charts, including horizontal bar charts (HBAR), pie charts (PIE), star charts (STAR) and block charts (BLOCK). See the *SAS Procedures Guide, Version 6, Third Edition* or *SAS/GRAPH Software: Reference, Version 6, First Edition, Volume 1* and *Volume 2* for more information. While PROC CHART and PROC GCHART are quite similar, the ability to annotate with PROC GCHART offers greater flexibility for statistical graphics, so the examples here use PROC GCHART.

Example: Baseball Data

Output 2.1 shows a bar chart of the variable HITS, the number of hits scored by each of 322 major league players in the 1986 baseball season from the BASEBALL data set (see Section A2.3, "The BASEBALL Data Set: Baseball Data"). The horizontal axis in the chart is labeled with the midpoints of the intervals into which the chart variable, HITS, has been divided. The chart shows that there were 62 players with hits in the interval from 37.5 to 62.5, whose midpoint is 50. Output 2.1 is produced with no options or other specifications using these statements:

```
%include BASEBALL;
proc gchart data=baseball;
   vbar hits;
```

Output 2.1
Bar Chart of HITS

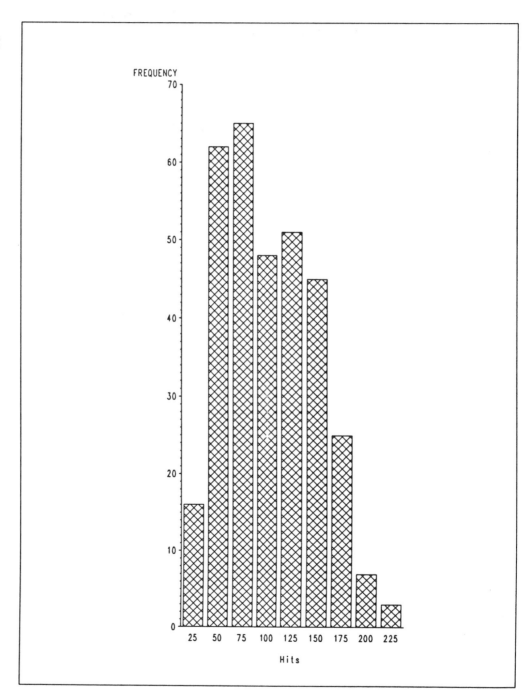

By default, PROC GCHART automatically calculates the midpoint values and number of intervals, fills the bars with a cross-hatched pattern, and separates the bars by a small space. The default midpoints and pattern may not be the best for data analysis display. You can change all these features of the bar chart display (and many others not discussed here) with options in the VBAR statement and with PATTERN and AXIS statements in the PROC GCHART step. For example, Output 2.2 is a bar chart of the same data produced by the following statements:

```
    /* specifying midpoints, pattern and axes */
proc gchart data=baseball  ;
   pattern v=empty  c=black;
   vbar hits  / midpoints=0 to 240 by 20
                raxis=axis1 maxis=axis2 space=0 width=5
                name='GB0202' ;
   axis1 label=(a=90 h=1.5) value=(h=1.2);
   axis2 label=(h=1.5) value=(h=1.2);
```

Output 2.2
Bar Chart of HITS,
Specifying
MIDPOINTS,
PATTERN, and
AXIS

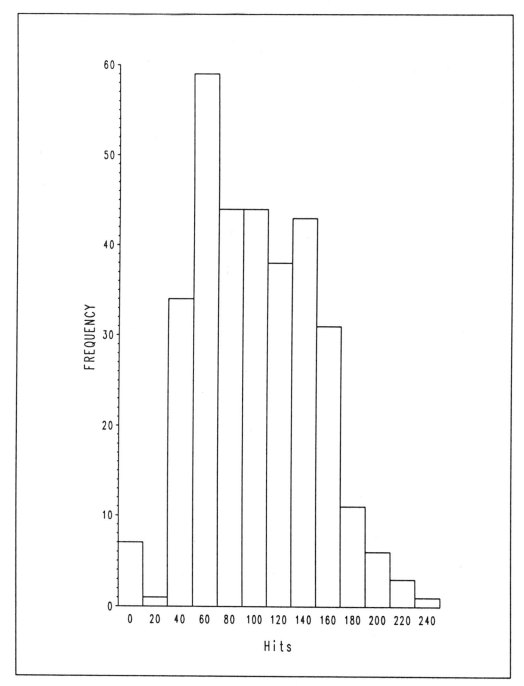

While the cross-hatched, separated bars may be preferred for presentation purposes, some people find the cross-hatched pattern produces a distracting, vibrating moiré effect that Tufte (1983) calls "chartjunk." For data analysis, many people prefer the empty bars specified by V=EMPTY in the PATTERN statement.

Because the variable HITS is continuous, it is more appropriate to join the bars, specifying SPACE=0 in the VBAR statement.*

The difference in midpoints used in these two displays highlights a limitation of the bar chart for showing the shape of the data. In Output 2.2, with intervals of width 20, the frequencies drop from the first interval to the second before rising sharply in the interval labeled 40. This gap, which reflects players who had very few times at bat, is not apparent in Output 2.1.

In general, the shape of the data displayed in the bar chart depends on the number and location of the intervals. It is usually better to choose these values with the MIDPOINT= option or to choose the number of intervals with the LEVELS= option, rather than let PROC GCHART choose for you. While the choice is somewhat arbitrary, smaller intervals show more detail, but larger intervals are generally smoother. Various rules of thumb for choosing the number of intervals have been suggested. Emerson and Hoaglin (1983) compare a number of these and recommend using the formula

$$\text{LEVELS} \leq \text{integer part of} \begin{cases} 2\sqrt{n}, & n < 100 \\ 10\log_{10}n, & n \geq 100 \end{cases} \tag{2.1}$$

for an upper bound on the number of intervals, where n is the number of observations in the data. However, in Version 6 of the SAS System, the CHART and GCHART procedures calculate the default number of intervals according to the results of Terrell and Scott (1985), who suggest the lower bound:

$$\text{LEVELS} \geq \text{integer part of } (2n)^{1/3} \quad . \tag{2.2}$$

For the baseball data, with $n=322$, these two rules suggest specifying $8 \leq \text{LEVELS} \leq 25$ in the VBAR statement. The default bar chart in Output 2.1 uses 9 bins, which for these data is slightly too coarse. But the upper bound of 25 (which is about double the number shown in Output 2.2) appears too ragged (see Output 2.3). Some graphical methods for smoothing histograms and choosing a reasonable interval width are described in Section 3.4.

PROC CAPABILITY Histograms

The previous examples all use intervals of equal width on the horizontal midpoint axis. If the midpoints of intervals are not equally spaced, the heights of the bars in a PROC GCHART display will not reflect the same data density (number of observations per horizontal data unit) across the scale. PROC CAPABILITY, however, scales the heights of bars properly whether the intervals are equally or unequally spaced. PROC CAPABILITY also uses the Terrell and Scott (1985) formula for choosing the number of intervals to display by default.

The example below uses the HISTOGRAM statement to request a PROC CAPABILITY histogram for the HITS variable in the baseball data. For the sake of comparison, this example uses the MIDPOINTS= option to specify 25 intervals in the display. The output from this program is shown in Output 2.3.

---

* When you specify SPACE=0, you may also need to use the WIDTH= option to allow enough space for printing the values on the midpoint axis. Otherwise, the axis values are rotated and printed vertically.

```
%include BASEBALL;              /* On MVS, use %include ddname(file);*/
title ' ';
proc capability data=baseball graphics  ;
   var hits;
   histogram hits /
       midpoints = 0 to 240 by 10
       vaxis=axis1 haxis=axis2
       vscale=count                              /* V6 */
       noframe name='GB0203';
   axis1 label=(h=1.5 a=90) value=(h=1.2);
   axis2 label=(h=1.5)      value=(h=1.2)
        order=(0 to 240 by 20);
```

Output 2.3
CAPABILITY
Procedure
Histogram of HITS

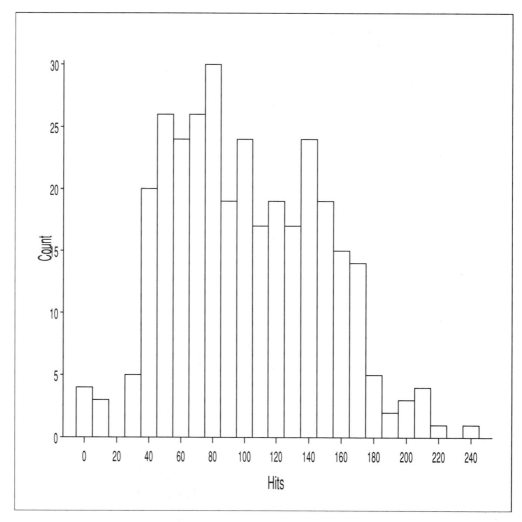

The vertical axis in the PROC CAPABILITY histogram is expressed in percent by default. In Version 6, the VSCALE= option allows the vertical axis to be scaled in terms of frequency (VSCALE=COUNT) or proportion (VSCALE=PROPORTION) as well.

Using 25 intervals, as suggested by equation 2.1, gives a display that is quite ragged. However, this display, like Output 2.2, does pick up the curious gap in the distribution at HITS=20, which turns out to be due to a small number of substitute batters who played in relatively few games. Comparing Output 2.1, 2.2, and 2.3 serves to illustrate how the appearance of the histogram depends on the number and location of class intervals.

Example: Infant Mortality Data

Output 2.4 shows a bar chart of infant mortality rates (per 1,000 live births) for 101 nations around the world. The data set NATIONS (Leinhardt and Wasserman 1979) is described in Section A2.11, "The NATIONS Data Set: Infant Mortality Data," and will appear again in later chapters in this book. The frequency distribution is highly skewed in the positive direction. From Output 2.4, you can see that more than half of the values are concentrated in the range 0—100, but a few data points trail off to surprisingly large values.

Output 2.4
Bar Chart of Infant
Mortality Rate

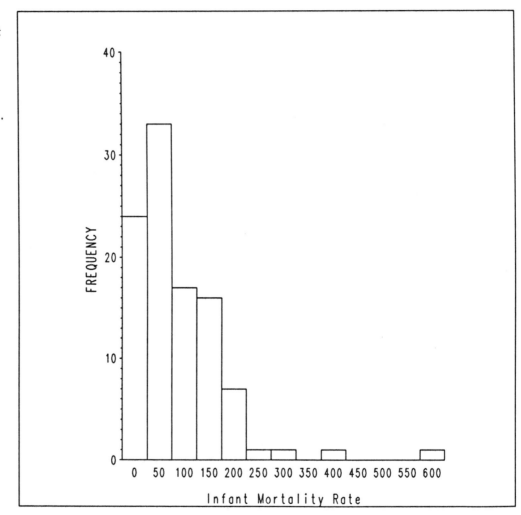

The PROC GCHART step to produce Output 2.4 is very similar to that used for the baseball data.

```
proc gchart data=nations  ;
   vbar imr / midpoints=0 to 600 by 50
              raxis=axis1
              maxis=axis2
              space=0 width=5
              name='GB0204';
   axis1 label=(a=90 h=1.5) value=(h=1.3);
   axis2 label=(h=1.5) value=(h=1.3);
   pattern v=empty color=black;
```

2.2.2 Bar Charts of Means

When the observations in a batch are grouped by some classification variable, you might want to see how the values of a primary variable depend on or change with the classification variable. For example, you might want to know how the number of hits per player varies over teams in the major leagues or how infant mortality rate varies over geographic regions.

One common way to display this information with a bar chart is to use the height of the bar to represent the mean value of the quantitative variable (HITS or IMR), for each level of the classification variable (TEAM or REGION). A bar chart of means is produced with PROC GCHART (and PROC CHART) with the TYPE= and SUMVAR= options using the following statements:

```
PROC GCHART DATA=SASdataset;
   VBAR groupvariable / TYPE=MEAN SUMVAR=variable;
   other statements
```

Note that the VBAR statement specifies the *groupvariable* as the VBAR midpoint variable, and the variable whose means are to be displayed as the *variable* in the SUMVAR= option. For example, Output 2.5 shows a bar chart of the mean number of hits for each of the 26 teams in the major leagues. This chart is produced by the following statements:

```
proc gchart data=baseball;
   vbar team / type=mean sumvar=hits;
```

Output 2.5   *Bar Chart of Mean Hits by Team*

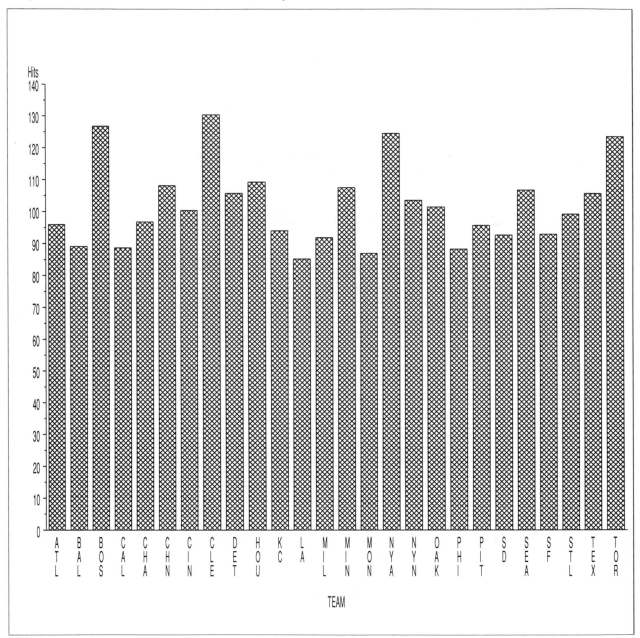

Adding Error Bars to the Bar Chart

The bar chart in Output 2.5 summarizes all of the values in each team by one number—their mean. This assumes that the data values have an approximately symmetric distribution, so that the mean is a reasonably typical value for each group. But it ignores the variability of the scores in each group. If some groups are more variable than others, or there are different numbers of observations in different groups, you can display the standard error of the mean or a confidence interval for the mean with an error bar.

For example, Output 2.6 shows a one-standard error range, $\bar{x} \pm s/\sqrt{n}$, by the colored line centered at the mean hits for each team. (A color version of this output appears as Output A3.6 in Appendix 3, "Color Output.") The length of each error bar is a measure of the amount of variability associated with the mean number of hits for each team and gives an approximate 68% confidence interval for the mean. Because the number of players per team is roughly the same, the different lengths of the bars reflect variability in number of hits among the players on each team.

Output 2.6   *Bar Chart of Mean Hits with Standard Error Bars*

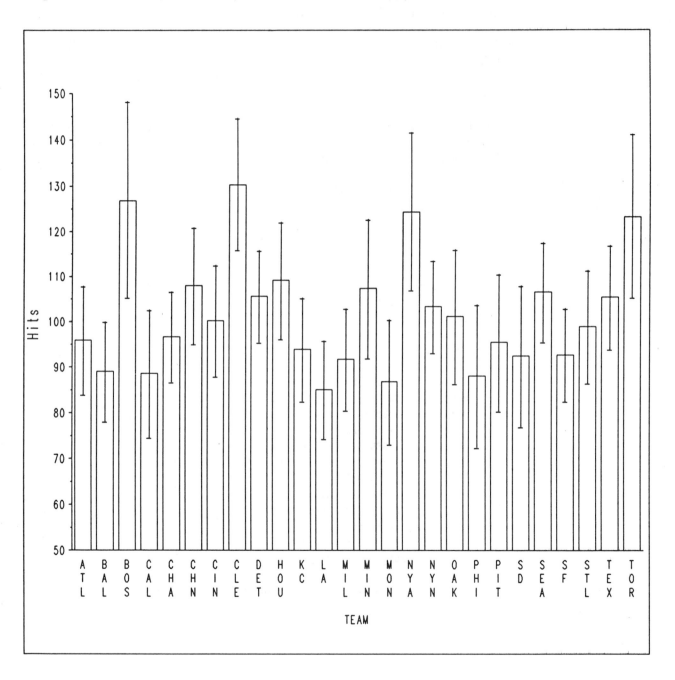

The error bars are drawn with an Annotate data set as shown in the program below. First, the mean and standard errors are found by summarizing the data with a PROC SUMMARY step, producing the data set MEAN that contains the MEAN and STDERR variables. The NWAY option suppresses the output observation for the grand mean. (The MEANS procedure could also be used, but it would require the data set to be sorted; the SUMMARY procedure is more efficient for large data sets.)

Adding annotations to a PROC GCHART bar chart is a little different from adding annotations with other SAS/GRAPH procedures. For PROC GCHART, the midpoint variable in an Annotate data set is referred to by the special Annotate variable MIDPOINT.* This variable refers to the horizontal axis in a VBAR chart and the vertical axis in an HBAR chart. The response axis in an Annotate data set for a VBAR chart is referenced by the Annotate variable Y (or by X in an HBAR chart).

The DATA step that produces the data set ERRBAR below sets MIDPOINT=TEAM to locate the center of the proper bar. The error bar is drawn by manipulating the variable Y, using Annotate MOVE and DRAW commands, and the ends of the bars are drawn with a '—' character.

```
proc summary data=baseball nway;
   class team;
   var hits;
   output out=stats mean=mean stderr=stderr;

data errbar;
   set stats;
   xsys='2'; ysys='2'; color='RED';
   length function $8 ;
   midpoint=team;
   midpnt=team;                    /* Version 5 */
   t = 1;
   y = mean + t * stderr;
   function = 'MOVE ';              output;
   function = 'LABEL';  text='-'; output;
   y = mean - t * stderr;
   function = 'LABEL';  text='-'; output;
   function = 'DRAW';              output;
proc gchart data=baseball  ;
   vbar team / type=mean sumvar=hits
            anno=errbar name='GB0206'
            raxis=axis1;
   axis1 order=(50 to 150 by 10) label=(a=90 r=0 h=1.3);
   pattern v=empty  c=black;
```

To show other confidence intervals, change the statement

```
t = 1;
```

---

*   Prior to Release 5.18 of SAS software, this variable was named MIDPNT. Later releases support both the old and new spellings. See *SAS/GRAPH Software: Reference.*

in the DATA step. For a two-tailed interval with confidence level $1-\alpha$, the standard error multiple can be determined with the TINV function (see Section 3.2.4, "Plotting Percentiles: Student's *t* Distribution"):

```
t = tinv(1-α/ 2, nobs-1 );
```

2.3 Stem and Leaf Displays

The stem and leaf display is a form of histogram that retains the actual data values while showing the shape of their distribution. Tukey (1977) invented the stem and leaf display as a quick means of writing down a set of data values, producing a data display from which summary values can be easily calculated. While it is especially suited to hand work with sets of data that are not large, the stem and leaf plot has become a useful component of standard data summary procedures such as the UNIVARIATE procedure.

Output 2.7 shows a stem and leaf display of the variable ATBAT, the number of times a player was at bat in 1986, from the BASEBALL data set. The plot is produced by PROC UNIVARIATE with the PLOT option:*

```
proc univariate data=baseball plot;
    var atbat;
    id name;
```

Output 2.7 *The UNIVARIATE Procedure Display for Times at Bat*

```
                                    UNIVARIATE

    VARIABLE=ATBAT          Times at bat in 1986


                  MOMENTS                            QUANTILES(DEF=4)

    N                  322   SUM WGTS        322    100% MAX      687    99%    673.78
    MEAN           380.929   SUM          122659     75% Q3    512.25    95%     613.4
    STD DEV        153.405   VARIANCE     23533.1     50% MED    379.5    90%     585.7
    SKEWNESS    -0.0780609   KURTOSIS   -0.889432     25% Q1    254.75    10%     193.3
    USS           54278439   CSS         7554121      0% MIN        16     5%    160.15
    CV            40.2713    STD MEAN    8.54893                           1%     19.23
    T:MEAN=0      44.5586    PROB>|T|     0.0001     RANGE         671
    SGN RANK      26001.5    PROB>|S|     0.0001     Q3-Q1       257.5
    NUM ¬= 0          322                            MODE          209
```

(continued on next page)

---

*   PROC UNIVARIATE decides whether to display a stem and leaf plot or a histogram based on the maximum number of observations in any class interval. If this value is no more than 48, a stem and leaf display is produced; otherwise, a histogram is produced. For the HITS variable, PROC UNIVARIATE produces a histogram because there are 51 observations in the most frequent class interval.

```
(continued from previous page)

                                 EXTREMES

                       LOWEST     ID         HIGHEST    ID
                          16(Tony Arm)          642(Tony Gwy)
                          19(Terry Ke)          663(Joe Cart)
                          19(Cliff Jo)          677(Don Matt)
                          20(Mike Sch)          680(Kirby Pu)
                          22(Bob Boon)          687(Tony Fer)

    STEM LEAF                                      #            BOXPLOT
       6 6889                                      4               |
       6 000011122233333444                       18               |
       5 555555666666777777788888888889999999999  39               |
       5 00000011111111122222223333334444         32            +-----+
       4 556666667777777888899999999              27            |     |
       4 000000011111122222222223334444444444     33            |     |
       3 55556666777788888889999                  22            *--+--*
       3 0000111111111111111122222223333333344444444  39        |     |
       2 5555666667778888888888888899999          31            +-----+
       2 00000000001111111111111222222222223333344444444  43        |
       1 5556666778888888999999                   22               |
       1 33344                                     5                |
       0                                                            |
       0 2222223                                   7                |
       ----+----+----+----+----+----+----+----+---
    MULTIPLY STEM.LEAF BY 10**+02
```

In this display, the *stem* represents the leading digits of each data value, so the stems labeled "6" in Output 2.7 represent ATBAT values from 600 to 699. In the PROC UNIVARIATE display, each data value is rounded to the next digit after the stem (the leaf), and that digit is written down to the right of the stem. The top line

```
    6  6889
```

thus represents the observations 663, 677, 680, and 689. The scale of the values is noted as "MULTIPLY STEM.LEAF BY 10**+02" in Output 2.7. Thus the stem, 6, and the first leaf, 6, represent $6.6 \times 10^2 = 660$, the rounded value of 663.*

The stem and leaf display in Output 2.7 uses two stems for each 100s digit in the ATBAT values, with leafs beginning 0—4 on the first and 5—9 on the second, so this display is equivalent to a histogram with an interval width of 50. The column of numbers after the stem and leaf display shows the number of scores represented on each stem. You can see there were seven players with fewer than 50 at bats in the 1986 season.

Depending on the range of data values, stem and leaf displays may use 1, 2, or 5 stems for the initial digits, corresponding to interval widths of 1, 5, or 2 times some power of 10. The same rules of thumb for choosing the number of intervals or interval width (for example, equation 2.1) also apply to the stem and leaf display.

Note that the gap between the seven lowest values and the remaining ones is also apparent in the stem and leaf display for ATBAT as it was in Output 2.2 for HITS; they are the same players. When an ID statement is used with PROC

---

* A value exactly halfway between two leaves is rounded to the nearest leaf with an even value. The value 675 would be represented as a stem of 6 and a leaf of 8. In Version 6 of the UNIVARIATE procedure, the ROUND= option controls how the data values are rounded in the stem-leaf display. For the ATBAT data, specifying ROUND=5 would cause the values of ATBAT to be rounded to the nearest multiple of 5.

UNIVARIATE, the extreme values are labeled with the value of the ID variable, which helps to identify outliers and determine why they may be discrepant.

Output 2.8 shows the stem and leaf portion of the PROC UNIVARIATE output for the infant mortality data. Note that although the stem and leaf display is based on the same number of intervals as Output 2.4, the stems in Output 2.8 are the lower boundaries of the intervals, while the axis values in Output 2.4 are midpoints, so there are 38 observations on the lowest stem, but 24 in the lowest interval of the histogram.

Output 2.8
Stem and Leaf
Display for Infant
Mortality Data

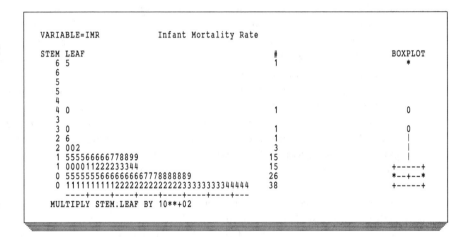

```
VARIABLE=IMR              Infant Mortality Rate

STEM LEAF                                    #              BOXPLOT
  6 5                                        1                 *
  6
  5
  5
  4
  4 0                                        1                 0
  3
  3 0                                        1                 0
  2 6                                        1                 |
  2 002                                      3                 |
  1 555566666778899                         15                 |
  1 000011222233344                         15              +-----+
  0 55555556666666667778888889              26              *--+--*
  0 11111111112222222222222233333333344444  38              +-----+
    ----+----+----+----+----+----+----+---
MULTIPLY STEM.LEAF BY 10**+02
```

When the number of intervals is the same, the shape of the distribution depicted in a histogram and stem and leaf display is the same. However, the stem and leaf display lets us see the distribution of data values within an interval and makes it easier to go from a leaf in the display to the data value it represents.

2.4 Boxplots

Sometimes it is more useful to display a summary of the data than to display all of the values individually or the shape of the entire distribution. But as pointed out in Section 1.2.1, "Means and Standard Deviations: Potentially Misleading Measures," the mean and standard deviation are not always the best summary measures, particularly if the data distribution is skewed or contains outliers.

For exploratory purposes, the *schematic plot* or *boxplot* invented by Tukey (1977; McGill, Tukey, and Larsen 1978) provides a high degree of resistance to outlying points and focuses attention on five important features of a data set:

□ typical or central value

□ spread or variability

□ shape—symmetry or skewness

□ outlying data points

□ behavior of the tails.

This section describes the ideas behind the boxplot and explains how to produce them on the printer with PROC UNIVARIATE. Because the boxplot is particularly useful for comparing different groups of data and for selecting transformations of the data to achieve homogeneity of variance, we return to this technique in Section 6.3, "Comparative Boxplots," where a high-resolution SAS/GRAPH version of the plot is described.

When the PLOT option is specified, PROC UNIVARIATE produces a boxplot display beside the stem and leaf or histogram. The two displays have the same vertical scale, so the column of frequencies in the middle applies to both displays. The basic features of the boxplot display can be seen in Output 2.7 and 2.8.

The central box extends from the lower quartile (Q1) to the upper quartile (Q3) in the UNIVARIATE QUANTILES output found in Output 2.7. The length of the box is the interquartile range (IQR≡Q3−Q1) that depicts the spread of the middle 50 percent of the observations. So in Output 2.7, the lower boundary of the box is on the same line as the stem for 250−299, corresponding to Q1=254.75. The location of the median is shown by the horizontal line in the middle of the box, and the mean is marked by a plus sign.

Whether the distribution is symmetric or skewed can normally be seen from the location of the median relative to the two quartiles. In a positively skewed distribution, the distance between the median and upper quartile will be greater than the distance between the median and lower quartile. Unfortunately, the vertical resolution in the boxplots produced by PROC UNIVARIATE is too small for this difference to be seen in Output 2.8.

The behavior of the tails and another indication of symmetry versus skewness is shown by the dashed whisker lines, which extend from the quartiles out to what are termed *adjacent values*. These are defined as the most extreme observations not outside $1.5 \times$ IQR beyond the upper and lower quartiles, that is, the largest value less than $Q3 + 1.5$ IQR and the smallest value greater than $Q1 - 1.5$ IQR. In a normal distribution, this range corresponds to the interval $\mu \pm 2.698\sigma$, which would include about 99 percent of the observations in the middle.

Any observations outside the adjacent values are called *outside points* and are plotted individually with a 0 in the PROC UNIVARIATE boxplots. A pair of more extreme limits, 3 IQR beyond the quartiles, define *far out points* that are plotted with an asterisk in these figures.

This section has shown how to produce full page boxplots on the printer with PROC UNIVARIATE in Version 6 of the SAS System. Section 6.3 describes how to produce high-resolution boxplots in Version 6 by specifying I=BOX in the SYMBOL statement, with the SHEWHART procedure in SAS/QC software, and with a BOXPLOT macro (Section A1.4, "The BOXPLOT Macro") for Version 5 or Version 6.

2.4.1 Boxplots with the UNIVARIATE Procedure

In Version 5, the SPLOT procedure (see the *SUGI Supplemental Library User's Guide, Version 5 Edition*) will produce full page boxplots on the printer. You can print a single boxplot for all observations with these statements:

```
PROC SPLOT DATA=SASdataset;
VAR variables;
```

For side-by-side boxplots of several groups defined by one or more classification variables, use the CLASS statement to list the classification variable(s):

```
PROC SPLOT DATA=SASdataset;
CLASS classvariables;
VAR variables;
```

The data set must first be sorted by the CLASS variables. Note that PROC SPLOT uses a different (and less typical) definition for outside points than the boxplot from PROC UNIVARIATE. PROC SPLOT marks all outside and far-out points with an asterisk; points that are between 1 and 1.5 IQR beyond the quartiles are plotted with a 0.

In Version 6, PROC UNIVARIATE gives full page boxplots when used with the PLOT option and with a BY statement. To produce a boxplot for a single group, define a dummy BY variable that has a constant value for all observations. For example, Output 2.9 shows the boxplot of infant mortality rate produced by PROC UNIVARIATE with these statements:

```
%include NATIONS;              /* On MVS, use %include ddname(file);*/
title ' ';
data nations;
   set nations;
   all=1;           * constant BY variable;
proc univariate plot data=nations;
   var imr;
   by all;
```

In this plot, the signs of positive skewness are more apparent than they were in the version produced beside the stem and leaf display (Output 2.8).

Output 2.9
UNIVARIATE
Procedure Boxplot
for Infant
Mortality Data

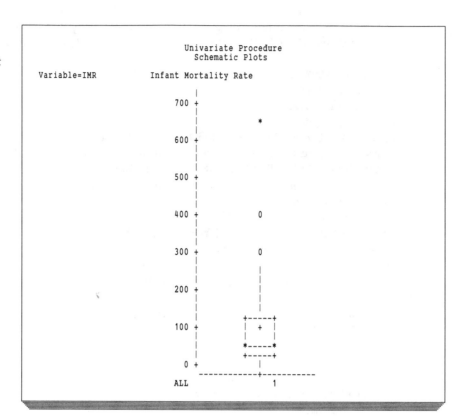

```
                         Univariate Procedure
                           Schematic Plots

    Variable=IMR          Infant Mortality Rate

                            |
                      700 + |
                            |
                            |              *
                      600 + |
                            |
                            |
                      500 + |
                            |
                            |
                      400 + |              0
                            |
                            |
                      300 + |              0
                            |
                            |              |
                      200 + |              |
                            |              |
                            |           +-----+
                      100 + |           |  +  |
                            |           |     |
                            |           *-----*
                            |           +-----+
                        0 + |              |
                            -----------+-----------
                      ALL               1
```

Boxplots are even more useful for comparing several sets of data. If the boxplots for each set are drawn side by side against a common scale, we can see the similarities and differences in central value, spread, shape, and unusual values. Output 2.10 shows this display for the infant mortality data, grouped by REGION. This plot is produced when PROC UNIVARIATE is invoked with the statement

```
by region;
```

as shown below:

```
    /* side-by-side boxplots, by region */
proc sort data=nations;
   by region;
proc univariate plot data=nations;
   var imr;
   by region;
```

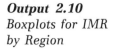

Output 2.10
Boxplots for IMR
by Region

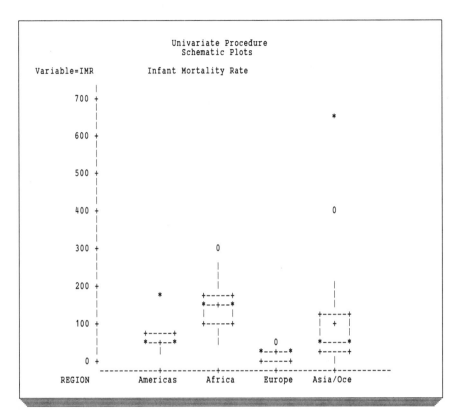

In Output 2.10, the few very large values have the effect of compressing most of the data values into the bottom quarter of the plot. Nevertheless, we can see that the infant mortality rates are very different in the four geographic regions, with considerably higher typical values in Africa and Asia/Oceania than in the Americas and Europe. The spread of the values is also greater in the former two regions than in the latter. In a grouped boxplot display, the outside values are determined from the interquartile range of each group separately. As a result, one value in the Americas (Brazil, at 170) stands out where it did not appear to be unusual in the total sample.

Sometimes it is useful to see the boxplot for all of the data together with the plots for each group, as shown in Output 2.11. You can produce this display by duplicating each observation in the data set NATIONS.

Output 2.11
Boxplots for IMR
for All Nations
and by Region

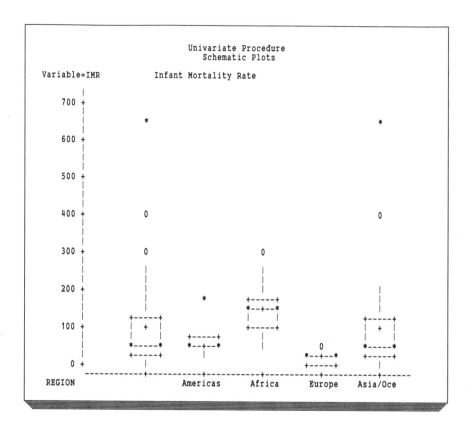

The second copy of each observation is given a constant value of the BY variable, different from any of the actual values (such as REGION=0), as shown in the DATA step below. Thus, each observation appears once in its own region and once in the 0 region.

```
/* side-by-side boxplots, including all regions */
data nation2;
   set nations;
   output;                  /* individual region */
   region=0; output;        /* all regions      */
proc sort data=nation2;
   by region;
proc univariate plot data=nation2;
   var imr;
   by region;
```

2.5 Dot Charts

We often need to display values of a single quantitative variable where each value has an identifying label associated with it. To interpret the display—and understand which values are high and which are low—we need to plot the labels along with the data values.

Output 2.12 is an example of a graphical method called a *dot chart* or *dot plot* developed by Cleveland (1984b) for this purpose. The figure displays a measure of occupational prestige, on a scale of 0 to 100, for 45 occupations from a classic sociological study by Duncan (1961). The goal of the study was to predict occupational prestige from more readily available measures, such as census data on income and education. The data from the Duncan study are described more fully in Section 5.4, "Validity of Assumptions: Plotting Residuals," where regression methods are used to predict occupational prestige.

Output 2.12
Dot Chart for Occupational Prestige

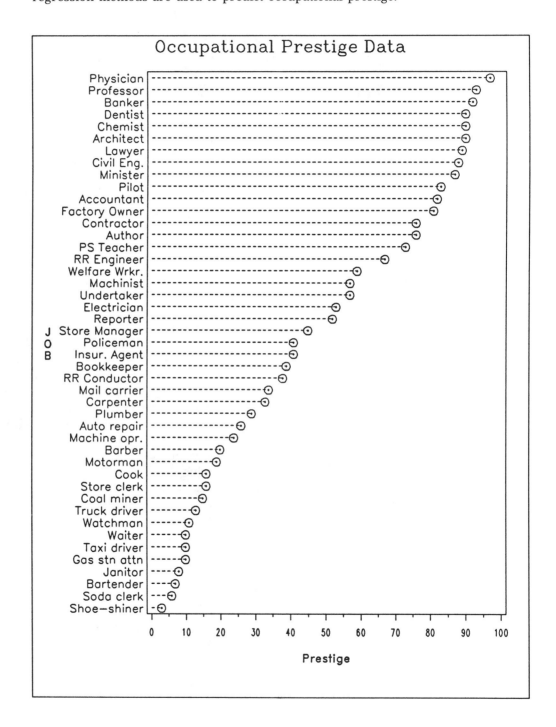

The data values plotted in Output 2.12 are listed in Output 2.13 together with short names for the 45 occupational categories. The complete data set is contained in the file DUNCAN SAS, listed in Appendix 2, "Data Sets."

Output 2.13
Occupational
Prestige Data

JOB	PRESTIGE	JOB	PRESTIGE
Accountant	82	Store clerk	16
Pilot	83	Carpenter	33
Architect	90	Electrician	53
Author	76	RR Engineer	67
Chemist	90	Machinist	57
Minister	87	Auto repair	26
Professor	93	Plumber	29
Dentist	90	Gas stn attn	10
Reporter	52	Coal miner	15
Civil Eng.	88	Motorman	19
Undertaker	57	Taxi driver	10
Lawyer	89	Truck driver	13
Physician	97	Machine opr.	24
Welfare Wrkr.	59	Barber	20
PS Teacher	73	Bartender	7
RR Conductor	38	Shoe-shiner	3
Contractor	76	Cook	16
Factory Owner	81	Soda clerk	6
Store Manager	45	Watchman	11
Banker	92	Janitor	8
Bookkeeper	39	Policeman	41
Mail carrier	34	Waiter	10
Insur. Agent	41		

The same sort of information can be displayed in a bar chart or pie chart, but the dot chart is more effective, especially as the number of observations grows, because the data-ink ratio is larger than for the other displays. When the observations are arranged in order, as they are in Output 2.12, the graph is, in effect, a cumulative distribution function for the sample, and the pattern of dots can be seen more easily than the same pattern shown with bars.

It is usually more effective to display the observations sorted by the response variable than alphabetically or in their order in the data set. Output 2.14, for example, shows the same data with the job labels arranged in alphabetical order. (The GCHART and GPLOT procedures arrange a character variable in alphabetical order by default.) This makes it easier to locate a particular occupation, but it is harder to see any patterns.

Output 2.14
Occupational
Prestige with
Observations in
Alphabetical Order

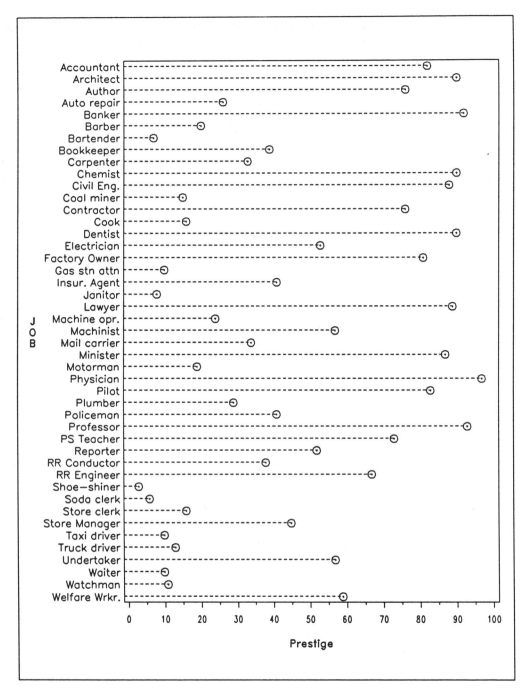

Simple dotplots can be produced on a printer with the TIMEPLOT procedure. The observations are arranged on the vertical axis in their order in the data set, so the data set should be sorted in descending order of the response variable. For example, the statements below produce the printer plot shown in Output 2.15.

The specification REF=0 and the JOINREF option produce the dashed line connecting the plotting symbol to the vertical axis.

```
%include DUNCAN;                    /* On MVS, use %include ddname(file);*/
proc sort data=duncan;
   by descending prestige;
proc timeplot data=duncan;
   plot prestige = '*' / ref=0 joinref;
   id job;
```

Output 2.15
Dot Chart of Occupational Prestige from the TIMEPLOT Procedure

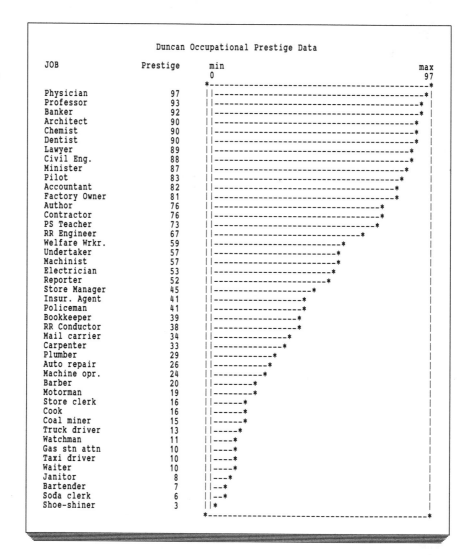

A high-resolution dot chart, such as Output 2.12, can be produced with PROC GPLOT. The dotted lines are drawn with an Annotate data set. It takes a bit of work, however, to arrange the observations in order of the response variable. With a character ID variable such as JOB, for example, the statements below produce a plot with the JOB values arranged in alphabetical order:

```
proc sort data=duncan;
   by prestige;
```

```
proc gplot data=duncan;
   plot job * prestige;
```

To reorder the observations for this plot, create an index variable, SORT_KEY, in a DATA step. SORT_KEY has the values 1, 2, . . . , 45, corresponding to the observations arranged in order of PRESTIGE values. Then use the FORMAT procedure to construct a format that associates the job labels with the SORT_KEY values.

This method is much more flexible than using PROC GPLOT (or PROC GCHART) directly because it enables the observations to be sorted in any desired order. In the program below, the format YNAME. is created by a small SAS macro, MAKEFMT. The job labels are assigned to macro variables VAL1—VAL45 using the SYMPUT function.

```
%include DUNCAN;                /* On MVS, use %include ddname(file);*/
 /*----------------------------------------------------*
   | Sort observations in the desired order on Y axis |
   *--------------------------------------------------*/
proc sort;
   by prestige;
 /*-------------------------------------------------------*
   | Add sort_key variable and construct macro variables |
   *-----------------------------------------------------*/
data duncan;
  set duncan end=eof;
  sort_key = _N_;
  call symput( 'val' || left(put( sort_key, 2.)), trim(job) );
  if eof then do;
     call symput('nobs', put(sort_key, 2.));
     end;
run;

  /*---------------------------------------------*
    |  Macro to generate a format of the form |
    |    1 = &val1  2=&val2 ... 45=val45      |
    |  for observation labels                 |
    *-----------------------------------------*/
%macro makefmt(nval);
  proc format;
      value yname
    %do i=1 %to &nval ;
       &i = "&&val&i"
       %end;
       ;
%mend makefmt;
%makefmt(&nobs);
```

The dotted lines are drawn by a simple Annotate data set with MOVE and DRAW functions. In the GPLOT step, we plot SORT_KEY against PRESTIGE and use the format YNAME. to write the job label. The macro variable NOBS, containing the number of observations in the data set, is used to define the tick marks in the AXIS1 statement.

```
   /*--------------------------------------------------*
    | Annotate data set to draw horizontal dotted lines |
    *--------------------------------------------------*/
data dots;
   set duncan;
   xsys = '2'; ysys='2';
   y   = sort_key;
   line=2;
   x = 0;          FUNCTION='MOVE'; output;
   x = prestige; FUNCTION='DRAW'; output;
   /*------------------------------------------------*
    | Draw the dot plot, plotting formatted Y vs. X |
    *------------------------------------------------*/
title h=1.5 'Occupational Prestige Data';
goptions vpos=55;
proc gplot data= duncan  ;
   plot sort_key * prestige / name='GB0212'
         vaxis=axis1 vm=0 haxis=axis2 hm=1
         frame annotate=dots;
   label   sort_key='JOB';
   format  sort_key yname.;
   symbol1 v=- h=1.4 c=black;
   axis1   order=(1 to &nobs by 1) label=(f=duplex)
           major=none value=(j=r f=simplex);
   axis2   order=(0 to 100 by 10)  label=(f=duplex)
           offset=(1);
run;
```

2.5.1 DOTPLOT Macro

Because the format for observation labels is constructed from the data itself, it is not difficult to generalize the program to work with any data set. A SAS macro, DOTPLOT, listed in Section A1.8, "The DOTPLOT Macro," prepares grouped or ungrouped dot charts, with an optional error bar for each observation. The dotted lines can be omitted or drawn in several ways. The macro takes the following arguments (with default values shown after the equal sign):

```
%macro dotplot(
        data=_LAST_,          /* input data set              .        */
        xvar=,                /* horizontal variable (response)       */
        xorder=,              /* plotting range of response           */
        xref=,                /* reference lines for response variable */
        yvar=,                /* vertical variable (observation label) */
        ysortby=&xvar,        /* how to sort observations             */
        ylabel=,              /* label for y variable                 */
        group=,               /* vertical grouping variable           */
        gpfmt=,               /* format for printing group variable   */
                              /* value (include the . at the end)     */
        connect=DOT,          /* draw lines to ZERO, DOT, AXIS, or NONE */
        dline=2,              /* style of horizontal lines            */
        dcolor=BLACK,         /* color of horizontal lines            */
        errbar=,              /* variable giving length of error bar  */
                              /* for each observation                 */
        name=DOTPLOT);        /* Name for graphic catalog entry       */
```

2.5.2 Grouped Dot Charts

If the observations are grouped by some classification variable, it may be useful to sort the observations first by groups and then by the response variable within each group. Output 2.16 is a dot chart of infant mortality rates for the 74 nations in the Americas, Africa, and Europe grouped by region. The same data are plotted in an ungrouped dot chart in Output 2.17. With 74 observations, these plots are quite legible. (The complete data set contains data on 105 nations; with this many observations, it becomes difficult to read the observation labels.)

Output 2.16   *Grouped Dot Chart for Infant Mortality Data*

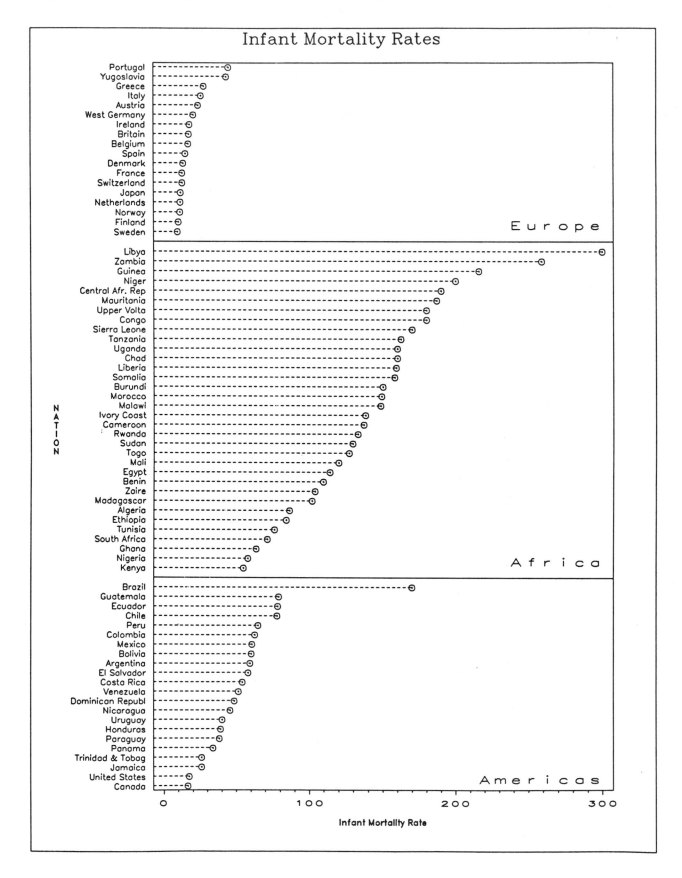

Output 2.17   *Ungrouped Dot Chart for Infant Mortality Data*

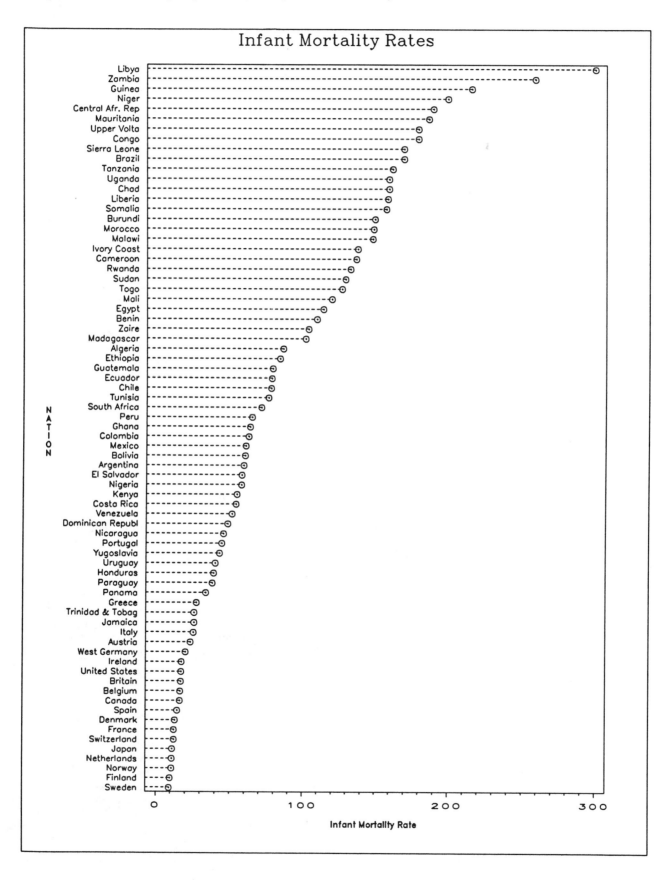

These two plots were produced with the DOTPLOT macro as shown below. Note that the GOPTION statement sets the VPOS= option to allow a sufficient number of vertical character cells for the labels.

```
%include NATIONS;           /* On MVS, use %include ddname(file);*/
%include dotplot;           /* On MVS, use %include ddname(file);*/
data nations;
   set nations;
   if region<4;    * exclude Asia for these plots;
   if imr ¬= . ;   * exclude missing data;
title 'Infant Mortality Rates';
goptions vpos= 94 hpos=60;
%dotplot(data=nations,
         xvar=imr, yvar=nation,
         group=region, gpfmt=region.,
         name=GB0216);
run;
%dotplot(data=nations,
         xvar=imr, yvar=nation,
         name=GB0217);
```

The ungrouped chart displays the entire distribution. The pattern of dots in Output 2.17 is characteristic of a positively skewed distribution. We can see that most of the nations have infant mortality rates less than 100, and a few have values greater than 200. The grouped chart displays the data values sorted within each geographical region. The regional differences, in typical value as well as in the shape of the distribution, are striking in Output 2.16.

2.5.3 Other Variations

Cleveland (1984b) cautions that bar charts and dot charts in the style of Output 2.12, 2.16, and 2.17 may be misleading when the origin of the horizontal response scale does not correspond to a meaningful baseline value. For then the eye tends to visually decode the length of the bar or the dotted line as measures whose ratios can be compared. For the occupational prestige and the infant mortality data, there is a natural zero point to the response scale, so the dot charts are appropriate.

When there is no meaningful baseline value on the scale, Cleveland (1984b) suggests that the dotted lines should go across the scale, or be omitted if there are only a small number of observations. The CONNECT parameter of the DOTPLOT macro provides the values of AXIS and NONE for the CONNECT= option for these possibilities.

Example: U.S. Public School Expenditures

For example, Output 2.18 displays per capita expenditures on public school education in the U.S., which range from a low of $112 in Alabama to a high of $372 for Alaska. Because the lowest value in the plot, $100, is not a meaningful reference point, this plot uses CONNECT=AXIS to draw the dotted lines across the whole scale. (A bar chart would give a misleading impression.) The public school expenditures data are discussed more fully and analyzed in relation to state demographic variables in Section 5.8, "Regression with Geographic Data."

Output 2.18 is produced by the statements below. The data set SCHOOLS is included in the program from a separate program, SPENDING SAS (see Section A2.13, "The SPENDING Data Set: School Spending Data").

```
%include SPENDING;          /* On MVS, use %include ddname(file);*/
%include dotplot ;          /* On MVS, use %include ddname(file);*/
title ' ';
goptions vpos=65;
%dotplot(data=schools,xvar=spending,
         yvar=st,ylabel=STATE,
         dcolor=GREEN,connect=AXIS,
         name=GB0218);
```

Note that the other dot charts actually use a dashed line (drawn with the DRAW Annotate function and LINE=2 by default) rather than a dotted line, because there is no line style for dotted lines in Version 5 of SAS/GRAPH software (a variety of dotted lines is available in Version 6). But a dashed line from side to side is too imposing visually when there are more than 15 or 20 observations, so the AXIS specification for the CONNECT= option draws dots with the POINT Annotate function. The other CONNECT= options would be more difficult to draw with dots because the data values would have to be scaled explicitly to achieve the right spacing.

Output 2.18   *Dot Chart for Public School Expenditures*

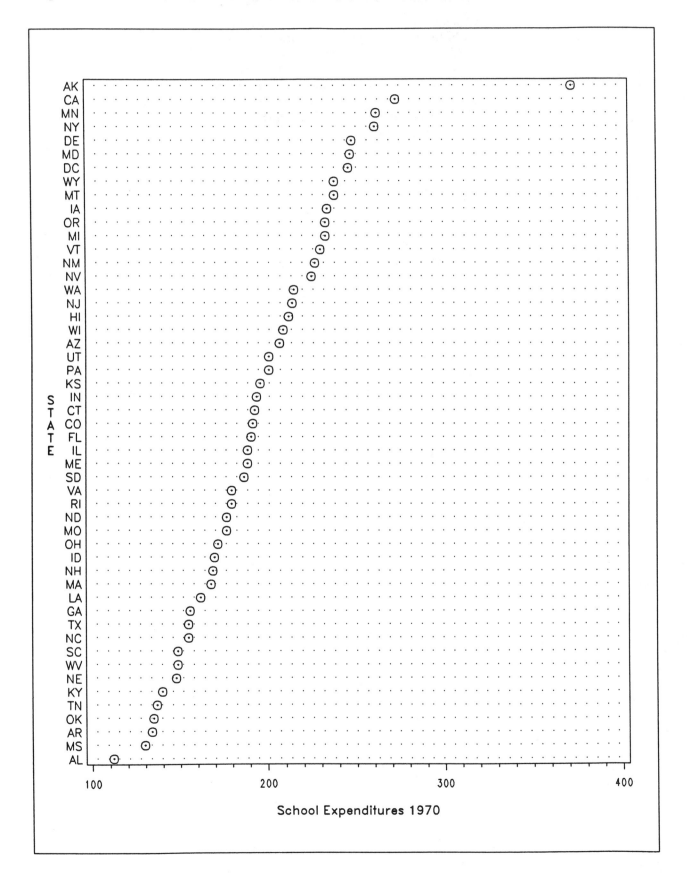

Another possibility is to sort the observations by some additional variable. Output 2.19 shows the occupational prestige data sorted by level of education for each job title.

Output 2.19
Occupational Prestige Ranked by Education

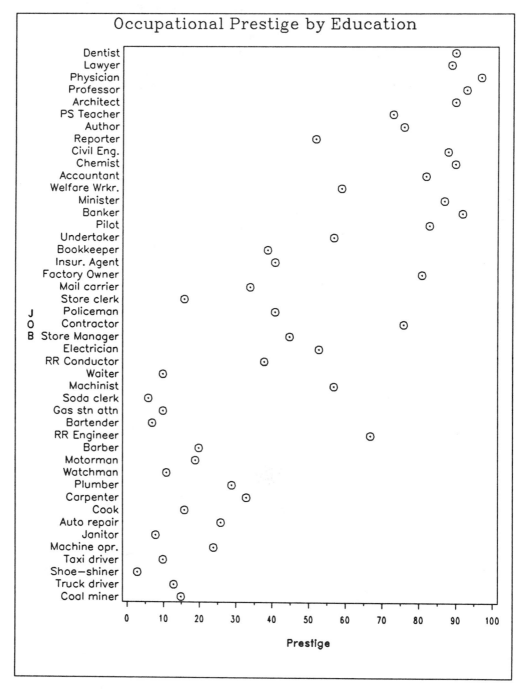

This plot uses the value of NONE for the CONNECT= option to suppress the dashed lines in this call to the DOTPLOT macro:

```
%dotplot( data=duncan, yvar=job, xvar=prestige,
          ysortby=educ, connect=NONE);
```

The result is like a scatterplot of EDUC * PRESTIGE, but the ordinate shows the rank order of the values of education. An advantage of this display is that it shows the observation labels without putting them into the data region where they would obscure the pattern of the points. A disadvantage is that the metric spacing of education is lost.

Adding Error Bars

The dot charts shown so far plot a single measure for each observation, but other information can be displayed as well. If there is a measure of uncertainty for each observation, this can be shown as an error bar around each point. Output 2.20 shows an example, using made-up values for the uncertainty of each data value in the school expenditures example. The fact that Alaska differs from the other states is made more apparent by the error bars.

Output 2.20　　　　*Dot Chart with Error Bars*

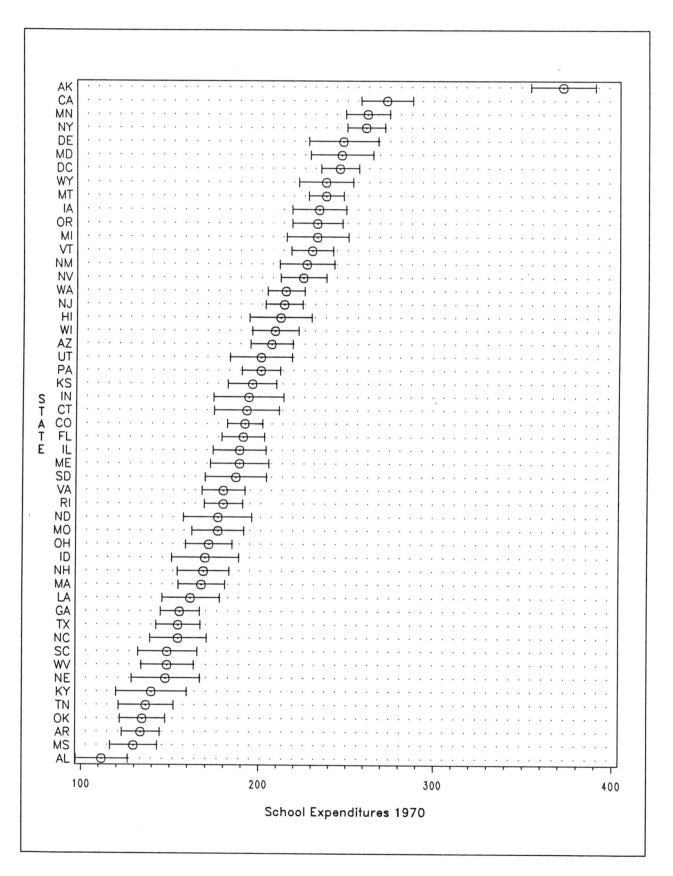

School Expenditures 1970

Output 2.20 is produced with the DOTPLOT macro using the program below. The error bars are produced when a variable in the data set is specified for the ERRBAR= option of the macro program.

```
%include SPENDING;              /* On MVS, use %include ddname(file);*/
%include dotplot ;             /* On MVS, use %include ddname(file);*/
data schools;
   set schools;
   error = 10 + 10*uniform(32653);
title ' ';
goptions vpos=65;
%dotplot(data=schools, xvar=spending,
         yvar=st, ylabel=STATE,
         dcolor=green, connect=axis,
         errbar=error, name=GB0220);
```

When the response is highly skewed, plotting the abscissa on a log scale, using log base 2 or natural logs as well as log base 10, can be a useful way to reduce the skewness. The infant mortality data, for example, would be better displayed on a log scale. If base 2 logs were used, unit steps on the scale of $\log_2$(IMR) would correspond to a doubling of the infant mortality rate.

Dot charts with log or other transformed scales for the response can be done with the DOTPLOT macro by calculating the transformed variable in a DATA step. The lines below, for example, give a dot chart of $\log_2$(IMR):

```
%include NATIONS ;
data nations;
   set nations;
   limr = log2(imr);
   label limr = 'log base 2 Infant Mortality Rate';
%dotplot(data=nations,
         xvar=limr,yvar=nation);
```

Just as the basic histogram can use the bar height to display a mean or other summary statistic, the DOTPLOT macro can do the same if the summary value is calculated in a separate PROC step. Output 2.21 shows the mean number of hits for each team grouped by league, together with error bars representing a 50% confidence interval about the mean.

Output 2.21   *Grouped Dot Chart for Mean Hits by Each Team*

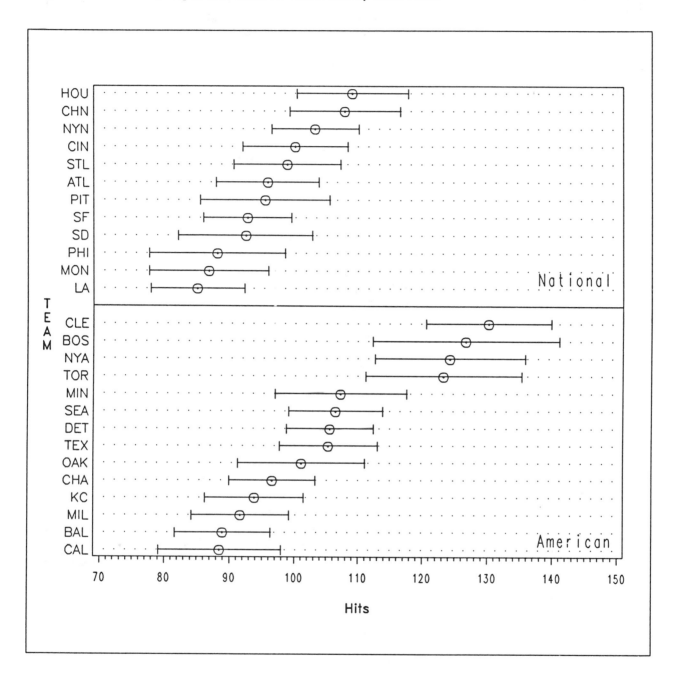

To produce this plot, use PROC SUMMARY to calculate the means and standard errors for each LEAGUE—TEAM combination. The NWAY option specifies that these statistics are to be output only for the LEAGUE—TEAM combinations (the highest _TYPE_ value). A subsequent DATA step calculates the width of the error bar, which is .67 times the standard error and gives a 50% confidence interval.

```
proc summary data=baseball nway;
   class league team;
   var hits;
   output out=mean mean=hits stderr=stderr;
data mean;
   set mean;
   err = .67 * stderr;   /* 50% conf. interval */
run;
%dotplot(data=mean, xvar=hits, yvar=team,
         errbar=err, connect=axis,
         dcolor=green,group=league,
         xorder=70 to 150 by 10,
         name=GB0221);
```

This display shows the same information portrayed by bar charts in Output 2.5 and 2.6, but the dot chart is easier to comprehend. Ordering the teams by mean hits helps, as does the visual emphasis given to the mean and error bar compared to the bar chart bar. In the dot chart, the mean and error information is the visual foreground, whereas this information is in the background in a bar chart of means. The grouped version in Output 2.21 also shows that four teams in the American league stand out in average hitting performance, both in their own league and in relation to all of the teams.

Cleveland (1984b) illustrates other possibilities for dot charts that are not provided by the DOTPLOT macro. Dot charts can also be used to display two or three different response variables for each observation using a different plotting symbol or color, or both, for each response. For example, if the infant mortality data included separate variables for infant mortality rates among male and female babies, these two values could be shown along a single line for each nation. Such data might provide indications of the reasons for the high infant mortality rates in some of these nations.

Chapter **3** Plotting Theoretical and Empirical Distributions

3.1 Introduction

Parametric statistical methods assume that data follow a probability distribution of known form, perhaps with some unknown parameters. For example, in the simple linear regression model, $y_i = \beta_0 + \beta_1 x_i + \varepsilon_i$, it is assumed that the error, ε_i, is normally distributed with mean 0 and unknown variance, σ^2, $\varepsilon_i \sim N(0, \sigma^2)$. The large number of mathematical and probability functions available in the SAS System makes it relatively simple to plot all of the common theoretical probability distributions.

The second section of this chapter, "Plotting Theoretical Statistical Distributions," shows how to construct plots of the normal, χ^2, t, and F distributions using different methods. Section 3.3, "Bivariate Normal Distribution," describes how to plot the bivariate normal distribution. You can use these techniques to plot other distributions as well. While these are plots of mathematical functions rather than plots of data, the plotting techniques offered by the SAS System are useful for constructing theoretical quantile plots (Section 3.5, "Quantile Plots") and power curves for statistical tests (Section 3.2.5, "Plotting Power Curves: Student's *t* Distribution" and Section 7.7, "Plotting Power Curves for ANOVA Designs"). These examples also serve to illustrate features of SAS/GRAPH software that will be used in later chapters.

For real data, a variety of plots (Section 3.4, "Histogram Smoothing and Density Estimation") are useful for studying the shape of a distribution. Parametric techniques fit a particular theoretical distribution to the data and provide a χ^2 test for goodness of fit. A nonparametric technique for estimating a smooth density function, without assuming a particular form for the distribution, is also described and applied to several data sets. Quantile plots are also used to determine if a set of data follows a particular distribution. Section 3.5 illustrates the construction and interpretation of normal quantile plots.

Finally, some graphical methods for determining if data have a reasonably symmetric distribution are described in Section 3.6, "Plots for Assessing Symmetry," together with methods for transforming the data if they are asymmetrical.

3.2 Plotting Theoretical Statistical Distributions

This section explains how to plot a statistical distribution that can be expressed as a mathematical function, $y = f(x)$, of the value, x, of a random variable. Constructing such a plot typically involves a DATA step to calculate the values of $f(x)$ as x is varied over its domain, and a PROC GPLOT step to plot the resulting values. The most commonly used probability distributions, the normal, χ^2, Student's *t*, and *F* distributions, can all be plotted in this way.

These plots illustrate various techniques for calculating and plotting the values of the function:

□ using SAS probability functions

□ plotting a family of parametric curves with different symbols

□ approximating a complex density function by difference quotients

□ using logarithmic axes to make a set of curves more nearly linear

□ using macro variables to generalize a program.

The CAPABILITY procedure in SAS/QC software provides the ability to fit a number of theoretical distributions to a set of data, including the normal, lognormal, gamma, beta, and Weibull distributions, which are commonly used in quality control applications. These distributions are easier to plot with PROC CAPABILITY because this procedure calculates the values and plots the result.

Two types of plots of a probability distribution are commonly used:

□ For a discrete distribution, the *probability density function* gives the probability, $y = f(x) = Pr\{X=x\}$, that a random variable, X, distributed according to that distribution, takes the value x. Density functions for continuous distributions are defined in terms of the probability of a value in a minute interval around the value x.

□ The *cumulative distribution function* (CDF) shows the probability, $y = F(x)$, that a random variable X, distributed according to that distribution, is less than or equal to a given value, x:

$$y = F(x) = Pr\{X \le x\} = \int_{-\infty}^{x} f(x)\, dx \quad . \tag{3.1}$$

3.2.1 Plotting the Normal Distribution

SAS functions are available to calculate the cumulative probabilities for all of the common statistical distributions. The example below plots the CDF for a standardized random variable X distributed $N(0, 1)$ (that is, normally with mean=0 and variance=1) according to the normal distribution. The value of the cumulative probability is calculated directly with the PROBNORM function, which computes an approximation to

$$y = F(x) = Pr\{X \le x\} = \int_{-\infty}^{x} \frac{1}{\sqrt{2\pi}} e^{-x^2/2} dx \quad . \tag{3.2}$$

The DATA step below uses the PROBNORM function to calculate the value of y for a range of x values from $x = -3.5$ to $x = 3.5$. (The range of x is actually $-\infty$ to $+\infty$, but over 99.9 percent of the probability is contained within the interval $[-3.5, 3.5]$.)

```
title 'Cumulative Normal Distribution plot';
data normal;
   do x = -3.5 to 3.5 by 0.01;
      y = probnorm(x);
      output;
   end;
```

The PROC GPLOT step plots y against x and draws a dashed vertical line at the horizontal value 0 (HREF=0), the mean of the standard normal distribution. The SYMBOL statement uses I=JOIN and V=NONE to draw a line connecting each pair of points without any plotting symbol at the points.

```
proc gplot data=normal ;
   plot y * x / href=0 chref=green lhref=20
                vaxis=axis1 haxis=axis2
                name='GB0301' ;
   symbol1 i=join v=none c=black;
```

```
axis1 order=0 to 1 by .1
      label=(f=titalic h=1.7 a=90 r=0 'Cumulative Probability')
      value=(h=1.3)
      minor=none;
axis2 label=(f=titalic h=1.7 'Normal Deviate (x)')
      value=(h=1.3)
      minor=none;
format y 3.1;
title1 c=black f=swisse h=2 'Normal Distribution Function';
```

The plot produced is shown in Output 3.1. The dashed line at *x*=0 locates the mean of the distribution. The slope of the curve at that point equals the standard deviation.

Output 3.1
*Plot of the
Cumulative Normal
Distribution
Function*

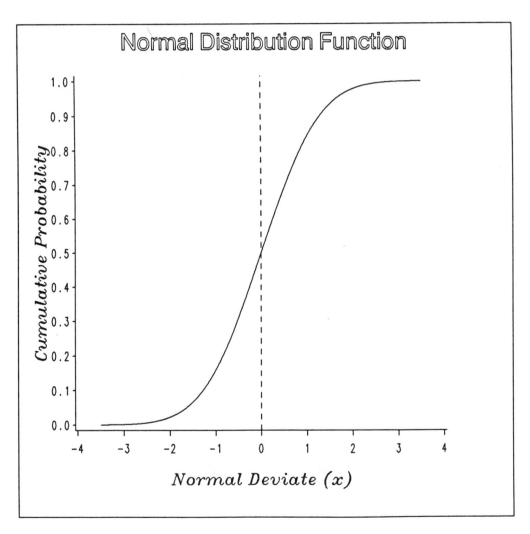

In a similar way, the SAS functions PROBCHI (χ^2 distribution), PROBF (*F* distribution), PROBT (Student's *t* distribution), and others can be used to compute and plot the cumulative probability distribution functions for other statistical distributions. The CDFPLOT statement in PROC CAPABILITY will produce a cumulative probability plot for any of the distributions fit by this procedure.

PROC CAPABILITY plots of the normal density function are illustrated in Output 3.3.

Normal Density Function

Most probability distributions are more easily seen (and understood) as a plot of the density function, $f(x)$, than as a plot of the cumulative distribution function, $F(x)$. There is no SAS function for the density function of the normal distribution, shown below:

$$y = f(x) = \frac{1}{\sqrt{2\pi}} \, e^{-x^2/2} \quad . \tag{3.3}$$

But it is not difficult to calculate the values of y from equation 3.3. The example below plots this density function over the same range, from $x=-3.5$ to $x=3.5$. The plot is shown in Output 3.2.

```
title 'Density plot of Normal Distribution with GPLOT';
data normal;
   constant = 1 / sqrt (2 * 3.141592654);
   do x = -3.5 to 3.5 by 0.01;
      y = constant * exp(-(x**2)/2);
      output;
   end;
```

Again, the PROC GPLOT step plots y against x and uses a SYMBOL statement with I=JOIN and V=NONE to connect the points without any plotting symbol. The TITLE statements use the MATH and GREEK fonts to draw the equation of the normal density function.

```
proc gplot data=normal  ;
   plot y * x / href=0 chref=green lhref=20
               vaxis=axis1 haxis=axis2
               name='GB0302' ;
   symbol1 i=join v=none c=black;
   axis1 order=0 to .4 by .1
        label=(f=titalic h=2 a=90 r=0 'Normal Density f(x)')
        value=(f=duplex h=1.8) minor=none;
   axis2 label=(f=titalic h=2 'Normal Deviate (x)')
        value=(f=duplex h=1.8) minor=none;
   format y 3.1;
   title1 c=black f=swiss h=2.4 'Normal Density Function';
   title2 c=blue  f=duplex h=2.0 'f (x) = 1/' f=math h=1.5 'a'
       f=duplex h=2.0 '2' f=cgreek 'p' f=duplex ' * exp(-x*x/2)';
```

Output 3.2   *Plot of the Standard Normal Density Function*

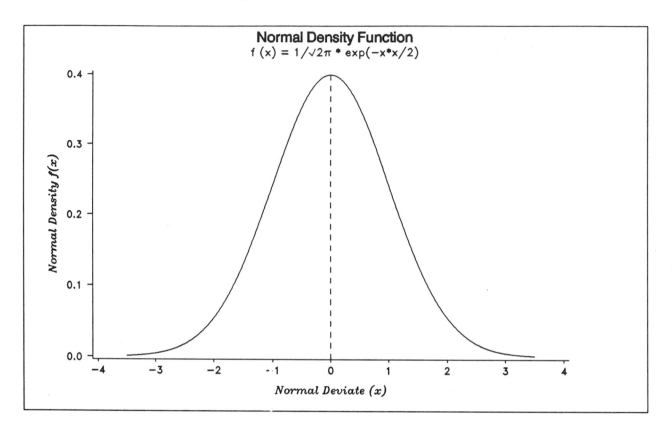

Normal Density Plot in the CAPABILITY Procedure

PROC CAPABILITY is typically used to compare an empirical data distribution with a theoretical distribution. The HISTOGRAM statement is used to draw a histogram of the data, and the NORMAL option requests a fitted normal curve. The example below constructs a set of equally spaced X values in a DATA step. The HISTOGRAM statement specifies a fitted normal curve with mean (MU) equal to 0 and standard deviation (SIGMA) of 1. The NOBARS option suppresses the histogram of X values (which have a uniform distribution). These statements produce the plot shown in Output 3.3.

```
title h=1.8 'Normal Density Function';
data xvalues;
   do x=-4 to 4;
      output;
      end;
proc capability graphics ;
   histogram x /
      normal(mu=0 sigma=1)
      midpoints=-4 to 4
      nobars nolegend noframe
      vaxis=axis1 haxis=axis2
      href=0 lhref=20
      name='GB0303' ;
```

```
axis1 label=(h=1.5 f=duplex) value=(h=1.5);
axis2 order=-4 to 4 value=(h=1.5)
        label=(h=1.5 f=duplex 'Normal Deviate (x)');
```

Output 3.3   *Normal Density Plot in the CAPABILITY Procedure*

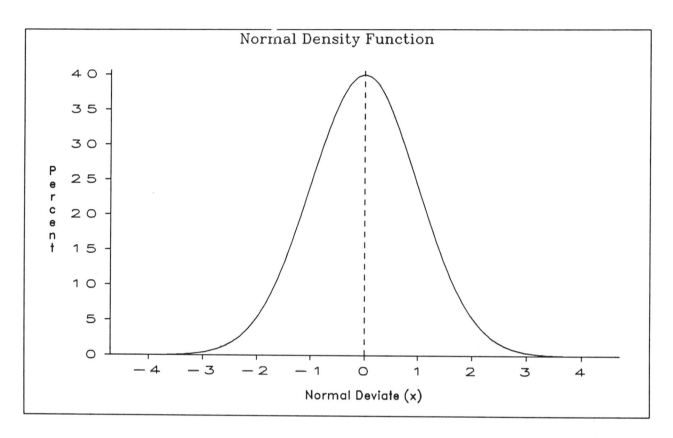

3.2.2 Parametric Plots: Chi Square Distribution

The theoretical χ^2 distribution is defined as the distribution of the sum of squares of v independent, normally distributed quantities with mean zero and unit standard deviation. In statistical applications, the χ^2 arises as the sampling distribution of a variance estimate, where v corresponds to the degrees of freedom of the variance estimate. The χ^2 density function is given by

$$y = f_v(x) = \frac{x^{(v/2)-1}}{2^{v/2}\Gamma(v/2)} e^{(-x/2)} \quad , \qquad x > 0 \tag{3.4}$$

where v is the degrees of freedom parameter and $\Gamma(\bullet)$ is the gamma function.

To plot the χ^2 density function, we proceed very much as we did with the normal distribution: calculate the values of $f_v(x)$ in a DATA step over a range of x values and plot with the GPLOT procedure. However, the χ^2 distribution changes its shape depending on the degrees of freedom parameter, v, and we might want to show the distribution for several different degrees of freedom values in a single *parametric plot*.

In the program below, the values of the χ^2 density are calculated for three different values of degrees of freedom, DF=2, 5, 10. Within the inner DO loop, values of Y are calculated for X=0 to 15. (The mean of the χ^2 distribution is equal to its degrees of freedom. Hence the range, 0 to 15, includes most of the distribution for 2 and 5 degrees of freedom. See the graph in Output 3.4.)

```
data chisq;
   keep x y df;
   do df = 2, 5, 10;              * Degrees of freedom ;
      k1  =   df/2;
      k2  =   1/((2**k1) * gamma(k1));
      do x =   0 to 15 by 0.05;
         h1 =   x ** (k1-1);
         h2 =   exp(-x/2);
         y  =   k2 * h1 * h2;
         output;
      end;
   end;
```

In the PROC GPLOT step, the PLOT statement

```
plot y * x = df;
```

causes PROC GPLOT to draw a different curve within the plot for each distinct value of the variable DF; each curve uses a different SYMBOL statement, and the value of the variable DF is identified in the plot legend. The PROC GPLOT step produces the three curves shown in Output 3.4. (A color version of this output appears as Output A3.7 in Appendix 3, "Color Output.")

```
proc gplot data=chisq  ;
   plot y * x = df /
           vaxis=axis1 haxis=axis2
           vminor=4 hminor=4
           legend=legend1
           name='GB0304'  ;
   symbol1 i=join v=none l=2  c=red;
   symbol2 i=join v=none l=3  c=blue;
   symbol3 i=join v=none l=20 c=green;
   axis1   label=(f=titalic h=2 j=r 'f(X)')
           value=(h=1.5)
           order=(0 to .4 by .1);
   axis2   label=(f=titalic h=2 j=r)
           value=(h=1.5)
           order=(0 to 15 by 5);
   legend1 label=(h=1.5) value=(h=1.5);
   title1 f=swiss h=2 'Chi-Square Density Function';
```

Output 3.4
Parametric Plot of Chi Square Density Function

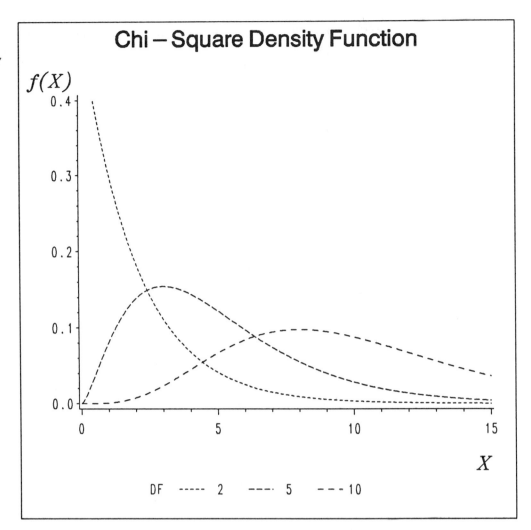

The graph shows how dramatically the shape of the χ^2 distribution changes with the degrees of freedom, going from an extreme J-shape at DF=2 to a nearly symmetric distribution as the degrees of freedom increase.

You can produce separate symbols, colors, interpolated lines, and so on (any of the parameters of the SYMBOL statement) for each level of the parameter variable after the equal sign in the PLOT statement. The parameter variable can be character or numeric. Use a series of SYMBOL*n* statements, one for each value of the parameter variable in the data set being plotted. SYMBOL*n* statements are associated with the sorted values. If there is a format defined for the parameter variable, the formatted values are used. Thus, in Output 3.4, the SYMBOL1 statement defines the line style and color for the DF=2 curve, SYMBOL2 is used for the DF=5 curve, and SYMBOL3 is used for DF=10. The appearance of the legend beneath the plot is controlled by the LEGEND statement.

3.2.3 Plotting Difference Quotients: *F* Distribution

The *F* distribution is the distribution of a ratio of two independent variance estimates. The cumulative distribution is given by the PROBF function, but again there is no built-in function for the density. You could use the same technique as in the χ^2 plot: look up the formula for the F density function and translate it into a DATA step calculation.

The formula for the F density function, however, is much more complex. Here is another technique you may find useful in some situations. If you have the cumulative distribution function, $F(x)$, for some distribution, the probability density function, $f(x)$, is the derivative of $F(x)$, which you can approximate by a difference quotient for small Δx:

$$f(x) \approx \frac{F(x + \Delta x) - F(x)}{\Delta x} \quad .$$

The program below uses this idea to plot the F distribution. It generates the values of $F(x)$ in one DATA step with the PROBF function. A second DATA step computes the difference quotient using the DIF function, defined as

```
DIF(X) = X - LAG(X)   .
```

LAG(X) returns the value of the variable X from the previous call to the function. When two or more parametric series (indexed by degrees of freedom) are generated in one DATA step, it is necessary to end each set of values with a missing observation so that the DIF function will not compute the difference between the last observation in one series and the first in the next.

In addition, the program illustrates how to generalize a plotting program slightly by use of SAS macro variables. In this case, the two degrees-of-freedom parameters, DF1 and DF2, are defined in the %LET statements at the top. The values entered there are used in the DATA step to define the ranges of the two DO loops and printed in the plot title. Note that literal strings in titles and other SAS statements that contain macro variables must be enclosed in double quotes. To adapt the program for different degrees of freedom, simply change the two %LET statements. This program produces the plot shown in Output 3.5. (A color version of this output appears as Output A3.8 in Appendix 3.)

```
title 'F density by differencing PROBF';
%let df1=2, 4, 40;
%let df2=4 ;
data fdist;
   do df1 = &df1 ;                              /* numerator df */
      do df2 = &df2 ;                           /* denominator df */
         df = put(df1,2.0)||','||trim(put(df2,2.0));
         do x = 0 to 4 by .05;
            yy = probf(x, df1, df2, 0);
            output;
         end;
         x = .; yy = .; output;     /* provide gap between series */
      end;
   end;
data fdist;
   set fdist;
   y = dif(yy) / dif(x);
```

Output 3.5
*Plot of the F
Density Function*

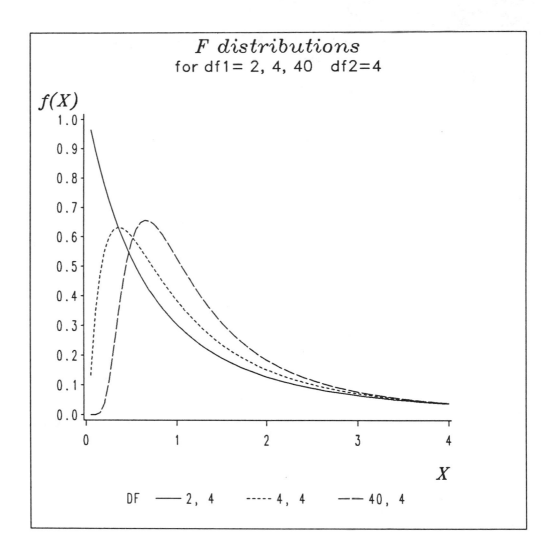

The graph is drawn with PROC GPLOT as a set of plots for each value of the variable DF. This variable was constructed as a character string containing the two degrees-of-freedom values, separated by a comma, so there is a unique value of the variable DF for each curve.

```
title1 c=black f=titalic h=2 'F distributions';
title2 c=red f=duplex h=1.5 "for df1= &df1    df2=&df2";
proc gplot data=fdist  ;
   plot y * x = df /
         vaxis=axis1 haxis=axis2
         legend=legend1
         name='GB0305'  ;
   axis1 label=(c=black f=titalic h=1.75 j=r 'f(X)')
         value=(h=1.5)
         minor=none;
   axis2 label=(c=black f=titalic h=1.75 j=r)
         value=(h=1.5)
         minor=none;
```

```
symbol1 i=join v=none l=1 c=red;
symbol2 i=join v=none l=2 c=green;
symbol3 i=join v=none l=4 c=blue;
legend1 label=(h=1.5) value=(h=1.5);
```

3.2.4 Plotting Percentiles: Student's *t* Distribution

In some situations, it is useful to construct plots of the *inverse functions* of probability distributions such as the normal, *t*, χ^2, and *F*. Such a plot shows the value *x* (on the ordinate) exceeded with probability *p* (on the abscissa). The values *x(p)* are the *quantiles* of the distribution. For example, such plots are the basis for power curves for well-known statistical tests, as illustrated in the following section.

The SAS function PROBIT gives the inverse of the standard normal CDF. Other inverse functions are also available.* Some of these are described below:

CINV (*p*, *df*, *nc*)
  is the inverse of the PROBCHI function. It returns the value *x*, exceeded with probability $1-p$ by a χ^2 random variable with *df* degrees of freedom and noncentrality parameter *nc*.

FINV (*p*, *df1*, *df2*, *nc*)
  is the inverse of the PROBF function. It returns the value *x*, exceeded with probability $1-p$ by an *F* random variable with *df1* and *df2* degrees of freedom and noncentrality parameter *nc*.

TINV (*p*, *df*, *nc*)
  is the inverse of the PROBT function. It returns the value *x*, exceeded with probability $1-p$ by a Student's *t* random variable with *df* degrees of freedom and noncentrality parameter *nc*.

The example below uses the TINV function to plot a set of curves of the critical values of the *t*-distribution function, for different numbers of degrees of freedom. When the noncentrality parameter, *nc*, is specified as 0, the values returned by TINV are those of the central *t* distribution, which are the ordinary critical values required for significance at the specified probability value.

The values of the horizontal variable, PVALUE, are better plotted on a logarithmic scale. To do this, the AXIS statement specifies LOGBASE=10 (base 10 logs) and LOGSTYLE=EXPAND so that tick marks are labeled with the actual values.

```
title h=1.6 'Percentiles of the t-distribution';
data ttable;
   label pvalue = 'Tail probability'
         t      = 'Critical t'
         df     = 'Degrees of Freedom';
```

---

* These are available in Version 6; prior to Version 6, they were part of the *SUGI Supplemental Library*. (See the *SUGI Supplemental Library User's Guide, Version 5 Edition*.)

```
do  df= 4 to 10 by 2, 20, 60;
  do prob =  .90, .95, .975, .98, .99, .995, .999;
     pvalue = 1 - prob;        /* upper tail probability  */
     t    = tinv (prob, df, 0); /* gives t-value whose cum. */
                                /* probability= prob       */
     output;
  end;
 end;
 run;
```

The SYMBOL statements below define a different plotting symbol (V=) and line style (L=) for each value of the third plotting variable, DF. I=SPLINE is used to draw a smooth curve through each of the points. (Splines and other interpolated smoothed curves are discussed in Section 4.4, "Interpolated Curves and Smoothings.")

```
goptions colors=(black,red,blue,green);
symbol1 i=spline v=star l=1;        * NB: no color= specified ;
symbol2 i=spline v=plus l=2;
symbol3 i=spline v=-   l=8;
symbol4 i=spline v=star l=12;
symbol5 i=spline v=plus l=16;
symbol6 i=spline v=-   l=20;
proc gplot data=ttable  ;
  plot t * pvalue= df
       / vaxis=axis1 haxis=axis2
         frame legend=legend1 name='GB0306' ;
  axis1 label=(a=90 r=0 h=1.4) value=(h=1.4);
  axis2 label=(h=1.4)          value=(h=1.4)
        logbase=10
        logstyle=expand;
  legend1 label=(h=1.3) value=(h=1.3);
```

Output 3.6 shows that for a given tail probability ($p=.05$, for example) the critical value of t is very large for small degrees of freedom. (A color version of this output appears as Output A3.9 in Appendix 3.) As the degrees of freedom increase, the critical t decreases, but at a decreasing rate. Using a log scale for the abscissa, PVALUE, helps straighten the curves and keeps them more widely separated.

Output 3.6
Percentiles of
Student's t
Distribution

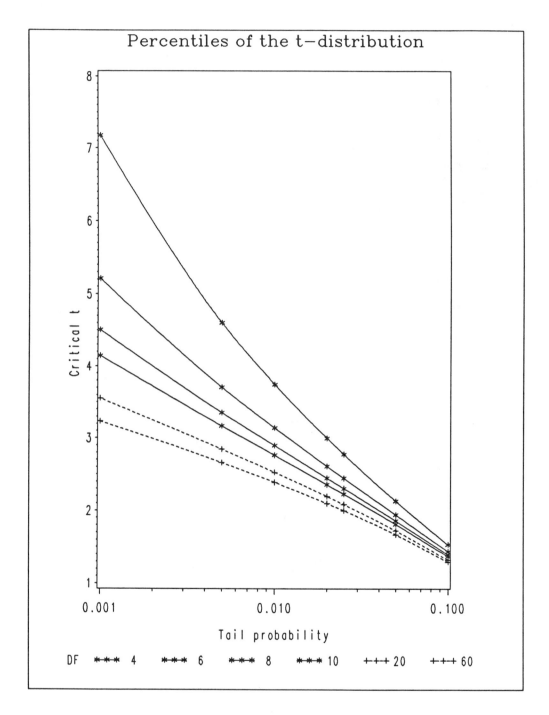

Defaults for SYMBOL Statements

If you are using a multicolor device and do not specify a color (C=*color*) option in a SYMBOL or PATTERN statement, SAS/GRAPH software cycles through all the available colors on the device before using the next SYMBOL or PATTERN statement. The available colors are determined by the COLORS= option in the GOPTIONS statement or by the default for the device.

For example, when the program above is plotted on a four-color device, the SYMBOL1 statement with L=1 is used four times—once with each color—before the SYMBOL2 statement is used. Then the SYMBOL2 statement is used for the remaining two values of the variable DF, and the other SYMBOL statements are ignored. Output 3.6 was in fact plotted on a four-color device, so the curves for DF=4, 6, 8, and 10 were plotted with stars joined by a solid line. Therefore, it is a good idea to specify the C= option in each SYMBOL statement.

In addition, SYMBOL statements are used globally; like TITLE statements, they stay in effect for all subsequent plots. See Chapter 16, "The SYMBOL Statement," in *SAS/GRAPH Software: Reference, Version 6, First Edition, Volume 1* for further details on canceling SYMBOL definitions and altering individual options in programs that produce more than one graph.

3.2.5 Plotting Power Curves: Student's *t* Distribution

This section illustrates the use of the inverse probability functions for plotting the power of the *t*-test in a two-group design. Such a plot would normally be used to determine the sample size required to give a specified power for testing the null hypothesis that the means of two populations are the same, $H_0: \mu_1 - \mu_2 = 0$, against the alternative hypothesis that the means have a particular nonzero difference, $H_1: \mu_1 - \mu_2 = d$. With more than two groups, the percentiles of the (noncentral) F distribution are used in a similar way in Section 7.7 to plot power curves for the F test in ANOVA designs.

The key idea is to compute the noncentrality parameter that corresponds to an effect of a given size. Power is then computed as the probability that the critical *t* value is exceeded in the noncentral distribution. This is illustrated in Output 3.7, which shows the distributions of the *t* statistic when H_0 is true ($nc=0$) and when H_1 is true ($nc=4$). The critical *t* value for a one-tailed test of size a is the *t* value in the central distribution exceeded with probability a. The power of the test is the area under the noncentral curve beyond this critical value.

Output 3.7
*Illustration of the
Power of the t-Test*

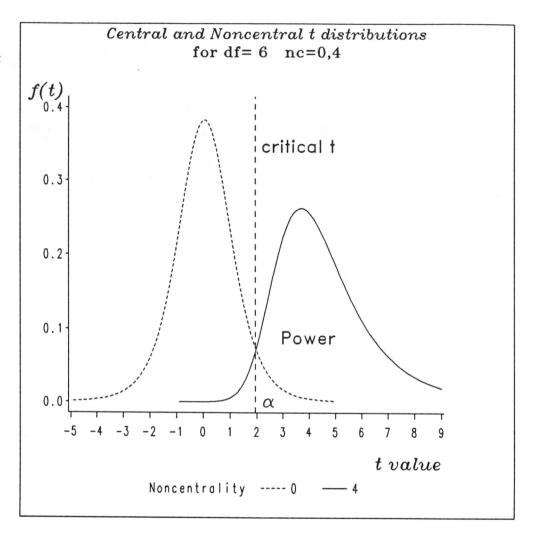

To calculate a power, you must specify the noncentrality value corresponding to the desired difference, *d*, to be detected, as well as the sample size, *n*, per group. When the variances of the two populations are the same value, σ^2, it is most convenient to specify this difference in units of the standard deviation, $\Delta = d/\sigma$. Cohen (1977) calls this an *effect size* index and suggests that the values $\Delta = 0.2$, 0.5, and 0.8 correspond to conventional small, medium, and large effects in behavioral science research.

The effect size can also be specified in terms of the proportion of variance of the scores predictable from group membership, using the relation

$$ r^2 = \frac{\Delta^2}{\Delta^2 + 4} \quad . \tag{3.5}$$

(The correlation here is the point-biserial correlation between the score and a dummy variable for group membership.) The small, medium, and large effects correspond to values of r^2 of .01, .06, and .14. See Cohen (1977) for other interpretive effect size measures.

The program below plots curves showing power as a function of sample size, for values of effect size ranging from 0.1 to 0.9. Macro variables are used to specify the significance level **(ALPHA)** and whether the test is 1-tailed or 2-tailed **(TAILS)**.

```
%let alpha = .05;
%let tails = 1;
title h=1.6 f=duplex "Power of the t-test, " f=cgreek "a = &alpha,"
         f=duplex " &tails tailed";
data tpower;
   label n    = 'Sample size per group'
         delta= 'Effect Size'
         df   = 'Degrees of Freedom';
   alpha = &alpha;
   do  n= 2, 3, 4 to 10 by 2, 15, 20;
     df = 2*(n-1);
     do delta = .1 to .9 by .1;
        nc   = sqrt(n/2) * delta;
        rsq  = delta**2 / (delta**2 + 4);
        t    = tinv (1-alpha/&tails, df, 0);        * critical t;
        power = 1 - probt(t, df, nc);
        output;
     end;
   end;
run;

symbol1 i=spline v=star     l=1   c=red;
symbol2 i=spline v=+        l=2   c=blue;
symbol3 i=spline v=-        l=5   c=green;
symbol4 i=spline v=plus     l=8   c=red;
symbol5 i=spline v=$        l=12  c=blue;
symbol6 i=spline v=:        l=16  c=green;
symbol7 i=spline v=a        l=20  c=red;
symbol8 i=spline v=square   l=22  c=blue;
symbol9 i=spline v=triangle l=25  c=green;
proc gplot data=tpower  ;
   plot power * n = delta
        / vaxis=axis1 haxis=axis2
          vref=.8 lvref=34 cvref=black          /* V6.06 */
          hminor=1 frame
          legend=legend1
          name='GB0308' ;
   axis1   order=(0 to 1 by .2)
           label=(a=90 r=0 h=1.4)
           value=(h=1.4);
   axis2   offset=(2)
           order=(2 to 20 by 2)
           label=(h=1.4)
           value=(h=1.4);
   legend1 label=(h=1.4 'Effect Size') value=(h=1.4);
```

The PROC GPLOT step uses different combinations of color, symbol, and line style and joins each set of points with a smooth curve using I=SPLINE in the SYMBOL statement. A dotted line is drawn at the value POWER=.80 to illustrate the use of the plot for determining sample size.* The plot produced is shown in Output 3.8. (A color version of this output appears as Output A3.10 in Appendix 3.)

Output 3.8
Power Curves for the Two-Sample t-Test

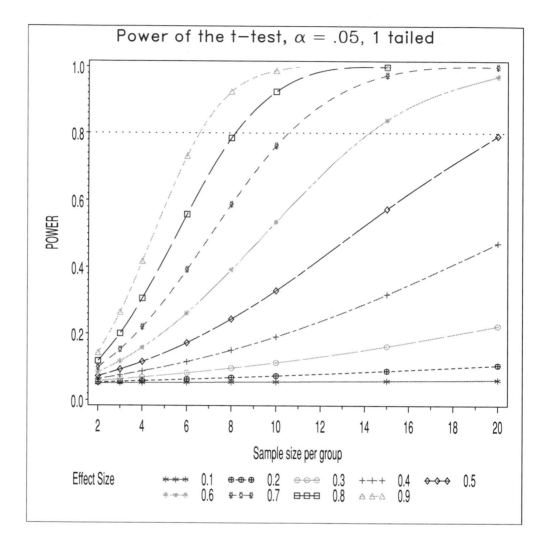

Once you have specified the value of Δ to be detected, you can use the plot to determine the power resulting from a given sample size or the sample size required to achieve a given power. For example, if a power of 0.80 is to be achieved when a difference between means of Δ=.6 standard deviations is to be detected, go across from the ordinate 0.8 as shown by the dotted line. Follow the intersection with the curve for Δ=.6 down to the abscissa to find that *n*=14 observations are required in each group.

---

*   The dotted line styles using L=33 and above are not available prior to Release 6.06 of the SAS System.

3.3 Bivariate Normal Distribution

A generalization of the normal distribution to two or more variables plays a fundamental role in multivariate analysis. The *multivariate normal distribution* of the p variables, $\mathbf{x} = \{x_1, x_2, x_3, \ldots, x_p\}'$ (centered so that the mean of each variable, $\mu_i = 0$), has the density function

$$z = f(\mathbf{x}) = \frac{1}{(2\pi)^{p/2} |\mathbf{\Sigma}|^{1/2}} e^{-1/2(\mathbf{x}' \mathbf{\Sigma}^{-1} \mathbf{x})} \tag{3.6}$$

where $\mathbf{\Sigma}$ is the variance-covariance matrix of $\mathbf{x}$. When $p = 2$ variables (x and y, for example) follow the *bivariate normal distribution*, the variance-covariance matrix in equation 3.6 can be specified in terms of three parameters: $\sigma_x^2 =$ variance of x, $\sigma_y^2 =$ variance of y, and $\rho =$ correlation between x and y.

$$\mathbf{\Sigma} = \begin{bmatrix} \sigma_x^2 & \rho\sigma_x\sigma_y \\ \rho\sigma_x\sigma_y & \sigma_y^2 \end{bmatrix}$$

The determinant $|\mathbf{\Sigma}|$ in equation 3.6 is a measure of generalized variance of the joint distribution. For two variables, $|\mathbf{\Sigma}| = \sigma_x^2 \sigma_y^2 (1 - \rho^2)$, so the generalized variance is greatest when the two variables are uncorrelated ($\rho = 0$) and shrinks to 0 as the variables become perfectly correlated ($\rho = \pm 1$).

These equations are complex enough that it is difficult to understand their properties, such as how the shape of the distribution depends on the correlation. Some of these properties can be better understood through a graphical presentation.

The program below uses the G3D procedure and the GCONTOUR procedure to plot the density of the bivariate normal, $z = f(x,y)$. The values of Z are calculated in a DATA step for values of X and Y over the range -3 to $+3$. For generality, the values of the parameters are assigned to the macro variables VX, VY (variances), and RHO (correlation). You can determine the effect of changing the correlation, for example, by varying the value of RHO in this program.

The PROC G3D step uses XYTYPE=2 to produce curves parallel to the y axis only. The spacing of the y values is made smaller to produce smoother curves. The plot in Output 3.9 shows that the (conditional) distribution of Y is a univariate normal distribution for any value of X. The program also uses PROC GCONTOUR to display the same data (Output 3.10).

Output 3.9
G3D Procedure
Plot of Bivariate
Normal
Distribution

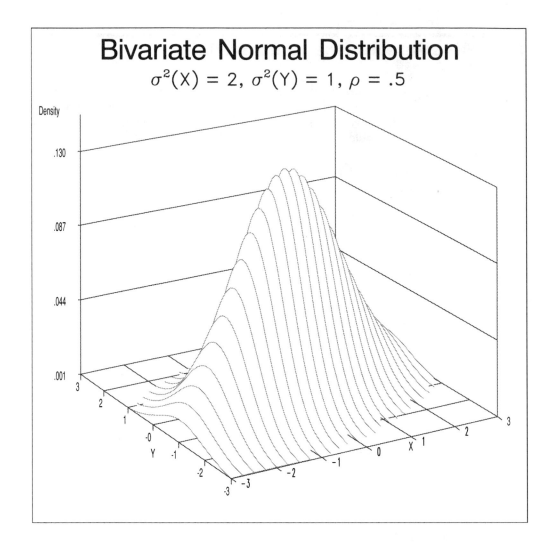

Note that the TITLE2 statement uses the characters 's' and 'r' in the CGREEK font to draw the symbols σ and ρ in the plot subtitle. To draw σ^2, it uses MOVE options (M=) that are assigned to a macro variable, SQUARED, so they can be used twice in the TITLE2 statement.

```
%let vx = 2;              /* variance of x */
%let vy = 1;              /* variance of y */
%let rho= .5;             /* correlation   */
data normal2;
   vx = &vx;
   vy = &vy;
   r  = &rho;
   keep x y z;
   label z='Density';
   constant = 1 / ( 2 * 3.141592654 * sqrt(vx*vy * (1-r**2)) );
```

```
    do x = -3 to 3  by 0.25;
       do y = -3 to 3  by 0.10;
          zx = x / sqrt(vx);      /* standardize */
          zy = y / sqrt(vy);
          hx = zx**2 + zy**2 - 2*r*zx*zy ;
          z = constant * exp(-hx/(2*(1-r**2)));
          if z > .001 then output;
       end;
    end;

proc g3d data=normal2  ;
    plot y * x = z /
        grid xytype=2              /* curves parallel to y axis */
        xticknum=7
        yticknum=7
        rotate =30
        name='GB0309';
    format x y 3.0 z 4.3;
title1 c=black f=swiss h=2.5
       'Bivariate Normal Distribution';
%let squared= m=(+.1, +.75) h=1 '2' m=(+.1, -.75) h=1.7;
title2 c=blue   f=duplex h=1.7
        f=cgreek 's' &squared f=duplex "(X) = &vx, "
        f=cgreek 's' &squared f=duplex "(Y) = &vy, "
        f=cgreek 'r' f=duplex " = &rho" ;
run;
```

It is useful to show the regression line predicting *y* from *x* on the contour plot. Because both variables are represented here in centered form (with means of zero), the regression line has an intercept of zero. The slope is $\rho(\sigma_y/\sigma_x)$. The line is drawn by the Annotate data set LINE, using the values of the macro variables for the parameters. You could plot the regression line predicting *x* from *y* by interchanging the roles of X and Y in the DATA step. The following statements produce Output 3.10. (A color version of this output appears as Output A3.11 in Appendix 3.)

```
    *-- Draw the regression line in the contour plot;
    data line;
       retain xsys ysys '2';
       vx = &vx; vy = &vy; r = &rho;
       x = -3;
       y =  r * sqrt(vy / vx) * x;
       function = 'MOVE';   output;
       x =   3;
       y =  r * sqrt(vy / vx) * x;
       function = 'DRAW';   output;
```

```
proc gcontour data=normal2  ;
   plot y * x = z /
      annotate= line
      levels =  .025 .05 .075 .10 .125
      llevels=  20   21  22   23  1
      clevels=  green blue red green blue
      caxis  = black
      vaxis  = axis1
      haxis  = axis1
      legend = legend1
      name='GB0310';
   axis1 label=(h=1.5) value=(h=1.3)
         order=(-3 to 3) offset=(1);
   legend1 label=(h=1.5) value=(h=1.3);
```

Output 3.10
Contour Plot of
Bivariate Normal
Distribution

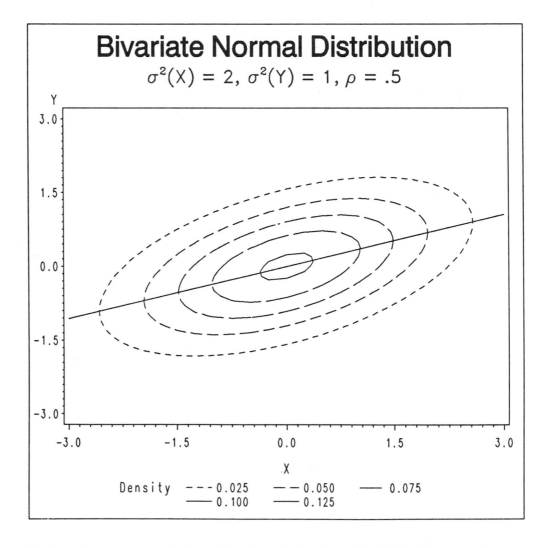

Each contour corresponds to a slice through the bivariate distribution at a given height above the X, Y plane. The plot shows that the contours of constant density in a bivariate normal distribution are a family of concentric ellipses. Note that the regression line passes through the points where the tangents to the ellipses are vertical. Section 4.5, "Enhanced Scatterplots," examines other properties of these elliptical contours and shows you how to draw a *data ellipse*, which includes a specified proportion of the data, assuming the variables have a bivariate normal distribution.

3.4   Histogram Smoothing and Density Estimation

For empirical data, a histogram provides the simplest display of the shape of the density function across the range of the variable. By its nature, however, the histogram is discrete, and more useful displays are based on some smoothed version of a histogram. Another problem with the histogram is that the shape of the distribution depends, more strongly than we would like, on the size and location of the class intervals used.

If you are willing to assume that the data follow a particular probability distribution, to a reasonable approximation, that distribution can be fit to the data and the smooth fit superimposed on the histogram. PROC CAPABILITY in SAS/QC software will do this for a variety of distributions.

On the other hand, in situations where there is no clear reason for preferring one theoretical distribution to another, it may be useful to estimate the shape of the data distribution by nonparametric means.

3.4.1   Density Displays in the CAPABILITY Procedure

Process capability analysis is a method of quality control designed to compare the distribution of results from some process with its design specifications. PROC CAPABILITY in SAS/QC software can produce a variety of graphic displays, including histograms superimposed with fitted density curves representing any of the normal, lognormal, exponential, Weibull, and gamma distributions. Though many of the features of PROC CAPABILITY are tuned for quality control applications, it can be used as a simple method for fitting these probability distributions to empirical data.

Chapter 2, "Graphical Methods for Univariate Data," examines histogram displays of variables in the BASEBALL data set. This section uses the number of runs scored variable (RUNS) to illustrate PROC CAPABILITY. Of the distributions available, the lognormal and gamma distributions are most reasonable to try to fit the skewed distribution of the RUNS variable. Alternatively, we might try to transform the data first to make them symmetrical. For the RUNS variable, methods described in Section 3.6.4, "Finding Transformations for Symmetry," suggest that $\sqrt{\text{RUNS}}$ is reasonably symmetrical.

The program below performs these analyses for the raw and transformed variables. The first HISTOGRAM statement produces the graph shown in Output 3.11 for the raw data with the fitted lognormal and gamma curves. The smooth curves show the fitted distributions.

Output 3.11
CAPABILITY
Procedure Plot for
RUNS

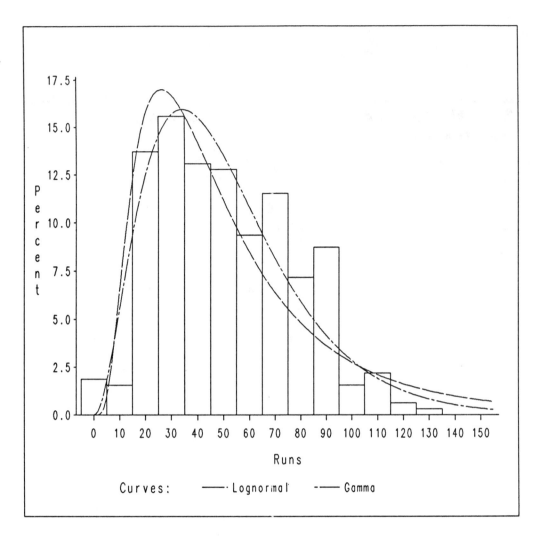

The lognormal and gamma distributions both require that the data values are strictly greater than the threshold parameter $\theta=0$, so values of RUNS=0 are deleted in a DATA step.

```
%include BASEBALL;           /* On MVS, use %include ddname(file);*/
title ' ';
data baseball;
   set baseball;
   if runs > 0;              * delete zero values for lognormal;
   sqrtrun = sqrt(runs);
   label sqrtrun='Square root of runs';

symbol1 v=none  c=black l=1 r=3;     /* histogram        */
symbol4 v=none  c=black l=3;         /* normal curve     */
symbol5 v=none  c=red   l=5 r=3;     /* lognormal curve  */
symbol8 v=none  c=green l=9;         /* gamma curve      */
```

```
proc capability data=baseball graphics  ;
   var runs sqrtrun;
   histogram runs/
      midpoints = 0 to 150 by 10
      lognormal gamma
      haxis=axis1 vaxis=axis2
      noframe legend=legend1
      name='GB0311';
   histogram sqrtrun/
      midpoints = 0 to 15
      normal
      haxis=axis1 vaxis=axis2
      noframe legend=legend1
      name='GB0312';
   axis1   label=(h=1.5) value=(h=1.3);
   axis2   label=(h=1.5) value=(h=1.3);
   legend1 label=(h=1.5) value=(h=1.3);
```

Note that PROC CAPABILITY uses the SYMBOL statements differently than they are used in SAS/GRAPH procedures. In the program above, the SYMBOL1 statement controls the histogram bars; the SYMBOL4, SYMBOL5, and SYMBOL8 statements control the appearance of the fitted normal, lognormal, and gamma curves, respectively. See "Using SYMBOL Statements" in Chapter 2, "The CAPABILITY Procedure," in *SAS/QC Software: Reference, Version 6, First Edition* for details.

The second HISTOGRAM statement in the program above fits the normal distribution to the square root variable and results on the plot shown in Output 3.12. On the plot, the smooth curve shows the fitted normal distribution.

Output 3.12
CAPABILITY
Procedure Plot for
Square Root of
RUNS

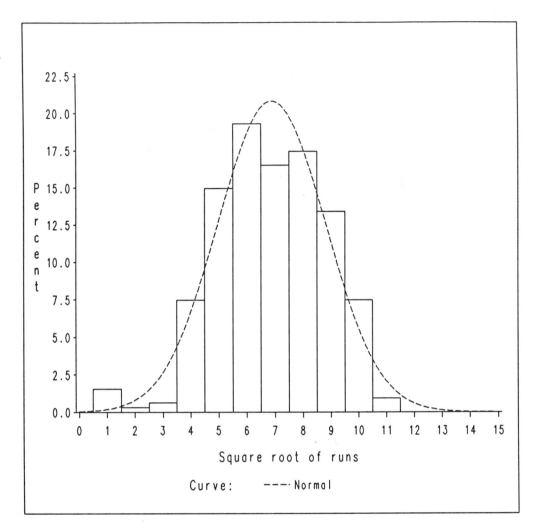

In fitting a parametric model, it is important to examine the goodness of fit between the observed and fitted frequencies. The output from PROC CAPABILITY includes the χ^2 tests shown in Output 3.13. While the best fit (the smallest value of χ^2) is achieved with the normal distribution for the transformed data, the χ^2 tests indicate that none of these provides an acceptable fit—all of the p values (PR>CHI-SQUARE) are highly significant.

Output 3.13
CAPABILITY
Procedure
Goodness-of-Fit
Chi Square Tests
for RUNS

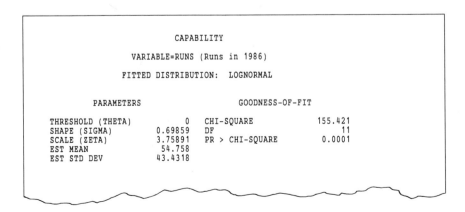

```
                        CAPABILITY

              VARIABLE=RUNS (Runs in 1986)

          FITTED DISTRIBUTION:  LOGNORMAL

          PARAMETERS                GOODNESS-OF-FIT

    THRESHOLD (THETA)        0   CHI-SQUARE           155.421
    SHAPE (SIGMA)      0.69859   DF                        11
    SCALE (ZETA)       3.75891   PR > CHI-SQUARE       0.0001
    EST MEAN            54.758
    EST STD DEV        43.4318
```

```
                    VARIABLE=RUNS (Runs in 1986)

               FITTED DISTRIBUTION:  GAMMA


            PARAMETERS                    GOODNESS-OF-FIT

    THRESHOLD (THETA)         0   CHI-SQUARE          60.4078
    SCALE (SIGMA)       16.8759   DF                       11
    SHAPE (ALPHA)       3.02613   PR > CHI-SQUARE      0.0001
    EST MEAN            51.0685
    EST STD DEV         29.3569


               VARIABLE=SQRTRUN (Square root of runs)

               FITTED DISTRIBUTION:  NORMAL


            PARAMETERS                    GOODNESS-OF-FIT

    MEAN (MU)           6.88533   CHI-SQUARE          44.4785
    STD DEV (SIGMA)     1.91629   DF                        8
                                  PR > CHI-SQUARE      0.0001
```

3.4.2 Nonparametric Density Estimation

Nonparametric techniques attempt to estimate the density function, $f(x)$, of a data distribution without assuming a particular form for that distribution. The estimate produced, $\hat{f}(x)$, can be thought of as a smoothed histogram. Smoothing a histogram involves a trade-off between fidelity to the data (low bias) and smoothness (low variance). The methods described below use a smoothing parameter, h, called the bandwidth or (half) window width that allows you to control the degree of smoothness. The description below follows Chambers et al. (1983) and Fox (1990).

The histogram display is discrete for two reasons:

□   Each observation is grouped into exactly one class interval.

□   All the observations in each interval are treated as if they were spread uniformly throughout that interval.

A solution to this problem is to compute the local density in some neighborhood around each of a set of equally spaced values in the range of x. Let h be the half-width of a window around x. Then the local density in the interval $[x-h, x+h]$ can be defined as

$$\hat{f}(x) = \frac{\text{number of observations in } [x - h, x + h]}{2h\, n} \ . \tag{3.7}$$

This idea, proposed by Rosenblatt (1956), generalizes the histogram to a density estimate that can be computed at any point x. The importance of this idea is that the local density can be formulated as spreading each observation, x_i, over the interval $x_i - h$ to $x_i + h$ with a weight function, $W(\bullet)$. That is, equation 3.7 is equivalent to

$$\hat{f}(x) = \frac{1}{h\, n} \sum_{i=1}^{n} W\left(\frac{x - x_i}{h}\right) \ , \tag{3.8}$$

using the uniform weight function,

$$W(z) = \begin{cases} 1/2 & \text{if } |z| < 1 \\ 0 & \text{otherwise} \end{cases} \quad .$$

(3.9)

Silverman (1986) calls this the *naive density estimator*. In effect, the fixed intervals of the histogram are replaced by a moving window of width $2h$, within which each observation is smeared uniformly. Output 3.14 shows the density plot for the RUNS data computed with equation 3.8 at 200 equally spaced points between the minimum and maximum values, using a window half-width $h=10$.

Output 3.14
Naive Density Plot
for RUNS Data

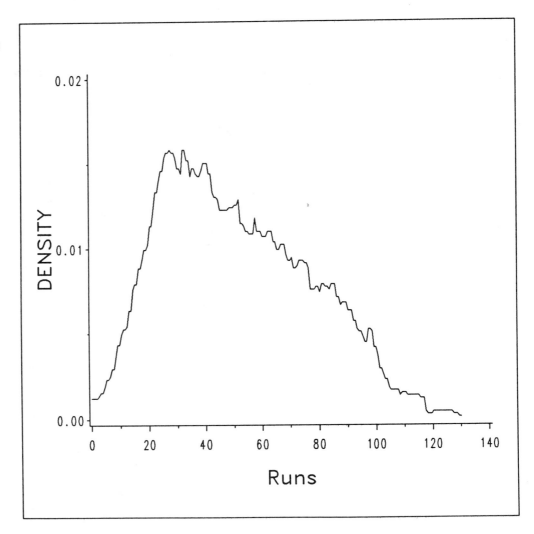

The density plot in Output 3.14 is more detailed than the histogram, but it is still jagged. The density plot can be made smoother by replacing the uniform function (equation 3.9) with something that decreases gradually to zero as we get closer to the edges of the window. The *kernel density estimator* uses a weighted average of the distances $(x - x_i)/h$,

$$\hat{f}(x) = \frac{1}{h\,n} \sum_{i=1}^{n} K\left(\frac{x - x_i}{h}\right) \quad , \tag{3.10}$$

where the weights are determined by the kernel function, $K(\bullet)$, and the window half-width, h. The kernel determines the shape of the distribution of weights; the window width determines the range of influence of a given observation and controls the trade-off between smoothness and fidelity.

One simple choice for the kernel function is the Gaussian kernel:

$$K_G(z) = \frac{1}{\sqrt{2\,\pi}} e^{-z^2/2} \quad . \tag{3.11}$$

Another possibility is the Biweight (or quartic) kernel:

$$K_B(z) = \begin{cases} \dfrac{15}{16}(1 - z^2)^2 & \text{if } |z| < 1 \\ 0 & \text{otherwise} \end{cases} \quad .$$

The uniform function is compared graphically to the Gaussian and quartic kernels in Output 3.15. All three functions are defined so that they enclose an area of 1.0. This means that as each observation x_i is spread over the window, it still contributes a total frequency of one to the density estimate. The Gaussian and Biweight functions provide a smoother density estimate than the uniform.

Output 3.15
Weight Functions
Used in Kernel
Density Estimates

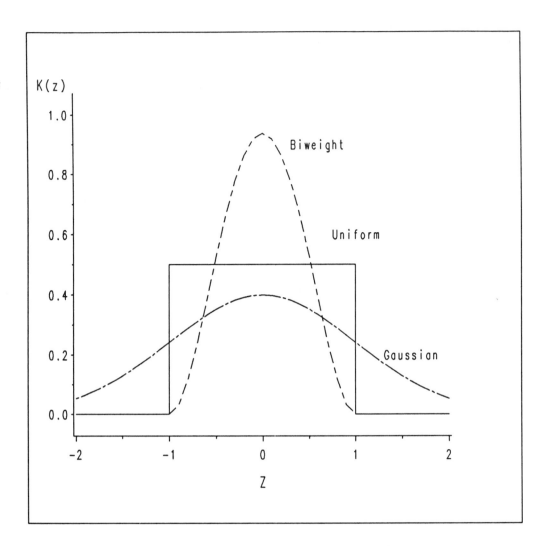

As long as the kernel itself is smooth, it turns out that the choice of the window width has a greater impact on the smoothness of the density estimate than the particular kernel function used. The examples shown here all use the Gaussian kernel. Output 3.16 shows the smoothed densities for the RUNS data using $h=$ 2, 8, 14, and 20. With the smallest window width, $h=2$, the density trace follows the data distribution closely (giving $\hat{f}(x)$ with small bias), but has large variance. The largest window width has the opposite effect (small variance, but large bias). The panel for $h=8$ is close to the optimal, as described below. (Output 3.16 was prepared using the DENSITY macro, described below. The four panels were assembled with the GREPLAY procedure.)

Output 3.16   *Density Estimates for the RUNS Data for Four Window Widths*

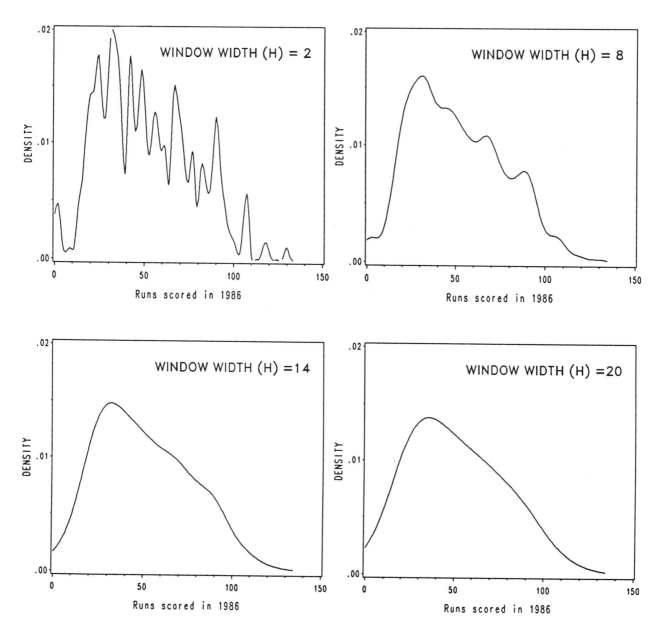

A reasonable choice, suggested by Silverman (1986), is based on minimizing a sum of variance and squared bias when the underlying distribution is Gaussian but performs well as long as the distribution is not too highly skewed or has widely separated modes. Silverman's formula sets

$$h = 0.9\,\hat{s}\,/\,n^{1/5} \quad,\qquad\qquad\qquad\qquad (3.13)$$

where $\hat{s}$ is the smaller of two estimates of σ in a normal distribution, the usual standard deviation, and the approximation based on the interquartile range or H-spread:

$$\hat{s} = \min \left\{ \frac{\Sigma(x_i - \bar{x})^2}{(n-1)}, \frac{\text{H-spread}}{1.349} \right\} \quad . \tag{3.14}$$

(In a standard normal distribution, the interquartile range is 1.349, and the robust estimate $\hat{\sigma}$=H-spread/1.349 uses only the data in the middle of the distribution.) For the RUNS data, the standard deviation is 26.02, H-spread/1.349 is 29.09, so equation 3.13 gives h=7.38. Thus, the panel for h=8 in Output 3.16 is close to the optimal.

3.4.3 Computing Density Estimates

Several SAS procedures use nonparametric density estimation as part of a larger process. In Version 5, the MODECLUS procedure calculates density estimates with a uniform kernel in order to cluster the observations into regions of high concentration (density). For single variables, however, the density estimates produced by the uniform kernel have the jagged look of Output 3.14. In Version 6, the DISCRIM procedure provides density estimates using the Gaussian, Biweight, and other kernels as nonparametric methods of classifying multivariate observations. The use of PROC DISCRIM, however, requires more than one group in the data and produces a separate smoothed density for each group. See Chapter 20, "The DISCRIM Procedure," Examples 1 and 4, in the *SAT/STAT User's Guide, Version 6, Fourth Edition, Volume 1* for univariate density estimation and plotting using the iris data (Section A2.10, "The IRIS Data Set: Iris Data").

The calculation of kernel density estimates can be done directly using the IML procedure. A general SAS macro, DENSITY, provided in Section A1.7, "The DENSITY Macro," is a simplified version of the program presented by Longbotham (1987). The essential parts of the program are shown below. The module WINDOW calculates the window half-width using Silverman's formula (equation 3.13). The observations are assumed to be previously sorted in the one-column matrix XA.

```
start WINDOW;    *-- Calculate default window width;
   mean = xa[+,]/n;
   css = ssq(xa - mean);
   stddev = sqrt(css/(n-1));
   q1 = floor(((n+3)/4) || ((n+6)/4));
   q1 = (xa[q1,]) [+,]/2;
   q3 = ceil(((3*n+1)/4) || ((3*n-2)/4));
   q3 = (xa [q3,]) [+,]/2;
   quartsig = (q3 - q1)/1.349;
   h  = .9*min(stddev,quartsig) * n##(-.2);  * Silvermans formula;
   finish;
```

The density values, using the Gaussian kernel, are calculated by the module DENSITY for the values in the matrix XA, for NX values of *x* in increments of DX.

```
start DENSITY;
   fnx = j(nx,3,0);
   vars = {"DENSITY" "&VAR" "WINDOW"};
   create &out from fnx [colname=vars];
   sigmasqr = .32653;                    * scale constant for kernel ;
   gconst = sqrt(2*3.14159*sigmasqr);
   nuh = n*h;
   x = xf - dx;
   do i = 1 to nx;
      x = x + dx;
      y = (j(n,1,x) - xa)/h;
      ky = exp(-.5*y#y / sigmasqr) / gconst;   * Gaussian kernel;
      fnx[i,1] = sum(ky)/(nuh);
      fnx[i,2] = x;
      end;
   fnx[,3] = round(h,.001);
   append from fnx;
   finish;
```

This direct method of evaluating the density estimate at, for example, *k* points involves computation proportional to *k n*, which can be slow in large samples. Silverman (1982) gives a much more efficient algorithm based on the fast Fourier transform.

3.5 Quantile Plots

A common problem in statistical analysis is to determine how well a given set of data or results from some analysis (such as residuals) follows a particular theoretical distribution. While statistical tests exist for many distributions (such as the normal), a general, graphical approach is to plot the ordered data values against the expected values of those observations if they followed the particular reference distribution. To the extent that the data follow the reference distribution, the points in such a plot will plot as a straight line; departures from the line show how the data differ from the assumed distribution.

Choice of a graphical rather than a hypothesis-testing approach is not simply a matter of taste. If the data do not agree with the theoretical distribution, we need to know *how* they differ. For example, standard least squares techniques perform relatively well with compressed, light-tailed data distributions but lose efficiency rapidly with elongated, heavy-tailed data.

To describe this plot, let x_i, $i=1, \ldots, n$ refer to the data values, and let $x_{(1)}$, $x_{(2)}, \ldots, x_{(i)}, \ldots, x_{(n)}$ denote the data values sorted from smallest to largest. The $x_{(i)}$ are called the *order statistics* of the data values. The *quantiles* of the distribution of *x* are a generalization of the idea of percentiles: The 25th and 75th percentiles, for example, are the values of *x* below which 25 percent and 75 percent of the data values fall, respectively. Analogously, a quantile of the distribution of *x* is defined as the value of *x* below which a proportion, *p*, of data values fall.

In particular, if we let $p_i = (i - .5)/n$, for $i = 1$ to n, then the corresponding quantile, $Q(p_i)$, is just the ordered value $x_{(i)}$.* For example, if there are $n = 100$ observations, then the smallest observation, $x_{(1)}$, is the quantile $p_1 = .005$, $x_{(2)}$ is the quantile $p_2 = .015$, and so forth.

The corresponding quantiles for the reference distribution, such as the normal, come from the cumulative distribution function, $F(z)$, of that theoretical distribution. That is, corresponding to the data value, $x_{(i)} = Q(p_i)$, that exceeds the proportion p_i of the data values, find the value $z_i = F^{-1}(p_i)$ that exceeds the same proportion in the theoretical distribution. For the normal distribution, z_i is the normal deviate corresponding to the cumulative proportion p_i. Computationally, z_i can be obtained with the SAS function PROBIT (p).

The plot of the empirical quantiles $x_{(i)}$ against the theoretical quantiles z_i is called a *theoretical quantile-quantile plot* or a *Q-Q plot* for short. When the reference distribution is the normal, the plot is commonly called a *normal probability plot* or *normal Q-Q plot*.

The following account of the properties of Q-Q plots and their interpretation is necessarily brief. This topic is described in detail by Chambers et al. (1983). Quantile plots and related techniques from the perspective of exploratory data analysis are discussed by Hoaglin (1985a).

3.5.1 Interpreting Q-Q Plots

Output 3.17 shows normal Q-Q plots of four sets of artificial data, each with $n = 100$ observations.

---

* Taking $p_i = (i - .5)/n$ avoids problems with cumulative proportions of zero and one that have quantiles at infinity in the normal distribution. This choice splits each observation at its value, assigning half the observation below, and half above $x_{(i)}$. Other approximations are commonly used. For the normal distribution, Blom (1958) suggests using $p_i = (i - 3/8)/(n + 1/4)$, which gives a better approximation to expected order statistics in the tails for small samples. In samples of $n \geq 100$, the difference between these adjustment values is negligible.

Output 3.17   *Normal Q-Q Plots for Simulated Data*

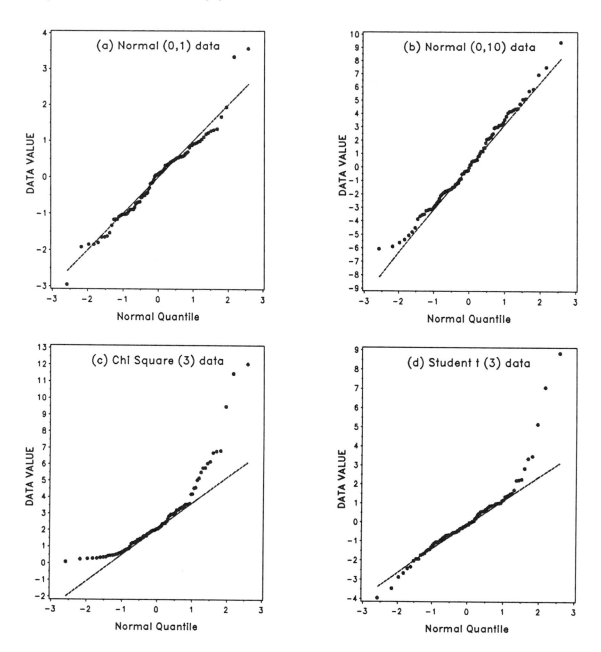

In the top two panels, the data are drawn from normal distributions: a standard normal, $N(0, 1)$, on the left (a) and a $N(0, 10)$ distribution with $\sigma^2 = 10$ on the right (b). The comparison line is $x_{(i)} = \mu + \sqrt{\sigma}\, z_i$, which is $x_{(i)} = z_i$ on the left and $x_{(i)} = \sqrt{10}\, z_i$ on the right. These examples illustrate that when the distribution of x has the same shape as the reference distribution, the Q-Q plot is linear. For the standard normal reference distribution, the slope of the line estimates σ, and the intercept estimates μ of the distribution of x.

The bottom panels show corresponding plots of data from two nonnormal distributions: a positively skewed χ^2 distribution with three degrees of freedom on the left (c) and an elongated, heavy-tailed Student's t distribution with three degrees of freedom on the right (d). As these examples show, skewness causes both ends of the plot to deflect in the same direction, while kurtosis causes them to deflect in opposite directions. Negatively skewed data would look like the panel for χ^2 reflected about the comparison line, and data from a compressed, short-tailed distribution, like the uniform distribution, would look like the reflection of the Student t panel. But note that the deviations from the line usually occur in the tails; in the middle, the observations appear fairly close to the line.

When the data deviate substantially from the normal, robust estimators of the mean and standard deviation of the corresponding normal distribution may be preferable to using $\bar{x}$ and $s^2 = \Sigma(x_i - \bar{x})^2/(n-1)$. The lines in Output 3.17 use the median x to estimate μ and the estimate of σ based on the interquartile range, or H-spread, given in equation 3.14. This choice gives better comparison lines for the bottom two panels.

Drawing the comparison line on the Q-Q plot helps you judge how well the data fit the shape of the comparison distribution. To see whether the distribution of the data differs from the comparison distribution, look for a systematic pattern of departure from the comparison line. When the pattern of deviation from the line is large, as in the bottom panels of Output 3.17, the diagnosis is fairly easy. A positively skewed distribution, like the χ^2 in panel (c), bends above the line at both ends; a negatively skewed distribution would bend down at both ends. A distribution with heavy tails (relative to the normal) dips below the line at the low end and rises above it at the high end, as does the t data in panel (d).

However, medium and small departures are harder to detect. Untilting the plot, by subtracting the values of the comparison line from the points, makes the comparison distribution plot as a horizontal line, and patterns of departure from a horizontal line are much easier to see. Thus, we plot $x_{(i)} - z_i$ against z_i. This *detrended* version of the Q-Q plot is shown in Output 3.18 for the same data plotted in Output 3.17. Note that the vertical scale in the detrended plot is much expanded, so that the departures from the line are viewed as if under a microscope.

Normal Q-Q plots can also be used to detect outliers, which appear as stragglers, separated from the comparison line at either end of the distribution. In Output 3.17(a), for example, two points seem to stray from the top end of the line. In Output 3.18, which displays the same data as Output 3.17, these two points appear as large deviations in the corresponding panel. Plotting the deviation from the comparison line makes it easier to see the pattern of disagreement.

Output 3.18   *Detrended Normal Q-Q Plots for Simulated Data*

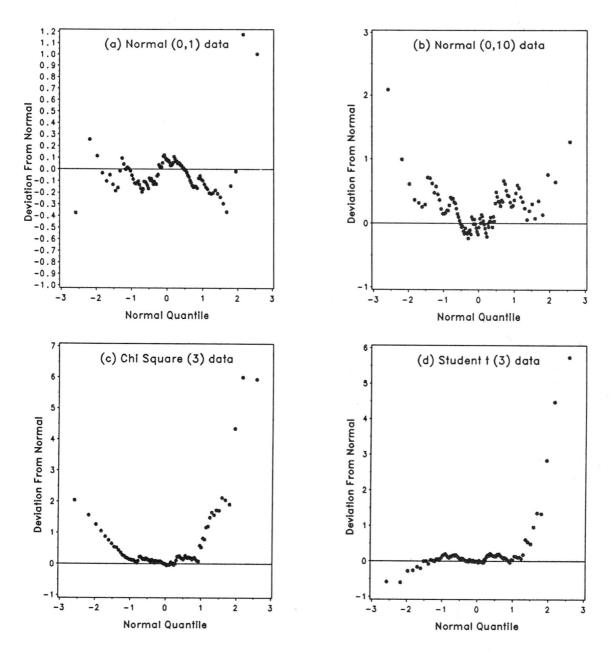

We should be cautious, however, in declaring points to be outliers solely on the basis of appearances. These two observations are in fact generated from the same $N(0, 1)$ distribution as the rest, and as the points get further from the mean of x we should expect greater variability. Adding confidence limits to the Q-Q plot, illustrated in Section 3.5.3, "Standard Errors for Normal Q-Q Plots," helps us judge how big a discrepancy we should worry about.

The normal Q-Q plots in Output 3.17 and 3.18 are produced with the NQPLOT macro program listed in Section A1.10, "The NQPLOT Macro." The construction of these plots is described in Section 3.5.2, "Constructing Normal Q-Q Plots." The panels of these figures are assembled using PROC GREPLAY.

Examples: Baseball Data

Some of the features of normal Q-Q plots are illustrated with data on major league baseball players from the BASEBALL data set (described in Section 2.2, "Histograms and Bar Charts"). Output 3.19 shows the standard and detrended versions of the Q-Q plot for ATBAT, the number of times a player was at bat in 1986.

Output 3.19   *Q-Q Plots for Times at Bat*

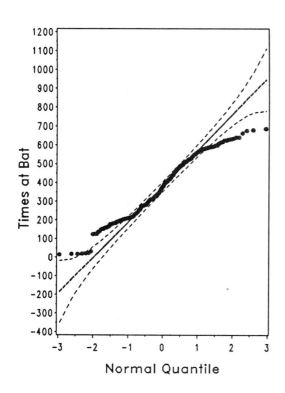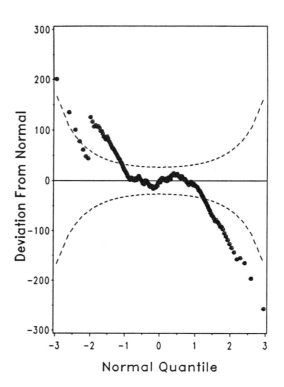

The distribution is quite short-tailed compared to the normal, and the pattern is like a reflected version of the plot of Student $t(3)$ in Output 3.17(d) and Output 3.18(d). There is a curious gap in the points at the left, which turns out to correspond to seven pinch hitters who played very little in 1986.

Output 3.20 displays the RUNS variable, number of runs scored in 1986. This variable appears to be slightly skewed and short-tailed relative to the normal.

Output 3.20   *Q-Q Plots for Runs in 1986*

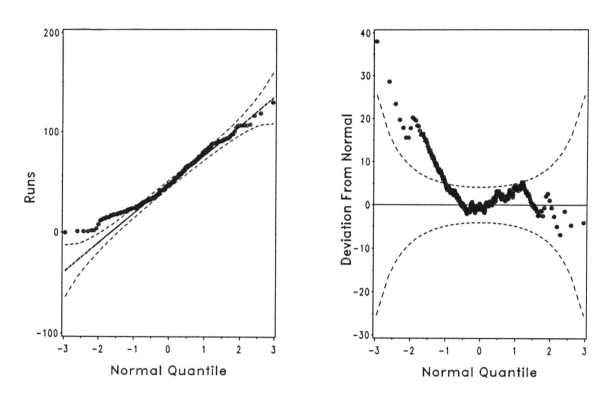

Output 3.21 shows the plots for the variable YEARS, years in the major leagues. The ordinary Q-Q plot shows a set of horizontal rows of points that appear as diagonal lines in the detrended plot. This pattern stems from the fact that YEARS is quite discrete—the values are all integers—while the normal quantiles are continuous. In addition, the upward bowing of the points indicates a positively skewed distribution.

Output 3.21   *Q-Q Plots for Years in the Major Leagues*

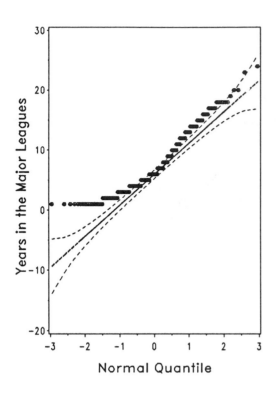
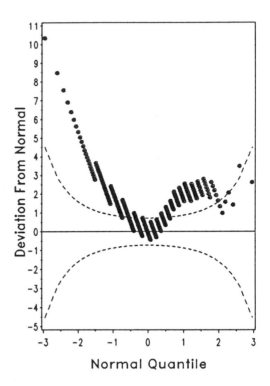

3.5.2 Constructing Normal Q-Q Plots

Normal Q-Q plots can be drawn in the SAS System in several ways. The simplest (and least versatile) plot is given by the PLOT option in the UNIVARIATE procedure. The normal quantiles can be calculated by the RANK procedure or in a DATA step with the PROBIT function, and these values can be plotted against the data quantiles with the PLOT procedure or the GPLOT procedure. Finally, PROC CAPABILITY in SAS/QC software can produce normal Q-Q plots (and Q-Q plots for several other theoretical distributions) directly.

This section illustrates the construction of these plots with data on infant mortality rates in nations in the world. The data set NATIONS is described in Section 4.1. The variable IMR contains the infant mortality rate. The normal Q-Q plot from PROC UNIVARIATE shown in Output 3.22 is produced by the PLOT option, with the following statements:

```
%include NATIONS;
proc univariate plot data=nations;
   var imr;
```

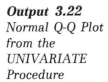

Output 3.22
Normal Q-Q Plot
from the
UNIVARIATE
Procedure

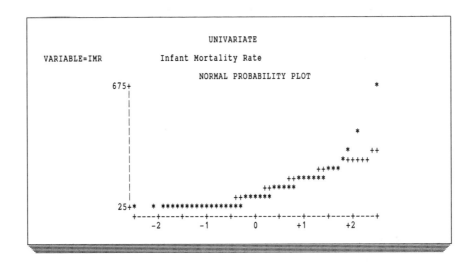

The PROC UNIVARIATE plot is fine for a quick look at the data, but because the vertical resolution of the plot is determined internally, you have no control over the scaling of the vertical (data) axis. As a result, the plot is sometimes squeezed into a few lines.

With PROC CAPABILITY in SAS/QC software, you can use the QQPLOT statement to generate Q-Q plots for the normal distribution and five other families: the beta, exponential, gamma, lognormal, and Weibull distributions. The statements below are used to draw a normal Q-Q plot, shown in Output 3.23, for the infant mortality data:

```
proc capability noprint graphics data=nations  ;
   qqplot imr /
      normal(mu=89 sigma=90.8)      /* sample mean and sigma */
      rankadj = -.5     nadj = 0    /* p = (i - .5)/n         */
      ctext=black
      font=duplex
      name='GB0323';
      symbol1 i=none h=1.1 v=- c=black r=3;  /* data points     */
      symbol4 i=join l=20 v=none c=blue;     /* comparison line */
```

Output 3.23
Normal Q-Q Plot for IMR from the CAPABILITY Procedure

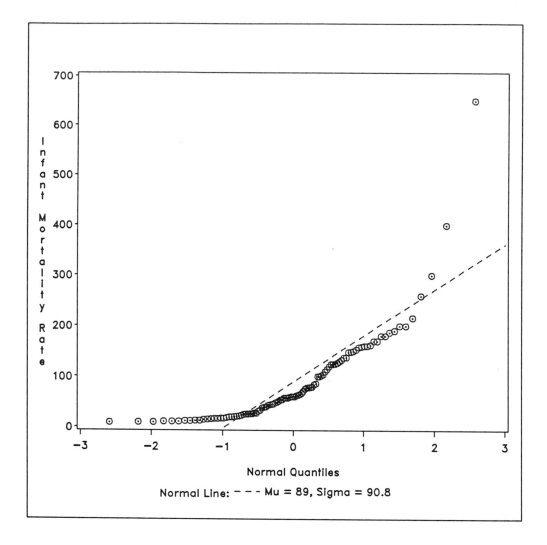

In the statements above, the NORMAL option specifies the mean and standard deviation of the distribution and causes the procedure to plot the comparison line. The RANKADJ and NADJ options allow different choices for the adjustment constants in calculating the proportions, $p_i = (i + \text{RANKADJ})/(n + \text{NADJ})$. See *SAS/QC Software: Reference, Version 6, First Edition* for further details.

To produce your own plot with SAS/GRAPH software, you can obtain normal quantiles with PROC RANK as shown below. If you specify NORMAL=BLOM, the variable Z named in the RANKS statement will be calculated as the normal quantiles corresponding to proportions $p_i = (i - 3/8)/(n + 1/4)$. The comparison line for the normal curve is calculated by linearly transforming the Z variable to the same mean and standard deviation as the data values. Alternatively, when the data are highly skewed or contain outliers (as in the IMR data), a robust comparison line may be obtained by using the median and hinge spread estimates instead of mean and standard deviation.

```
**- Normal Q-Q plot using PROC RANK --;
proc rank data=nations normal=blom out=quantile;
   var   imr;          * the input variable;
   ranks z ;           * the output variable;
   label z  = 'Normal Quantile';
```

```
proc univariate noprint data=nations;
   var imr;
   output out=stats n=nobs median=median qrange=hspr
                         mean=mean std=std;
data quantile;
   set quantile;
   if _n_=1 then set stats;
   normal = mean    + z * std;
   robust = median + z * hspr/1.349;

proc gplot data=quantile  ;
   plot imr    * z = 1
        normal* z = 2
        robust* z = 3
        / overlay frame
          hminor=1 vminor=1
          vaxis=axis1 haxis=axis2
          name='GB0324' ;
   symbol1 i=none h=1.1 v=-    color=black;
   symbol2 i=join l=20  v=none color=blue;
   symbol3 i=join l=1   v=none color=red;
   axis1  label=(h=1.5 a=90 r=0) value=(h=1.3);
   axis2  label=(h=1.5)          value=(h=1.3);
```

Output 3.24 shows the PROC GPLOT version with both normal comparison lines. The dashed comparison line uses the mean and standard deviation of the data. The solid comparison line uses robust estimates based on the median and H-spread. Because the infant mortality data are both positively skewed and have several high outliers, the robust estimators (solid line) produce a better comparison line.

Output 3.24
*Normal Q-Q Plot
from RANK
Procedure
Quantiles*

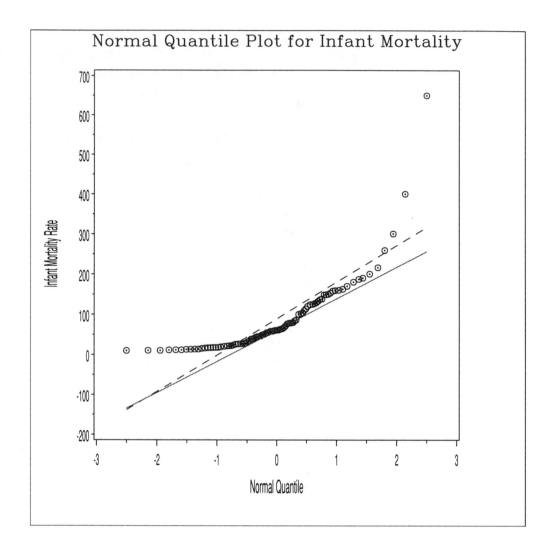

Normal Quantile Plot for Infant Mortality

If you prefer to use the proportions defined by $p_i=(i-.5)/n$, you can sort the data and calculate the normal quantiles in a DATA step with the PROBIT function as shown below. This is the most general method, because you can easily adjust the calculation of p_i, use a different F^{-1} function for Q-Q plots based on other distributions, and add calculations for standard errors, as described below. The PROC GPLOT step is the same as used with the PROC RANK method, so it is not shown.

```
**- Normal Q-Q plot using PROBIT --;
proc sort data=nations;
   by imr;
proc univariate noprint data=nations;
   var imr;
   output out=stats n=nobs median=median qrange=hspr
                    mean=mean std=std;
```

```
data quantile;
   set nations;
   if _n_=1 then set stats;
   if imr ¬= . ;                        *-- remove missing data!;
   i + 1;
   p = (i - .5) / nobs;
   z = probit(p);
   normal = mean   + z * std;
   robust = median + z * hspr/1.349;
```

3.5.3 Standard Errors for Normal Q-Q Plots

Determining when the pattern of points deviates from the comparison line sufficiently to declare that the distributions differ, or when a few stragglers are far enough away to call them outliers is, at least in part, a matter of judgment, as it should be. But points on the Q-Q plot are not equally variable—we should expect observations in the tails to vary most—and it can be quite helpful to display variability information on the plot.

One way to do this is to calculate an estimated standard error, $\hat{s}(z_i)$, of the ordinate z_i and plot curves showing the interval $z_i \pm 2 \, \hat{s}(z_i)$ to give approximate 95% confidence intervals. Chambers et al. (1983) give formulas for estimating the standard errors for most of the common comparison distributions. For the normal Q-Q plot, the standard error is estimated by

$$\hat{s}(z_i) = \frac{\hat{\sigma}}{f(z_i)} \sqrt{\frac{p_i(1 - p_i)}{n}} \tag{3.15}$$

where $f(z_i)$ is the normal density function given in equation 3.3, and $\hat{\sigma}$ estimates the slope of the comparison line, the standard deviation of x. Chambers et al. (1983) stress using a robust estimate of $\hat{\sigma}$, such as the measure based on the Hinge-spread, so that the standard errors are not distorted by outliers or straggling tails.

The program below shows the calculation of the estimated standard errors for the normal Q-Q plot of the infant mortality data. Quantities for the detrended version of the plot are also calculated, and both plots are displayed in one PROC GPLOT step. The two plots produced are shown in Output 3.25.

```
*-- Normal Q-Q plot with standard errors--;
data quantile;
   set nations;
   if _n_=1 then set stats;
   if imr ¬= . ;                        *-- remove missing data!;
   i + 1;
   p = (i - .5) / nobs;
   z = probit(p);
   sigma  = hspr / 1.349;
   normal = median + z * sigma;         *-- use robust estimates;
   se = (sigma/((1/sqrt(2*3.1415926))*exp(-(z**2)/2)))
        *sqrt(p*(1-p)/nobs);
   lower = normal - 2*se;
   upper = normal + 2*se;
```

```
            resid = imr - normal;                *-- deviation from normal;
            reslo  = -2*se;                       *    +/- 2 SEs ;
            reshi  = 2*se;
            label z   ='Normal Quantile'
                  resid='Deviation From Normal';
         proc gplot data=quantile;
            plot imr   * z = 1
                 normal* z = 2
                 lower  * z = 3
                 upper  * z = 3
                 / overlay frame vaxis=axis1 haxis=axis2;
            plot resid * z = 1
                 reslo  * z = 3
                 reshi  * z = 3
                 / overlay frame vaxis=axis1 haxis=axis2 vref=0;
            symbol1 i=none h=1.1 v=- color=black;
            symbol2 i=join l=3  v=none color=red ;
            symbol3 i=join l=20 v=none color=green;
            axis1  label=(a=90 r=0 h=1.5) value=(h=1.3);
            axis2  label=(h=1.5)          value=(h=1.3);
```

Output 3.25   *Normal Q-Q Plot with Standard Errors*

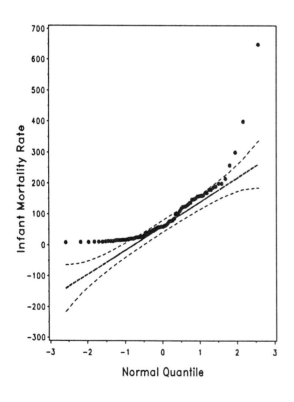
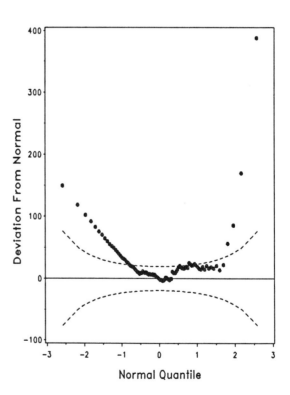

3.5.4 NQPLOT Macro

A SAS macro program, NQPLOT, which prepares normal Q-Q plots in various forms, is listed in Section A1.10, "The NQPLOT Macro." By default, the program plots both the standard and detrended versions together with standard error curves. You can select the classical or robust estimates for the comparison line with the MU= and SIGMA= parameters. The macro program uses the following syntax:

```
%macro nqplot (
        data=_LAST_,     /* input data set                   */
        var=x,           /* variable to be plotted           */
        out=nqplot,      /* output data set                  */
        mu=MEDIAN,       /* est of mean of normal distribution: */
                         /*  MEAN, MEDIAN or literal value   */
        sigma=HSPR,      /* est of std deviation of normal:  */
                         /*  STD, HSPR, or literal value     */
        stderr=YES,      /* plot std errors around curves?   */
        detrend=YES,     /* plot detrended version?          */
        lh=1.5,          /* height for axis labels           */
        anno=,           /* name of input annotate data set  */
        name=NQPLOT,     /* name of graphic catalog entries  */
        gout=);          /* name of graphic catalog          */
```

Only the name of the data set and the variable to be plotted need be supplied. Output 3.25 would also be produced with this statement:

```
%nqplot(data=nations,var=imr);
```

3.6 Plots for Assessing Symmetry

Sometimes we may be interested not so much in the detailed shape of the distribution of data, but merely in whether the data have a reasonably symmetric distribution. Or if our data are far from symmetric, it may be satisfactory to find some transformation that makes the variables more nearly symmetric.

While histograms, boxplots, density traces, and Q-Q plots all enable you to make some judgment about symmetry, displays designed particularly to reveal asymmetry can be more revealing in difficult cases. These plots are described and illustrated in Section 3.6.1, "Plotting the Upper against the Lower Value" and Section 3.6.2, "Untilting with Mid-Spread Plots." Section 3.6.3, "Constructing the Upper-Lower and Mid-Spread Plots," shows how these plots are constructed.

For these purposes, the simplest, most useful plots are based on the idea that in a symmetric distribution, the distances of points at the lower end to the median should match the distances of corresponding points in the upper end to the median. "Corresponding" here means counting in from either end, so $x_{(1)}$ and $x_{(n)}$ at depth$=1$ should be equally distant from the median, M, as should $x_{(2)}$ and $x_{(n-1)}$ at depth$=2$:

$$M - x_{(1)} = x_{(n)} - M$$

$$M - x_{(2)} = x_{(n-1)} - M \quad .$$

In general, if the distribution were perfectly symmetric, the observations at depth i should satisfy

$$M - x_{(i)} = x_{(n+1-i)} - M \quad , \qquad \text{for } i = 1 \text{ to } \begin{cases} n/2 \, , & n \text{ even} \\ (n+1)/2, & n \text{ odd} \end{cases} \quad . \quad (3.16)$$

3.6.1 Plotting the Upper against the Lower Value

If the data values are reasonably close to symmetric, this relationship should hold approximately from the extremes in toward the middle. Chambers et al. (1983) and others suggest plotting the right side of equation 3.16, the upper value, against the left side, the lower value. For a symmetric distribution, these points lie on a line with a slope of 1. In skewed distributions, the points tend to rise above the line (positive skew) or fall below (negative skew).

Example: Baseball Data

Output 3.26 shows the Upper vs. Lower plot for the distribution of number of runs in the baseball data, which we saw (Output 3.11) has a positively skewed distribution. The points near the origin are those closest to the median of the distribution; those on the upper right of the plot reflect the corresponding observations in the tails. Here we see that the points rise above the line as we go out toward the tails, indicating that the upper tail stretches out farther than the lower.

Output 3.26
Upper Vs. Lower Plot for Number of Runs in Baseball Data

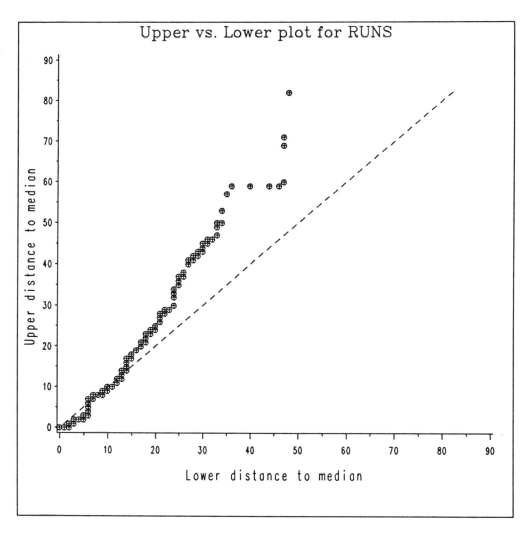

3.6.2 Untilting with Mid-Spread Plots

On the Upper vs. Lower plot, we judge departure from symmetry by divergence from the line $y=x$. A simple modification of this plot suggested by Hoaglin (1985a), following Tukey (1977), changes the coordinates so that the reference line for symmetry becomes horizontal; it is always easier to judge departure from a flat line than from a tilted one. On this plot, called the *mid-spread plot*, we plot

$$\text{mid} \equiv (x_{(n+1-i)} + x_{(i)}) \, / \, 2 \qquad \text{vs.} \qquad x_{(n+1-i)} - x_{(i)} \equiv \text{spread} \quad .$$

The reference line is horizontal because in a symmetric distribution each mid value equals the median,

$$(x_{(n+1-i)} + x_{(i)}) \, / \, 2 = M \quad . \tag{3.17}$$

Because the spreads increase as we go from the center out to the tails, this plot depicts how the pattern of mid values changes from the middle of the distribution to the extremes. Because the plot is untilted when the distribution is symmetric, expansion of the vertical scale allows us to see systematic departures from flatness far more clearly.

For the baseball RUNS data, Output 3.27 plots the mid value against spread. The flat dashed line is the reference line of median=mid value, and the tilted solid line is the linear regression line through the points. Again we see a clear deviation from symmetry, but on an expanded vertical scale.

Output 3.27
Mid-Spread Plot for Baseball Data, Number of Runs

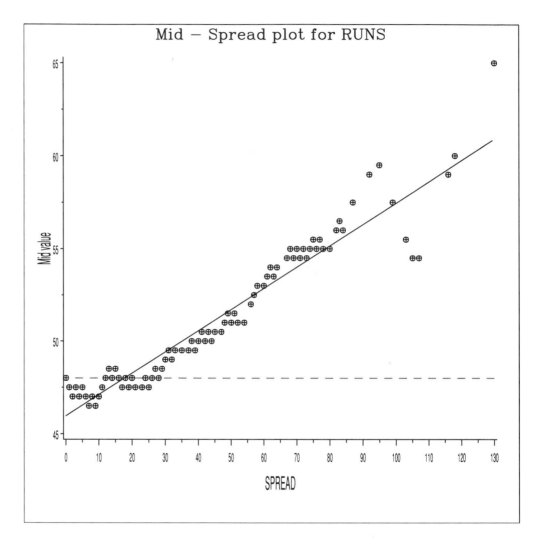

Other variations of the mid-spread plot are possible. With extreme skewness, the spread values for the extremes can be very large, compressing the other points to the left. (Note that in Output 3.27 the points are spaced more widely on the right. If the largest spread value were doubled, the remaining points would be squeezed to the left to fit the same plot frame.) Hoaglin (1985a) suggests using z_i^2, the square of the corresponding normal quantile, as the horizontal coordinate. This keeps the order of the points the same, affecting only the horizontal spacing. This plot, called the mid vs. z^2 plot, has desirable theoretical properties, discussed in Hoaglin (1985b).

Another variation (Emerson and Stoto 1982) is to plot the centered mid value, $mid-M$, against a squared measure of spread, the sum of squares of the upper and lower distance-to-median values. This plot, as we will see in Section 3.6.4, makes it easy to choose a power transformation to make a variable symmetrical.

3.6.3 Constructing the Upper-Lower and Mid-Spread Plots

The method for all these plots is to fold the distribution of scores around the middle value. A simple way to do this is to sort the scores twice—once in ascending order and once in descending order—and then to merge the two sets.* This operation pairs ($x_{(1)}$ with $x_{(n)}$), ($x_{(2)}$ with $x_{(n-1)}$), and so on.

The program below shows these steps for the data plotted in Output 3.26 and 3.27.

```
%include BASEBALL;          /* On MVS, use %include ddname(file);*/
data analyze;
   set baseball;
   keep runs;
   if runs=. then delete;

proc univariate data=analyze noprint;
   var runs;
   output out=stats n=nobs median=median;
proc sort data=analyze out=sortup;
   by runs;
proc sort data=analyze out=sortdn;
   by descending runs;
```

The two sorted data sets are merged (by depth, which is implicit in the order of the values) in the following DATA step. The STATS data set from PROC UNIVARIATE is used to get values for the number of observations (NOBS) and the median. The DATA step executes the STOP statement when the middle of the distribution is reached.

```
     /* merge x(i) and x(n+1-i)    */
data symplot;
   merge sortup(rename=(runs=frombot))      /* frombot = x(i)     */
         sortdn(rename=(runs=fromtop));      /* fromtop = x(n+1-i) */
   if _n_=1 then set stats;                  /* get nobs, median   */
   depth = _n_ ;
   mid = (fromtop + frombot) / 2;
   spread = fromtop - frombot;
   lower = median - frombot;
   upper = fromtop - median;
   if _n_ > (nobs+1)/2 then stop;
   label mid =   'Mid value'
         lower= 'Lower distance to median'
         upper= 'Upper distance to median';
run;
```

---

*   You can do this more efficiently with just one sort. After sorting, use the POINT option in the SET statement to read observation i and observation $n+1-i$ within a DO loop for $i=1$ to $(n+1)/2$.

Both plots and the reference lines are drawn using the data in data set SYMPLOT. The first PROC GPLOT step plots the UPPER value against itself to give the reference line with slope 1 for the Upper vs. Lower plot. The second PROC GPLOT step plots the constant MEDIAN for the horizontal reference line and draws a linear regression line through the points to show the actual trend of the mid values.

```
title h=1.5 'Upper vs. Lower plot for RUNS';
proc gplot data=symplot ;
   plot upper * lower = 1
        upper * upper = 2
      / overlay
        vaxis=axis1 haxis=axis2 vm=1 hm=1
        name='GB0326' ;
   symbol1 v=+ c=black;
   symbol2 v=none i=join c=red l=20;
   axis1 label=(h=1.5 a=90 r=0);
   axis2 label=(h=1.5);
run;
title h=1.5 'Mid - Spread plot for RUNS';
*-- Mid vs. Spread plot;
proc gplot data=symplot ;
   plot mid   * spread = 1
        median* spread = 2
      / overlay
        vaxis=axis1 haxis=axis2 vm=1 hm=1
        name='GB0327' ;
   symbol1 v=+    i=rl  c=black;
   symbol2 v=none i=join c=red l=20;
   axis1 label=(h=1.5 a=90 r=0) order=(45 to 65 by 5);
```

A more general SAS macro version of this program is included in Section A1.15, "The SYMPLOT Macro." This macro constructs any of the symmetry plots described in this section. The SYMPLOT macro takes the following parameters:

```
%macro SYMPLOT(
        data=_LAST_,    /* data to be analyzed               */
        var=,           /* variable to be plotted            */
        plot=MIDSPR,    /* Type of plot(s): NONE, or any of  */
                        /* UPLO, MIDSPR, MIDZSQ, or POWER    */
        trim=0,         /* # or % of extreme obs. to be trimmed */
        out=symplot,    /* output data set                   */
        name=SYMPLOT);  /* name for graphic catalog entry    */
```

Specifying PLOT=POWER produces a version of the mid-spread plot designed to select a power transformation that makes the data more nearly symmetrical. This plot is described in Section 3.6.4.

3.6.4 Finding Transformations for Symmetry

You can use symmetry plots not only as a diagnostic display, but also as a tool to find a transformation that makes a distribution more symmetric. One idea is simply to try several transformations until you find one that makes the plot reasonably flat. But we need not search blindly, for the mid-spread plot points us in the right direction, and the version of the mid-spread plot suggested by Emerson & Stoto (1982) helps us choose a good transformation easily.

We will consider a simple family of power transformations called the ladder of powers (Tukey 1977), where a variable x is transformed to $t_p(x)$ by a power p according to

$$t_p(x) = \begin{cases} x^p, & p > 0 \\ \log_{10} x, & p = 0 \\ -x^p, & p < 0 \end{cases} . \tag{3.18}$$

The values of x must all be positive; if they are not, the transformation can be applied to $(x+c)$, where c is greater than the minimum value of x. Although the family of transformations given by equation 3.18 is defined for all powers p, it is common to use only the simple integer and half-integer powers listed in Table 3.1. (In some situations, we may add the cube root, $p=1/3$, to this list.)

Table 3.1
Simple Transformation in the Ladder of Power

Power	Transformation	Re-expression
2	Square	x^2
1	Raw	x
1/2	Square root	$\sqrt{x}$
0	Log	$\log_{10} x$
−1/2	Reciprocal root	$-1/\sqrt{x}$
−1	Reciprocal	$-1/x$

The most important properties of this family of transformations for our purposes are as follows:

□ They preserve the order of data values. Larger data values on the original scale will be larger on the transformed scale.

□ They change the relative spacing of the data values. For powers less than 1, such as $\sqrt{x}$ and $\log x$, the transformation has the effect of compressing or pulling in values in the upper tail of the distribution relative to low values; powers $p > 1$, such as x^2, have the opposite effect, expanding the spacing of values in the upper end relative to the lower end.

□ The effect on the shape of the distribution changes systematically with p. If $\sqrt{x}$ pulls in the upper tail, $\log x$ will do so more strongly, and negative powers will be stronger still.

Therefore, if the mid-spread plot (or mid vs. z^2 plot) shows the mid value increasing with spread, indicating positive skewness, we know to go down the ladder of powers and try $\sqrt{x}$, log x, and so on; if the points on the plot decrease with spread, showing negative skewness, we go up the ladder of powers. A transformation that makes the plot reasonably flat is a good choice.

Output 3.28 and Output 3.29 show the mid vs. z^2 plots for $\sqrt{\text{RUNS}}$ and $\log_{10}(\text{RUNS})$. You can see that the plot for the square root is mostly flat, indicating that the square root transformation makes the distribution symmetric. Only the eight points that are most extreme fall off the line of mid value=median. (The linear regression line slopes downward, for it is strongly influenced by these extreme points. A robust fitted line, not influenced by these extreme values, would be much better here.)

Output 3.28
Mid Vs. z^2 Plot for Sqrt(RUNS)

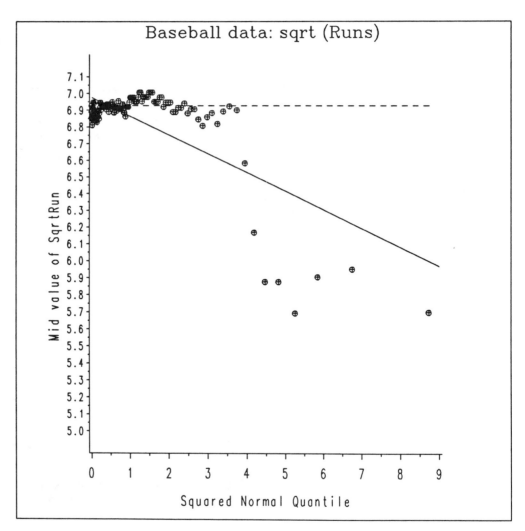

The plot for $\log_{10}(\text{RUNS})$ in Output 3.29 shows that the log transformation has gone too far because the points now clearly drift downward, indicating that the distribution is negatively skewed.

Output 3.29
Mid Vs. z^2 Plot for Log(RUNS)

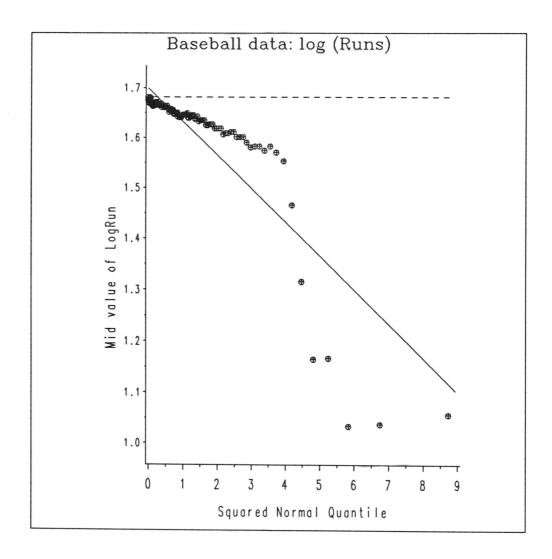

In fact, the process of choosing a transformation from the ladder of powers can be made even easier with a variation of the mid-spread plot suggested by Emerson and Stoto (1982). In this display, produced by using PLOT=POWER in the SYMPLOT macro, we plot the centered mid value,

$$\frac{x_{(i)} + x_{(n+1-i)}}{2} - M \quad ,$$

as the vertical coordinate, against a squared measure of spread,

$$\frac{\text{Lower}^2 + \text{Upper}^2}{4\,M} = \frac{(M - x_{(i)})^2 + (x_{(n+1-i)} - M)^2}{4\,M}$$

as the vertical coordinate. Emerson and Stoto show that, if this graph is approximately linear, with a slope, b, then $p = 1 - b$ is the indicated power for a transformation to approximate symmetry. If the line is calculated by least squares, it is useful to trim some proportion of the extreme observations as a substitute for a resistant line.

For the RUNS data, with 5 percent of the highest and lowest observations trimmed, this plot is shown in Output 3.30. The pattern of the points is very similar to the mid-spread plot in Output 3.27. Because the mid values have been centered, the dashed line (indicating symmetry) is now at the vertical value of zero. The slope of the least squares line, calculated with the REG procedure, is shown on the plot to be $b=.55$. The suggested power is therefore $p=1-.55 =.45$, which, rounded to the nearest .5, is $p=.5$, indicating the square root transformation.

Output 3.30
Symmetry Transformation Plot for Number of Runs

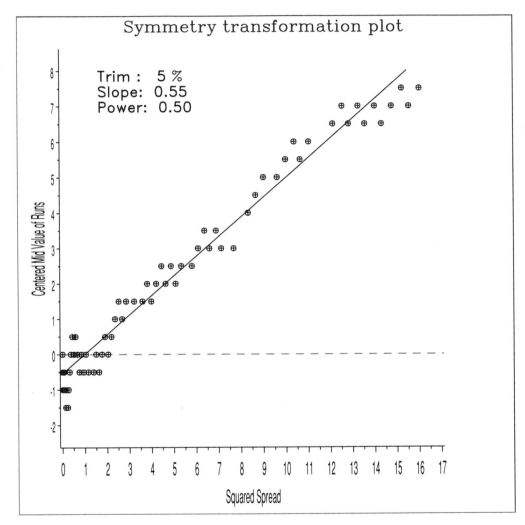

It turns out that the square root transformation is most often selected for data that represent counted values or frequencies. In the BASEBALL data set, the numeric variables such as RUNS, HITS, HOMER, and so on, are all of this type. The symmetry transformation plots for these variables all result in slopes ranging from about .3 to .7, for which the square root is a reasonable transformation.

Output 3.28 to 3.30 were all drawn with the SYMPLOT macro. For example, these statements produce Output 3.30:

```
%include BASEBALL ;
%include symplot  ;

title h=1.6 'Symmetry transformation plot';
%symplot(data=baseball,var=Runs,
         plot=POWER,
         trim=5 pct);
```

Specifying TRIM=5 PCT causes the highest and lowest 5 percent of the observations to be excluded from the plot and the slope calculation, reducing the impact of extreme observations.

Part 3

Bivariate Displays

Chapter 4 Scatterplots

4.1   Introduction

In the realm of statistical graphics, the scatterplot is the basic tool on which most methods are based. Indeed, there is probably no statistical tool more powerful or more generally useful for understanding the relationship between two variables. Chapter 8, "Displaying Multivariate Data," considers techniques designed for multivariate data that often use the scatterplot as a building block.

This chapter discusses both simple and enhanced scatterplots for data concerned with the relationship between two quantitative variables, *x* and *y*. The next two sections focus on the basic tools for plotting, which include the following:

□   constructing basic scatterplots

□   plotting on logarithmic axes

□   using different plotting symbols for different observations

□   labeling observations.

Section 4.4, "Interpolated Curves and Smoothings," and Section 4.5, "Enhanced Scatterplots," describe several techniques for adding information to a scatterplot to provide a visual summary of the relation or to help interpret the display. Plotting methods for more specialized situations are discussed in the final sections of this chapter, including techniques for plotting discrete data (Section 4.6, "Plotting Discrete Data") and displaying a third variable in a scatterplot (Section 4.7, "Displaying a Third Variable on a Scatterplot" and Section 4.8, "Three-Dimensional Plots").

4.2   Simple Scatterplots

Even the simplest scatterplots are made for some particular purpose, and that purpose generally determines what we plot and how we do it. This section begins with examples showing different types of data and the goals graphical display might serve. Later sections show how most of the plots were constructed and how they can be enhanced.

4.2.1   Automobiles Data

In some situations, scatterplots are intended to show how one variable, the *response*, plotted on the vertical scale as *y*, depends on another variable, the *factor*, plotted on the horizontal scale as *x*. Output 4.1, for example, shows the relation between gas mileage in miles per gallon and weight for 74 automobiles in the 1979 model year (data from Chambers et al. 1983; see Section A2.2, "The AUTO Data Set: Automobiles Data"). It is natural to think of gas mileage as determined, at least in part, by weight, so weight is the factor and mileage is the response. As we would expect, Output 4.1 shows that heavier cars are less fuel efficient. However, even such a simple graph can reveal things we might not expect. It appears, for example, that mileage decreases nonlinearly with weight, dropping steeply at first and then leveling off from left to right.

Output 4.1
Gas Mileage Plotted against Weight for Automobiles Data

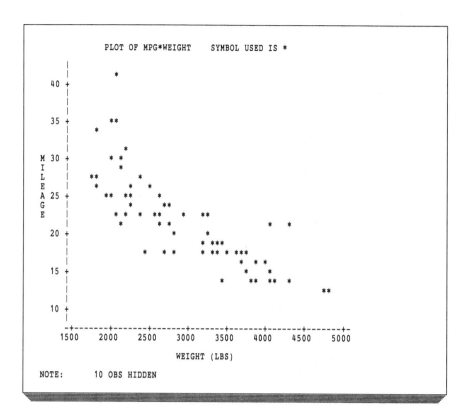

In other situations, neither variable is distinguished, and we may make a scatterplot merely to show how one variable varies compared with the other. The automobile data also contain the base price for each model, and Output 4.2 shows the relation between price, on the ordinate, and weight, on the abscissa. While we might argue that price is determined at least in part by weight, we could equally well reverse the axes and argue that weight is fixed by design and materials and this, in part, determines selling price. We can see that price generally increases with weight. There is also a slightly peculiar pattern here, with a hole right in the middle where we might otherwise expect the points to be most concentrated.

Output 4.2
Price Plotted
against Weight for
Automobiles Data

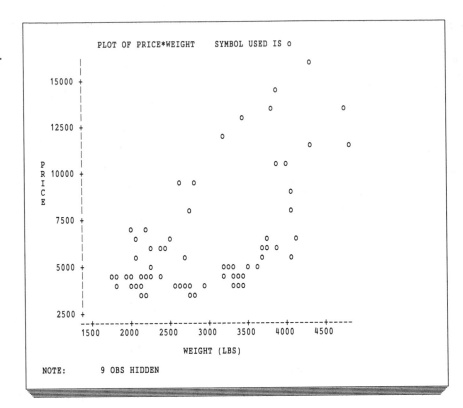

4.2.2 Infant Mortality Data

Output 4.3 shows the data on infant mortality rate (per 1,000 live births) in relation to per capita income for 101 nations around the world (see Section A2.11, "The NATIONS Data Set: Infant Mortality Data"). We might consider infant mortality rate to be a response that we would like to understand in relation to per capita income as a factor (perhaps among others). The distributions of these variables make this a rather uninformative display (most of the data are squeezed into the bottom left corner), but two or three points stand out as different from the rest.

Output 4.3
*Infant Mortality
Rate and Per
Capita Income*

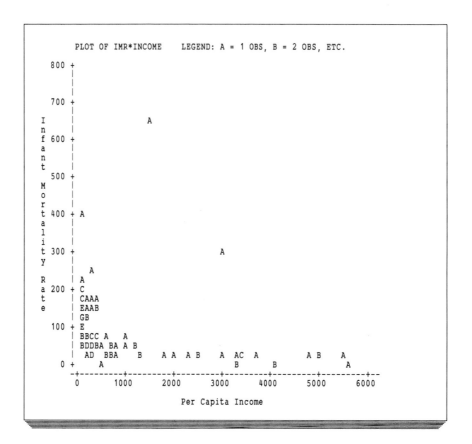

4.2.3 Fisher's Broadbalk Wheat Data

Another kind of scatterplot shows how a variable changes over time. Output 4.4
displays data used by Fisher (1925) to illustrate the calculation of the slope in
linear regression. The data come from records of wheat yields, in bushels per
acre, at Broadbalk over a 30-year period, 1855—1884 (see Section A2.15, "The
WHEAT Data Set: Broadbalk Wheat Data"). Fisher lists the yields from two plots,
labeled 9a and 7b, which had been treated identically, except that the fertilizers
applied contained nitrogen in different forms. He noted that the separate yields
were very variable from year to year; but the difference in yields (plot9a—plot7b)
was much less variable. Fisher commented that plot 9a appeared to be gaining in
yield on plot 7b over time, and illustrated the calculation of the least squares slope
of yield difference on time as a way to determine if this apparent gain was
significant.

Output 4.4

Differences in Yield between Two Plots in the Broadbalk Wheat Experiment

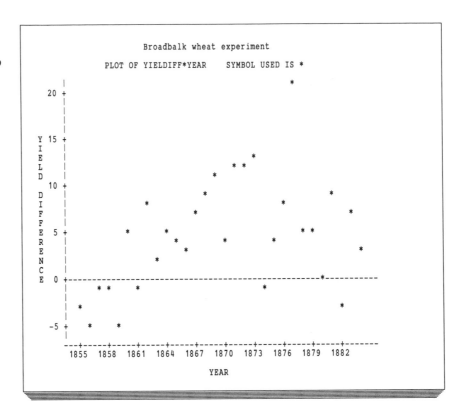

Output 4.4 does, indeed, show a trend of yield differences going from negative values in the early years to mostly positive values after 1860. (Note how this perception is aided by the horizontal reference line at zero.) The general decline of values after 1873, however, suggests a curvilinear relation rather than a linear one.

4.2.4   Basic Scatterplots with the PLOT and GPLOT Procedures

For simple plots like those in Output 4.1 to 4.4, the PLOT procedure (for printer plots) and the GPLOT procedure (for high-resolution plots) are used in very much the same way. Nevertheless, it is usually easiest to begin with PROC PLOT. Output 4.1 and 4.2 are produced using the following statements:

```
options ls=64 ps=40 nodate nonumber;
%include AUTO;               /* On MVS, use %include ddname(file);*/
title ' ';
proc plot data=auto;
   plot mpg * weight = '*';
   plot price * weight = 'o';
```

PLOT statements of the form

```
plot y * x = 'char'
```

use the character specified as the plotting symbol. If several points fall in the same location, a note is printed indicating the number of hidden observations, as in Output 4.1 and 4.2.

Output 4.3 is drawn with the following statements:

```
options ps=45 ls=70 nodate nonumber;
proc plot;
   plot imr * income ;
```

Here, no plotting symbol is specified, so the characters A, B, C, and so on are used to indicate the number of points at a given location. In these cases, the size of the plot is determined by the PAGESIZE and LINESIZE options. With PROC PLOT, you could use the HPOS= and VPOS= options equally well in the PLOT statement to control the size of the plot.

To do a plot with PROC GPLOT, it is best to start very simply, preview the plot on a video display, and then add a few options to specify the plotting symbol(s) and axis style according to your taste. Output 4.5 shows the PROC GPLOT version of Output 4.1 after one iteration.

Output 4.5
The GPLOT Procedure Version of Gas Mileage Plotted against Weight

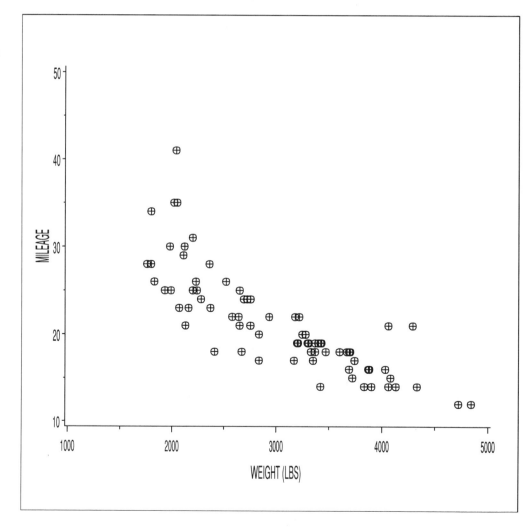

The size of the plot is controlled by the HSIZE and VSIZE options. I often prefer the plotted points to appear slightly larger than the default H=1 size (one character cell), and the vertical axis looks better to me when the axis label is rotated, with ANGLE=90 and ROTATE=0 specified in the AXIS statement.* Output 4.5 is drawn using these statements:

```
proc gplot data=auto ;
   plot mpg * weight / vminor=1 hminor=1
                       vaxis=axis1 haxis=axis2
                       name='GB0405' ;
   symbol v=+ h=1.4 c=black;
   axis1 label=(h=1.5 a=90 r=0) value=(h=1.3);
   axis2 label=(h=1.5)          value=(h=1.3);
```

I might also specify fonts for labels and axis values, but I'll leave that for other examples (Output 4.7, for instance).

4.2.5 Logarithmic Axes

We usually want to show all the data in raw form, but that is sometimes hard to do. On the plot of the infant mortality data (Output 4.3), it is difficult to see much detail because most of the observations are located in the lower-left corner. In fact, 72 percent of the points appear in about 4 percent of the area of the plot window.

When the compression of the data to a small corner of the plot is due to a few extreme observations, we may sometimes opt to redefine the plotting range so that the outlying points are excluded. In this case, defining the plot range for infant mortality from 0 to 300 would exclude two points (Afghanistan and Saudi Arabia) and increase the resolution of the remaining points.

The VAXIS and HAXIS options in the PLOT statement in PROC PLOT and PROC GPLOT can be used to limit the plotting range to a subset of the data, so the two points with infant mortality values greater than 300 are excluded with these statements:

```
proc plot data=nations;
   plot imr * income= 'O' / vaxis=0 to 300 by 100 ;
   footnote 'Note: Vertical axis truncated artificially';
```

Excluding extreme observations can sometimes improve resolution of the remaining points. Output 4.6 shows the resulting plot, a slight improvement but still unsatisfactory. PROC PLOT prints a note indicating that two observations are out of range.

---

* In Version 6, ROTATE=0 need not be specified because it is the default for any ANGLE= specification.

Output 4.6
*Infant Mortality
Rate and Per
Capita Income*

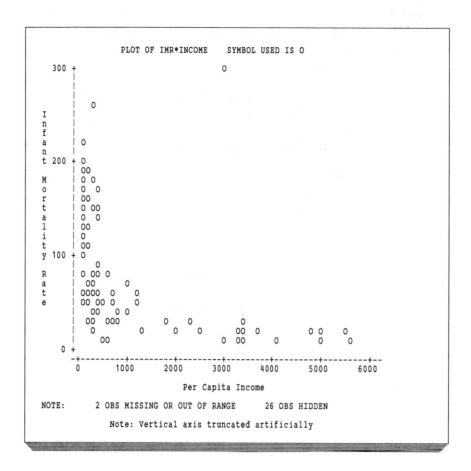

However, a more satisfactory solution to this problem is to transform the scales of *x* or *y* or both so the data are spread more evenly throughout the range. In this case, infant mortality is highly skewed in the positive direction; plotting the data on a log scale has the desired effect.* With PROC PLOT, we can transform the data in a DATA step and plot the transformed variables. (This plot is not shown. It looks like the PROC GPLOT version in Output 4.7, except that the scales are labeled with the log values.)

```
data nations;
   set nations;
   limr = log10(imr);
   lincome = log10(income);
   label limr    = 'Log Infant Mortality Rate'
         lincome = 'Log Per Capita Income';
proc plot;
   plot limr * lincome = region ;
```

---

* Choosing a good transformation to straighten a bivariate relationship is discussed in Emerson and Stoto (1982).

PROC GPLOT can plot the data on log scales without transforming it first, by using the LOGBASE= option in the AXIS statement. The log-log plot from PROC GPLOT shown in Output 4.7 is produced by these statements:

```
title h=1.75 'Infant mortality rate and income';
proc gplot data=nations  ;
   plot imr * income = region
         / frame vaxis=axis1 haxis=axis2
           name='GB0407' ;
   axis1 logbase=10
         order=(3 10 30 100 300 1000)
         value=(h=1.5 f=duplex)
         label=(a=90 r=0 h=1.5 f=duplex);
   axis2 logbase=10
         value=(h=1.5 f=duplex)
         label=(h=1.5 f=duplex);
   symbol1 h=1.6 color=black v=- ;
   symbol2 h=1.6 color=black v=star;
   symbol3 h=1.6 color=black v=square;
   symbol4 h=1.6 color=black v=triangle;
   format region region.;
```

Output 4.7
Log-Log Plot of
Infant Mortality
Data

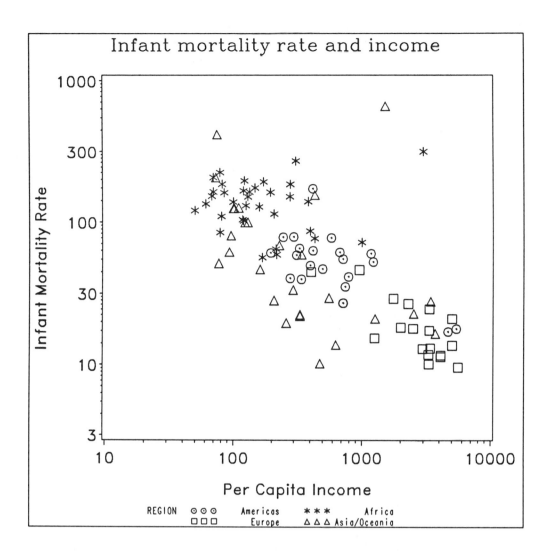

When LOGBASE=10 is specified, the values are spaced in equal steps of $\log_{10}(x)$, but the value labels are written in the scale of the original variable. This is often easier to understand than a plot with transformed data.

In the first version of this plot, the vertical axis is scaled from 1 to 1,000 in log units. Because the smallest value of infant mortality is 9.6, the bottom third of the plot (from 1 to 10) is empty. Output 4.7 uses the ORDER= option in the AXIS1 statement to define the plotting range starting at $\log_{10}(IMR)=.5$, or IMR=3 for the vertical axis. Another possibility would be to scale the vertical axis by specifying LOGBASE=2, so equal steps would correspond to a doubling of the infant mortality rate. (Using a different base only multiplies the log values by a constant.)

4.2.6 Using Different Plotting Symbols for Different Observations

Frequently the observations we are plotting fall into different groups or categories. In order to understand the relation between the *x* and *y* variables, and how this relationship varies from group to group, it is helpful to show the group membership of each point in the scatterplot.

The plot in Output 4.7 uses different plotting symbols for countries, classified in four geographic regions shown in the plot legend. The PLOT statement instructs PROC GPLOT to associate a different SYMBOL*n* statement with each different value of the REGION variable:

```
plot imr * income = region ;
```

REGION is a numeric variable with values 1, 2, 3, and 4. The descriptive labels printed in the legend come from a user-defined format constructed with the FORMAT procedure:

```
proc format;
   value region 1='Americas'
                2='Africa'
                3='Europe'
                4='Asia/Oceania';
```

4.3 Labeling Observations

When observations have individual identities, like the automobiles or nations of the world, our ability to understand patterns or suspicious cases is often helped by plotting an identification (ID) label for each point. For example, we noticed that a few points in Output 4.3 are quite detached from the rest, but it took a look through the data values to find out that they were Saudi Arabia and Afghanistan. Moreover, the results of a multidimensional scaling, principal components analysis (Section 9.4, "Plotting Principal Components and Component Scores"), correspondence analysis (Section 10.3, "Correspondence Analysis"), and similar techniques are extremely difficult to interpret without a labeled scatterplot.

However, unless the number of points is quite small, labels for data points in the data region can interfere with our ability to see shape and patterns and can make the plot quite crowded. There is thus a delicate balance between the advantage of identifying points and the disadvantage of obscuring the data. Even worse, when there are more than a small number of points, the labels of nearby points tend to collide, obscuring each other. There are two strategies for dealing with these problems:

Visual thinning
> Reduce the visual impact of the labels relative to the points so they interfere as little as possible with the data pattern.

Collision avoidance
> If you are drawing a graph by hand or with an interactive system, you can minimize the tendency for labels to overlap. This is much harder to do with a procedure-based graphic system.

These problems and some solutions are illustrated below.

4.3.1 Labeling Printer Plots

With PROC PLOT, you can label each point with the first character of an ID variable using a PLOT statement of the form

```
plot y * x = id ;
```

In a future release of the SAS System, a planned feature of PROC PLOT is the ability to use character labels of any length, with provisions to reduce collisions. Because this feature is not available at this time, the examples here use the IDPLOT procedure from Version 5.

PROC IDPLOT prints scatterplots with each point on the plot labeled by the value of an ID variable.* The ID variable can be a number or a character string of any length. PROC IDPLOT offers several different plotting methods (Kuhfeld 1986) that vary the trade-off between completely presenting the values of the point identification variable and accurately presenting the metric distance information. The METRIC method produces ordinary scatterplots. It makes some attempt to avoid overlapping labels, but portions of the labels will be obscured in dense regions. The RANK method completely avoids collisions by using the ranks of the variables as the plot coordinates. The distances between points are not represented accurately, however. The default MODRANK (modified rank) method also guarantees that no point labels are obscured. It retains some metric information by varying the spacing between the plotting positions of successive points in proportion to the distance between those points. A fourth method, GENRANK (generalized rank), enables the user to control the combination of metric and rank information. See the *SUGI Supplemental Library User's Guide, 1983 Edition* for further details.

Usually, it is most important to appreciate the metric information in the plot, so the METRIC method should be the method of choice. You can control the degree of overlap between point labels and use visual thinning by abbreviating the ID variable, labeling only some of the points, or enlarging the plot so that more print positions are available.

Example: Occupational Prestige Data

In Section 2.5, "Dot Charts," we examine the measure of occupational prestige for 45 occupational titles in the DUNCAN data set. The goal of the study is to predict occupational prestige from more readily available measures, and here we examine the relation between occupational prestige and income.

Output 4.8 and 4.9 show the METRIC and the MODRANK plots from PROC IDPLOT. The points on both plots are labeled with the value of the JOB variable, a 12-character abbreviation of the occupational title. These plots are produced by these statements:

```
%include DUNCAN ;
proc idplot data=duncan ;
    plot prestige * income = job  / metric modrank;
    format prestige income 3.;
```

---

* PROC IDPLOT is part of the Version 5 SUGI Supplemental Library. It is not available in Version 6 of the SAS System.

Output 4.8
METRIC Plot of PRESTIGE by INCOME from the IDPLOT Procedure

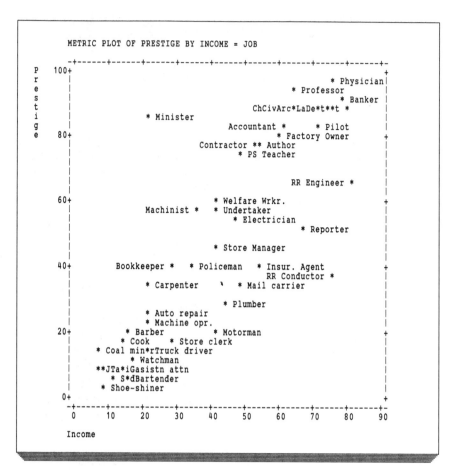

```
METRIC PLOT OF PRESTIGE BY INCOME = JOB

      -+------+------+------+------+------+------+------+------+-
P  100+                                                          +
r      |                                         * Physician|
e      |                                  * Professor
s      |                                          * Banker  |
t      |                           ChCivArc*LaDe*t**t *
i      |       * Minister                                   |
g   80+                        Accountant *    * Pilot       +
e      |                              * Factory Owner        |
       |                   Contractor ** Author              |
       |                       * PS Teacher                  |
       |                                                     |
       |                                  RR Engineer *      |
    60+                                                      +
       |                       * Welfare Wrkr.               |
       |           Machinist *  * Undertaker                 |
       |                        * Electrician                |
       |                                  * Reporter         |
       |                     * Store Manager                 |
    40+        Bookkeeper *   * Policeman   * Insur. Agent    +
       |                              RR Conductor *          |
       |           * Carpenter   `  * Mail carrier           |
       |                     * Plumber                       |
       |             * Auto repair                           |
       |             * Machine opr.                          |
    20+         * Barber      * Motorman                      +
       |        * Cook     * Store clerk                     |
       |    * Coal min*rTruck driver                         |
       |          * Watchman                                 |
       |    **JTa*iGasistn attn                              |
       |       * S*dBartender                                |
     0+    * Shoe-shiner                                     +
      -+------+------+------+------+------+------+------+------+-
       0    10    20    30    40    50    60    70    80    90

      Income
```

Output 4.9
MODRANK Plot of
PRESTIGE by
INCOME from the
IDPLOT Procedure

```
          MODIFIED RANK PLOT OF PRESTIGE BY INCOME = JOB

P   97|--------------------------------------------------------
r   93|                                          Physician *  |
e   92|                                        * Professor    |
s   90|                                              Banker * |
t   90|                                           Dentist *   |
i   90|                                      * Chemist        |
g   89|                                   Architect *         |
e   88|                                             * Lawyer| |
    87|                               Civil Eng. *            |
    83|             * Minister                                |
    82|                                      * Pilot          |
    81|                                   * Accountant         |
    76|                                  * Factory Owner      |
    76|                              * Contractor             |
    73|                               * Author                |
    67|                            * PS Teacher               |
    59|                                       RR Engineer *|  |
    57|                        * Welfare Wrkr.               |
    57|                     * Machinist                      |
    53|                        * Undertaker                  |
    52|                           * Electrician              |
    45|                                 * Reporter           |
    41|                     * Store Manager                  |
    41|                             * Insur. Agent           |
    39|                * Policeman                           |
    38|              * Bookkeeper                            |
    34|                               RR Conductor *         |
    33|                           * Mail carrier             |
    29|          * Carpenter                                 |
    26|                      * Plumber                       |
    26|            * Auto repair                             |
    24|          * Machine opr.                              |
    20|       * Barber                                       |
    19|                    * Motorman                        |
    16|               * Store clerk                          |
    16|      * Cook                                          |
    15| * Coal miner                                         |
    13|           * Truck driver                             |
    11|        * Watchman                                    |
    10|        * Gas stn attn                                |
    10|     * Taxi driver                                    |
    10| * Waiter                                             |
     8|* Janitor                                             |
     7|        * Bartender                                   |
     6|     * Soda clerk                                     |
     3|   * Shoe-shiner                                      |
      |--------------------------------------------------------
        1 11111 22222 22  3 3  4444 4 444  5 55  6 6 66 6  77 7777 7 88
     77899 2 45667 11112 99  4 6  1222 4 788  3 55  0 2 44 7  22 5666 8 01

      Income
```

Note that in the METRIC plot (Output 4.8), most of the labels are distinct, but some are obscured in the bottom-left and upper-right corners where the density of points is larger. Though the relationship between prestige and income is fairly strong ($r=0.837$), the dispersion of the labels in the METRIC plot gives a visual impression of a weaker relation. On the MODRANK plot (Output 4.9), none of the labels overlap, but the spacing of values on the axes has been distorted. One way to improve these plots with PROC IDPLOT is to abbreviate the observation labels. For example, if IDLENGTH=4 is specified in the PLOT statement, only the first four characters of the JOB name are plotted.

4.3.2 Labeling Points with the Annotate Facility

The SAS/GRAPH Annotate facility offers great flexibility in labeling observations on a scatterplot. The first example below illustrates the technique for constructing an Annotate data set for situations where the points are not so crowded that overplotting occurs. The second example illustrates some additional steps to reduce collisions.

To plot point labels, create an Annotate data set containing one observation for each observation in the PLOT data set. Specify FUNCTION='LABEL'. The Annotate variables X and Y give the plot coordinates, and the TEXT variable gives the label for the point. Then the Annotate data set name is specified as the value of the ANNOTATE= option in the PLOT statement in a PROC GPLOT step.

These steps are illustrated below using the occupational prestige data. For the purpose of the example, a randomly selected subset of about 25 percent of the data set DUNCAN is chosen in the DATA step.

The labels are created with the Annotate data set LABELS. The Annotate variables X and Y are assigned the values of INCOME and PRESTIGE, respectively, and the variable TEXT is assigned the value of JOB. The length of the label is determined by the length attribute of the TEXT variable. The Annotate facility can interpret the X and Y variables in different coordinate systems, as specified by the variables XSYS and YSYS. When the values of X and Y refer to data coordinates of the plot variables, as in this example, both XSYS and YSYS are assigned the value '2'. By default, the labels are centered at the X, Y coordinates; in this example the labels are drawn centered, one character cell below the point, by specifying POSITION='8'. The data set LABELS constructed by this DATA step is shown in Output 4.10.

Output 4.10
Annotate Data Set
LABELS to Label
Observations

OBS	FUNCTION	XSYS	YSYS	X	Y	TEXT	POSITION
1	LABEL	2	2	62	82	Accountant	8
2	LABEL	2	2	76	38	RR Conductor	8
3	LABEL	2	2	53	76	Contractor	8
4	LABEL	2	2	29	39	Bookkeeper	8
5	LABEL	2	2	48	34	Mail carrier	8
6	LABEL	2	2	55	41	Insur. Agent	8
7	LABEL	2	2	21	33	Carpenter	8
8	LABEL	2	2	44	29	Plumber	8
9	LABEL	2	2	7	15	Coal miner	8
10	LABEL	2	2	21	24	Machine opr.	8
11	LABEL	2	2	7	8	Janitor	8

In the PROC GPLOT step, specifying ANNOTATE=LABELS causes the Annotate instructions to be applied to the plot. The AXIS statements use the OFFSET= option to provide some additional space at the plot boundaries for the point labels. These statements produce Output 4.11:

```
data subset;
   set duncan;
   if uniform(5621729) < .25;
```

```
data labels;                    /* Annotate data set */
   set subset;
   retain xsys '2' ysys '2';
   length text $12 function $8 ;
   text = job;
   y    = prestige;
   x    = income;
   size = 1.2;
   position = '8';              /* centered below the point */
   function = 'LABEL';

proc print data=labels;
   var function xsys ysys x y text position ;

title ' ';
footnote j=l h=1.3 '  Note: Subset of data points';
symbol v=+ h=1.2 color='BLACK';
proc gplot data=subset  ;
   plot prestige * income /
      annotate=labels
      vaxis=axis1 haxis=axis2
      hminor=1 vminor=1 frame
      name='GB0411' ;
   axis1 label=(h=1.5 a=90 r=0 f=duplex)
        value=(h=1.3 f=duplex) offset=(1);
   axis2 label=(h=1.5 f=duplex)
        value=(h=1.3 f=duplex) offset=(4);
```

Output 4.11   *Annotated Plot of PRESTIGE by INCOME*

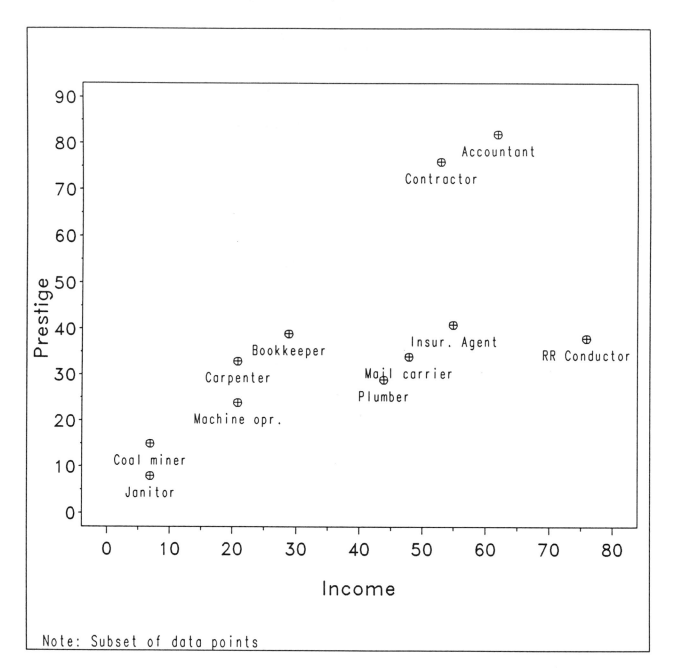

Reducing Overplotting

In the previous example, point labels do not overlap; but even with as few as 11 points, this was mostly fortuitous. With a larger number of observations, a general solution is difficult. See Noma (1987) for a heuristic approach that minimizes collisions by examining plot locations around each point. However, we can attempt to reduce the effects of label overlap in a PROC GPLOT scatterplot by using the following strategies:

□ plotting labels in a different color

□ choosing the positions of labels more carefully

□ making labels smaller or shorter.

The program below plots all the observations in the DUNCAN data set. Because the labels are rather long, we can choose to abbreviate them, either uniformly or selectively. In this case, we decide to identify almost half of the points with the full JOB label, and the remaining points by the sequential CASE number in the data set. To do this, the program below first assigns the TEXT variable to the value of CASE number, converted to a character string of length 2. Then the INDEX function is used to determine if the current CASE value is one of the list to be identified by the full JOB name.

In addition, the program alternates the position of the labels, plotting them either on the left (POSITION='4') or right (POSITION='6') of the point symbols. For the alternating order to be effective, the observations are first sorted by the ordinate variable, PRESTIGE.

This program produces the plot shown in Output 4.12. (A color version of this output appears as Output A3.12 in Appendix 3, "Color Output.") Overplotting is reduced by using color, alternating the positions of labels, and making some labels shorter. Because the density of points is high along the diagonal, it is difficult to do much better without a good deal more effort. Nevertheless, the simple techniques used here are useful generally and enable more labels to be plotted than would otherwise be possible.

```
proc sort data=duncan;
   by prestige;
data labels;
   set duncan;
   retain xsys '2' ysys '2' color 'RED' size .9;
   length text $12 function $8 ;
   text = put(case,2.);
   if index(' 6 7 9 11 13 15 16 18 19 20 22 24 25 27 28 30 32 33 36 37 39 44 ',
           trim(text)||' ') > 0
     then text = job;
   y = prestige;
   x = income;
   if mod(_n_,2)=0 then do; position ='4'; x=x-1; end;
                    else do; position ='6'; x=x+1; end;
   function = 'LABEL';
title ' ';
symbol v=+ color='black';
proc gplot data=duncan  ;
   plot prestige * income /
      annotate=labels
      vaxis=axis1 haxis=axis2
      hminor=1 vminor=1 frame
      name='GB0412'  ;
```

```
axis1 label=(h=1.5 a=90 r=0 f=duplex)
      value=(h=1.3 f=duplex) offset=(1);
axis2 label=(h=1.5 f=duplex)
      value=(h=1.3 f=duplex) offset=(4);
```

Output 4.12   *Annotated Plot of PRESTIGE by INCOME*

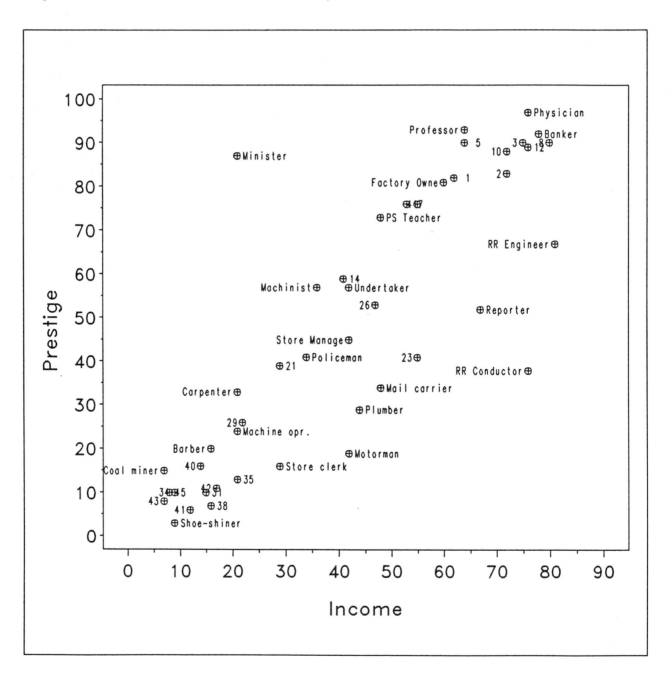

4.4 Interpolated Curves and Smoothings

When y is a response variable and x is a factor variable, scatterplots are often used to study how y depends on x. Is the relationship linear or nonlinear? Does the variability of y change with x, or is it constant? With noisy data, it may be hard to tell just by looking at the points. Even with fairly regular, well-behaved data, adding smooth curves that summarize the relation between y and x makes it easier to understand and communicate.

More generally, adding a smooth curve to the scatterplot may be viewed as a means to highlight the pattern or signal in the data. The smooth curve is easier for the eye to follow, reducing the distracting effects of noise. When you have a particular pattern in mind, one idea is to fit that model to the data and display the fitted relation along with the data. For example, the regression methods discussed in Chapter 5, "Plotting Regression Data," fit a linear or polynomial pattern to the data, and the I=R series of options in the SYMBOL statement enable you to display a linear, quadratic, or cubic on a PROC GPLOT scatterplot.

In other cases, you may prefer to fit a smooth curve without specifying a preconceived model. If the data form an essentially functional, though possibly complex, relation, the specification I=SPLINE can be used to draw a smooth interpolating curve that passes through all the points. The spline methods fit a series of continuous cubic polynomials to adjacent points. For example, plots of the Student's t critical values (Section 3.2.4, "Plotting Percentiles: Student's t Distribution") and power curves (Section 3.2.5, "Plotting Power Curves: Student's t Distribution") use spline curves.

This section describes several smoothing methods that are useful for noisy data. The simplest methods use the specification I=SMnn in the SYMBOL statement, either with the raw data or with some presmoothed summary values such as medians. The SMnn curve differs from the SPLINE interpolation options in that the SPLINE methods pass through all the points and will not be smooth if the data are noisy. The nn value in the SMnn curve enables the user to balance smoothness against closeness of the points to the curve. The method minimizes a weighted sum of two terms: the sum of squares of residuals (measuring fidelity) and the integrated second derivative of the curve (smoothness). The larger the weight for the second term, the smoother the curve, but smoothness may be achieved by passing far from some points. A value of $nn=0$ gives no weight to smoothness and is equivalent to specifying I=SPLINE.

A second approach is to fit a robust, locally weighted regression curve through the points with the method known as *lowess* (Cleveland 1979). The method finds a smooth value at each x by fitting a regression line that gives the most weight to nearby points. Robust means that points far from the rest are discounted or ignored.

There are many other approaches to scatterplot smoothing. Tukey (1977) discusses a variety of methods based on running medians; some of these are illustrated in the program TUKEY SAS in the SAS Sample Library. The TRANSREG procedure in Version 6 of SAS/STAT software offers a large variety of optimal data transformations, of which the SPLINE and MSPLINE (monotone spline) methods provide reasonable smoothings. Various forms of weighted moving averages are also widely used; see Anscombe (1973) and Hamming (1977). A recent general survey of smoothing methods is given by Goodall (1990).

4.4.1 SYMBOL Statement
Smoothers for Noisy Data

This example shows two ways in which smoothing can be used to tame noisy data in order to see a trend that may not be at all apparent in the raw data.

The data (Section A2.7, "The DRAFTUSA Data Set: Draft Lottery Data"), are from the U.S. draft lottery, conducted in 1969 to determine the order in which draft-eligible men would be called up to fight in Vietnam. A random lottery was held assigning draft priority numbers to birth dates (days of the year, from 1 to 366). The number 001 was the highest priority, and 366 was the lowest priority. So someone born on the date assigned 001 was almost certain to be drafted into the army, while someone born on the day assigned 366 was almost certain not to be drafted. However, some observers questioned the validity of the randomization procedure used to assign priority values and suggested that flaws in the procedure made those born later in the year more likely to be drafted (Fienberg 1971).

The first display (Output 4.13) plots the raw data, using a strongly smoothed line (I=SM80) to join the points. Because the x values for a given month are all the same, the spline curve passes approximately through the mean y value for each month.

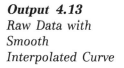

Output 4.13
Raw Data with
Smooth
Interpolated Curve

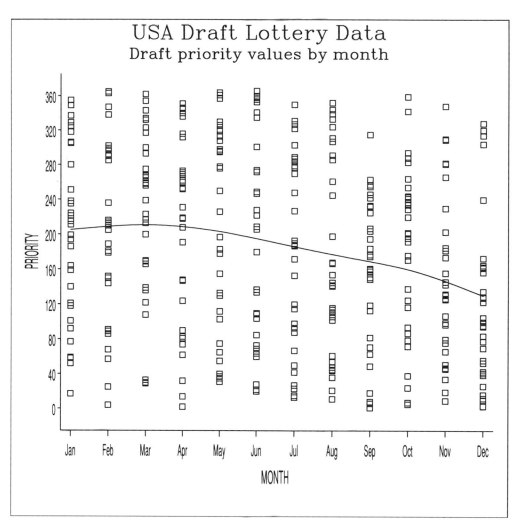

The decreasing trend from January to December is very apparent in the smoothed curve, but not in the raw data (see Output 1.23).

```
%include DRAFTUSA;              /* On MVS, use %include ddname(file);*/
 /*--------------------*
  |   Plot raw data    |
  *--------------------*/
title 'USA Draft Lottery Data';
title2 f=complex h=1.5 'Draft priority values by month';
symbol1 v=square
       i=sm80   c=black;                /* sm80 for very noisy data */
proc gplot data=draftusa  ;
    plot priority * month=1
         / vaxis=axis1 vminor=0
           haxis=axis2 hminor=0
           name='GB0413' ;
    axis1 order=(0 to 380 by 40)
          offset=(2) label=(h=1.5 a=90 r=0) value=(h=1.3);
    axis2 offset=(2) label=(h=1.5)          value=(h=1.3);
    format month mon.;
```

The second idea, from Tukey (1977), is to plot the trace of the median and quartiles of the set of priority values of each month as shown in Output 4.14. This shows the behavior of the data in three middle portions of the plot, and so shows more detail while ignoring excessively high or low values.

These summary values are found using the UNIVARIATE procedure with an OUTPUT statement to store them in a SAS data set. The SYMBOL statements specify a moderately smooth interpolated line (I=SM60) for the quartiles and medians. Note that the AXIS1 statement specifies the label for the ordinate, which otherwise would be labeled MEDIAN, the first of the variables plotted. The statements below produce Output 4.14.

```
/*--------------------*
 |   Plot trace lines |
 *--------------------*/
*-- Get medians & spreads;
proc univariate noprint data=draftusa;
    var priority;
    by  month;
    output out=sumry median=median q1=q1 q3=q3;
run;

title2 f=complex h=1.5 'Median and Quartile Trace lines';
symbol1 v=square h=1.5 i=sm60 c=black; /* median    */
symbol2 v=star   h=1.2 i=sm60 c=red;   /* quartiles */
proc gplot data=sumry  ;
    plot median * month = 1
         q1      * month = 2
         q3      * month = 2
         / overlay vaxis=axis1 haxis=axis2 hminor=0
           name='GB0414' ;
    axis1 label=(a=90 r=0 h=1.5 'PRIORITY')
          value=(h=1.3) order=(0 to 380 by 40)
          offset=(2);
    axis2 offset=(2) label=(h=1.5) value=(h=1.3);
    format month mon.;
```

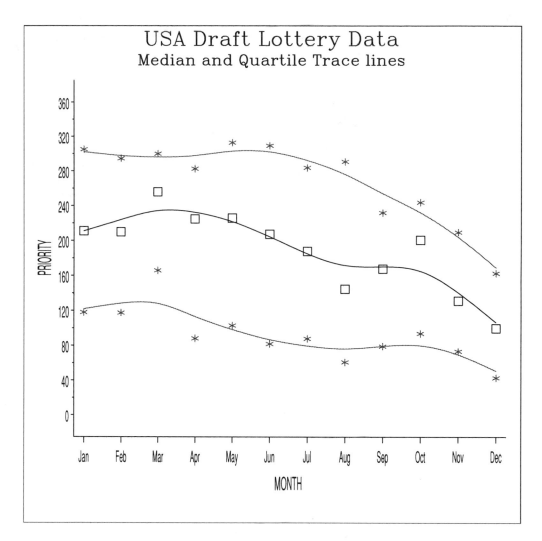

Because the *x* values, MONTH, are replicated in this example, another possibility would be to plot parallel boxplots of the *y* value, PRIORITY * MONTH. Such a plot (not shown here) would have the advantage of showing any individual outliers as well as the smooth middle traces. In this instance, the priority values are bounded above and below, so the display of Output 4.14 is preferable, but in other situations a boxplot display may be more informative.

4.4.2 Lowess Smoothing

In the plot of draft lottery data, the *x* values (MONTH) are regularly spaced, there are quite a few (28 to 31) observations at each *x* value, and none of the *y* values are outliers. In these special circumstances, both the SYMBOL statement I=SM*nn* smoother and the trace lines for median and quartiles produce reasonably smooth curves showing the trend of the data.* In other cases, one or the other of these techniques will not work as well, and it is worthwhile to have a robust smoothing technique that will work for any set of data.

A useful technique, which performs well under much more general conditions (particularly with outliers) is robust, locally weighted regression smoothing, or *lowess* (Cleveland 1979; Chambers et al. 1983). The procedure finds a smoothed fitted value, $\hat{y}_i$, for each x_i value by fitting a weighted regression to the points in the neighborhood of x_i. The points closest to the value x_i receive the greatest weight. This is the locally weighted regression part of the procedure, and the weights are called *neighborhood weights*. Because the fitted values are not constrained to a prespecified functional model, the procedure can also be construed as a form of *nonparametric regression*.

The robust part works as follows. Once the fitted values, $\hat{y}_i$, have been found, the residuals, $r_i = y_i - \hat{y}_i$, are used to determine a new set of weights (*robustness weights*) so that points that have large residuals are down-weighted and the locally weighted regression is repeated.

In practice, lowess depends on the choice of a smoothing parameter, f, $0 < f \leq 1$, the fraction of the data points to be considered in the calculation of the value $\hat{y}_i$. Choosing $f = .5$ means that only the $[f\ n]$ points closest to the value x_i have nonzero weights. Increasing the value of f makes the fitted curve smoother, analogous to the effect of increasing nn in the I=SMnn smoother in the SYMBOL statement. Values of f between ⅓ and ⅔ usually work well in practice. For the statistical details, see Cleveland (1979); Chambers et al. (1983, Section 4.6) give a nontechnical description of lowess smoothing.

Implementing Lowess Smoothing in the IML Procedure

The lowess procedure involves a series of weighted regressions, one for each data point. While this might seem like a lot of work, the process can be carried out easily with the matrix operations of the IML procedure.

The program below implements lowess smoothing and applies it to the automobile data on the variables MPG and WEIGHT (see Output 4.1 and 4.5). The resulting plot with the lowess smooth for $f = .5$ is shown in Output 4.15. The main calculations are carried out in the module LOWESS.

```
title h=1.8 'Auto data with lowess smoothing';
%include AUTO;                  /* On MVS, use %include ddname(file);*/
proc sort data=auto;
   by weight;
proc iml;

start wls( x, y, w, b );        *-- weighted least squares;
   x = j(nrow(x), 1, 1) || x;
   xpx = x` * diag( w ) * x;
   xpy = x` * diag( w ) * y;
   b  = inv(xpx) * xpy;
finish;
```

---

* When the spacing between successive ordered *x* values is highly irregular, the parametric versions of smoothing splines, specified by I=SM*nn*P, will generally give better results.

```
start median( w, m);          *-- calculate median ;
   n = nrow( w );
   r = rank( w );
   i = int((n+1)/2);
   i =  i || n-i+1;
   m = w[ r[i] ];
   m = .5 # m[+];
finish;

start robust( r, wts);        *-- calculate robustness weights;
   run median(abs(r), m);
   w = r / (6 # m);           * bisquare function;
   wts = (abs(w) < 1) # (1 - w##2) ## 2;
finish;

start lowess( x, y, f, steps, yhat);
   n = nrow(x);
   if n < 2 then do;
      yhat = y;
      return;
      end;
   q = round( f * n);         * # nearest neighbors;

   res  = y;
   yhat = J(n,1,0);
   delta= J(n,1,1);           * robustness weights;
   if steps <= 0 then steps=1;
   do it = 1 to steps;
      do i = 1 to n;
         dist = abs( x - x[i] );      * distance to each other pt;
         r = rank( dist );
         s = r; s[r]=1:n;
         near =  s[1:q] ;             * find the q nearest;
         nx = x [ near ];
         ny = y [ near ];
         d  = dist[ near[q] ];        * distance to q-th nearest;
         if d > 0 then do;
            u = abs( nx - x[i] ) / d ;
            wts = (u < 1) # (1 - u##3) ## 3; * neighborhood wts;
            wts = delta[ near ] # wts;
            if sum(wts[2:q]) > .0001 then do;
               run wls( nx, ny, wts, b );
               yhat[i] = (1 || x[i]) * b;      * smoothed value;
               end;
            else yhat[i] = y[i];
         end;
         else do;
            yhat[i] = ny [+] /q;
         end;
      end;
      res = y - yhat;
      run robust(res,delta);
```

```
            If it < steps then do;
                print "Iteration" it;
                print "Robustness weights", delta;
            end;
        end;
        xyres =x || y || yhat || res;
        print "Data, smoothed fit & residuals",
            xyres[ colname={"X" "Y" "YHAT" "RESIDUAL"}];
    finish;
```

The main routine reads the variables WEIGHT and MPG from the AUTO data set, calls the LOWESS module, and outputs the smoothed data to the data set SMOOTH. The LOWESS procedure here is run with $f = .5$ and two iterations.

```
    use AUTO;
    read all var{WEIGHT MPG} into xy[ colname=vars ];
    close AUTO;
    x = xy[,1];
    y = xy[,2];
    run lowess(x, y, .50, 2, yhat);

*-- Output results to data set SMOOTH ;
    xys = x || y || yhat;
    cname = vars || {"YHAT"};
    create smooth from xys [ colname=cname ];
    append from xys;
```

PROC GPLOT is then used to plot the original data together with the smoothed $\hat{y}$ values. The SYMBOL1 statement specifies unconnected points for the data; in the SYMBOL2 statement, I=JOIN is specified to connect the smoothed values.

```
    proc gplot data=smooth  ;
      plot mpg  * weight = 1
           yhat * weight = 2
           / overlay frame
             vaxis=axis1 haxis=axis2
             name='GB0415'  ;
      symbol1 v=+ h=1.5  i=none c=black;
      symbol2 v=none w=2 i=join c=red;
      axis1   label=(h=1.5 f=duplex) label=(h=1.4);
      axis2   label=(h=1.5 f=duplex) label=(h=1.4);
```

Output 4.15
Gas Mileage Plotted against Weight with Robust Lowess Smoothing

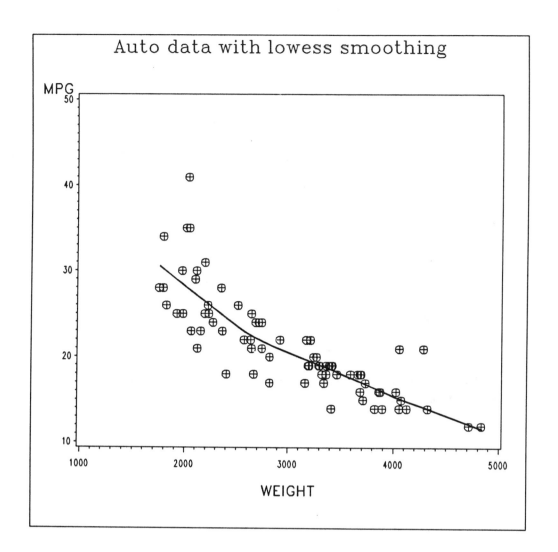

Note that there are quite a few points in Output 4.15 that are far from the curve, particularly near WEIGHT of 2,000 and 4,000. In the robust estimation iteration, these points receive robustness weights of zero and have no influence in determining the fitted curve. Near values of 4,000, there are not many points, and without the robustness step, the smooth curve would be deflected upwards by the two relatively high points.

LOWESS Macro

A macro program for lowess smoothing is listed in Section A1.9, "The LOWESS Macro." The program calculates the smoothed value of the variable named by the Y= parameter over the range of the X= variable. The smoothed values are returned as the variable _YHAT_ in the OUT= data set. To have the program plot the data points and the smoothed lowess curve (as shown in the PROC GPLOT step for the MPG * WEIGHT example), specify PLOT=YES.

```
%macro LOWESS(
        data=_LAST_,        /* name of input data set         */
        out=SMOOTH,         /* name of output data set        */
        x = X,              /* name of independent variable   */
        y = Y,              /* name of Y variable to be smoothed */
        id=,                /* optional row ID variable       */
        f = .50,            /* lowess window width            */
        iter=2,             /* total number of iterations     */
        plot=NO,            /* draw the plot?                 */
        name=LOWESS);       /* name for graphic catalog entry */
```

For example, the plot shown in Output 4.15 is also produced by these statements:

```
%include AUTO;
%include lowess;
%lowess(data=auto,
        x=weight,
        y=mpg,
        id=model,
        plot=YES);
```

4.5 Enhanced Scatterplots

Scatterplots can be enhanced by adding information to help you interpret the display or understand aspects of the data that are not shown directly. This section demonstrates two such enhancements:

□ The first adds boxplots for the X and Y variable to the plot margins. These help show the shape of the distributions of the individual variables and detect univariate outliers.

□ The second adds an elliptical confidence region around the mean to highlight the joint relationship. With data for several groups, this helps show the extent to which the relationship is the same for all groups.

Together these examples illustrate some ways that features of the SAS System can be combined to create custom scatterplot displays.

4.5.1 Annotating a Scatterplot with Boxplots for X and Y Variables

The ideas used to construct the boxplot display (see Section 2.4, "Boxplots" and Section 6.3, "Comparative Boxplots") can be used to enhance a scatterplot with a boxplot for each of the X and Y variables with PROC GPLOT or for the X, Y, and Z variables with the G3D procedure. In a two-variable display, the boxplot for X (or Y) is drawn parallel to the *x* axis (*y* axis), at one extreme of the other axis, so as not to obscure the points.

Such a display is shown in Output 4.16 for the WEIGHT and PRICE variables in the automobiles data. The boxplot for PRICE at the top of the figure shows clearly that the distribution of PRICE is positively skewed and that most of the car models are in the under $9,000 price range, with a handful over that. In a regression application, this dense packing of the observations toward the low end of PRICE might suggest the need for a transformation of PRICE to log(PRICE).

Output 4.16
Scatterplot of WEIGHT Plotted against PRICE Annotated by BOXANNO Macro

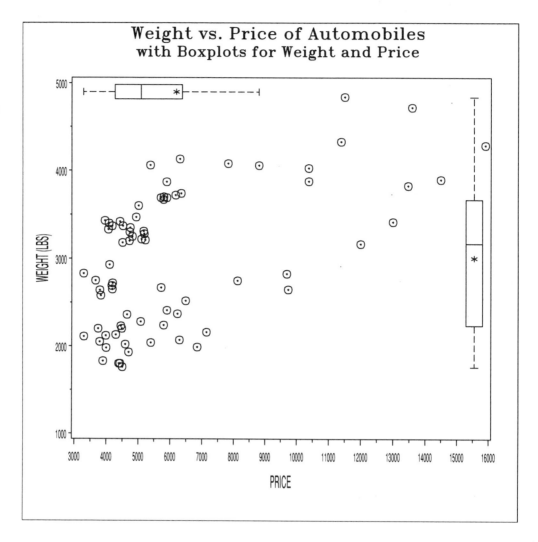

For generality, the program was designed as a SAS System macro application that constructs an Annotate data set to draw a boxplot on any one axis in a scatterplot. The basic macro is called BOXAXIS. It takes the parameters shown below:

```
%macro BOXAXIS(
        data=_LAST_,          /* Input data set               */
        out=_DATA_,           /* Output Annotate data set     */
        var=,                 /* Variable to be plotted       */
        baxis=x,              /* Axis on which it goes- X, Y, or Z */
        oaxis=y,              /* The other axis in the plot   */
        paxis=z,              /* The 3rd axis (ignored in GPLOT) */
        boxwidth=4,           /* Width of box in data percent */
        pos=98);              /* Position of box on OAXIS 0<POS<100*/
```

The complete macro is listed in Section A1.3, "The BOXANNO Macro." To draw boxplots on both axes in a scatterplot such as Output 4.16, %BOXAXIS is called twice, once with the specifications BAXIS=X and OAXIS=Y, and once with the specifications BAXIS=Y and OAXIS=X. Each call produces an Annotate data set (named by the OUT= parameter). These two data sets can be concatenated and used as the Annotate input to PROC GPLOT. These two steps are packaged together in another macro, BOXANNO, which produces one Annotate data set containing the instructions to draw the boxplots for both axes.

```
/*-----------------------------------------------------------*
 | BOXANNO macro - creates Annotate data set for both X & Y |
 *-----------------------------------------------------------*/
%macro boxanno(
    data=_last_,            /* Data set to be plotted  */
    xvar=,                  /* Horizontal variable     */
    yvar=,                  /* Vertical variable       */
    out=boxanno             /* Output Annotate data set*/
    );

%boxaxis(
    data=&data, var=&xvar,
    baxis=x,    oaxis=y,    out=xanno);

%boxaxis(
    data=&data, var=&yvar,
    baxis=y,    oaxis=x,    out=yanno);
/*-----------------------------------------*
 | Concatenate the two Annotate data sets |
 *-----------------------------------------*/
data &out;
    set xanno yanno;
%mend boxanno;
```

These macros do not plot the graph; it is up to the user to call PROC GPLOT after using the BOXANNO macro. The program below shows how the BOXANNO macro is used to plot Output 4.16.

```
%include AUTO;                 /* On MVS, use %include ddname(file);*/
%include boxanno;              /* On MVS, use %include ddname(file);*/
 /*-------------------------------------------------------------*
  | Plot WEIGHT vs. PRICE for AUTO data with marginal boxplots |
  *-------------------------------------------------------------*/
%boxanno(
    data=auto,              /* Data set to be plotted  */
    xvar=price,             /* Horizontal variable     */
    yvar=weight,            /* Vertical variable       */
    out=boxanno             /* Output Annotate data set*/
    );
```

```
proc gplot data=auto  ;
    symbol1 v=- h=1.5;
    axis1 label=(h=1.4 a=90 r=0);
    axis2 label=(h=1.4);
    plot weight * price /
        frame
        vaxis=axis1 haxis=axis2
        vminor=1 hminor=1
        annotate = boxanno
        name='GB0416' ;
    title  h=1.6 f=triplex 'Weight vs. Price of Automobiles';
    title2 h=1.4 f=triplex 'with Boxplots for Weight and Price';
```

The following shows how to produce an annotated boxplot for each axis
of a three-dimensional scatterplot with PROC G3D. In a plot of
WEIGHT*PRICE=MPG, we need an Annotate data set for each of the three axes.
The call to the BOXAXIS macro below constructs the boxplot for the *z* axis in the
y-z plane. The output data set ZANNO is appended to the data set BOXANNO for
the *x* and *y* axes before calling PROC G3D. The three-dimensional plot is shown
in Output 4.17. (A color version of this output appears as Output A3.13 in
Appendix 3.)

```
%boxaxis(
    data=auto,  var=mpg,
    baxis=z,    oaxis=y,    paxis=x, pos=99, out=zanno);
data boxanno;
    set boxanno zanno;

 /* Distinguish autos by region of origin */
data auto;
    set auto;
    if origin='A' then do;
        shape ='SQUARE'; color='GREEN'; end;
    if origin='E' then do;
        shape ='BALL  '; color='RED  '; end;
    if origin='J' then do;
        shape ='FLAG  '; color='BLUE '; end;
title h=1.5 'Weight, Price and Gas Mileage';
proc g3d data=auto  ;
    scatter weight * price = mpg /
        caxis = black
        shape = shape
        color = color
        size = .6
        annotate = boxanno
        name='GB0417';
```

Output 4.17   *Plot of Auto Data with Boxplots for All Axes from the G3D Procedure*

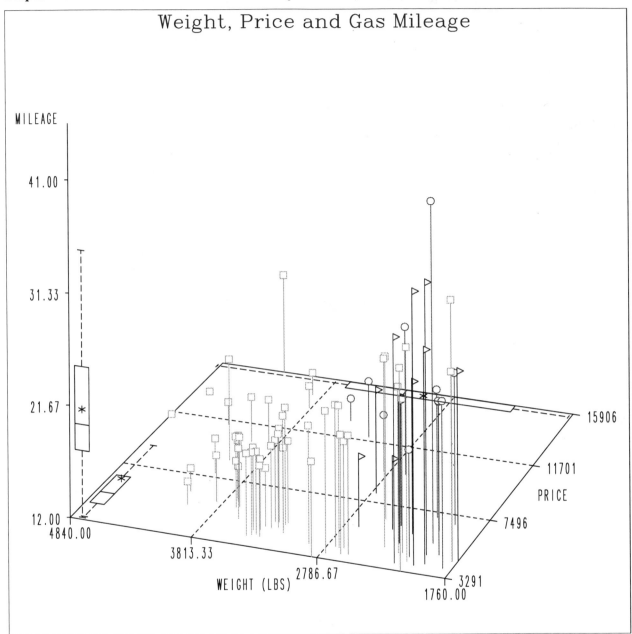

Displaying the shape of the distribution of each variable with a boxplot in the margin helps determine whether and how the variables should be transformed. The plot of the infant mortality data in Output 4.18 highlights the positive skewness of both variables and again suggests that the variables IMR and INCOME should be displayed on a log scale. Output 4.18 is produced by the following statements:

```
%include NATIONS;           /* On MVS, use %include ddname(file);*/
%include boxanno;           /* On MVS, use %include ddname(file);*/
 /*----------------------------------------------------------*
  | Plot IMR vs. INCOME for nations data with marginal boxplots |
  *----------------------------------------------------------*/
%boxanno(
     data=nations,          /* Data set to be plotted  */
     xvar=income,           /* Horizontal variable     */
     yvar=imr,              /* Vertical variable        */
     out=boxanno            /* Output Annotate data set*/
     );

proc gplot data=nations  ;
     symbol1 v='-' h=1.4;
     axis1 label=(a=90 r=0 h=1.5) value=(h=1.4);
     axis2 label=(        h=1.5) value=(h=1.4);
     plot imr * income /
          frame name='GB0418'
          vaxis=axis1 haxis=axis2 vminor=1 hminor=1
          annotate = boxanno;
     title  h=1.7 f=triplex 'Infant mortality rate and income';
run;
```

Output 4.18
Infant Mortality Data with Boxplots for IMR and INCOME

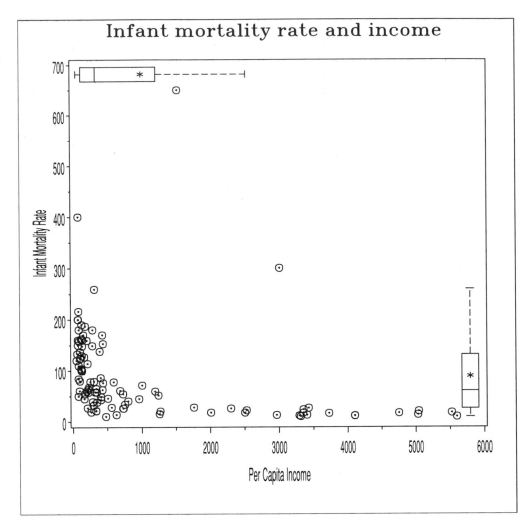

4.5.2 Adding Confidence Ellipses

When you have (*x,y*) data for several groups, you may want to examine how the means, variances, and correlations differ from group to group and how these relate to the data for the total sample. This example uses the IRIS data set (Section A2.10, "The IRIS Data Set: Iris Data") to illustrate the point that the within-group relationships between *x* and *y* may be very different from the relationship in the total sample.

The statements below use the CORR procedure to calculate the correlation between petal width and sepal width, once for the total sample of 150 observations and separately for each of the three species of 50 observations. The two sets of correlations are combined in a DATA step for display purposes and shown in Output 4.19.

```
%include IRIS;              /* On MVS, use %include ddname(file);*/
proc corr data=iris noprint out=total;
   var petalwid sepalwid;
proc sort data=iris;
   by species;
```

```
proc corr data=iris noprint out=groups;
    var petalwid sepalwid;
    by species ;

title 'Total-sample and Within-group correlations';
data corr;
    set total groups;
    drop _type_ petalwid;
    if _type_='CORR' & _name_= 'PETALWID';
    if species=' ' then species='_All_';
proc print;
    id species;
```

Output 4.19

Total-Sample and
Within-Cell
Correlations for
Iris Data

```
            Total-sample and Within-group correlations

        SPECIES        _NAME_        SEPALWID

        _All_          PETALWID      -.366126
        Setosa         PETALWID      0.232752
        Versicolor     PETALWID      0.663999
        Virginica      PETALWID      0.537728
```

From Output 4.19, you can see that petal width and sepal width are positively correlated within each of the species. However, the means of these two variables are inversely related across the three species; this produces an overall negative correlation in the total sample. Therefore, a correlational analysis of this data that ignored group membership would be misleading. (A correct analysis should use the pooled within-sample correlations.) One way to avoid being misled is to plot an elliptical region for each group, and for the total sample to show the within-group and total-sample relationships visually. Such a plot is shown in Output 4.20. The large ellipse depicts the 75% confidence region for the total sample. The three smaller ellipses show the same confidence regions for the separate species.

Output 4.20
Contour Regions
for Separate
Groups and Total
Sample, Iris Data

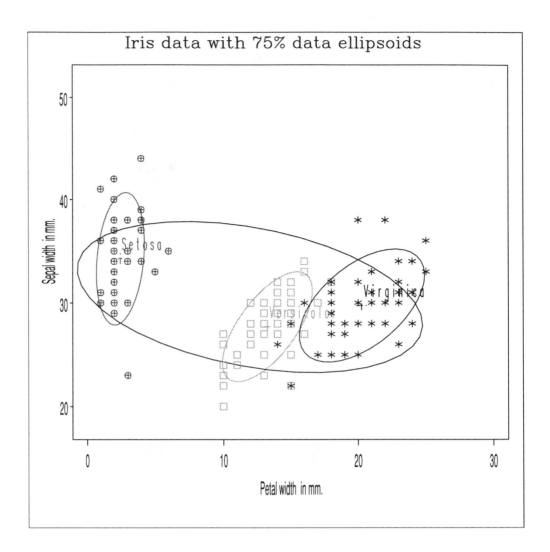

The idea of a confidence interval for a single variable generalizes to an elliptical joint-confidence region for two variables (and to a p-dimensional ellipsoid for p variables). Let $\mathbf{x}_i = (x_i, y_i)$ be a pair of scores, with sample means $\bar{\mathbf{x}} = (\bar{x}, \bar{y})$. For data that have a bivariate normal distribution, the joint $(1-\alpha)$ confidence region for the population mean, $\boldsymbol{\mu} = (\mu_x, \mu_y)$, can be shown to be given by the values $\boldsymbol{\mu}$ satisfying

$$(\bar{\mathbf{x}} - \boldsymbol{\mu})' \, \mathbf{S}^{-1} (\bar{\mathbf{x}} - \boldsymbol{\mu}) \le \frac{2}{n-1} F_{2,\,n-1}(1-\alpha) \quad , \tag{4.1}$$

where $\mathbf{S}$ is the (2×2) variance-covariance matrix of (x, y), and $F_{2,\,n-1}(1-\alpha)$ is the $(1-\alpha)$ percentage point of the F distribution with 2 and $n-1$ degrees of freedom (Johnson and Wichern 1982). (The left side of equation 4.1 is like the square of a t statistic for a single mean, $t = (\bar{x} - \mu)/s$). For the individual observations, $\mathbf{x}_i$, an analogous relationship holds, giving a data ellipse that includes $\mathbf{x}_i$ with probability $(1-\alpha)$:

$$(\mathbf{x}_i - \bar{\mathbf{x}})' \, \mathbf{S}^{-1} (\mathbf{x}_i - \bar{\mathbf{x}}) \le 2 \, F_{2,\,n-1}(1-\alpha) \quad . \tag{4.2}$$

Points on the boundary of the ellipse (where equality holds in equations 4.1 and 4.2) can be calculated from the eigenvalues and eigenvectors of **S** (see Johnson and Wichern 1982, Section 5.5).

When the data are, at least roughly, bivariate normal, the data ellipse has a number of remarkable properties (Monette 1990), some of which are illustrated in Output 4.21.

Output 4.21
Data Ellipse,
Regression Line,
and Correlation

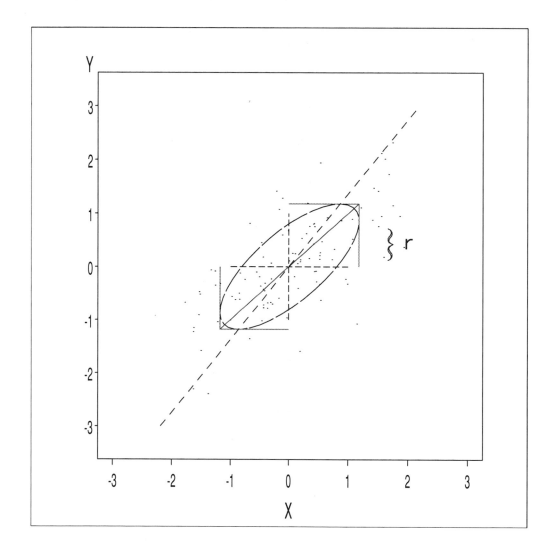

Output 4.21 has the following features:

□   The center of the ellipse shows the location (mean) of x and y; the arms of the central cross show the spread (standard deviation) of each variable.

□   With several groups, if the error bars are scaled to the standard error of each mean, they give a visual indication of which groups differ on each of the x and y variables.

□   The regression line predicting y from x goes through the mean $(\bar{x}, \bar{y})$ and through the point where a tangent to the ellipse is vertical.

□ The other regression line—predicting *x* from *y*—can be constructed by drawing a line from the mean through the point where a tangent to the ellipse is horizontal (the dashed line in Output 4.21).

□ The correlation between *x* and *y* can be estimated visually from the fraction of the vertical tangent line below the regression line. (As the correlation approaches ± 1, the ellipse becomes thinner and the regression line approaches the major axis of the ellipse. As the correlation approaches 0, the regression line becomes horizontal.)

The program in the next section uses PROC IML to calculate the elliptical regions for the observations in each group in the data set IRIS. The value of $(1-\alpha)$ for the ellipse is assigned to the macro variable PVALUE. The plot is drawn with PROC GPLOT, using the PROC IML result to construct an Annotate data set. A general version of the program is the CONTOUR macro, listed in Section A1.5, " The CONTOUR Macro."

Calculating the Contours with the IML Procedure

The PROC IML program contains two modules (subroutines):

ELLIPSE   calculates points on the elliptical contour for a single set of (X, Y) data. The variable FORMEAN determines whether the ellipse gives a confidence region for the mean vector or a data ellipse.*

DOGROUPS   loops through the groups in the data set, calling ELLIPSE for each. The points on the boundary of the ellipse are output to a data set CONTOUR.

```
%include IRIS;                    /* On MVS, use %include ddname(file);*/
title h=1.6 'Iris data with 75% data ellipsoids';

%let pvalue=.75;
proc iml;
start ellipse(c, x, y, npoints, pvalues, formean);
   /*------------------------------------------------------------*
    |  Computes elliptical contours for a scatterplot            |
    |    C       returns the contours as consecutive pairs of columns |
    |    X,Y     coordinates of the points                       |
    |    NPOINTS scalar giving number of points around a contour  |
    |    PVALUES column vector of confidence coefficients         |
    |    FORMEAN 0=contours for observations, 1=contours for means |
    *------------------------------------------------------------*/
```

---

* The program is based on an example from Chapter 3, "Applications," in the *SAS/IML User's Guide, Version 5 Edition.*

```
xx = x||y;
n  = nrow(x);
*-- Correct for the mean --;
mean = xx[+,]/n;
xx = xx - mean @ j(n,1,1);

*-- Find principal axes of ellipses --;
xx = xx` * xx / (n-1);
call eigen(v, e, xx);

*-- Set contour levels --;
c =  2*finv(pvalues,2,n-1,0);
if formean=1 then c = c / (n-1) ;
print 'Contour values',pvalues c;
a = sqrt(c*v[ 1 ] );
b = sqrt(c*v[ 2 ] );

*-- Parameterize the ellipse by angles around unit circle --;
t = ( (1:npoints) - (1)) # atan(1)#8/(npoints-1);
s = sin(t);
t = cos(t);
s = s` * a;
t = t` * b;

*-- Form contour points --;
s = ( ( e*(shape(s,1)//shape(t,1) )) +
      mean` @ j(1,npoints*ncol(c),1) )` ;
c = shape( s, npoints);
*-- C returned as NCOL pairs of columns for contours--;
finish;

start dogroups(x, y, gp, pvalue);
   d = design(gp);
   d = d || j(nrow(x),1,1);     * unit vector for total sample;
   do group = 1 to ncol(d);
      Print group;
      *-- select observations in each group;
      col = d[, group ];
      xg = x[ loc(col), ];
      yg = y[ loc(col), ];

      *-- Find ellipse boundary ;
      run ellipse(xyg,xg,yg, 40, pvalue, 0 );
      nr = nrow(xyg);
```

```
            *-- Output contour data for this group;
            cnames = { X Y PVALUE GP };
            col = 1:2 ;
            xygp = xyg[,col] || j(nr,1,pvalue) || j(nr,1,group);
            if group=1
               then create contour from xygp [colname=cnames];
            Append from XYGP;
       end;
   finish;
```

The following PROC IML statements read the data set IRIS and call the module DOGROUPS to calculate 40 points on the boundary of the data ellipse for each group and for the total sample. The X, Y points on all four ellipses are output to a SAS data set, CONTOUR. The CONTOUR data set is used later to plot the contours with PROC GPLOT.

```
   *-- Get input data: X, Y, GP;
      use iris;
      read all var {petalwid} into x  ;
      read all var {sepalwid} into y  ;
      read all var {species}  into gp ;
      close iris;

   *-- Find contours for each group;
      run dogroups(x, y, gp, { &pvalue} );
   quit;
```

Plotting the Contours with the Annotate Facility

The elliptical contours can be plotted with PROC GPLOT using the Annotate facility. The DATA step below reads the CONTOUR data set produced by PROC IML. The ellipse is drawn using FUNCTION= 'POLY' (to start a new polygon for each group) and FUNCTION= 'POLYCONT' (to continue the polygon).

```
    /*-----------------------------------*
    |  Plot the contours using Annotate  |
    *-----------------------------------*/
data contour;
   set contour ;
   by gp ;
   length function $8;
   xsys='2'; ysys='2';
   if first.gp then function='POLY';
               else function='POLYCONT';
   color=scan("RED GREEN BLUE BLACK",gp);
   line = 5;
symbol1 v=+      h=1.2 c=RED;
symbol2 v=square h=1.2 c=GREEN;
symbol3 v=star   h=1.2 c=BLUE;
```

```
proc gplot data=iris;
   plot sepalwid * petalwid = species
        /annotate=contour nolegend frame
         vaxis=axis1 vminor=0
         haxis=axis2 hminor=0
   ;
   axis1 offset=(3) value=(h=1.5) label=(h=1.5 a=90 r=0);
   axis2 offset=(3) value=(h=1.5) label=(h=1.5);
```

(The crosses at the means and group labels in Output 4.20 are drawn with another Annotate data set, not shown here. See the listing for the CONTOUR macro in Section A1.5.)

CONTOUR Macro

The CONTOUR macro listed in Section A1.5 produces a scatterplot with elliptical contours for any data set. One ellipse is drawn for each value of the variable specified for the GROUP= parameter. If the parameter ALL=YES is specified, an ellipse for the total sample is drawn as well. The macro takes the following parameters:

```
%macro CONTOUR(
        data=_LAST_,          /* input data set                   */
        x=,                   /* X variable                       */
        y=,                   /* Y variable                       */
        group=,               /* Group variable (optional)        */
        pvalue= .5,           /* Confidence coefficient (1-alpha) */
        std=STDERR,           /* error bar metric: STD or STDERR  */
        points=40,            /* points on each contour           */
        all=NO,               /* include contour for total sample?*/
        out=CONTOUR,          /* output data set                  */
        plot=YES,             /* plot the results?                */
        i=none,               /* SYMBOL statement interpolate opt */
        name=CONTOUR,         /* Name for graphic catalog entry   */
        colors=RED GREEN BLUE BLACK PURPLE YELLOW BROWN ORANGE,
        symbols=+ square star -    plus    :      $     = );
```

For example, the contour plot for the iris data shown in Output 4.20 is also produced by these statements:

```
%include IRIS;              /* On MVS, use %include ddname(file);*/
%include contour;           /* On MVS, use %include ddname(file);*/
Title h=1.6 'Iris data with 75% data ellipsoids';
%contour(data=iris,
        x=petalwid, y=sepalwid,
        group=species,
        pvalue=.75,all=yes,
        name=GB0420);
```

Example: Automobiles Data

The observations in the AUTO data set are classified by region of origin. To help see how the relationship between weight and price of an automobile is moderated by region of origin, the data shown in Output 4.2 and 4.16 are redrawn in Output 4.22, with a 50% data ellipse for each region. (A color version of this output appears as Output A3.14 in Appendix 3.)

The original plots appear to show an empty space in the middle of the plot. In Output 4.22, it is apparent that there are separate linear relations, with approximately the same slope, for the three regions. The points at the top are all the American models, which are substantially heavier for a given price.

Output 4.22
WEIGHT Plotted against PRICE of Automobiles, with Data Ellipse for Each Region of Origin

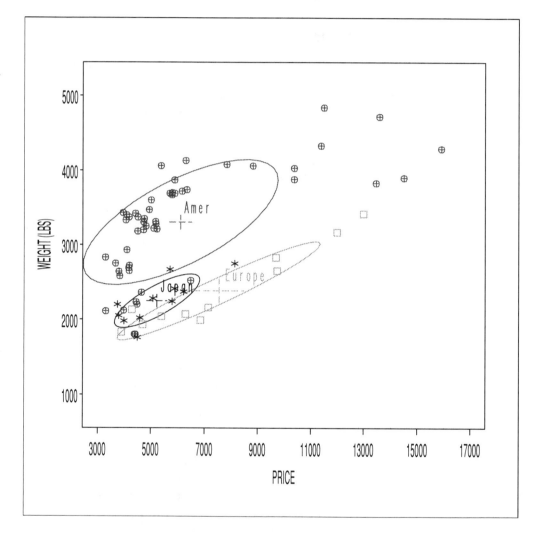

Nonparametric Contours

It should be noted that the adequacy of the data ellipse as a bivariate visual summary depends on means, standard deviations, and correlations being reasonable summary measures for the data. The data ellipse can be distorted by outliers, highly skewed distributions, or nonlinear relations. For example, each of Anscombe's data sets (Section 1.2, "Advantages of Plotting Data") has an identical data ellipse. In the automobiles data, the American cars in Output 4.22 do not appear well represented by a linear relation. For such cases, a robust or nonparametric version of the data ellipse would be preferable.

Cleveland and McGill (1984b) describe a clever procedure using lowess to compute a robust bivariate central oval. The (x, y) data are first scaled to have equal spread and an origin at median(x), median(y). Then the idea is to transform the data to polar coordinates, (r, θ), use lowess to smooth the distance, r, of each point from the origin as a function of the angle, θ, and finally transform the smoothed result back to original (x, y) coordinates.

4.6 Plotting Discrete Data

When the x or y values are highly discrete, or when there are a lot of data, it may be difficult to see patterns in scatterplots because of *overplotting*—many points may fall at the same (x, y) plot locations.

In PROC PLOT, with a PLOT statement of the form

```
PLOT y * x;
```

multiple points are shown by the characters A, B, C, . . . representing 1, 2, 3, . . . observations at a given plot position. This may be suitable for archival use, but when the data are highly discrete, the resulting plot does not convey density visually. For example, Output 4.23 is a plot of data generated from a mixture of three bivariate normal distributions. The letters A, B, C, and so on are poor visual codes for numerosity. The plot looks like just one mass of points, and it is very hard to see that there are actually three regions of high density. PROC GPLOT gives a much finer resolution, but with large data sets, the visual impression of regions of high and low density still may be misleading, especially if the data are discrete or have been rounded.

Output 4.23
Mixture of Three
Bivariate Normal
Distributions

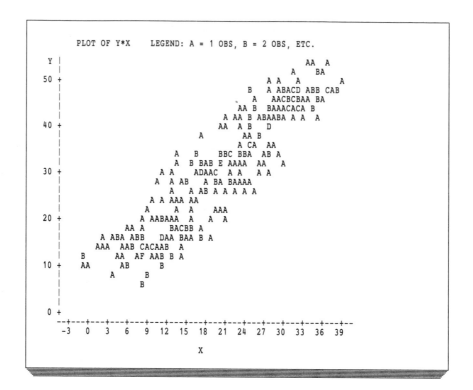

```
        PLOT OF Y*X     LEGEND: A = 1 OBS, B = 2 OBS, ETC.
   Y |                                            AA   A
     |                                         A    BA
  50 +                                      A  A   A        A
     |                                 B   A ABACD ABB CAB
     |                                   A   AACBCBAA BA
     |                                AA B  BAAACACA B
     |                              A AA B ABAABA A A  A
  40 +                                AA  A B   D
     |                           A       AA B
     |                                 A CA  AA
     |                      A    B   BBC BBA  AB A
     |                    A  B BAB E AAAA  AA   A
  30 +                 A A    ADAAC A A   A A
     |                  A  A AB   A BA BAAAA
     |                   A   A AB A A A A A
     |                 A A AAA AA
     |               A   A  A    AAA
  20 +               A AABAAA  A   A   A
     |                AA A    BACBB A
     |            A ABA ABB   DAA BAA B A
     |             AAA  AAB CACAAB  A
     |         B        AA  AF AAB B A
  10 +      AA      AB      B
     |               A    B
     |                    B
     |
   0 +
     --+---+---+---+---+---+---+---+---+---+---+---+---+---+---+--
      -3   0   3   6   9  12  15  18  21  24  27  30  33  36  39

                                  X
```

4.6.1 Sunflower Plots

Cleveland and McGill (1984b) propose the *sunflower plot* as a solution to this problem. Their idea is to divide the plotting region into cells of equal size and count the number of observations that fall into each cell. For the data of Output 4.23, the counts from a 20×20 cellulation are displayed in Output 4.24. However, even this display requires careful examination to see the regions where observations are most numerous.

Output 4.24
Mixture Data
Displayed as
Counts in Cells

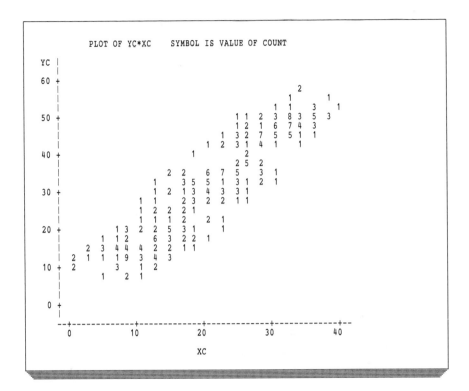

Then the counts are displayed by sunflower symbols, as shown in Output 4.25. Thus, the number of observations in a given cell is shown by the number of petals. A single observation is shown as a small square. A square with two radial lines (petals) represents two observations in a cell, a square with three lines is a count of three, and so on.

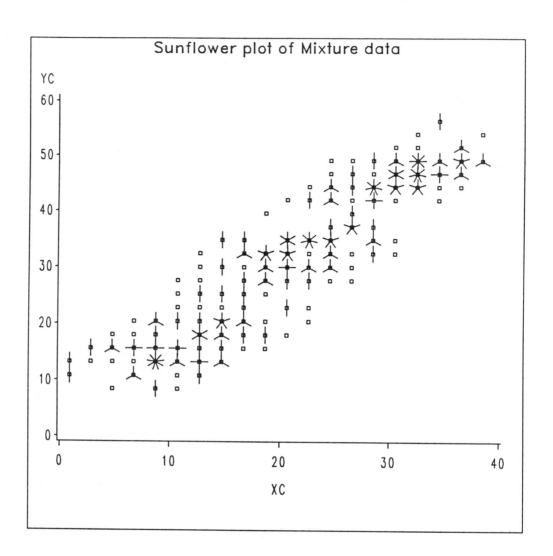

The sunflower plot is a kind of two-dimensional histogram, because the sunflower symbol shows the frequency in each (*x*, *y*) cell. Visually, the display of frequency is direct in Output 4.25, whereas in Output 4.23 and 4.24 we must interpret letters and digits to understand where the points are the most concentrated. The sunflower symbols are an interesting application of the psychology of graphical display. Visually, we are best able to discriminate differences among the symbols for small counts (cell frequencies from 1 to about 7), where discrimination is most important. That is, it is easier to distinguish between the symbols for three and four observations than it is to distinguish between those for ten and eleven observations, and the smaller counts are typically most prevalent.

Constructing Sunflower Symbols with the GFONT Procedure

The sunflower symbols are constructed as a SAS/GRAPH font with PROC GFONT.* The data set SUNSYMB in the program below is used as input to PROC GFONT to construct 26 characters in a font called SUN. These characters are the sunflower symbols for counts 1 to 26 and are identified with the keyboard characters A through Z. That is, the symbol for a count of two could be drawn by a SYMBOL statement as follows:

```
SYMBOL FONT=SUN V=B;
```

PROC GFONT uses the variables named CHAR, SEGMENT, X, and Y. The variables CHAR and SEGMENT identify the keyboard character and segment within the character. For each segment, the variables X and Y list the points that are connected by straight lines. The SUN font is shown in Output 4.26.

```
title 'Construct a font for making Sunflower Plots';
data sunsymb;
  alpha = 'ABCDEFGHIJKLMNOPQRSTUVWXYZ';
  do n=1 to 26;
     char=substr(alpha,n,1);
     segment=n;
     x = .2; y = .2; output;          /* Draw small box at center */
     x =-.2; y = .2; output;          /* of each symbol           */
     x =-.2; y =-.2; output;
     x = .2; y =-.2; output;
     x = .2; y = .2; output;
     if n>1 then
        do i=1 to n;                  /* draw n radial lines      */
          x=0; y=0; output;
          x=cos(2*atan(1) + i/n*(8*atan(1)));
          y=sin(2*atan(1) + i/n*(8*atan(1)));
          output;
        end;
  end;
proc gfont data=sunsymb deck                /* name=GB0426 */
          name=sun showroman h=6 romht=3;
   title ' ';
```

---

* A method for drawing sunflower plots with the Annotate facility rather than PROC GFONT is described by Benoit (1985b).

Output 4.26   *Sunflower Symbols Character Set*

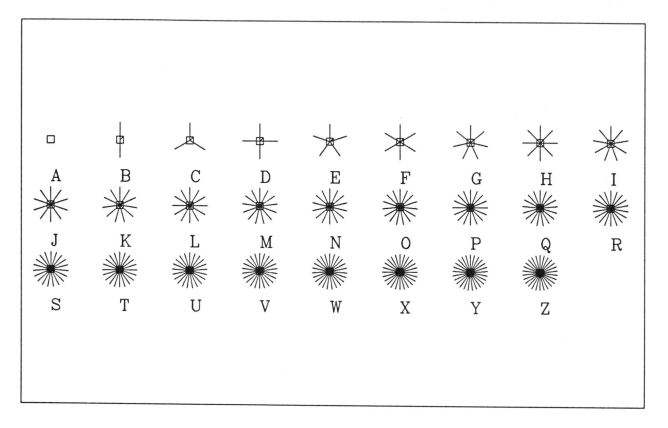

Making a Sunflower Plot

The sunflower plot is constructed using the following steps:

1. The range of *x* and *y* values is found, and these values are output to a SAS data set using the MEANS or the UNIVARIATE procedure. The *x* and *y* values are then grouped into discrete cells of width δ=range/number of cells, giving cellulated (grouped) values, XC, YC.

2. The frequencies of the cellulated values are found with the FREQ procedure, and an output data set containing the variables XC, YC, and COUNT is produced.

3. The data are plotted with PROC GPLOT with a PLOT statement of the following form, so PROC GPLOT will use a different SYMBOL statement for each value of the variable COUNT:

```
plot yc * xc = count ;
```

The following SYMBOL statements produce the sunflower symbols for each value of the variable COUNT:

```
symbol1  font=sun v=A;
symbol2  font=sun v=B;
symbol3  font=sun v=C;
symbol4  font=sun v=D;
  ...
symbol26 font=sun v=Z r=100;
```

Note that the last SYMBOL statement specifies R=100 so that if there happen to be cells with more than 26 observations, the SYMBOL26 statement will continue to be repeated up to 100 times.

Listed below are two SAS macros, CELL and SUNSYMB. The CELL macro carries out the steps (1—3 above) of dividing the data points into discrete cells and finding the number of observations in each. The SUNSYMB macro determines the highest value of COUNT and generates the SYMBOL statements.

```
/*------------------------------------*
 | Macro to cellulate an X-Y data set |
 *------------------------------------*/
%macro CELL( data=_LAST,    /* input data set */
             out=_DATA_,
             x = X, y = Y,    /* input X, Y variables */
             xc=xc, yc=yc,    /* output X, Y variables */
             nx=15, ny=15);   /* number of bins in X & Y directions */

 /*--------------------------------------------------*
  | Quantize the X and Y values into discrete cells  |
  *--------------------------------------------------*/
proc means data=&data noprint;
   var &x &y;
   output out=_temp_ range=rx ry;

data _cell_;
   if _n_=1 then set _temp_;
   set &data;
   drop rx ry;
   deltax = rx / &nx;
   deltay = ry / &nx;
   &xc = deltax * (round( &x/deltax )+.5);
   &yc = deltay * (round( &y/deltay )+.5);
 /*----------------------------------------*
  | Find number of points in each X-Y cell |
  *----------------------------------------*/
proc freq data=_cell_;
   tables &xc * &yc / noprint out=&out ;
%mend CELL;
```

```
/*----------------------------------------------------*
 |  Macro to generate SYMBOL statement for each COUNT |
 *----------------------------------------------------*/
%macro SUNSYM( data=_LAST_,ht=1.5, color=black );
%local alpha ch repeat;
%let alpha=ABCDEFGHIJKLMNOPQRSTUVWXYZ;

proc means data=&data noprint;
   var count;
   output out=_temp_ max=maxcount;
data _null_;
   set _temp_;
   maxcount = min( maxcount, 26 );
   call symput('HIGH',put(maxcount,2.));
run;

  %let repeat=;
  %do i=1 %to &HIGH %by 1;
     %let ch = &i;
     %if &i = 26 %then %let repeat = r=100;
     symbol&i h=&ht f=sun c=&color v=%substr(&alpha,&ch,1) &repeat ;
  %end;
%mend SUNSYM;
```

The plot in Output 4.25 is generated by the statements below. The DATA step creates the three groups of bivariate normal observations. The CELL macro is called to divide the observations into a 20×20 grid, which produces the output data set CELLED with the variables XC, YC, and COUNT. These variables are plotted in the PROC GPLOT step, using the SYMBOL statements generated by the SUNSYM macro.

```
data suntest;
    do gp=1 to 3;
        mx = 10*gp;
        my = mx;
        do i = 1 to 100;
            x = round((mx + 5 * normal(1242421)), .5);
            y = round((my + 3 * normal(2424243) + .5*x), .5);
            output;
        end;
    end;

%cell(data=suntest, out=celled, nx=20, ny=20);
%sunsym(data=celled);
```

```
title1 h=1.5  f=duplex 'Sunflower plot of Mixture data';
proc gplot data=celled  ;
   plot yc * xc = count /
        nolegend
        vaxis=axis1 haxis=axis2
        name='GB0425' ;
   axis1 order = 0 to 60 by 10  label=(h=1.5) value=(h=1.5)
        minor = none;
   axis2 order = 0 to 40 by 10  label=(h=1.5) value=(h=1.5)
        minor = none;
```

Example: Baseball Data

The relationship between years in the major leagues and salaries of baseball players is examined in Section 1.3, "The Roles of Graphics in Data Analysis." Because the variable YEARS is discrete, and most of the players are concentrated in the range 3—8 years, it is very hard to see the relationship between salary and years of experience. Output 4.27 shows the sunflower plot of log (SALARY) against the variable YEARS. Although the plot has been made somewhat coarse, the more populous regions stand out more clearly than in the plot of the raw data (see Output 1.2).

Output 4.27
Sunflower Plot of
Baseball Salaries
against Years

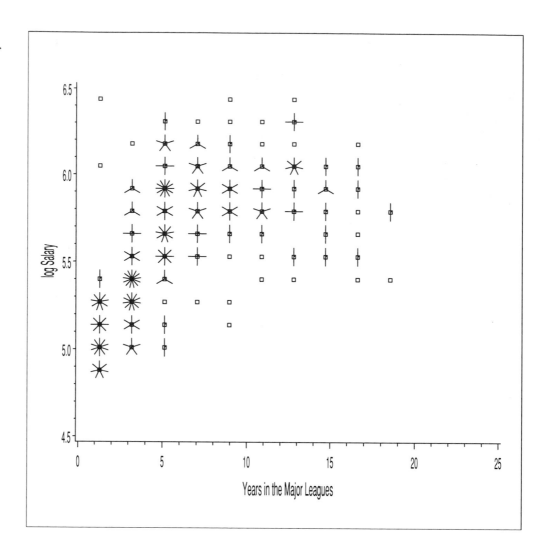

4.6.2 Jittering

Another idea is to add a small amount of random noise to each point to reduce overplotting. This technique, called *jittering*, is particularly useful when the data are highly discrete or have been coarsely rounded. Whereas the sunflower plot is most effective with large data sets, jittering seems to work well with any discrete data.

To jitter the values of a variable, add a uniformly distributed random quantity, scaled so that it breaks up the overlap but does not corrupt the data. Let u_i be a random value uniformly distributed from -1 to 1. Then, to jitter x, calculate

$$x_i' = x_i + s \ u_i$$

where the value s is the scale factor, chosen as some small fraction (for example, .02) of the range of the data, or one half the rounding interval if the data have been rounded. If the y variable is also discrete, the same process can be applied to jitter y, producing a jittered value y'. Then, in the scatterplot of y (or y') * x' it should be easier to see the density of points in different regions. See Muhlbaier (1987) for SAS macros for jittering.

Example: Baseball Data

For the baseball data, the variable YEARS can be jittered in the following DATA step:

```
data baseball;
   set baseball;
   if salary ¬= .;
   logsal = log10(1000 * salary);
   yearj= years + (uniform(15732789)-.5);
```

The UNIFORM function returns a random number in the range (0, 1). Subtracting .5 gives a uniform value in the interval (-0.5, 0.5), which is half of the rounding interval of the variable YEARS. Adding a small random number to YEARS helps to reduce the discreteness of this variable. The plot of log(SALARY) against YEARS is shown in Output 4.28.

Output 4.28
Baseball Salaries against Years, Jittered

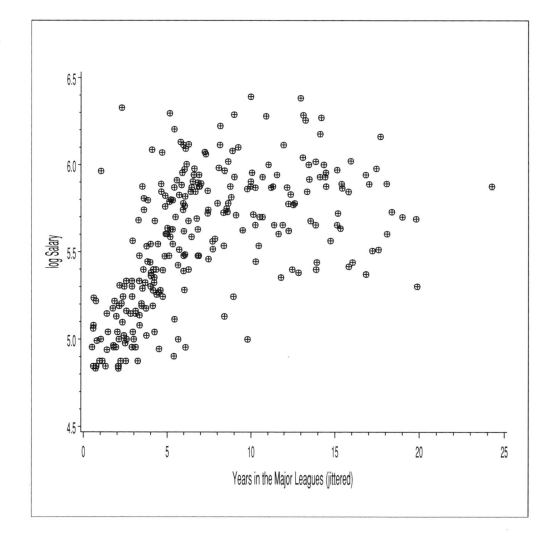

4.7 Displaying a Third Variable on a Scatterplot

There are situations when it is useful to display an additional variable in a scatterplot to help interpret the two primary ones. Two techniques are described in this section:

□ plotting an additional *y* variable against a separate axis

□ producing a bubble plot that displays a third variable by the size of a circle drawn at each point.

4.7.1 Same X, Different Y

With PROC PLOT, you can plot any number of *y* variables against a single *x* with the OVERLAY option in the PLOT statement. You can do this with PROC GPLOT as well. With the OVERLAY option, all the variables should be measured on the same scale because there is just one vertical axis.

With the PLOT2 statement in PROC GPLOT, you can plot other *y* variables on a separate, independently scaled, vertical axis that is drawn on the right side of the plot. Fisher (1925) used an interesting example of this type of plot in connection with the Broadbalk wheat experiment. Data were also available on the amount of rainfall in the same set of years. To determine if there is an association between the yield differences shown in Output 4.4, the most immediate idea would be to make a scatterplot of yield difference plotted against rainfall, which might be called y_i against r_i.

Instead, Fisher sorted the pairs of (yield difference, rainfall) by yield difference. An index plot of the ordered values, $y_{(i)}$ plotted against *i*, shows the empirical cumulative distribution of the value y_i. Now on the same plot, he displayed the corresponding values of rainfall, $r_{y(i)}$, on a separate axis, giving a plot like that shown in Output 4.29. (A color version of this output appears as Output A3.15 in Appendix 3.) The axis displaying rainfall is scaled independently by the PLOT2 statement. Yield difference must increase from left to right because the observations were sorted; the rainfall values are quite variable, but the fact that they increase as well is a clear indication that the increasing advantage of plot 9a over plot 7b is associated with an increase in rainfall.

Output 4.29
Index Plot of Yield Difference Accompanied by Rainfall

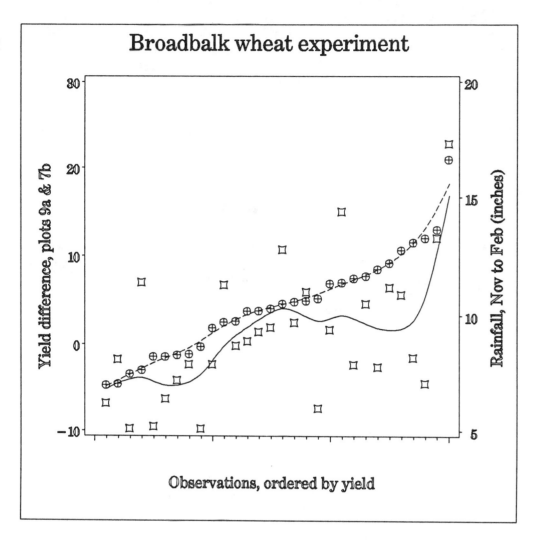

The program below is used to construct the plot. Note the use of different SYMBOL and AXIS statements to draw the two vertical axes and corresponding points in different colors. The values on the horizontal axis are not particularly useful, so they are suppressed by specifying VALUE=NONE in the AXIS2 statement.

```
%include WHEAT;              /* On MVS, use %include ddname(file);*/
title f=centx 'Broadbalk wheat experiment';
proc sort data=wheat;
   by yieldiff;
data ordered;
   set wheat;
   index = _n_;
   label index='Observations, ordered by yield';

proc gplot data=ordered  ;
   plot yieldiff * index = 1 / vaxis=axis1 haxis=axis2
                               vminor=0 name='GB0429' ;
   plot2 rain    * index = 2 / vaxis=axis3 vminor=0 frame;
```

```
axis1 label=(f=centxe h=1.4 a=90 r=0)
      value=(f=centxe h=1.2);
axis2 label=(f=centxe h=1.4) offset=(2)
      value=none;
axis3 label=(f=centxe h=1.4 a=90 r=0 c=red)
      value=(f=centxe h=1.2 c=red)
      order=(5 to 20 by 5);
symbol1 v=+ h=1.4 i=sm50 l=3 color=black;
symbol2 v=_ h=1.4 i=sm50 l=1 color=red;
```

4.7.2 Bubble Plots

Bivariate scatterplots can sometimes be enhanced by coding additional variables
into the plotting symbol. When a quantitative variable is used in this way, it is
important to use a coding scheme that

□ conveys magnitude graphically

□ enables positive and negative values of equal magnitude to have similar visual
   impact.

The BUBBLE statement in PROC GPLOT produces a scatterplot of *y* against *x* on
which each point is plotted with a circle whose size (area or radius) is
proportional to the magnitude of *z*, the bubble variable:

```
BUBBLE y * x = z ;
```

Positive values are drawn with solid circles, and negative values are drawn with
dashed circles, so they have approximately the same impact. The BSCALE option
in the BUBBLE statement allows you to choose whether the size of the bubble
variable is represented by the area of the circle (by specifying BSCALE=AREA) or
by the radius (by specifying BSCALE=RADIUS). For plots of this sort, the AREA
scale is usually most appropriate.

    Bubble plots are often used to display the residuals from some analysis, such
as regression, where one can use the size of the residual as the bubble variable.
The example below illustrates this idea in an analysis of weight, price, and
gasoline mileage from the automobiles data (see Chambers et al. 1983, Fig. 5.35).

    The first step in this analysis is a regression of the variable MPG on the
variable WEIGHT. The output data set from the REG procedure contains the
residual variable, RESID, which can be thought of as excess mileage, that is, the
amount by which gas mileage differs from what is predicted by a linear relation
on the variable WEIGHT. Positive values of the variable RESID therefore
correspond to cars that get more miles to the gallon than would be predicted from
their weight. The plot produced by PROC GPLOT shows the variable WEIGHT in
relation to the variable PRICE, using RESID as the bubble variable.

```
%include AUTO;              /* On MVS, use %include ddname(file);*/

proc reg data = auto;
    model mpg = weight;
    id model;
    output out=resid r=resid p=predict ;
```

```
data resid;                  /* scale & add labels */
    set resid;
    weight = weight/1000;
    price  = price /1000;
    label weight = 'WEIGHT (1000 LBS)'
          price  = 'PRICE ($1000)'
          resid  = 'Residual MPG';
```

The next DATA step constructs an Annotate data set to label the larger residuals with the name of the automobile model. There is a large cluster of low weight, low price cars, however, so a larger value of the absolute residual is used to select the points in this group that will be labeled. We use the color black for positive residuals and the color red for negative ones. In the PROC GPLOT step, ANNOTATE=LABELS is specified in the BUBBLE statement.

```
data labels;
    set resid;
    if weight > 2.2 & abs(resid) > 5
     | weight <=2.2 & abs(resid) > 8 ;
    retain xsys ysys '2';

    x = price ; y = weight +.15;
    text = model;
    size = .9;
    if resid > 0 then color = 'BLACK';
                 else color = 'RED';
    function = 'LABEL';

proc gplot data=resid   ;
    bubble weight * price = resid
        / bsize=12
          bscale=area
          annotate=labels
          vaxis=axis1 haxis=axis2
          vminor=0 hminor=0
          name='GB0430'
          ;
    axis1  label=(h=1.3 a=90 r=0)
           order=(1.5 to 5 by .5)
           offset=(2);
    axis2  label=(h=1.3)
           offset=(2);
title  h=1.5 f=triplex
    'Bubble plot of Weight against Price';
title2 h=1.5 f=triplex
    'Bubble area shows MPG adjusted for Weight';
```

The plot, shown in Output 4.30, shows a generally positive relationship between the variables WEIGHT and PRICE, though the points appear to follow two branches. (A color version of this output appears as Output A3.16 in Appendix 3.) The size of the bubble symbol shows excess mileage, the part of gas mileage that cannot be predicted from weight. According to Chambers et al., "although there are several negative values of excess mpg (dashed circles) on the upper branch and two or three positive values on the lower branch (solid circles), the preponderance of points on the upper branch (which turn out to be U.S. cars) have positive excess mpg, and most of the points on the lower branch have negative excess mpg (and they are mostly foreign)" (1983, p. 180). In other words, the most economical cars in this data are found in the cluster of light, inexpensive cars at the bottom left that get greater gas mileage than is predicted from weight (large black bubbles).

Output 4.30   *Bubble Plot of Weight Plotted against Price for Automobiles Data*

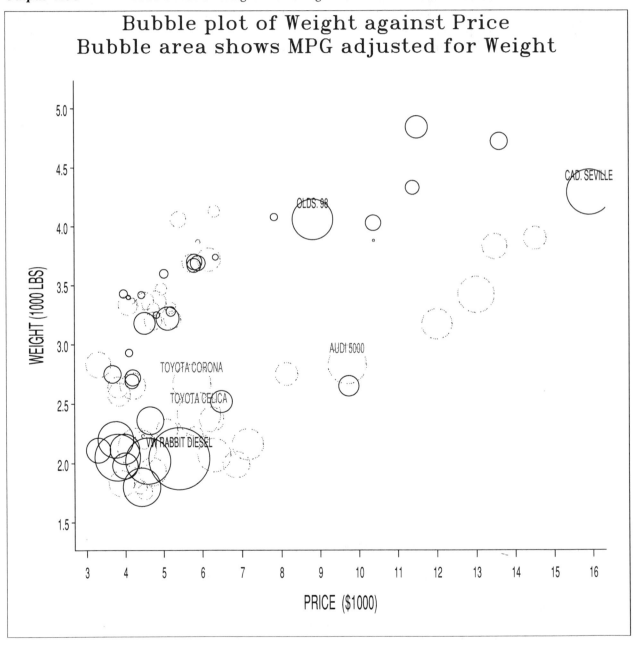

4.8 Three-Dimensional Plots

The G3D procedure produces three-dimensional surface graphs and scatterplots. A three-dimensional *surface graph* plots the value of one variable, *z*, as a function of two other variables, $z = f(x, y)$. The plot of the bivariate normal density function in Section 3.3, "Bivariate Normal Distribution," is an example of a surface plot. The surface graph requires data where the *x*, *y* values form a rectangular grid with a *z* value present for every possible (*x*, *y*) combination. A three-dimensional *scatter graph* does not require a complete grid of (*x*, *y*) values and is used to portray a statistical relation among three variables.

4.8.1 Three-Dimensional Scatterplots

The SCATTER statement of PROC G3D will produce a three-dimensional scatterplot. Options in the SCATTER statement let you assign the size, shape, and color used to plot the points.

This example displays a well-known set of data consisting of measures of size of three varieties of irises from the Gaspé Peninsula (Anderson 1935) used by Fisher (1936) in a classic paper on group classification. The data set IRIS is listed in Section A2.10. The DATA step below reads the iris data and assigns additional variables SHAPEVAL and COLORVAL based on the species of iris.

```
%include IRIS;                  /* On MVS, use %include ddname(file);*/
data iris;
   set iris;
   length colorval $8. shapeval $8.;
   sizeval = sepalwid / 30;
   select;
      when (spec_no  = 1 ) do;
         shapeval = 'CLUB';       colorval = 'RED';     end;
      when (spec_no  = 2 ) do;
         shapeval = 'BALLOON';    colorval = 'GREEN';   end;
      when (spec_no  = 3 ) do;
         shapeval = 'SPADE';      colorval = 'BLUE';    end;
   end;
```

The PROC G3D scatterplot of these data is shown in Output 4.31, which is produced by these statements:

```
title1 f=swiss c=black 'Iris Species Classification';
title2 f=swiss c=black 'by Physical Measurement';
title3 f=swiss c=black 'Source: Fisher (1936) Iris Data';
footnote1 j=l ' PETALLEN: Petal length in MM.'
          j=r 'PETALWID: Petal width in MM. ';
footnote2 j=l ' SEPALLEN: Sepal length in MM.'
          j=r 'Sepal width is not shown      ';
```

```
proc g3d data = iris  ;
   note;
   note j=r c=black    'SPECIES: ' c=blue    'VIRGINICA   '
        j=r c=green    'VERSICOLOR '
        j=r c=red      'SETOSA      ';
   scatter petallen * petalwid = sepallen /
           color = colorval
           shape = shapeval
           name='GB0431' ;
```

In the SCATTER statement, the options SIZE=, SHAPE=, and COLOR= can be given literal values (for example, COLOR=RED) or can refer to a variable in the data set (for example, COLOR=COLORVAL). In Output 4.31, both the shape and color of the symbol are determined by the species. (A color version of this output appears as Output A3.17 in Appendix 3.) The size of the symbol could have been used to display the sepal width if SIZE=SIZEVAL had been specified. These variables were assigned in the DATA step. Also note the use of the NOTE statements in the PROC G3D step to produce a color-coded legend identifying the species inside the plot display area.

Output 4.31
Three-Dimensional Scatterplot of Iris Data

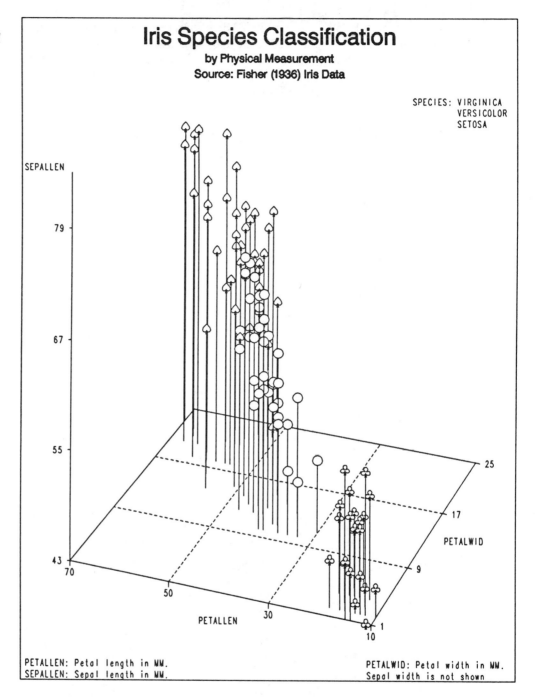

4.8.2 Annotated Three-Dimensional Scatterplots

The process for annotating a PROC G3D plot is similar to that used for an ordinary two-dimensional scatterplot. The main difference is that to perform an Annotate function in a PROC G3D plot, you must specify a Z variable in the Annotate data set in addition to X and Y.

This example plots a three-dimensional scatterplot of income, education, and prestige in the DUNCAN data set, shown in Output 4.32. A bivariate data ellipse for the income and education variables is constructed with the CONTOUR macro (Section 4.5, "Enhanced Scatterplots") and drawn in the X, Y plane. A few unusual observations are labeled as well.

Output 4.32   *Three-Dimensional Scatterplot of Duncan Data*

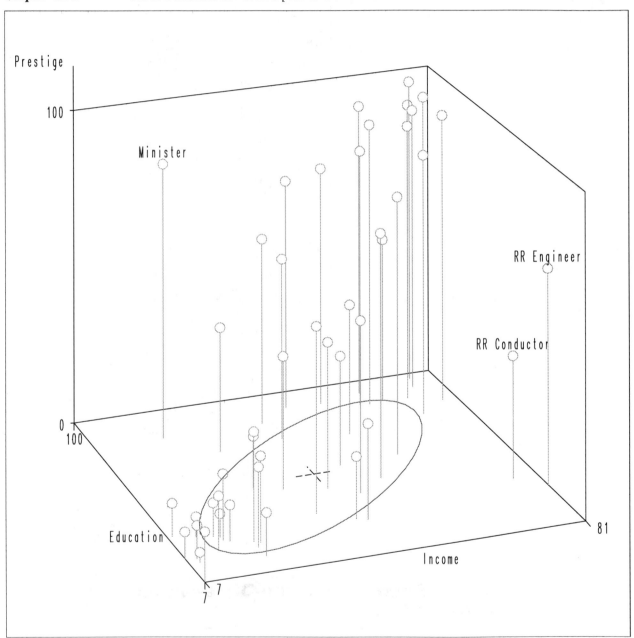

The CONTOUR macro is used to produce an Annotate data set named ELLIPSE for the variables INCOME and EDUC. (The bivariate scatterplot should be suppressed by specifying PLOT=NO because we only want the data set ELLIPSE.) To use this data set with PROC G3D, the variables Z and ZSYS are added in the next DATA step. Setting ZSYS=2 specifies the data coordinate system for the variable Z, and Z=0 causes the ellipse to be drawn in the X, Y plane where the value of the variable PRESTIGE is zero.

The initial version of this plot showed that three observations appeared to deviate from the generally positive relationship between income and education summarized by the data ellipse. The CASE numbers of these observations were identified and used to construct a data set LABELS to label these points with the JOB name on the plot. The Annotate observations in the LABELS data set are then appended to the end of the ELLIPSE data set. These statements produce Output 4.32:

```
%include DUNCAN;              /* On MVS, use %include ddname(file);*/
%include contour;            /* On MVS, use %include ddname(file);*/
title ' ';

   /* get ANNOTATE= data set to draw ellipse */
%contour(data=duncan,
         x=income, y=educ,
         pvalue=.50, out=ellipse, plot=no);
run;
   /* add Z variable for G3D */
data ellipse;
   set ellipse;
   zsys='2';
   z=0;

   /* label 3 interesting cases */
data labels;
   set duncan;
   if case=6 | case=16 | case=27 ;
   length text $12 function $8;
   xsys='2'; ysys='2'; zsys='2';  /* data system coordinates */
   function = 'LABEL   ' ;
   position = '2';                /* centered above the point */
   x = income;
   y = educ;
   z = prestige;
   text=job;
data ellipse;
   set ellipse labels;
```

```
proc g3d data=duncan  ;
   scatter educ * income = prestige /
      annotate=ellipse
      shape='balloon'
      size=.8
      grid
      zmin=0 zmax=100
      zticknum=2
      xticknum=2
      yticknum=2
      caxis   =black
      tilt    =70
      rotate  =20
      name='GB0432';
```

In Output 4.32, you can see that occupational prestige generally increases with both income and education and that these two variables are positively related. The use of multiple regression methods to predict prestige from income and education is discussed in Section 5.4.4, "Leverage and Influential Observations." It turns out that the three cases identified as unusual in this plot are potentially troublesome because their distance from the centroid (shown by the cross) gives them a larger potential to influence (termed *leverage*) the regression results than other observations have. The data ellipse for education and income is precisely a contour of constant distance from the mean that takes the variability and correlation of these variables into account.

Chapter **5** Plotting Regression Data

5.1 Introduction

Regression methods naturally lead to plots of the fitted relationship between the response measure and one or more predictor variables. Beyond this, it has become standard practice to check whether the assumptions of the analysis are met by plotting residuals from the fitted model.

This chapter describes techniques for plotting the data and fitted regression function for linear, polynomial, and multiple-predictor regression models. Section 5.2, "Plotting Data, Regression Curves, and Confidence Limits," explains how to plot the data, regression curves, and confidence limits for single-predictor linear and polynomial models. Section 5.3, "Regression with Several Groups," illustrates graphic displays of the data and residuals for situations in which data are available from several groups. A variety of special-purpose plots for diagnosing violations of assumptions and detecting unduly influential observations are described in Section 5.4, "Validity of Assumptions: Plotting Residuals." The final sections discuss graphical methods particularly useful for multiple regression models, including partial regression residual plots (Section 5.5, "Partial Regression Plots"), response surface plots (Section 5.6, "Multiple Regression Response Surface"), plots for selecting variables to be included in the model (Section 5.7, "Plots of C_p against P for Variable Selection"), and plots for geographic data (Section 5.8, "Regression with Geographic Data").

5.1.1 The Linear Regression Model

In the model for *simple linear regression*, we assume that a dependent, response variable, y, is to be explained as a linear function of an independent, explanatory variable, x, according to

$$y = \beta_0 + \beta_1 x + \varepsilon \tag{5.1}$$

where β_0 and β_1 are the intercept and slope of the linear regression line and ε is the (unobservable) error, that is, the part of y unexplained by model 5.1. In addition to the assumption that model 5.1 correctly specifies the relation between y and x, the usual statistical inference requires the following assumptions on the errors (ε):

normality
    The errors have a normal distribution (with mean zero).

homogeneity of variance
    The variance of ε is a constant value, σ^2.

independence
    The errors associated with distinct observations are independent.

Polynomial and multiple regression models are straightforward generalizations of model 5.1. If the relation between y and x is not linear in x, we may be able to fit y by a quadratic, cubic, or higher-degree polynomial in x:

$$y = \beta_0 + \beta_1 x + \beta_2 x^2 + \beta_3 x^3 + \ldots + \varepsilon \tag{5.2}$$

Model 5.2 is still a *linear model*, which means that it expresses y as a linear function of the parameters $\beta_0, \beta_1, \ldots$. Or, with several explanatory variables (p), $x_1, x_2, \ldots, x_p$, each linearly related to the variable y, the *multiple regression* model is as follows:

$$y = \beta_0 + \beta_1 x_1 + \beta_2 x_2 + \beta_3 x_3 + \ldots + \beta_p x_p + \varepsilon \tag{5.3}$$

We can express each of these models in matrix form for a sample of n observations by the general linear model

$$\mathbf{y} = \mathbf{X}\boldsymbol{\beta} + \boldsymbol{\varepsilon} \tag{5.4}$$

or

$$
\begin{bmatrix} y_1 \\ y_2 \\ \cdots \\ y_n \end{bmatrix} = \begin{bmatrix} 1 & x_{11} & x_{12} & \cdots & x_{1p} \\ 1 & x_{21} & x_{22} & \cdots & x_{2p} \\ \cdots & \cdots & \cdots & \cdots & \cdots \\ 1 & x_{n1} & x_{n2} & \cdots & x_{np} \end{bmatrix} \begin{bmatrix} \beta_0 \\ \beta_1 \\ \beta_2 \\ \cdots \\ \beta_p \end{bmatrix} + \begin{bmatrix} \varepsilon_1 \\ \varepsilon_2 \\ \cdots \\ \varepsilon_n \end{bmatrix} \tag{5.5}
$$

Detailed discussion of regression methods is beyond the scope of this book. See Freund and Littell (1991) for a thorough discussion of fitting regression models with the SAS System. See Fox (1984), Neter, Wasserman, and Kutner (1985), or Rawlings (1988) for the theory of linear models. Bowerman and O'Connell (1990) give particular emphasis to using the SAS System for the analysis of linear models.

5.2 Plotting Data, Regression Curves, and Confidence Limits

In plotting regression data, you often want to show the fitted regression curve and sometimes the confidence limits for the regression curve along with the data. Specifying I=RL, I=RQ, or I=RC in the SYMBOL statement requests the GPLOT procedure to fit a linear, quadratic, or cubic regression curve to the data and draw the curve on the plot. Adding the characters CLMnn to the I= option will draw confidence limits as well.*

In Chapter 4, "Scatterplots," we examine Fisher's data from the Broadbalk wheat experiment (1925). Although Fisher uses that data to illustrate linear regression calculations, Output 4.4 indicates that a quadratic relation might provide a better description of the change in y (yield difference) over time.

To determine whether a quadratic model provides a significant improvement in goodness of fit, we fit the model**

$$
\hat{y} = \beta_0 + \beta_1 (\text{YEAR} - 1870) + \beta_2 (\text{YEAR} - 1870)^2 \quad . \tag{5.6}
$$

To fit this model using the REG procedure, first construct the variables YEAR1 and YEAR2 in a DATA step as shown below. The regression output, shown in Output 5.1, indicates that the quadratic model achieves an $R^2 = 0.42$ and that the addition of the quadratic term provides a significant improvement in fit, as indicated by the t value for β_2 ($t = -3.497$, $p = .0017$).

---

* The same regression curves can be plotted by obtaining the predicted values from the REG procedure in an output data set and overlaying these on the plot of the data. This method must be used for regressions other than linear, quadratic, or cubic, or when points are weighted unequally.

** Subtracting the mean, 1870, from the variable YEAR has no effect on the goodness of fit or on predicted values, but helps to avoid problems of multicollinearity in polynomial models. In addition, centering makes the coefficients more interpretable because the values for β_1 and β_2 measure the slope and curvature of the regression function over the region of the data, rather than over the origin (YEAR=0).

```
%include wheat ;
data wheat2;
   set wheat;
   year1 =  year - 1870;
   year2 = (year - 1870)**2;
proc reg data=wheat2;
   model yieldiff = year1 year2 / ss1;
```

Output 5.1   *REG Procedure Output for Broadbalk Wheat Data*

```
                        Broadbalk wheat experiment

DEP VARIABLE: YIELDIFF Yield difference, plots 9a & 7b
                          ANALYSIS OF VARIANCE

                         SUM OF         MEAN
          SOURCE    DF   SQUARES       SQUARE      F VALUE      PROB>F

          MODEL      2   428.24557    214.12279     9.760       0.0006
          ERROR     27   592.31826   21.93771319
          C TOTAL   29  1020.56383

               ROOT MSE     4.683771    R-SQUARE     0.4196
               DEP MEAN        4.467    ADJ R-SQ     0.3766
               C.V.        104.8527

                          PARAMETER ESTIMATES


                 PARAMETER    STANDARD    T FOR H0:
VARIABLE   DF     ESTIMATE      ERROR    PARAMETER=0    PROB > |T|   TYPE I SS

INTERCEP    1    7.93817130   1.28246576     6.190       0.0001      598.62267
YEAR1       1    0.22208964   0.09962103     2.229       0.0343      159.97248
YEAR2       1   -0.04470235   0.01278314    -3.497       0.0017      268.27309
```

The plot of the data in Output 5.2 with linear and quadratic regression curves shows these effects graphically. In order to show both regression curves, we plot the data twice, using two SYMBOL*n* statements. In the program below, the PLOT statement causes PROC GPLOT to use the SYMBOL1 statement for the first plot and the SYMBOL2 statement for the second:

```
proc gplot data=wheat;
   plot yieldiff * year = 1
        yieldiff * year = 2 / overlay ... ;
```

SYMBOL1 draws the points and the quadratic regression (I=RQ); SYMBOL2 draws the linear regression (I=RL) but suppresses the points (V=NONE).

```
%include WHEAT;                /* On MVS, use %include ddname(file);*/
title1 f=duplex h=1.5 'Broadbalk wheat experiment';
title2 f=duplex h=1.3 'with linear and quadratic regressions';
proc gplot data=wheat  ;
   plot yieldiff * year = 1
        yieldiff * year = 2
     / overlay frame vminor=0
       vaxis=axis1 haxis=axis2
       name='GB0502' ;
```

```
axis1 label=(f=duplex h=1.4 a=90 r=0) value=(h=1.4);
axis2 label=(f=duplex h=1.4) value=(h=1.4) offset=(2) ;
symbol1 v=+ h=1.4 i=rq  l= 1 color=black;
symbol2 v=none   i=rl  l=20 color=red;
```

Output 5.2
Broadbalk Wheat Data with Linear and Quadratic Regressions

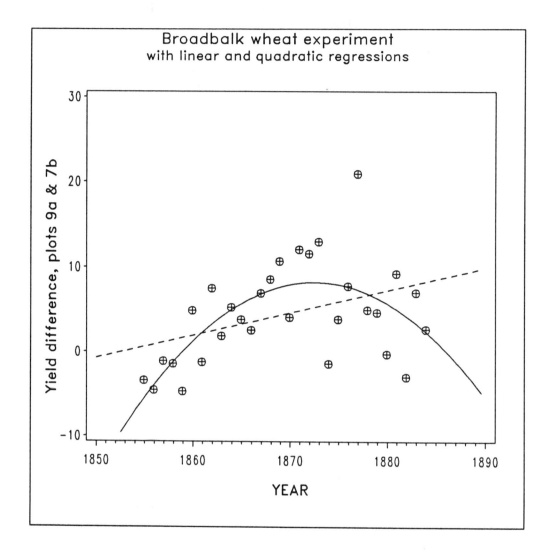

The lowess smoothing technique (Section 4.4.2, "Lowess Smoothing") can often be used to help determine the form of a polynomial regression model. Output 5.3 displays the same data together with the lowess smooth fitted curve. The need for a quadratic term in the model is clearly indicated by the downward turn in the data after 1870.* Output 5.3 is produced by the LOWESS macro with these statements:

```
%include lowess;              /* On MVS, use %include ddname(file);*/
title2 f=duplex h=1.3 'with lowess smooth';
%lowess(data=wheat,
        x=year, y=yieldiff,
        plot=YES, name=GB0503);
```

Output 5.3
*Broadbalk Wheat
Data with Lowess
Smoothed Curve*

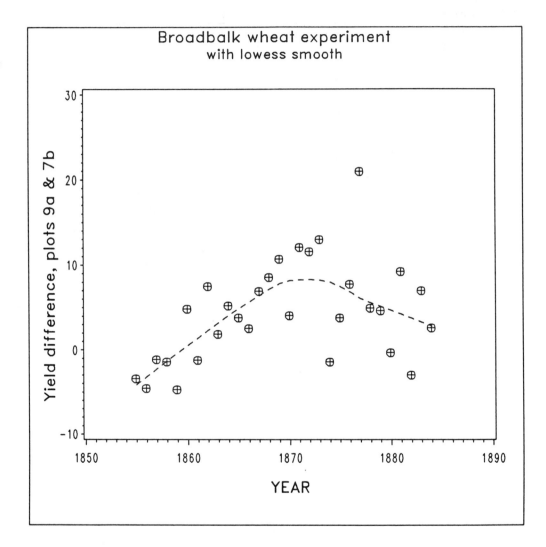

---

* Another possibility would be to fit a periodogram model, with terms composed of sine and cosine functions whose period is the length of the sequence (30 years here), to account for cyclic effects over time. Anscombe (1981) suggests this analysis provides an equally good fit to Fisher's data.

5.2.1 Confidence Intervals

In addition to the fitted regression curve, you might also want to plot confidence bands for the mean response or for individual observations. Specifying I=RLCLM95 for the interpolate option in the SYMBOL statement draws a linear regression line and a 95% confidence band for the mean response, $E(y_i)$; specifying I=RLCLI95 gives a linear regression line and a confidence band for an individual y_i. See Freund and Littell (1991, Section 2.4, "Various MODEL Statement Options") for a discussion of these options and guidelines for their use.

Example: Salary Survey Data

This example plots data from a salary survey (fictitious) of computer professionals (Chatterjee and Price 1977) designed to investigate the roles of experience, education, and management responsibility as determinants of salary. The complete data set is listed in Section A2.12, "The SALARY Data Set: Salary Survey Data," and shown below in abbreviated form. Education (EDUC) and management responsibility (MGT) are qualitative variables that are used in later examples. As shown in the PROC FORMAT statements, the variable EDUC has values 1, 2, and 3, corresponding to high school, B.S. degree, and advanced degree, respectively. The variable MGT is coded 0 and 1, corresponding to nonmanagement and management responsibility. These formats are used for labeling the plots described below. The DATA step reads the raw data and defines a GROUP variable with values 1—6 for the six combinations of the variables EDUC and MGT.

```
Title 'Salary survey data';
* Formats for group codes;
proc format;
     value glfmt    1='HS' 2='BS' 3='AD' 4='HSM' 5='BSM' 6='ADM';
     value edfmt    1='High School' 2='B.S. Degree' 3='Advanced Degree';
     value mgfmt    0='Non-management' 1='Management';

data salary;
   input case exprnc educ mgt salary;
   label exprnc = 'Experience (years)'
         educ   = 'Education'
         mgt    = 'Management responsibility'
         salary = 'Salary (in $1000s)';

   salary = salary / 1000;
   group = 3*(mgt=1)+educ;
   format group glfmt.;
```

```
cards;
 1   1   1   1   13876
 2   1   3   0   11608
 3   1   3   1   18701
 4   1   2   0   11283
  ... (cases 5 - 42 omitted)
43  16   2   0   18838
44  16   1   0   17483
45  17   2   0   19207
46  20   1   0   19346
;
```

Output 5.4 plots the variable SALARY against the variable EXPRNC with confidence bands for individual observations and for the mean response. (A color version of this output appears as Output A3.18 in Appendix 3, "Color Output.") The PLOT and SYMBOL statements are used as in the previous example, overlaying three plots of the same data, each with a different SYMBOL statement.

```
proc gplot data=salary ;
   plot salary * exprnc              /* points      */
        salary * exprnc              /* cl - mean   */
        salary * exprnc /            /* cl - individual */
        overlay frame
        vaxis=axis1 haxis=axis2
        vminor=5 hminor=5
        name='GB0504' ;
   symbol1 v=-      h=1.5 i=none     c=black;  /* points */
   symbol2 v=none         i=rlclm95 c=red;     /* cl m   */
   symbol3 v=none         i=rlcli95 c=blue;    /* cl i   */
   axis1   order=(10 to 30 by 5)
           offset=(2)
           label=(h=1.5 a=90 r=0 f=duplex);
   axis2   order=(0 to 20 by 5)
           offset=(2)
           label=(h=1.5 f=duplex);
   title   h=1.5 'Salary Survey Data';
   title2  h=1.2 f=duplex
      'Common regression line with 95% confidence intervals';
   footnote1 f=duplex j=l 'Source: Chatterjee & Price (1977)'
             f=italic j=r 'Regression by Example';
```

The points in the plot appear to be separated into several parallel groups. This pattern, as we will see, is due to salary differences among those with different levels of education and management responsibility. The inner (red) curves give the confidence interval for the expected value of y given x. The outer (blue) curves give the confidence interval for an individual y_i given x.

Output 5.4   *Plot of Salary Survey Data with Confidence Bands*

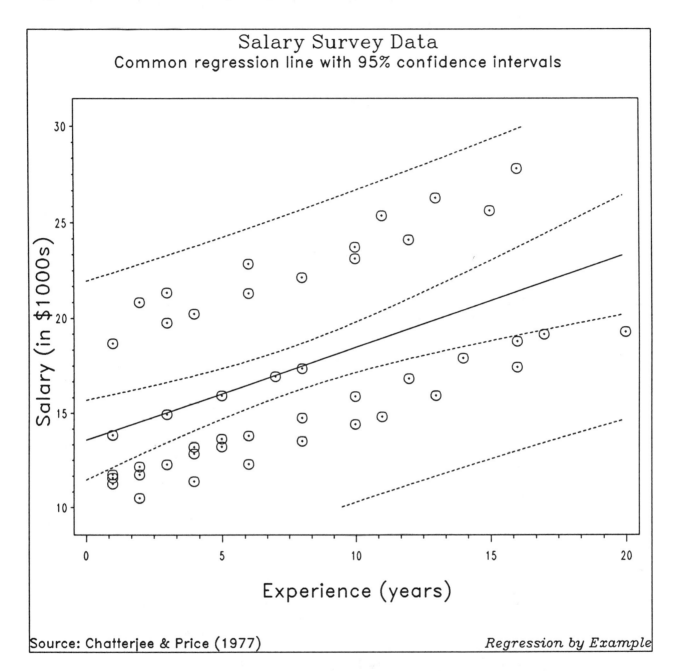

The plot of the data together with regression line and confidence intervals can also be done with the PLOT procedure. To do this, use the OUTPUT statement in PROC REG to produce a data set that contains the predicted values and upper and lower confidence values, in addition to the original data. The limits for the mean response are stored in the variables given in the U95M= and L95M= options. The limits for individual observations are stored in the variables named in the U95= and L95= options. As shown below, each of these variables can be plotted against the variable EXPRNC in an overlay plot (not shown) that is equivalent to Output 5.4.

```
proc reg data=salary;
   model  salary = exprnc ;
   output out=limits p=pred u95m=up95m l95m=lo95m
                            u95 =up95  l95 =lo95;
proc plot data=limits;
   plot salary * exprnc = 'o'
        pred   * exprnc = '*'
        up95m  * exprnc = '-'
        lo95m  * exprnc = '-'
        up95   * exprnc = '='
        lo95   * exprnc = '=' / overlay;
```

This method can also be used with PROC GPLOT to plot a fitted regression and 95% confidence intervals for any regression model, in particular for models not handled by the series of specifications I=R in the SYMBOL statement.

5.3 Regression with Several Groups

When observations belong to different groups, classified by factors such as sex, level of education, marital status, and so forth, it is important to determine if the relationship between quantitative regression variables is the same for all groups. The statistical question, called *homogeneity of regression,* is concerned with testing whether the regression slopes are equal for all groups. When the interest is focused on group differences in means (intercepts), assuming equal slopes, the questions are those of *analysis of covariance.* These topics are discussed in Chapter 6, "Covariance and the Heterogeneity of Slopes," in *SAS System for Linear Models, Third Edition* (Freund, Littell, and Spector 1991).

One way to examine whether group membership affects the relation is to fit separate regressions to the data from each group. We can do this graphically with PROC GPLOT with a PLOT statement of the form

```
plot y * x = group ;
```

together with one SYMBOL statement for each group. If the regression is linear, specify I=RL for an interpolated linear regression line.

Output 5.5 shows this plot for the salary survey data. (A color version of this output appears as Output A3.19 in Appendix 3.) It is clear from the plot that salary increases linearly with experience within each education—management group and that those with management responsibility earn considerably more than those without.

Output 5.5   *Salary Survey Data, with a Separate Regression Line for Each Group*

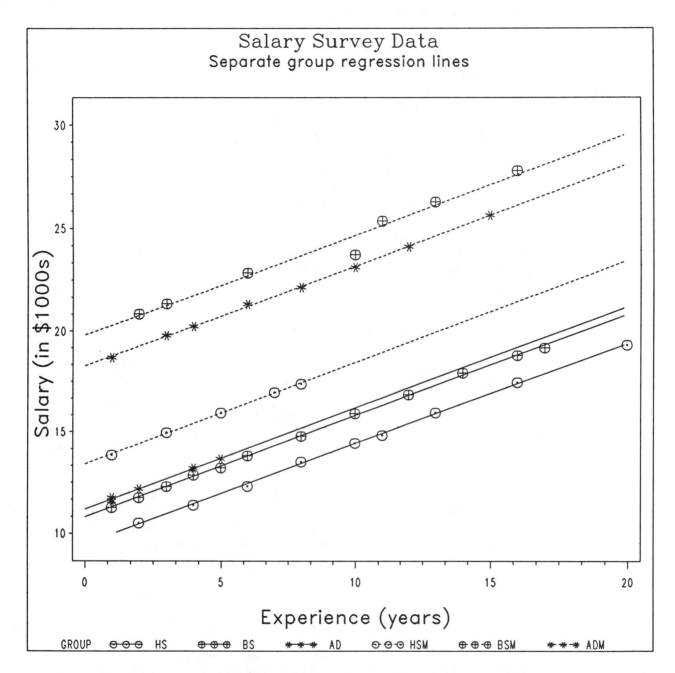

This plot is produced with the statements shown below. Note that the SYMBOL statements use color and line style to distinguish between the nonmanagement and management groups, and code the educational level with the plotting symbol. The legend for the GROUP variable uses the character codes defined by the GLFMT format.

```
proc gplot data=salary ;
   plot salary * exprnc = group
        / frame
          vaxis=axis1 haxis=axis2
          vminor=5 hminor=5
          name='GB0505' ;
```

```
symbol1 v=- h=1.5 i=rl c=blue l=1;
symbol2 v=+ h=1.5 i=rl c=blue l=1;
symbol3 v=: h=1.5 i=rl c=blue l=1;
symbol4 v=- h=1.5 i=rl c=red  l=2;
symbol5 v=+ h=1.5 i=rl c=red  l=2;
symbol6 v=: h=1.5 i=rl c=red  l=2;
axis1   order=(10 to 30 by 5)
        offset=(2)
        label=(h=1.5 a=90 r=0 f=duplex);
axis2   order=(0 to 20 by 5)
        offset=(2)
        label=(h=1.5 f=duplex);
title   h=1.5 'Salary Survey Data';
title2  h=1.2 f=duplex  'Separate group regression lines';
```

In this example, the different groups of individuals have the same slope (approximately) for the relation between salary and experience, but the groups differ in intercepts. This is the source of the parallel sets of points we see in Output 5.4.

5.3.1 Fitting Dummy Variables

In cases like this, we can account for the effects of group membership in a regression model by defining *indicator* or *dummy* variables that take on the values 0 or 1, such as the variable MGT in the salary survey data. For a qualitative variable with three values, such as EDUC, group membership can be represented by two dummy variables jointly. In general, for n groups we need $n-1$ dummy variables. For example, we will represent the three educational groups by defining two indicator variables,

$$E_{i1} = \begin{cases} 1 & \text{if person } i \text{ has HS education (high school education)} \\ 0 & \text{otherwise} \end{cases}$$

$$E_{i2} = \begin{cases} 1 & \text{if person } i \text{ has BS education (B.S. degree)} \\ 0 & \text{otherwise} \quad . \end{cases}$$

If we then fit the model

$$SALARY = \beta_0 + \beta_1 EXPRNC + \gamma_1 E_1 + \gamma_2 E_2 + \gamma_3\, MGT + \text{residual} \tag{5.7}$$

then β_1 is the common slope for all groups, and γ_1 and γ_2 represent the differences in salary for the HS and BS groups relative to the advanced degree group, while γ_3 estimates the difference in salary for those with management responsibility relative to those without.

To fit model 5.7 with PROC REG, we first define the dummy variables E1 and E2 in a DATA step as shown below.* The TEST statement in the PROC REG step tests whether the three dummy variables E1, E2, and MGT together contribute significantly to predicting salary. The OUTPUT statement produces an output data set containing the fitted values and residuals.

```
data salary;
   set salary;
   e1 = (educ=1);              /* dummy variables for education */
   e2 = (educ=2);
   label e1     = 'HS Ed.'
         e2     = 'BS Ed.';

proc reg data=salary;
   model  salary = exprnc e1 e2 mgt / ss1 ss2;
dummies: test   e1=0,e2=0,mgt=0;
   output out=resids r=residual p=pred;
   Title2 'Fitting dummy variables for Education and Management';
```

The printed output from PROC REG appears in Output 5.6. The joint test of the three dummy variables is highly significant, as shown by the PROB>F=0.0001 in the TEST: DUMMIES section. Note that the mean square for this joint hypothesis (NUMERATOR: 222.367) corresponds to the sum of the Type I SS for the variables E1, E2, and MGT, divided by the degrees of freedom. The Type II SS, on the other hand, shows the net (unique) contribution of each variable when added last to the model. Of the three dummy variables, management responsibility accounts for the greatest differences in salary. The parameter estimates show that management responsibility is worth an additional $6,883 in salary, while those with only HS education earn $2,996 less than those with advanced degrees; workers with a BS degree earn $148 more on average than those with advanced degrees, but this is not a significant difference.

---

* The same model can be fit with the GLM procedure without using dummy variables. To do this, the group variables EDUC and MGT are specified in a CLASS statement:
```
proc glm data=salary;
   class educ mgt;
   model salary = exprnc educ mgt;
```

Output 5.6
REG Procedure
Output for Salary
Survey Data with
Dummy Variables

```
                              Salary survey data
                 Fitting dummy variables for Education and Management
DEP VARIABLE: SALARY    Salary ($1000s)
                           ANALYSIS OF VARIANCE

                        SUM OF         MEAN
     SOURCE      DF      SQUARES        SQUARE      F VALUE       PROB>F

     MODEL       4      957.81686     239.45421     226.836      0.0001
     ERROR      41     43.28071949     1.05562730
     C TOTAL    45     1001.09758

          ROOT MSE      1.027437     R-SQUARE      0.9568
          DEP MEAN      17.2702      ADJ R-SQ      0.9525
          C.V.           5.949193

                           PARAMETER ESTIMATES


                     PARAMETER       STANDARD     T FOR H0:
     VARIABLE  DF     ESTIMATE        ERROR      PARAMETER=0    PROB > |T|

     INTERCEP   1    11.03180789    0.38321713      28.787       0.0001
     EXPRNC     1     0.54618402    0.03051919      17.896       0.0001
     E1         1    -2.99621026    0.41175271      -7.277       0.0001
     E2         1     0.14782495    0.38765932       0.381       0.7049
     MGT        1     6.88353101    0.31391898      21.928       0.0001

                                                 VARIABLE
     VARIABLE  DF    TYPE I SS      TYPE II SS     LABEL

     INTERCEP   1   13719.94426     874.81088    INTERCEPT
     EXPRNC     1     290.71672     338.09792    Experience (years)
     E1         1     141.73740      55.89619564 HS Ed.
     E2         1      17.79031800    0.15349903 BS Ed.
     MGT        1     507.57242     507.57242    Management responsibility


     TEST: DUMMIES      NUMERATOR:    222.367  DF:    3   F VALUE:  210.6489
                        DENOMINATOR:  1.05563  DF:   41   PROB >F :   0.0001
```

   Though model 5.7 provides a substantially better fit than the regression without dummy variables, we can still ask whether there is more systematic variance to be explained. Any regression analysis provides a breakdown of each observation into the fitted value, $\hat{y}_i$—the part explained by the model—and the residual, e_i—the part not yet explained:

   data = fit + residual

   $$y_i = \hat{y}_i + e_i$$

So any systematic variance remaining is contained in the residuals. Plots of the residuals can help us assess whether model 5.7 is correctly specified or whether it can be improved. For regression data with multiple groups, there are two types of plots we can make. First, a plot of residuals against *x* with group membership coded in the plotting symbol shows what is unexplained in the individual group regressions. Such a plot for residual against experience is shown in Output 5.7. (A color version of this output appears as Output A3.20 in Appendix 3.) The individual group regression lines have a small negative slope, indicating that for each group separately, the common slope for salary on experience is slightly too large. (The residuals as a whole have a slope of zero in Output 5.7.)

Output 5.7   *Residuals from Model Including Education and Management Dummy Variables*

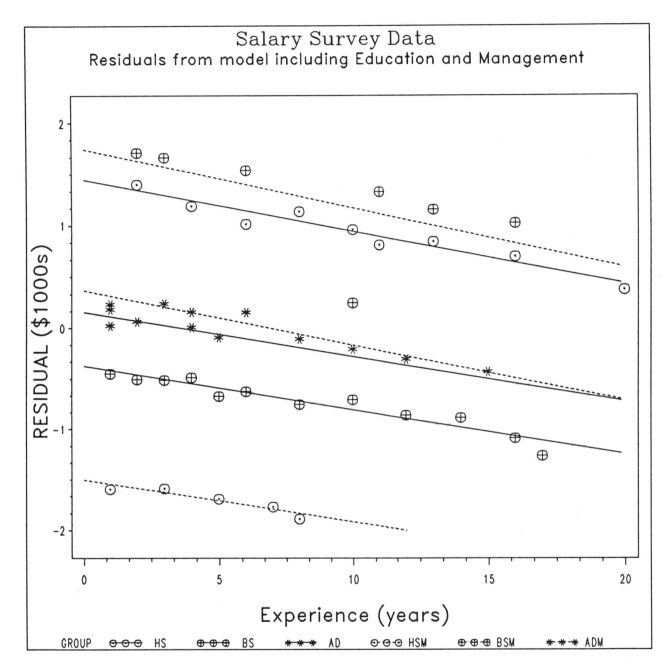

Second, plots summarizing the overall distribution of residuals within each group can help us understand the nature of group differences in salary that are not explained by the current model. Output 5.8 shows a mean and standard error plot of the residuals for different education—management groups. (A color version of this output appears as Output A3.21 in Appendix 3.) This plot shows clearly that the effects of education on salary (when experience has been taken into account) are not the same for management and nonmanagement groups; that is, these factors have interactive effects. For those with no management responsibility, workers with BS education earn considerably more than those with HS education, while the situation is reversed for those with management responsibility.

Output 5.8   *Residuals from Model Including Education and Management Dummy Variables*

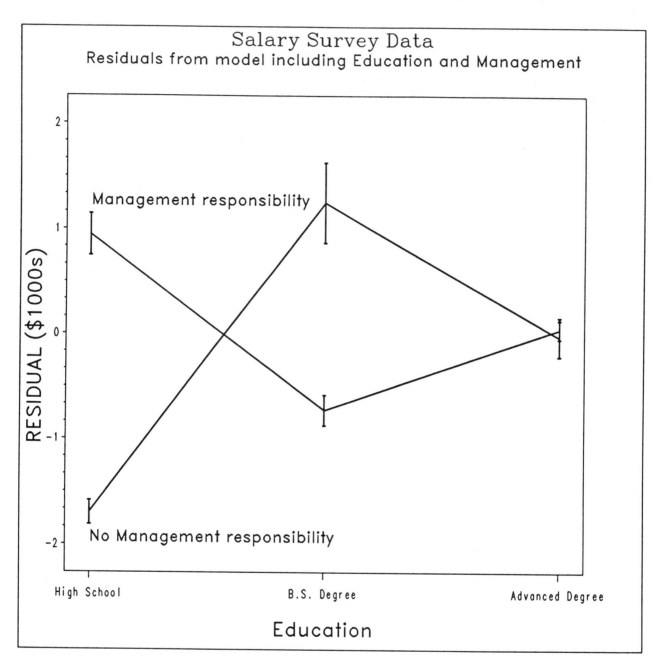

Output 5.7 results from specifying RESIDUAL * EXPRNC=GROUP. It uses the same SYMBOL statements that are used in Output 5.5 to draw a linear regression line for each group.

```
proc gplot data=resids ;
   plot residual * exprnc = group
        / frame
          vaxis=axis1 haxis=axis2
          vminor=5 hminor=5
          name='GB0507' ;
```

```
axis1   order=(-2 to 2)
        offset=(2)
        label=(h=1.5 a=90 r=0 f=duplex 'RESIDUAL ($1000s)' );
title   h=1.5 'Salary Survey Data';
title2 h=1.2 f=duplex
'Residuals from model including Education and Management';
```

For the plot of means and standard errors, a separate curve is drawn for each management group. We create an Annotate data set to label curves in the plot, rather than using a legend at the bottom of the plot.

```
data label;
   xsys='2'; ysys='2';
   x = 1.0;  size=1.2;
   position='6';
   function='LABEL';
   style='DUPLEX';
   y = 1.3;
   color='BLUE';
   text='Management responsibility   ';   output;
   y =-1.9;
   color='RED';
   text='No Management responsibility';   output;
```

The plot of means and standard errors is produced by specifying I=STD2MJT in the SYMBOL statement. The value STD2M specifies error bars extending ±2 standard errors of the mean (residual); the letters JT join the means and draw small horizontal lines at the end of the error bars. Specifying V=NONE suppresses plotting of the individual points. The two SYMBOL statements correspond to the two levels of the variable MGT.

```
proc gplot data=resids ;
   plot residual * educ = mgt
      / frame vaxis=axis1 haxis=axis2
        annotate=label nolegend
        vminor=5 hminor=0
        name='GB0508' ;
   symbol1 v=none i=std2mjt c=blue w=2;
   symbol2 v=none i=std2mjt c=red  w=2;
   axis1   order=(-2 to 2)
           offset=(2)
           label=(h=1.5 a=90 r=0 f=duplex 'RESIDUAL ($1000s)' );
   axis2   offset=(4)
           label=(h=1.5 f=duplex 'Education' );
   format educ edfmt.;
```

A boxplot of residuals by group can be used to see the same effects as are seen in Output 5.8. Output 5.9 shows the boxplot produced by the UNIVARIATE procedure with the following statements:

```
proc sort data=resids;
    by group;
proc univariate plot data=resids;
    by group;
        var residual;
```

Output 5.9   *Boxplot of Residuals from Model with Education and Management Effects*

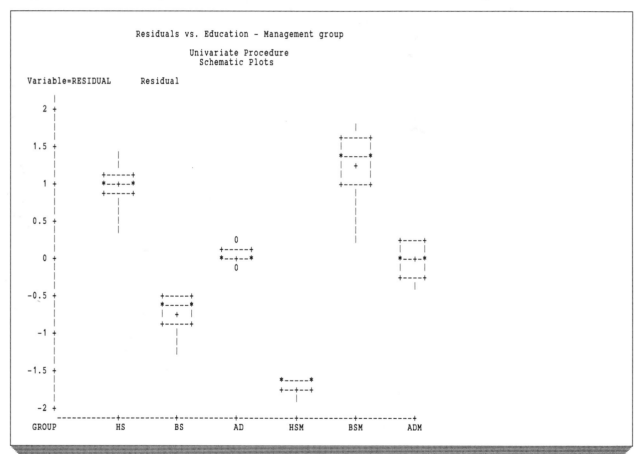

5.3.2 Interaction Effects

Both Output 5.7 and 5.8 indicate that the combined effects of education and management have not fully been taken into account in model 5.7. This model asserts that the effects of education on salary are the same for both management groups, but we've just seen that the difference between HS and BS education goes in opposite directions for the two levels of the variable MGT, so model 5.7 is incorrectly specified. The model may be amended by adding an interaction term, represented by the product of the dummy variables for education and management. The expanded model is

$$SALARY = \beta_0 + \beta_1 \, EXPRNC + \gamma_1 E_1 + \gamma_2 E_2 + \gamma_3 MGT$$
$$+ \gamma_4 E_1 \cdot MGT + \gamma_5 E_2 \cdot MGT + \text{residual} \quad . \tag{5.8}$$

The code below shows the statements used to fit the expanded model with interaction terms. To fit model 5.8 with PROC REG, we must define the interaction variables E1M and E2M in a DATA step as before.* The joint test of E1M and E2M is specified in a TEST statement, and the OUTPUT statement requests residuals and predicted values for the new model. The PROC REG printed output is shown in Output 5.10.

```
*-- Fit model with Education-Management interaction;
data salary;
   set salary;
   e1m = e1 * mgt;            /* interaction variables */
   e2m = e2 * mgt;

proc reg data=salary;
   model salary = exprnc e1 e2 mgt e1m e2m / ss1 ss2;
Interact: test    e1m=0,e2m=0;
   id case;
   output out=resids r=residual;
```

Output 5.10
REG Procedure Output for Model with Education— Management Interaction

```
                  Expanded model with Education - Mgt Interaction

       DEP VARIABLE: SALARY   Salary ($1000s)
                              ANALYSIS OF VARIANCE

                          SUM OF         MEAN
         SOURCE     DF    SQUARES        SQUARE      F VALUE      PROB>F

         MODEL       6    999.91941     166.65323    5516.596     0.0001
         ERROR      39    1.17816786    0.03020943
         C TOTAL    45    1001.09758

              ROOT MSE    0.1738086     R-SQUARE     0.9988
              DEP MEAN      17.2702     ADJ R-SQ     0.9986
              C.V.         1.006408

                                           (continued on next page)
```

---

* This model may also be fit, giving equivalent results, using PROC GLM with EDUC and MGT as CLASS variables. The test for an interaction is given by the result for the term EDUC*MGT in the MODEL statement below:
```
  proc glm data=salary;
     class educ mgt;
     model salary = exprnc educ mgt educ*mgt ;
```
Either way, these models assume that the slope for the variable EXPRNC is the same for all groups. See Freund, Littell, and Spector (1991) for methods for testing equality of slopes.

```
(continued from previous page)
                    PARAMETER ESTIMATES

                     PARAMETER      STANDARD     T FOR H0:
VARIABLE    DF        ESTIMATE         ERROR    PARAMETER=0    PROB > |T|

INTERCEP     1     11.20343377    0.07906545       141.698        0.0001
EXPRNC       1      0.49698701    0.005566415       89.283        0.0001
E1           1     -1.73074832    0.10533389       -16.431        0.0001
E2           1     -0.34907769    0.09756790        -3.578        0.0009
MGT          1      7.04741202    0.10258919        68.695        0.0001
E1M          1     -3.06603512    0.14933044       -20.532        0.0001
E2M          1      1.83648795    0.13116736        14.001        0.0001

                                                VARIABLE
VARIABLE    DF     TYPE I SS      TYPE II SS      LABEL

INTERCEP     1    13719.94426     606.55670     INTERCEPT
EXPRNC       1      290.71672     240.81384     Experience (years)
E1           1      141.73740    8.15594318     HS Ed.
E2           1       17.79031800  0.38669883    BS Ed.
MGT          1      507.57242     142.56032     Management responsibility
E1M          1       36.18056835  12.73503451
E2M          1        5.92198328   5.92198328


TEST: INTERACT    NUMERATOR:     21.0513  DF:     2    F VALUE:  696.8445
                  DENOMINATOR:  .0302094  DF:    39    PROB >F :   0.0001
```

In plots of residuals, it is useful to identify cases with large (absolute) residuals. In the SALARY data, observations are identified by CASE number. The Annotate data set below is used to label any observation with a residual greater than \$400 in absolute value. (The value \$400 is chosen arbitrarily.)

```
data label;
   set resids;
   xsys='2'; ysys='2';
   if abs(residual) > .4 then do;
      x = exprnc+.4;
      y = residual;
      function='LABEL';
      position='6';
      text = 'Obs.' ||put(case,3.);
      output;
      end;
```

The PROC GPLOT step uses new SYMBOL statements to plot the residuals with the same point symbols as before, but without interpolated lines (I=NONE).

```
proc gplot data=resids ;
   plot residual * exprnc = group
        / frame anno=label vref=0 lvref=3
          vaxis=axis1 haxis=axis2
          vminor=5 hminor=0
          name='GB0511' ;
```

```
symbol1 v=- h=1.5 i=none c=blue;
symbol2 v=+ h=1.5 i=none c=blue;
symbol3 v=: h=1.5 i=none c=blue;
symbol4 v=- h=1.5 i=none c=red ;
symbol5 v=+ h=1.5 i=none c=red ;
symbol6 v=: h=1.5 i=none c=red ;
Label   residual='RESIDUAL ($1000s)';
axis1   order=(-1 to .5 by .5)
        offset=(2)
        label=(h=1.5 a=90 r=0 f=duplex);
axis2   order=(0 to 20 by 5)
        offset=(2)
        label=(h=1.5 f=duplex);
title2 h=1.2 f=duplex
    'Residuals from Model with Education - Management Interaction';
```

Output 5.11 shows the plot of residuals for model 5.8. (A color version of this output appears in Output A3.22 in Appendix 3.) For the most part, the residuals seem unstructured and relatively small. Observation 33 appears to be an outlier; this individual is receiving roughly $900 less than would be expected for someone with the same experience, education, and management responsibility.

Because this observation is so far from the rest, one might be inclined to remove it from the data set and re-estimate the model. A more prudent course of action is to investigate that observation and eliminate it only if it appears to be in error or represents a person who does not belong to the same population. It may also turn out that the outlier suggests a crucial explanatory variable that has been omitted from the model, in which case such an observation may well be the most important in the data set. See Chatterjee and Price (1977) for further discussion of these data. Graphical methods for detecting influential observations are discussed in Section 5.4.4, "Leverage and Influential Observations."

Output 5.11   *Residuals from Model with Education—Management Interaction*

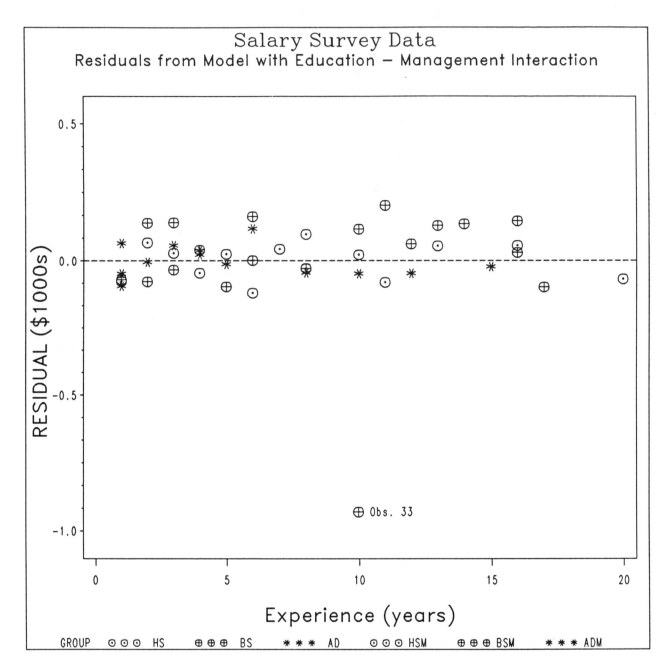

5.4 Validity of Assumptions: Plotting Residuals

We see in Section 5.3, "Regression with Several Groups," that plots of residuals from a fitted model can be useful in determining if the regression model has been correctly specified, that is, whether important explanatory variables have been omitted. This section considers plots of residuals and other quantities designed to assess whether the assumptions of the linear model 5.4 itself are satisfied. Many of the ideas and plots discussed here are considered in more detail in Atkinson (1987) and Cook and Weisberg (1982).

5.4.1 Example: Duncan Occupational Prestige Data

This section examines data presented by Duncan (1961) in a classic study of occupational prestige.* Because adequate measures of occupational prestige were not available for most occupational titles, Duncan (1961) attempted to estimate a substitute measure by predicting occupational prestige from census measures of income and education that were available for many more occupations. The variables are defined as follows:

INCOME — refers to the proportion of males in a given occupational category reporting income of $3,500 or more in the 1950 U.S. Census.

EDUC — refers to the proportion of males in each occupation with at least a high school education in that census.

PRESTIGE — refers to a survey of almost 3,000 people by the National Opinion Research Center (NORC) that obtained ratings on a five-point scale of the general standing of someone engaged in each occupation. Duncan's prestige measure is the percent of people giving a rating of good or excellent.

There were 45 occupations (out of 90) in the original NORC study that matched the census occupational categories. The data set DUNCAN is listed in Section A2.8, "The DUNCAN Data Set: Duncan Occupational Prestige Data." The data appear in Output 5.12.

Output 5.12
DUNCAN Data Set

OBS	JOB	INCOME	EDUC	PRESTIGE	OBS	JOB	INCOME	EDUC	PRESTIGE
1	Accountant	62	86	82	24	Store clerk	29	50	16
2	Pilot	72	76	83	25	Carpenter	21	23	33
3	Architect	75	92	90	26	Electrician	47	39	53
4	Author	55	90	76	27	RR Engineer	81	28	67
5	Chemist	64	86	90	28	Machinist	36	32	57
6	Minister	21	84	87	29	Auto repair	22	22	26
7	Professor	64	93	93	30	Plumber	44	25	29
8	Dentist	80	100	90	31	Gas stn attn	15	29	10
9	Reporter	67	87	52	32	Coal miner	7	7	15
10	Civil Eng.	72	86	88	33	Motorman	42	26	19
11	Undertaker	42	74	57	34	Taxi driver	9	19	10
12	Lawyer	76	98	89	35	Truck driver	21	15	13
13	Physician	76	97	97	36	Machine opr.	21	20	24
14	Welfare Wrkr.	41	84	59	37	Barber	16	26	20
15	PS Teacher	48	91	73	38	Bartender	16	28	7
16	RR Conductor	76	34	38	39	Shoe-shiner	9	17	3
17	Contractor	53	45	76	40	Cook	14	22	16
18	Factory Owner	60	56	81	41	Soda clerk	12	30	6
19	Store Manager	42	44	45	42	Watchman	17	25	11
20	Banker	78	82	92	43	Janitor	7	20	8
21	Bookkeeper	29	72	39	44	Policeman	34	47	41
22	Mail carrier	48	55	34	45	Waiter	8	32	10
23	Insur. Agent	55	71	41					

---

* I am grateful to John Fox for suggesting these data. Fox (1991) analyzes these data more fully in a monograph on regression diagnostics.

The regression, predicting PRESTIGE from the variables INCOME and EDUC, is carried out by PROC REG as follows. The OUTPUT statement creates an output data set, DIAGNOSE, containing predicted values (YHAT), residuals (RESIDUAL and RSTUDENT), and other diagnostic values that will be used later. The INFLUENCE option gives various measures for detecting influential observations described in Section 5.4.4. The PARTIAL option, for partial regression residual plots, is discussed in Section 5.5.

```
proc reg data=duncan;
    model prestige=income educ / influence partial;
    output out=diagnose
        p=yhat r=residual rstudent=rstudent h=hatvalue
        cookd=cookd dffits=dffits;
    id case;
```

5.4.2   Normal Probability Plot of Residuals

The standard way to assess the assumption of normality of errors graphically is to make a normal probability plot. As explained in Section 3.5, "Quantile Plots," our ability to detect deviations from normality is enhanced by plotting confidence intervals around the expected normal 45 degree line and by plotting a detrended version of the Q-Q plot.

Output 5.13 shows the standard normal Q-Q plot of the studentized (deleted) residuals, RSTUDENT.* The detrended version is shown in Output 5.14. Neither plot shows a tendency for the residuals to stray outside the confidence limits, though the detrended plot in Output 5.14 indicates that the largest residual is suspiciously close to the upper confidence boundary.**

---

*   The ordinary residuals do not have equal variance and are all influenced by outliers. Studentized residuals correct for these problems. See Section 5.4.4.

**   The standard confidence limits given in Section 3.5.3, "Standard Errors for Normal Q-Q Plots," assume that the observations are independent and identically distributed. Neither the least squares residuals nor the studentized residuals are independent because they must sum to zero. As a result, the confidence limits must be taken with a grain of salt, particularly in small samples. In large samples, the lack of independence tends to be negligible. Atkinson (1981; 1987) uses a simulation-based technique to produce an envelope showing the expected variability of regression residuals. Flores and Flack (1990) show how to implement this method in the SAS System.

Output 5.13
Normal Q-Q Plot of Studentized Residuals

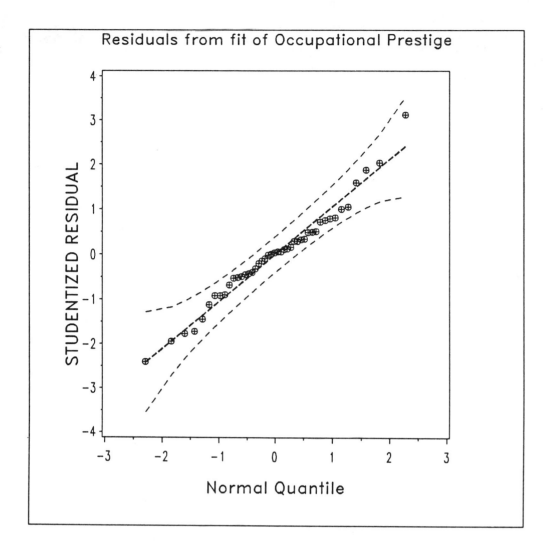

Output 5.14
Detrended Normal Q-Q plot of Studentized Residuals

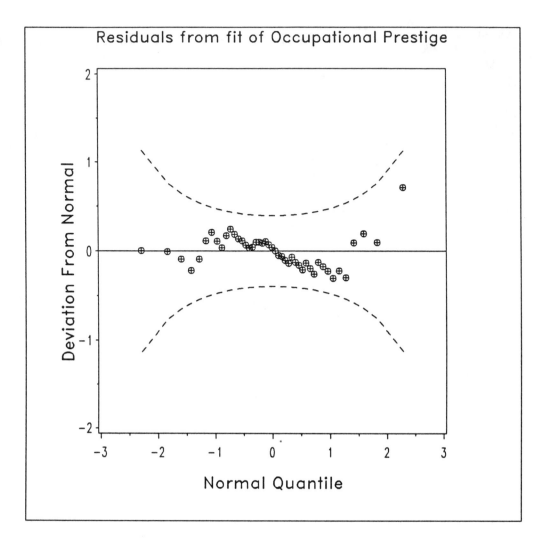

The calculation of the quantities for these plots is described in Section 3.5. The plots are drawn using the NQPLOT macro using these statements:

```
%include nqplot ;
title h=1.5 f=duplex 'Residuals from fit of Occupational Prestige';
%nqplot(data=diagnose, var=rstudent,
        mu=MEAN, sigma=STD);
```

5.4.3 Detecting Heterogeneity of Variance

The usual way to look for evidence that residual variance is not constant is to make a scatterplot of the residuals against fitted values.* In situations where residual variance increases systematically with *y* or with some *x*, you will see a fan-shaped pattern in the plot, as in Output 5.15 (a). Less clear cases, however,

---

* The SPEC option in the MODEL statement provides a formal χ^2 test of the assumption of homoscedasticity. See the *SAS/STAT User's Guide, Version 6, Fourth Edition, Volume 1* and *Volume 2*.

are more difficult to diagnose from a plot of residuals against fitted values. In Output 5.15 (b), the residuals are actually equally variable, but our perception of changes in vertical spread from left to right is confounded by changes in density of points.

A better way to plot residuals to assess changes in spread is to plot the absolute value of the residual, $|e_i|$, against $\hat{y}_i$ or x. This has the effect of folding the plot along a horizontal line at zero. Still, cases like Output 5.15 (b) are hard to judge visually. Adding a smoothed curve to the configuration of points helps considerably.

Output 5.15   *Detecting Changes of Spread*

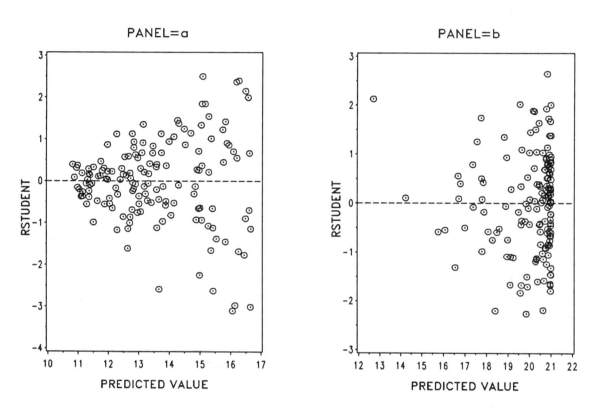

Output 5.16 shows a plot of the absolute value of RSTUDENT against fitted value for the Duncan data, with a lowess smoothed curve. The lowess smoothed curve shows that error variance is not constant. It is clear that residual variance is greatest in the middle of the range. This pattern is not due to the few large residuals, which are ignored in the robust smoothing procedure. Rather, it is due to the fact that the dependent variable, PRESTIGE, was measured as a proportion, and the standard error of a proportion, P, is $\sqrt{P\,(1-P)/n}$, which is maximal for $P=.5$. The heterogeneity of variance here can be corrected by use of a transformation for proportions, such as $P\rightarrow\text{logit}(P)=\log(P/(1-P))$ or $P\rightarrow \arcsin\sqrt{P}$.

Output 5.16
Plot of Absolute(Residual) against Fitted Value for Duncan Data

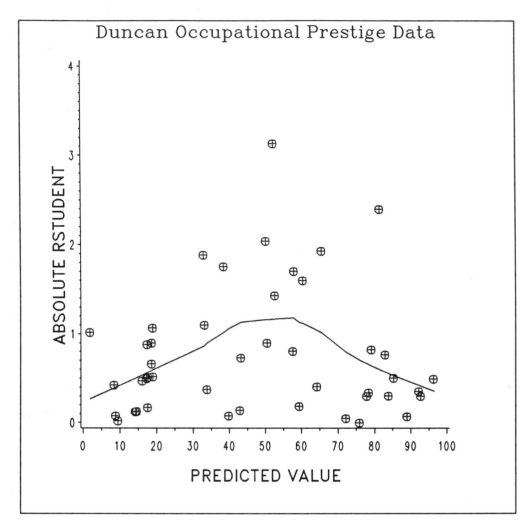

Transformations to Equalize Spread

Nonconstant error variance violates the assumptions of the linear regression model. When the pattern is one of systematic increase or decrease in spread with fitted value, a transformation of the response variable can often cure this defect.*

One idea is to make a diagnostic plot of $\log_{10}|\text{RSTUDENT}|$ against $\log_{10}(x_i)$. If the points in this plot are approximately linear with slope b, it can be shown that a power transformation, $y \to y^{(1-b)}$ where y^0 is interpreted to mean $\log(y)$, will make the residual variance more nearly equal (Leinhardt and Wasserman 1979). This plot thus has the same relation between slope and the ladder of powers as the diagnostic plot for symmetry discussed in Section 3.6, "Plots for Assessing Symmetry."

For example, the data in Output 5.15 (a) are generated so that $\sigma \sim x$, so standard theory would suggest the transformation $y \to \log(y)$. Output 5.17 shows the plot of $\log_{10}|\text{RSTUDENT}|$ against $\log_{10}10\ (x_i)$ for these data. The slope of the

---

* Alternatively, you can use weighted least squares regression to incorporate the change in residual variance into the model. See Freund and Littell (1991) or Chatterjee and Price (1977).

least squares line in this plot is $b = 1.03$, which suggests that log(y) is the appropriate transformation. (In general, if the slope is b, the value $1 - b$, rounded to the nearest .5, is the recommended power.) A related diagnostic plot, the *spread-level plot*, is described in Section 6.4, "Diagnostic Plot for Equalizing Variability."

Output 5.17
Diagnostic Plot of
Log|RSTUDENT|
against Log(x)

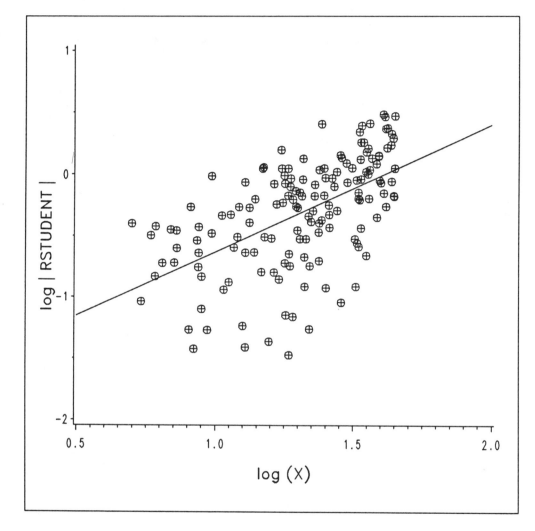

5.4.4 Leverage and Influential Observations

The method of least squares works very well when your data follow the assumptions of the linear model, but unusual observations can have a great impact on the fitted regression. In extreme cases, a single observation can completely determine a parameter estimate, as we saw in Anscombe's example (Section 1.2.2, "Anscombe's Data Set"). Fortunately, PROC REG has a number of diagnostic measures that help you detect such influential observations. Following is a discussion of some of these measures and how to plot them. For further details on influence measures and outlier detection, see Belsley, Kuh, and Welsch (1980), Cook and Weisberg (1982), Fox (1991), or Freund and Littell (1991).

Not All Unusual Observations Are Influential

Output 5.18 shows a plot of simulated data with three unusual observations marked. (A color version of this output appears as Output A3.23 in Appendix 3.) The observation marked O- is an outlier on the *y* axis, but near the center of the *x* values, so it does not affect the slope of the regression line. The potential of an observation to pull the regression line towards it, termed *leverage*, is directly related to its squared distance from the mean of the *x* axis. The observation marked -L is far from the mean of *x* and so has high leverage, but it is not an outlier on *y* (it is not far from the regression line). Observations far from the mean of *x* have high leverage, but an observation must have a moderate or large residual as well to be influential. Although both of these observations are unusual, they have little influence in determining the regression parameters.

Output 5.18
Illustration of Leverage and Influence

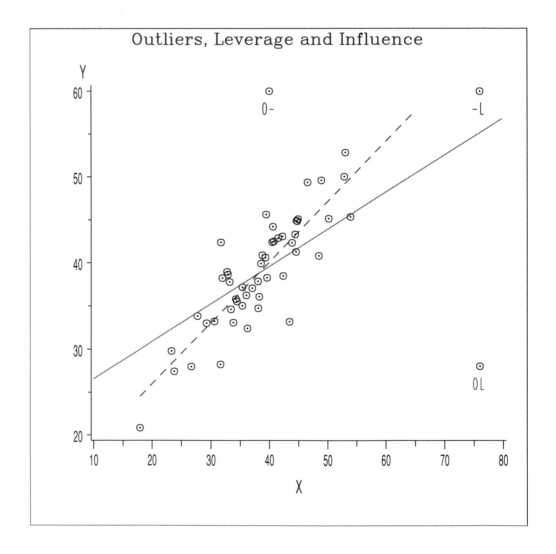

However, the observation in the lower right, marked OL, has both a large residual and high leverage. Such observations exert a great deal of influence on the fitted relationship. The regression line for the observations with the OL point excluded is shown by the dashed line in Output 5.18; with the addition of this single observation, the regression line is deflected to the position shown by the solid (red) line. Deleting either of the other two observations has little effect on the slope estimate.

From this example, you can see that there are two aspects to influence:

leverage

For a single predictor, leverage is proportional to the squared distance of an x value from $\bar{x}$. In multiple regression, leverage is the squared (Mahalanobis) distance from the centroid of the predictors, taking the correlations of the xs into account. Thus, the data ellipse for the predictors is a contour of constant leverage, as illustrated for the Duncan data in Output 4.32.

In general, leverage is computed from the diagonal elements, h_i, of the hat matrix,

$$\mathbf{H} = \mathbf{X}(\mathbf{X}'\,\mathbf{X})^{-1}\mathbf{X}' \tag{5.9}$$

so-called because it is the least-squares projection matrix that determines the fitted values, $\hat{\mathbf{y}} = \mathbf{H}\mathbf{y}$. When the model contains an intercept and p predictors, the hat values h_i must all lie between $1/n$ and 1 and the average hat value is $(p+1)/n$. In practice, observations are typically considered high leverage if $h_i > 2\,(p+1)/n$.

residual (studentized)

Residual is the (scaled) vertical distance between y_i and a regression line that ignores y_i. The ordinary residuals, $e_i = y_i - \hat{y}_i$ are not as useful here for these reasons:

□ The variance of the residuals varies inversely with leverage, $\mathrm{var}(e_i) = \sigma^2\,(1 - h_i)$

□ Outliers on the y axis tend to pull the regression line towards themselves.

To correct these difficulties, we commonly use what is called the *studentized residual* or (more properly) the *studentized deleted residual*, RSTUDENT, which is the standardized residual estimated for y_i without using that observation. Let $s_{(-i)}^2$ be the error mean square estimate of σ^2 omitting observation i. Then

$$\mathrm{RSTUDENT} \equiv e_i^* = \frac{e_i}{s_{(-i)}\sqrt{1 - h_i}} \quad . \tag{5.10}$$

Thus, if the errors in the population have equal variance, so will the studentized residuals, whereas the ordinary residuals do not. Moreover, when no outliers are present the studentized residuals follow a central t distribution with $n - p - 2$ degrees of freedom. Thus a single observation, specified in advance, can be considered unusually deviant if $|\mathrm{RSTUDENT}|$ exceeds the $1 - \alpha/2$ percentage point of the t distribution.

Influence Measures

Two influence measures based on these ideas are available from PROC REG. Both are functions of the product of leverage and residual. (The INFLUENCE option in the MODEL statement gives other influence measures as well. See the *SAS/STAT User's Guide*.)

DFFITS

is a scaled measure of the change in the predicted value, $\hat{y}_i$, when observation i is omitted:

$$\text{DFFITS}_i = \frac{\hat{y}_i - \hat{y}_{(-i)}}{s_{(-i)}\sqrt{h_i}} = \frac{e_i}{s_{(-i)}} \times \frac{\sqrt{h_i}}{1 - h_i} \ . \tag{5.11}$$

Belsley et al. (1980) suggest using the formula DFFITS $> 2\sqrt{(p+1)/n}$ as a rough criterion for declaring an influential outlier.

COOKD

is a measure of change in the parameter estimates when observation i is deleted:

$$\text{COOKD}_i = \frac{e_i^2}{(p+1)s^2} \times \frac{h_i}{1 - h_i^2} \ . \tag{5.12}$$

Comparing equations 5.11 and 5.12 shows that the Cook's D statistic is approximately $DFFITS^2/(p+1)$, so a cutoff of $4/n$ is sometimes used to detect influential outliers.

The cutoff values for declaring influential observations are simply rules-of-thumb, not precise significance tests, and should be used merely as indicators of potential problems.

Example: Duncan Occupational Prestige Data

The PROC REG step to fit PRESTIGE from the variables INCOME and EDUC uses the INFLUENCE option to print the influence statistics. Plots of these values use the output data set DIAGNOSE requested in the OUTPUT statement. Selected variables from this data set are shown in Output 5.19.

Output 5.19
OUTPUT Data Set from the REG Procedure for Duncan Data

CASE	JOB	PRESTIGE	YHAT	RESIDUAL	RSTUDENT	HATVALUE	COOKD	DFFITS
1	Accountant	82	77.9985	4.002	0.3039	0.0509	0.0017	0.0704
2	Pilot	83	78.5275	4.473	0.3409	0.0573	0.0024	0.0841
3	Architect	90	89.0570	0.943	0.0723	0.0696	0.0001	0.0198
4	Author	76	75.9907	0.009	0.0007	0.0649	0.0000	0.0002
5	Chemist	90	79.1960	10.804	0.8266	0.0513	0.0124	0.1923
6	Minister	87	52.3588	34.641	3.1345	0.1731	0.5664	1.4339
7	Professor	93	83.0168	9.983	0.7683	0.0645	0.0137	0.2017
8	Dentist	90	96.4174	-6.417	-0.4981	0.0879	0.0081	-0.1546
9	Reporter	52	81.5380	-29.538	-2.3970	0.0544	0.0990	-0.5749
10	Civil Eng.	88	83.9858	4.014	0.3062	0.0593	0.0020	0.0769
11	Undertaker	57	59.4738	-2.474	-0.1873	0.0468	0.0006	-0.0415
12	Lawyer	89	92.9308	-3.931	-0.3031	0.0793	0.0027	-0.0889
13	Physician	97	92.3849	4.615	0.3557	0.0777	0.0036	0.1032
14	Welfare Wrkr.	59	64.3334	-5.333	-0.4114	0.0783	0.0049	-0.1199
15	PS Teacher	73	72.3454	0.655	0.0505	0.0826	0.0001	0.0152
16	RR Conductor	38	57.9974	-19.997	-1.7040	0.1945	0.2236	-0.8375
17	Contractor	76	50.2307	25.769	2.0438	0.0433	0.0585	0.4346
18	Factory Owner	81	60.4260	20.574	1.6024	0.0433	0.0373	0.3407
19	Store Manager	45	43.0988	1.901	0.1424	0.0263	0.0002	0.0234
20	Banker	92	85.3949	6.605	0.5084	0.0722	0.0068	0.1419

```
21  Bookkeeper      39  50.5986  -11.599  -0.9024  0.0798  0.0236 -0.2657
22  Mail carrier    34  52.6954  -18.695  -1.4332  0.0241  0.0165 -0.2254
23  Insur. Agent    41  65.6198  -24.620  -1.9309  0.0313  0.0377 -0.3472
24  Store clerk     16  38.5903  -22.590  -1.7605  0.0327  0.0333 -0.3237
25  Carpenter       33  19.0629   13.937   1.0689  0.0455  0.0181  0.2334
26  Electrician     53  43.3633    9.637   0.7319  0.0409  0.0077  0.1511
27  RR Engineer     67  57.7160    9.284   0.8089  0.2691  0.0810  0.4908
28  Machinist       57  32.9564   24.044   1.8870  0.0363  0.0421  0.3663
29  Auto repair     26  19.1158    6.884   0.5227  0.0464  0.0045  0.1153
30  Plumber         29  33.9254   -4.925  -0.3780  0.0692  0.0036 -0.1031
31  Gas stn attn    10  18.7455   -8.746  -0.6666  0.0497  0.0079 -0.1524
32  Coal miner      15   1.9473   13.053   1.0185  0.0803  0.0302  0.3010
33  Motorman        19  33.2738  -14.274  -1.1045  0.0607  0.0261 -0.2806
34  Taxi driver     10   9.6948    0.305   0.0233  0.0645  0.0000  0.0061
35  Truck driver    13  14.6962   -1.696  -0.1292  0.0586  0.0004 -0.0322
36  Machine opr.    24  17.4254    6.575   0.4999  0.0496  0.0044  0.1142
37  Barber          20  17.7067    2.293   0.1738  0.0484  0.0005  0.0392
38  Bartender        7  18.7984  -11.798  -0.9024  0.0479  0.0137 -0.2023
39  Shoe-shiner      3   8.6031   -5.603  -0.4294  0.0657  0.0044 -0.1138
40  Cook            16  14.3259    1.674   0.1272  0.0537  0.0003  0.0303
41  Soda clerk       6  17.4951  -11.495  -0.8831  0.0570  0.0158 -0.2170
42  Watchman        11  17.7596   -6.760  -0.5135  0.0475  0.0045 -0.1146
43  Janitor          8   9.0431   -1.043  -0.0799  0.0687  0.0002 -0.0217
44  Policeman       41  39.9464    1.054   0.0788  0.0247  0.0001  0.0125
45  Waiter          10  16.1919   -6.192  -0.4760  0.0706  0.0058 -0.1312
```

A good way to understand how the effects of residual and leverage combine to measure influence is to make a bubble plot of RSTUDENT * HATVALUE, showing the influence measure by the size of the bubble. To identify influential observations in this plot, the DATA step below defines an Annotate data set named LABELS that draws the JOB title for those observations that are either high leverage ($h_i > 2(p+1)/n = .133$) or have large studentized residuals ($|\text{RSTUDENT}| > 2.02$, the critical value of $t(41)$ at $\alpha = .05$). This critical value, shown as the constant 2.02 below, can be obtained with the TINV function as `tinv(.975, 41)`.[*]

```
data labels;
  set diagnose;
  length xsys $1 ysys $1 function $8 position $1 text $12 color $8;
  retain xsys '2' ysys '2' function 'LABEL' color 'BLACK';
  x=hatvalue; y=rstudent;
  if abs(rstudent) > 2.02 or hatvalue > 2 * 3/45 then do;
    text=job; size=1.4;
    if rstudent > 0  then y = y+.4;
                     else y = y-.4;
    if hatvalue > .2 then position='4';
                     else position='6';
  end;

proc gplot data=diagnose  ;
  bubble rstudent * hatvalue = cookd /
      annotate=labels frame
      haxis=axis1 vaxis=axis2
      vref=0 bsize=5
      name='GB0520'  ;
```

---

[*] This critical value does not take the number of residuals tested into account. To allow for the fact that we are, in effect, simultaneously testing all n residuals, a Bonferroni adjustment would test each residual at the $\alpha/2n$ level to give an overall error rate of α.

```
axis1 label=(font=duplex height=1.5 'Hat Value')
      value=(font=duplex height=1.5)
      minor=none offset=(2,2)
      order=0 to .30 by .05;
axis2 label=(font=duplex height=1.5 a=90 r=0 'Studentized Residual')
      value=(font=duplex height=1.5)
      minor=none offset=(0,0)
      order=-4 to 4 by 1;
format hatvalue 3.2;
```

The BUBBLE statement in the PROC GPLOT step represents each observation
with a bubble whose size is proportional to the value of the measure COOKD. The
plot produced is shown in Output 5.20. Of the five points labeled, the two at the
left (Reporter and Contractor) have large residuals, but low leverage. Of the three
points with high leverage, RR Engineer has a small residual. The remaining two
points (Minister and RR Conductor) are the troublesome ones with both high
leverage and large residuals. The bubble size shows how leverage and residual
combine multiplicatively. The value of the Cook's D statistic is greatest for these
two observations with high leverage and large residuals and decreases as residual
or leverage gets smaller.

Output 5.20
Bubble Plot of RSTUDENT * HATVALUE Showing the Cook's D Statistic's Influence

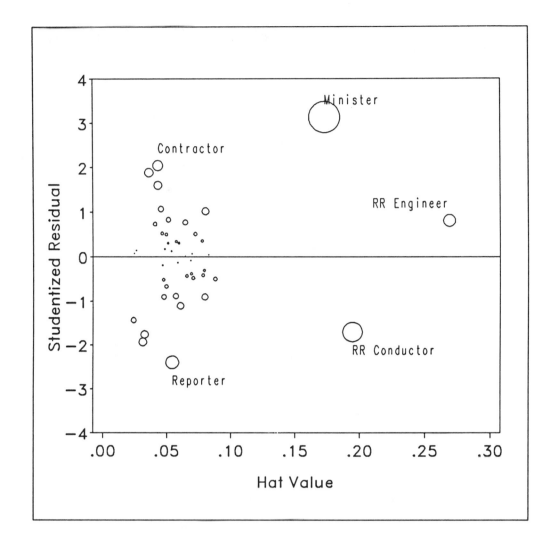

You can also make separate plots of the influence measures, DFFITS and COOKD. In this case, because each observation is identified by job title, dot charts of these measures provide an informative display. Output 5.21 and 5.22 show these plots for the Duncan data. The dashed reference lines correspond to the suggested cutoff values for declaring an influential residual. For example, for Cook's D, $(4/n)$ is 0.089.

Output 5.21
*Dotplot of DFFITS
Influence Measure
for Duncan Data*

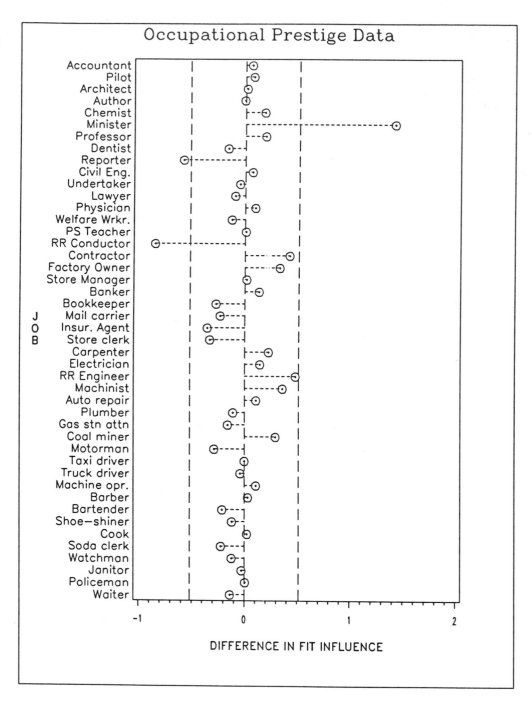

Output 5.22
Dotplot of COOKD Influence Measure for Duncan Data

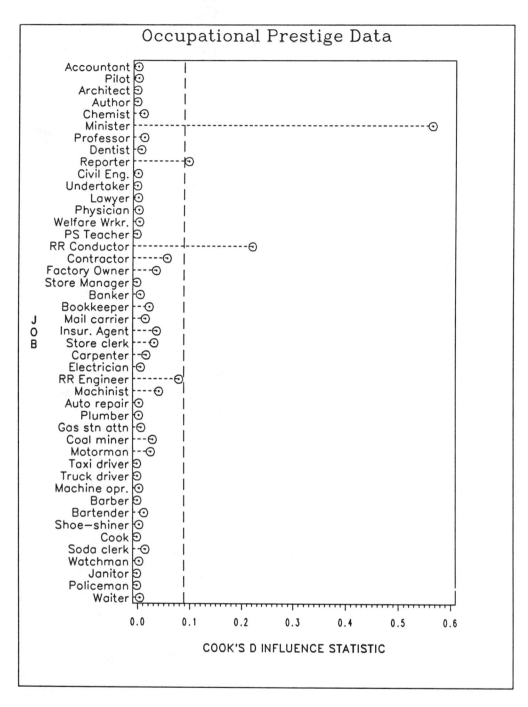

These figures are constructed from the data set DIAGNOSE with the DOTPLOT macro, using the following statements:

```
goptions vpos=55;
title h=1.5 'Occupational Prestige Data';
%include DOTPLOT ;              /* On MVS, use %include ddname(file);*/
```

```
%dotplot( data=diagnose,
          yvar=Job, ysortby=descending case,
          xvar=dffits,
          xref=%str(-.516 to .516 by .516),
          connect=ZERO,
          name=GB0521 );
run;
%dotplot( data=diagnose,
          yvar=JOB, ysortby=descending case,
          xvar=cookd,
          xref=.089,
          connect=DOT,
          name=GB0522 );
```

The measure DFFITS shows the direction of the residual, so you can see that Minister has a substantially higher prestige rating than is predicted by income and education, while Reporter and RR Conductor have much lower prestige than predicted. The COOKD measure, based on the squared residual, makes Minister stand out even more as an unusually influential observation.

5.4.5 Recursive Residuals and Cusum Plots for Time Dependencies

When the observations are ordered in time, you may wish to test whether the regression relation and regression parameters remain constant over time. In this situation, plots based on *recursive residuals* proposed by Brown, Durban, and Evans (1975) can be far more sensitive to changes in regression parameters than plots of ordinary residuals.

Assuming the observations y_i are ordered in time, let b_t be the least squares estimate of the regression coefficients based on the first t observations, and let $X_t = (x'_1, \ldots, x'_t)$ be the first t rows of $\mathbf{X}$. Then the recursive residual, w_t, for observation t is defined to be the (scaled) residual for y_t using the coefficients b_{t-1} obtained from the previous $t-1$ observations:

$$w_t = \frac{y_t - x'_t b_{t-1}}{\sqrt{1 + x'_t(X'_{t-1} X_{t-1})^{-1} x_t}} \quad , \qquad t = p+1, \ldots, n \quad . \tag{5.13}$$

The first p recursive residuals are defined to be zero. Brown, Durban, and Evans propose tests for the constancy of regression relation based on the cumulative sum (CUSUM) and the cumulative sum of squares (CUSUMSS) of the recursive residuals:

$$\text{cusum}_t = \sum_{i=p+1}^{t} w_i \quad , \tag{5.14}$$

$$\text{cusumss}_t = \sum_{i=1}^{t} w_i^2 \quad , \qquad t = p+1, \ldots, n \quad . \tag{5.15}$$

Unlike the ordinary residuals (and even the studentized residuals), the recursive residuals are statistically independent. If the regression relation remains constant over time, the plot of the cumulative sum versus time will vary randomly around the value zero. Conversely, a shift in the regression relation will appear in the cusum plot as a systematic drift toward positive or negative values after the change point. In the plot of the cumulative sum of squares, a shift will appear as a discontinuous jump after the change point.

While the calculation of recursive residuals may seem computationally intensive (involving $n-p$ regressions), a standard updating formula that allows observations to be added one-by-one to a regression model is calculated economically by the INVUPDT function in the IML procedure. The term recursive stems from this use of the updating method to calculate b_t from b_{t-1}.

The program below (based on IMLCUSUM from the SAS Sample Library, Version 5) calculates the recursive residuals, cusum values, and cumulative sum of squares (SS) for a given set of data.* It has been generalized slightly to use SAS macro variables for the names of the input data set and variables. In the program below, &DATA stands for the name of the input data set, &OUT names the output data set, &YVAR is the dependent variable, and &XVARS is the list of independent variables. An example of the use of the program follows the code.

```
proc iml;
   use &data;
   read all var {&xvars} into x[ colname=xname ];
   read all var {&yvar}  into y[ colname=yname ];
   close &data;
   *--------cusum----------;
   n=nrow(x);
   p=ncol(x);
   x = j(n,1,1) || x;
   p1=p+1;

   *---first p+1 observations---;
   z=x[ 1:p1, ];
   yy=y[ 1:p1, ];
   xy=z`*yy;
   ixpx=inv(z`*z);
   b=ixpx*xy;
   r=y[ p1, ]-z[ p1, ]*b;

   *---recursion until n---;
   start recurse;
      do i=p+2 to n;
         xi=x[ i, ]; yi=y[ i, ];
         xy=xy+xi`*yi;
         ixpx=invupdt(ixpx,xi);    /* update inverse matrix */
         b=ixpx*xy;                /* new coefficients      */
         r=r//(yi-xi*b);           /* current residual      */
      end;
   finish;
   run recurse;

   *---print estimates---;
   print "Final parameter estimates",  b /;
   print "Inverse XPX matrix", ixpx /;
```

---

* The CUSUM procedure in SAS/QC software can produce a wide variety of cusum charts, but it does not calculate the recursive residuals required in this application.

```
            *---form lower triangle of ones---;
            m=n-p;
            tri= (1:m)`*repeat(1,1,m)  >= repeat(1,m,1)*(1:m) ;

            *---form csum and csumss---;
            cusum=tri*r;
            cusumss=tri*(r#r);

            *---output to dataset---;
            rnames=xname || yname || { n  residual   cusum   cusumss };
            x= x[ p1:n, 2:p1  ]|| y[ p1:n ];
            x= x || (p1:n)` || r || cusum || cusumss;
            print  x  [ colname=rnames ] /;
            create &out from x [ colname=rnames ];
            append from x;
         quit;
```

The following example shows how you can apply the CUSUM program to test for change in regression relation. We examine the adequacy of the quadratic model 5.6 for the Broadbalk wheat data. The program produces plots of the recursive residuals, cusum values, and cumulative sum of squares, which are shown in Output 5.23, 5.24, and 5.25, respectively.

The plot of recursive residuals in Output 5.23 shows two large residuals (in 1874 and 1877). There is also a slight tendency for the variance of residuals to increase over time.

Output 5.23
*Recursive
Residuals from
Quadratic Model,
Broadbalk Wheat
Data*

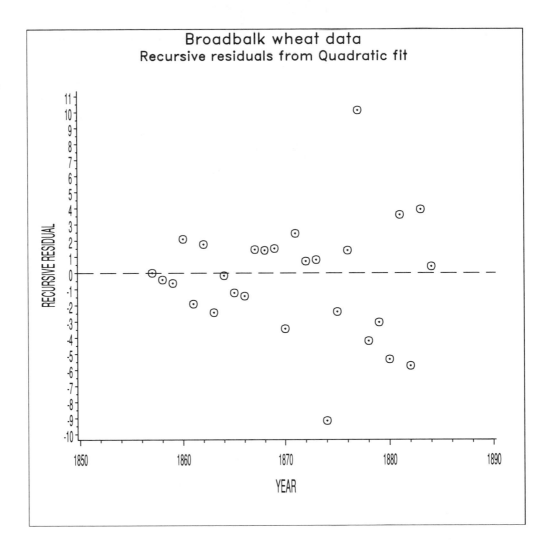

The cusum plot in Output 5.24 shows an essentially random pattern up to 1873 with a sharp decline thereafter, indicating that the quadratic relation appears to fit progressively less well after that date.

Output 5.24
*Cusum Plot of
Recursive
Residuals*

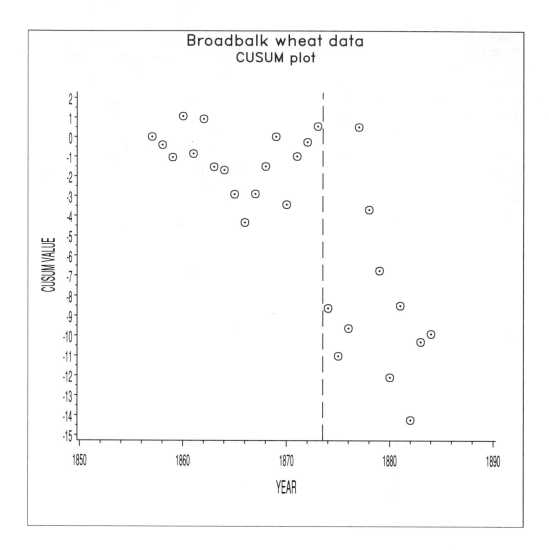

The cumulative SS plot (Output 5.25) confirms this pattern but suggests there may be an additional discontinuity after 1876.

Output 5.25
Cumulative SS Plot
of Recursive
Residuals

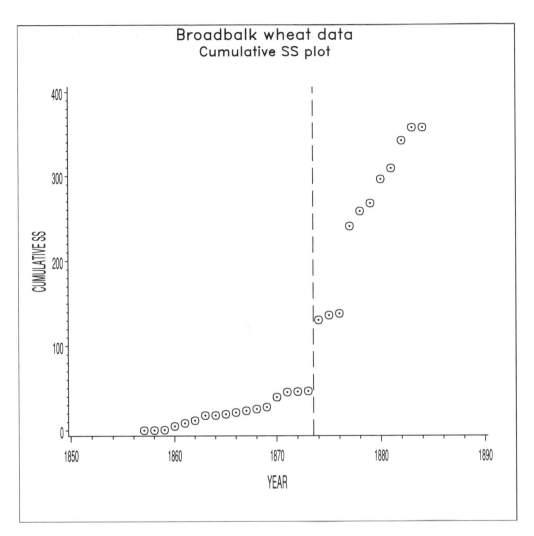

The DATA step below creates the squared year variable, YEAR2. The macro variables used in the CUSUM program are assigned in %LET statements.

```
%include WHEAT;              /* On MVS, use %include ddname(file);*/
title1 f=duplex h=1.5 'Broadbalk wheat data';

data wheat2;
   set wheat;
   year2 = (year - 1870)**2;
  /* set macro variables for CUSUM program */
%let data =wheat2;
%let out  =cus;
%let xvars=YEAR YEAR2;
%let yvar =YIELDIFF;

  /* get the CUSUM program */
%include cusum;              /* On MVS, use %include ddname(file);*/
```

The data set CUS produced by the CUSUM program now contains the original variables, plus the variables RESIDUAL, CUSUM, and CUSUMSS. These are plotted against year by the PROC GPLOT steps below. The vertical reference lines in the plot of the variables CUSUM and CUSUMSS help show the change in trend after 1873.

```
symbol1 v=- h=1.5 c=black;
proc gplot data=cus ;
   title2 f=duplex h=1.3 'Recursive residuals from Quadratic fit';
   plot residual * year
      / vaxis=axis1 haxis=axis2 hminor=4
        vref=0 lvref=21
        name='GB0523' ;
   label residual='RECURSIVE RESIDUAL';
   axis1   label=(h=1.5 a=90 r=0) value=(h=1.3);
   axis2   label=(h=1.5) value=(h=1.3);
run;
proc gplot data=cus ;
   title2 f=duplex h=1.3 'CUSUM plot';
   plot cusum     * year
      / vaxis=axis1 haxis=axis2 hminor=4
        href=1873.5 lhref=21
        name='GB0524' ;
   label cusum='CUSUM VALUE';
run;
proc gplot data=cus ;
   plot cusumss  * year
      / vaxis=axis1 haxis=axis2 hminor=4
        href=1873.5 lhref=21
        name='GB0525' ;
   label cusumss ='CUMULATIVE SS';
   title2 f=duplex h=1.3 'Cumulative SS plot';
run;
```

If you compare the plots of the data with the quadratic regression curve (Output 5.2) and the lowess smooth (Output 5.3), you will see that quadratic regression becomes an increasingly poor fit after 1873.

5.5 Partial Regression Plots

In multiple regression, the correlations among predictors often make it difficult to see the unique effects of one predictor, controlling for the others. A *partial plot* is a generalization of an $x-y$ scatterplot that displays information about the unique effects of a single predictor. This section describes *partial regression plots*, which are designed to show the relationship between y and each x_k, after the effects of all other predictors have been removed (Belsley, Kuh, and Welsch 1980). In addition, these plots depict the leverage of each observation on the partial relation between the variables y and x_k, and the effect of adding the variable x_k to the regression model, and so are also known as *partial regression leverage plots* or *added variable plots*. Another kind of partial plot, the *partial residual plot*, is described by Larsen and McCleary (1972).

The partial regression plot for the variable x_k is defined as follows. Let $\mathbf{X}[k]$ be the matrix of predictors omitting the variable k, and let y_k^* be the part of the variable y that cannot be predicted from $\mathbf{X}[k]$, that is, the residuals from regression of y on $\mathbf{X}[k]$. Similarly, let the variable x_k^* be the residuals from regression of the variable x_k on $\mathbf{X}[k]$. Then the partial regression plot for the variable x_k is the scatterplot of y_k^* against x_k^*.

The usefulness of the partial regression plot stems from the following remarkable properties:

□ The least squares slope of y_k^* on x_k^* is equal to b_k, the estimate of the (partial) regression coefficient, β_k, in the full model. (The simple scatterplot of y against x_k does not necessarily show the correct partial slope because the correlations of x_k with the other predictors is not taken into account.)

□ The residuals from the regression line in this plot are identically the residuals for y in the full model. That is, $y_k^* = b_k x_k^* + e$.

□ The simple correlation between y_k^* and x_k^* is equal to the partial correlation between y and x_k with the other x variables partialled out or controlled. Visually, this correlation indicates how precise the estimate of β_k is.

□ When the relation between y and x_k is nonlinear, the partial regression plot may show both the linear and nonlinear components of the relation, controlling for the other predictors. This helps determine if a term in x_k^2 or some other function of x_k needs to be added to the model.

□ The values x_{ik}^{*2} are proportional to the partial leverage added to h_i by the addition of x_k to the regression. High leverage observations on x_k are those with extreme values of x_k^*.

The plot of y_k^* and x_k^* therefore serves the same purposes that an ordinary scatterplot does in simple linear regression. See Belsley, Kuh, and Welsch (1980), Cook and Weisberg (1982), Fox (1991), or Chambers et al. (1983) for further discussion of the properties of partial regression plots.

The PARTIAL option of PROC REG produces partial regression plots, one for each predictor given in the MODEL statement (plus one for the intercept). In these plots, points are identified by the value of the variable (character or numeric) given in the ID statement.*

For the Duncan occupational prestige data, the partial regression plots are produced by the following statements:

```
proc reg data=duncan;
   model prestige=income educ / influence partial r;
   output out=diagnose
     p=yhat r=residual rstudent=rstudent h=hatvalue
     cookd=cookd          dffits=dffits;
  id case;
```

The plots for the variables INCOME and EDUC are shown in Output 5.26 and 5.27, respectively. The plot for the variable INCOME in Output 5.26 shows that three observations (6, 16, and 27) have great influence on the regression coefficient for income. Without these three observations, the partial effect (slope) of income on prestige would be a good deal greater. The plot for the variable EDUC, however, shows an opposite effect of the influence of the same three observations. This plot shows that the partial regression coefficient for education,

---

* In Version 6, points are labeled with the left-most nonblank character of the ID variable.

controlling income, is inflated by the presence of observations 6 (Minister), 16 (RR Conductor), and, to a lesser degree, 27 (RR Engineer).

Output 5.26
Partial Regression Plot for Income

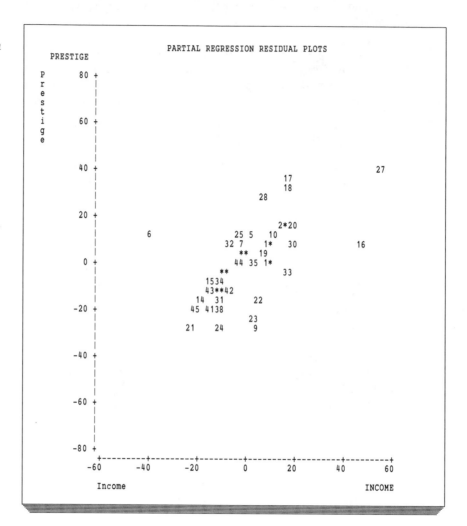

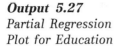

Output 5.27
Partial Regression
Plot for Education

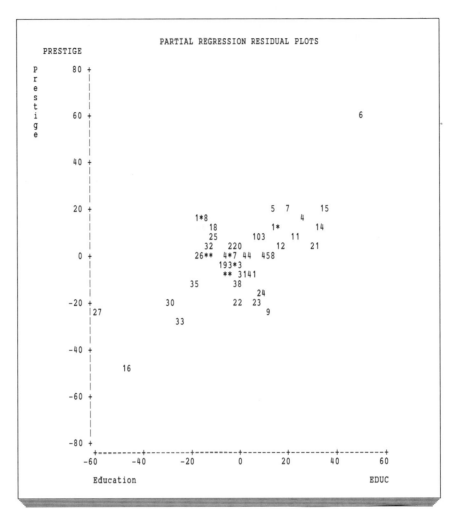

The partial regression plots from PROC REG could be enhanced in several ways. It would be useful, for example, to show the regression line with slope b_k explicitly and to provide some way of marking influential observations. PROC REG does not make the values for the plots available in an output data set, but the calculations are not difficult in PROC IML. A SAS macro, PARTIAL, given in Section A1.12, "The PARTIAL Macro," plots the partial residuals with the regression line. The macro program takes the following parameters:

```
%macro PARTIAL(
        data = _LAST_,   /* name of input data set              */
        yvar =,          /* name of dependent variable          */
        xvar =,          /* list of independent variables        */
        id =,            /* ID variable                         */
        label=INFL,      /* label ALL, NONE, or INFLuential obs */
        out =,           /* output data set: partial residuals  */
        gout=gseg,       /* name of graphic catalog             */
        name=PARTIAL);   /* name of graphic catalog entries     */
```

An example is shown in Output 5.28. The observations with large leverage or studentized residual are marked in red. (A color version of this output appears as

Output A3.24 in Appendix 3.) This plot, with all points labeled with case number, is produced by the following statements:

```
%include partial ;
%partial(data=duncan,
        yvar=prestige,
        xvar=educ income,
        id=case,label=ALL);
```

Output 5.28   *Partial Regression Plot for Income, from PARTIAL Macro*

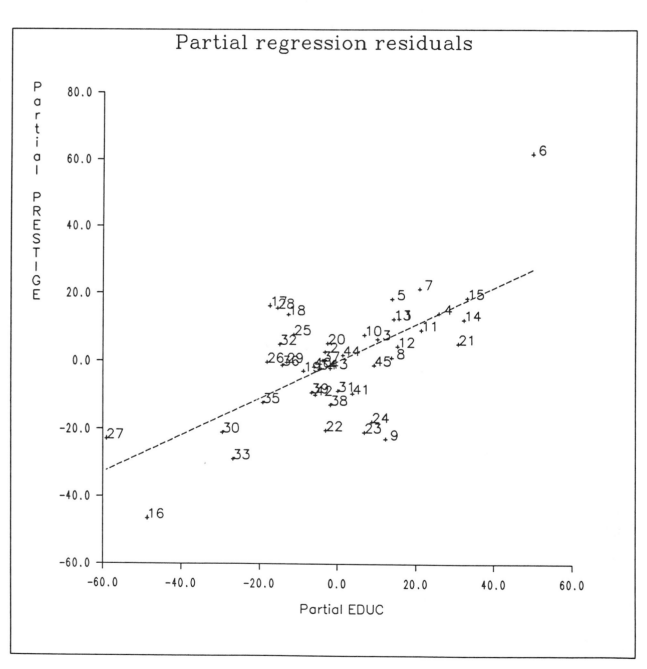

5.6 Multiple Regression Response Surface

When you have two predictor variables, x_1 and x_2, it is sometimes helpful to plot the fitted regression surface as a function of x_1 and x_2. With the G3D procedure, you can plot the actual or predicted values with the SCATTER statement. However, to plot the response surface using the PLOT statement, the values of the variables x_1 and x_2 must form a rectangular grid with one value of the variable y for each combination of x_1 and x_2.

As an example of one technique for constructing response surface plots, we plot the relation between prestige, income, and education from the Duncan data, using the first-order (linear) model

$$\text{PRESTIGE} = \beta_0 + \beta_1 \text{ INCOME} + \beta_2 \text{ EDUC} + \varepsilon \quad . \tag{5.16}$$

The plot, shown in Output 5.29, depicts each observation as a needle line from the observed PRESTIGE value to the fitted response plane estimated by regression:

$$\text{PRESTIGE} = -6.06466 + .59874 \text{ INCOME} + .54583 \text{ EDUC} \quad .$$

Thus, the length of each needle line is proportional to the residual. On the plot, positive residuals are drawn with solid black lines and negative residuals with dashed red lines. Each observation is labeled with the job title for potentially influential observations and case number otherwise.

The three-dimensional display in Output 5.29 makes clear why the three labeled observations are so influential in the prediction of PRESTIGE. (A color version of this output appears in Output A3.25 in Appendix 3.) Except for these points, the correlation between the variables INCOME and EDUC is quite high, and the fitted regression plane is not that firmly anchored. Graphically, the leverage of an observation appears as its ability to pull the fitted plane towards it. For example, Minister, in the upper-left corner, can pull the plane up, and the method of least squares obliges by trying to make that residual small. (In an interactive graphics system, it is possible to see the effect of individual cases by pointing to an observation and having the fitted regression surface redrawn without that case.)

Output 5.29   *Response Surface Plot of Occupational Prestige Data*

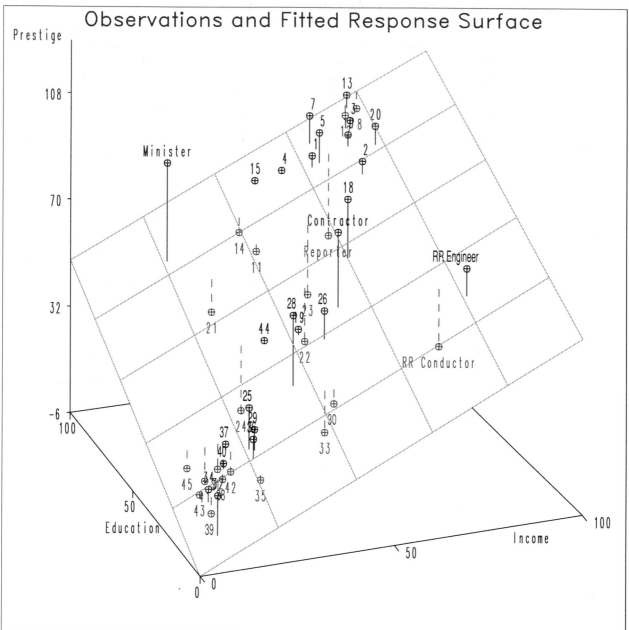

5.6.1 Constructing the Response Surface Plot

Output 5.29 combines aspects of both a surface plot and a three-dimensional scatterplot. The method used to construct this plot depends on the fact that all SAS regression procedures will calculate predicted values for observations in the input data set where the dependent variable is missing.

Thus, the first DATA step below generates a grid of 36 combinations of the variables INCOME and EDUC used for the regression surface, with PRESTIGE set to a missing value. The predicted values for these observations form the fitted regression plane. The second DATA step combines these observations with the Duncan data. The PROC REG step then uses only the actual data values (where PRESTIGE is not missing) to estimate and test the regression surface, but the output data set (RESULTS) contains predicted values (YHAT) for all the observations.

```
%include DUNCAN;                /* On MVS, use %include ddname(file);*/
data grid;
   prestige = .;
   do income = 0 to 100 by 20;
      do educ = 0 to 100 by 20;
         output;
         end;
      end;
data duncan2;
   set duncan grid;

proc reg data=duncan2 ;
   model prestige=income educ ;
   output out=results
      p=yhat r=residual rstudent=rstudent h=hatvalue;
   id case;
```

Then the points on the fitted response surface are obtained by selecting the grid observations in the RESULTS data set.

```
data fit;
   set results;
   if prestige = .;          *-- select predicted values on grid;
   prestige = yhat;
```

The needle lines, points, and case labels are drawn by an Annotate data set named NEEDLES. This is constructed from the output data set RESULTS from PROC REG, selecting the observations where PRESTIGE is not missing. The values of the studentized residual, RSTUDENT, and the variable HATVALUE are used to determine which observations should be identified by the full job name (see Section 5.4.4).

```
data needles;          /* draw residuals and label points */
   set results;
   if prestige ¬=.;              *-- select observed data;
   length caseid text $12;
   caseid = put(case,2.);
   if abs(rstudent) > tinv(.975, 45-4, 0)
      |  hatvalue  > 2 * 3/45 then caseid = job;

   xsys='2'; ysys='2'; zsys='2';
   x = income;
   y = educ;
```

```
z = yhat;     function = 'MOVE    '; output;
if residual > 0 then do;
   line=1;  color='BLACK'; position='2'; end;
else do;
   line=20; color='RED';   position='8'; end;

z = prestige; function = 'DRAW   '; output;        *-- needle line;

function = 'SYMBOL'; text='+';      output;        *-- point;
function = 'LABEL' ; text=caseid;   output;        *-- label;
```

Finally, the response surface in the FIT data set is drawn with the PLOT statement in PROC G3D. The Annotate data set NEEDLES draws the rest.

```
title h=1.6 f=duplex 'Observations and Fitted Response Surface';
proc g3d data=fit  ;
    plot educ * income = prestige /
          tilt=70
          rotate=20
          xticknum=3
          yticknum=3
          caxis=black
          anno=needles
          name='GB0529';
    format income educ prestige 5.0;
```

5.6.2 Quadratic Response Surface

The model 5.16 and the fitted regression plane shown in Output 5.29 assume that both income and education are linearly related to occupational prestige. None of the plots of these data have suggested that this assumption is not a reasonable one. However, for the sake of testing this assumption formally and illustrating plotting a quadratic response surface, we fit the second-order model

$$PRESTIGE = \beta_0 + \beta_1 \, INCOME + \beta_{11} \, INCOME^2 + \beta_2 \, EDUC$$
$$+ \beta_{22} \, EDUC^2 + \beta_{12} \, INCOME \times EDUC \quad . \tag{5.17}$$

In this model, the squared terms in the variables INCOME and EDUC represent potential curvature in the relation of these variables to PRESTIGE. The cross-product term represents an interaction of these variables. See Freund and Littell (1991, Section 5.4) for more discussion of fitting polynomial models with several variables with the SAS System.

For testing whether the higher-order terms in model 5.17 add significantly to the fit, the RSREG procedure is most convenient. In the step below, PROC RSREG builds the linear, quadratic, and cross-product terms for the variables INCOME and EDUC. The same model can be tested with PROC REG, but the squared and cross-product terms must first be generated in a DATA step.

```
%include DUNCAN ;
proc rsreg data=duncan;
    model prestige=income educ ;
    id case;
```

The printed output from PROC RSREG is shown in Output 5.30. The section headed REGRESSION gives the tests for the combined effects of each of the linear terms, the squared terms, and the cross-product term. It may be seen that none of the higher-order terms add significantly to the prediction of occupational prestige. Nevertheless, for the sake of illustration, we will plot the regression surface that has been fit by this model.

Output 5.30
Quadratic
Response Surface
Model Tested with
the RSREG
Procedure

```
                         Duncan Occupational Prestige Data
        RESPONSE SURFACE FOR VARIABLE PRESTIGE Prestige

            RESPONSE MEAN           47.68889
            ROOT MSE               13.53278
            R-SQUARE               0.8365141
            COEF OF VARIATION      0.2837723


            REGRESSION        DF     TYPE I SS    R-SQUARE    F-RATIO    PROB

            LINEAR            2    36180.94579      0.8282      98.78    0.0001
            QUADRATIC         2      147.90647      0.0034       0.40    0.6705
            CROSSPRODUCT      1      216.47833      0.0050       1.18    0.2836
            TOTAL REGRESS     5    36545.33059      0.8365      39.91    0.0001

            RESIDUAL          DF           SS    MEAN SQUARE

            TOTAL ERROR       39    7142.31386     183.13625


            PARAMETER         DF     ESTIMATE       STD DEV    T-RATIO    PROB

            INTERCEPT         1   -0.85620716    9.29959399     -0.09    0.9271
            INCOME            1    0.96745283    0.51364563      1.88    0.0671
            EDUC              1   -0.02437611    0.54223208     -0.04    0.9644
            INCOME*INCOME     1  -0.000869464   0.005835443     -0.15    0.8823
            EDUC*INCOME       1  -0.005773841   0.005310614     -1.09    0.2836
            EDUC*EDUC         1   0.007450869   0.005500430      1.35    0.1833


            FACTOR            DF           SS    MEAN SQUARE    F-RATIO    PROB

            INCOME            3     4833.694       1611.231       8.80    0.0001 Income
            EDUC              3     5655.535       1885.178      10.29    0.0001 Education
```

The method used to plot the regression plane in the previous example can be used for the quadratic model (and any two-predictor polynomial model) because the fitted values depend only on the input data and the MODEL statement. The program below constructs the same grid of INCOME and EDUC values, which are combined with the Duncan data. The DATA step that produces the DUNCAN2 data set also calculates the squared and cross-product terms in model 5.17.

```
data grid;
   prestige = .;
   do income = 0 to 100 by 20;
      do educ = 0 to 100 by 20;
         output;
         end;
      end;
data duncan2;
   set duncan grid;
   inc2 = income**2;
   edu2 = educ  **2;
   inc_edu = income * educ;
proc reg data=duncan2;
   model prestige=income educ inc2 edu2 inc_edu ;
   output out=results
      p=yhat r=residual rstudent=rstudent h=hatvalue;
   id case;
```

The remaining steps to plot the fitted quadratic together with needle lines and case labels are nearly identical to those for the linear case: the first DATA step extracts the fitted values, the second DATA step constructs the Annotate data set, and the PROC G3D step produces the plot shown in Output 5.31. (A color version of this output appears as Output A3.26 in Appendix 3.) The main difference between this plot and the linear fit (Output 5.29) is the upward bowing of the opposite corners, due to the influence of Minister and RR Engineer.

```
data fit;
   set results;
   if prestige = .;          *-- select predicted values on grid;
   prestige = yhat;

data needles;    /* annotate dataset to draw residuals */
   set results;
   xsys='2'; ysys='2'; zsys='2';
   length caseid text $12;
   if prestige ¬=.;   *-- select observed data;
   caseid = put(case,2.);
   if abs(rstudent) > 2.03 |
         hatvalue  > 2 * 6/45 then caseid = job;
   x = income;
   y = educ;
   z = yhat;     function = 'MOVE     '; output;
   if residual > 0 then do;
      line=1;  color='BLACK'; position='2'; end;
```

```
    else do;
       line=20; color='RED';   position='8'; end;
    z = prestige; function = 'DRAW   '; output;

    function = 'SYMBOL'; text='+';        output;
    function = 'LABEL' ; text=caseid;    output;

title h=1.6 f=duplex 'Quadratic Response Surface';
proc g3d data=fit  ;
    plot educ * income = prestige /
           tilt=70
           rotate=20
           xticknum=3
           yticknum=3
           caxis=black
           anno=needles
           name='GB0531';
    format income educ prestige 5.0;
```

Output 5.31   *Quadratic Response Surface for Occupational Prestige Data*

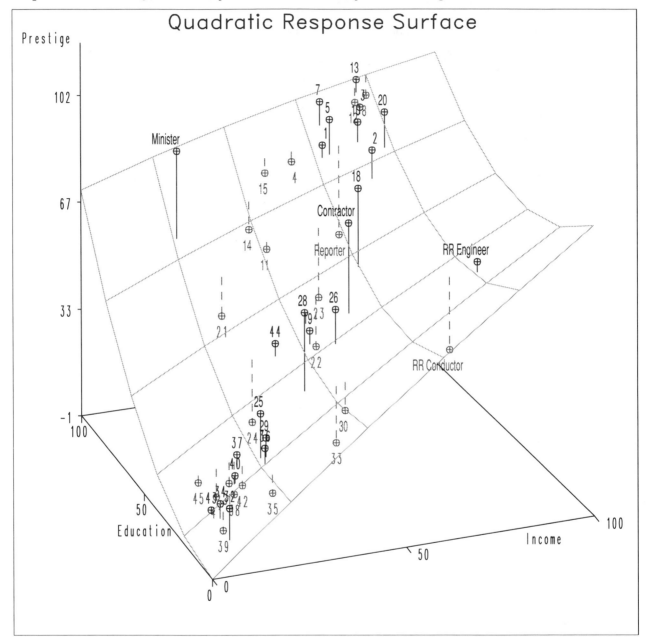

5.7 Plots of C_P against P for Variable Selection

In multiple regression, you often have a large number of potential predictor variables available. As you add more predictors, the R^2 variable for the model must increase (it can't decrease), but because the predictors are usually correlated, the standard errors of the regression parameters and predicted values will often increase too, reflecting a loss in precision of estimation. Thus, you usually want to choose the most parsimonious subset of predictors to provide a reasonable level of goodness of fit.

While people often use the STEPWISE or RSQUARE procedure to examine many or all possible subsets of the predictors, stepwise selection to maximize the R^2 variable by itself can involve a substantial risk of capitalization on chance and shrinkage (that is, the regression model will not validate well in a future sample; see, for example, Stevens 1986).* A useful graphical technique for model selection proposed by Mallows (1973) involves plotting a measure of total squared error of the predicted y values, which takes into account both error sum of squares and bias due to omission of important variables from the model.

Mallows suggests plotting the total squared error statistic C_P against P, where $P = p + 1$ is the number of the parameters (including the intercept) in the regression model. C_P is calculated as

$$C_P = \frac{\text{SSE}_P}{\text{MSE}_{full}} - (n - 2P) \qquad (5.18)$$

where SSE_P is the error sum of squares for a model with P parameters, and MSE_{full} is the mean square error estimate of σ_2 for the full model with all available predictors. Mallows (1973) shows that when the model is correctly specified (when there is no bias), the expected value of C_P is approximately equal to the value of P, so good models should have $C_P \approx P$.

The values of C_P are printed by PROC RSQUARE when the CP option is used in the PROC statement. The values are also included as the variables _CP_ and _P_ in the output data set requested in the OUTEST= option. It takes a bit of care, however, to construct a useful plot of C_P against P for these reasons:

□ To interpret the plot, you need to identify the particular variables in a model in the plotting symbol.

□ Large values of C_P will cause the C_P axis to be scaled so that the small values, where the models of interest reside, are highly compressed.

□ Drawing the line $C_P = P$ in the plot helps you to decide among models.

As a result, it is most useful to do the plots of C_P with PROC GPLOT and use the Annotate facility to label the points and draw the line.

The C_P plot can be illustrated using the data on fuel consumption in the United States (see Section 4.3.1, "Labeling Printer Plots," and Section A2.9, "The Fuel Data Set: Fuel Consumption Data"). The variables in the data set FUEL are the following:

FUEL   is the gasoline fuel consumption per capita, regarded as the response variable.

ROAD   is the length of federal highways, in miles.

INC   is the per capita personal income.

TAX   is the state tax rate on motor fuel, in cents per gallon.

---

* In Version 6, the model-selection methods of both PROC RSQUARE and PROC STEPWISE are carried out by the REG procedure. For compatibility across versions, if PROC STEPWISE or PROC RSQUARE is requested in Version 6, PROC REG with the appropriate model-selection method is actually used.

POP   is the state population, in thousands.

DRIVERS   is the proportion of licensed drivers.

The values of C_P are obtained from the following PROC RSQUARE step. The printed output is shown in Output 5.32. Note that for $p=5$ predictors, there are $2^p-1=31$ distinct models with different subsets of predictors.

```
proc rsquare data=fuel outest=models cp ;
    model fuel = tax drivers road inc pop ;
```

Output 5.32
RSQUARE
Procedure Output
for Fuel
Consumption Data

```
                    Fuel Consumption across the US

    N=48        REGRESSION MODELS FOR DEPENDENT VARIABLE: FUEL

    NUMBER IN    R-SQUARE      C(P)     VARIABLES IN MODEL
      MODEL

        1       0.00036260   95.290966   ROAD
        1       0.05995744   86.986932   INC
        1       0.20365389   66.964054   TAX
        1       0.21414550   65.502137   POP
        1       0.48855266   27.265835   DRIVERS
    ------------------------------------------------------
        2       0.06094116   88.849858   ROAD INC
        2       0.21779981   66.992941   INC POP
        2       0.26085413   60.993687   TAX INC
        2       0.26814438   59.977853   TAX ROAD
        2       0.38797374   43.280652   ROAD POP
        2       0.48957868   29.122867   TAX POP
        2       0.49264840   28.695128   DRIVERS ROAD
        2       0.53816217   22.353171   DRIVERS POP
        2       0.55668109   19.772717   TAX DRIVERS
        2       0.61750981   11.296753   DRIVERS INC
    ------------------------------------------------------
        3       0.31776329   55.063880   TAX ROAD INC
        3       0.39312108   44.563413   ROAD INC POP
        3       0.48990880   31.076868   TAX INC POP
        3       0.51069901   28.179929   TAX ROAD POP
        3       0.56697273   20.338665   TAX DRIVERS ROAD
        3       0.61968405   12.993792   DRIVERS INC POP
        3       0.62087362   12.828035   DRIVERS ROAD POP
        3       0.62492422   12.263618   DRIVERS ROAD INC
        3       0.65222138    8.459992   TAX DRIVERS POP
        3       0.67485834    5.305724   TAX DRIVERS INC
    ------------------------------------------------------
        4       0.51159447   30.055154   TAX ROAD INC POP
        4       0.65237216   10.438982   DRIVERS ROAD INC POP
        4       0.66872863    8.159847   TAX DRIVERS ROAD POP
        4       0.67868671    6.772273   TAX DRIVERS ROAD INC
        4       0.69558820    4.417193   TAX DRIVERS INC POP
    ------------------------------------------------------
        5       0.69858224    6.000000   TAX DRIVERS ROAD INC POP
    ------------------------------------------------------
```

The OUTEST data set MODELS is used to construct the C_P plot. In the following DATA step, a character variable INDEX is constructed, which labels the variables in a given model. The labels are the initial letters of the variable names; for example, the model with TAX and DRIVERS is labeled TD. The calculation of the variable INDEX uses the fact that the MODELS data set contains the regression coefficients for each model, or missing values when a given variable is not included in a particular model. The MODELS data set, with the addition of the INDEX variable, is shown in Output 5.33.

Output 5.33
MODELS Data Set
for C_P Plot

_P_	INDEX	_RSQ_	_CP_	TAX	DRIVERS	ROAD	INC	POP
2	R	0.00	95.29	.	.	0.00	.	.
2	I	0.06	86.99	.	.	.	-0.05	.
2	T	0.20	66.96	-53.106	.	.	.	.
2	P	0.21	65.50	.	.	.	.	-0.01
2	D	0.49	27.27	.	1409.84	.	.	.
3	RI	0.06	88.85	.	.	0.00	-0.05	.
3	IP	0.22	66.99	.	.	.	-0.01	-0.01
3	TI	0.26	60.99	-52.750	.	.	-0.05	.
3	TR	0.27	59.98	-71.402	.	-0.01	.	.
3	RP	0.39	43.28	.	.	0.02	.	-0.02
3	TP	0.49	29.12	-62.435	.	.	.	-0.01
3	DR	0.49	28.70	.	1418.14	0.00	.	.
3	DP	0.54	22.35	.	1233.46	.	.	-0.01
3	TD	0.56	19.77	-32.075	1251.49	.	.	.
3	DI	0.62	11.30	.	1525.04	.	-0.07	.
4	TRI	0.32	55.06	-69.998	.	-0.01	-0.04	.
4	RIP	0.39	44.56	.	.	0.02	0.02	-0.02
4	TIP	0.49	31.08	-62.260	.	.	-0.00	-0.01
4	TRP	0.51	28.18	-50.560	.	0.01	.	-0.02
4	TDR	0.57	20.34	-40.627	1193.32	-0.00	.	.
4	DIP	0.62	12.99	.	1473.33	.	-0.07	-0.00
4	DRP	0.62	12.83	.	1077.86	0.01	.	-0.01
4	DRI	0.62	12.26	.	1537.92	0.00	-0.07	.
4	TDP	0.65	8.46	-43.266	941.07	.	.	-0.01
4	TDI	0.67	5.31	-29.484	1374.77	.	-0.07	.
5	TRIP	0.51	30.06	-50.038	.	0.01	0.01	-0.02
5	DRIP	0.65	10.44	.	1290.50	0.01	-0.05	-0.01
5	TDRP	0.67	8.16	-33.015	928.45	0.01	.	-0.01
5	TDRI	0.68	6.77	-34.790	1336.45	-0.00	-0.07	.
5	TDIP	0.70	4.42	-36.434	1167.83	.	-0.05	-0.00
6	TDRIP	0.70	6.00	-32.451	1138.07	0.00	-0.04	-0.01

```
data models;
    set models;
    drop _model_ _depvar_ _type_ fuel _in_ _rmse_ ;
    * Calculate plotting symbol based on predictors in the model;
    length index $5;
    index =
      compress( substr(' T',1+(tax ¬=.),1)
             || substr(' D',1+(drivers¬=.),1)
             || substr(' R',1+(road¬=.),1)
             || substr(' I',1+(inc ¬=.),1)
             || substr(' P',1+(pop ¬=.),1) );
proc print data=models;
    id _p_ index;
    var _rsq_ _cp_ tax drivers road inc pop;
    format _cp_ 6.2 _rsq_ 5.2 road inc pop 5.2;
```

The plot of C_P for these data is shown in Output 5.34. (A color version of this output appears in Output A3.27 in Appendix 3.) The ordinate has been restricted to the range (0, 30) so that large values do not cause the smaller ones to be compressed. Models with large C_P values are shown above the arrows at the top of the figure. Each point represents one model, identified by the initial letter of each included variable. Points near the line $C_P=P$ represent satisfactory models. An alternative display might plot the minimum value of C_P for each number of parameters. This would not be as useful, however, because it would not show nearly competitive models.

From the plot, two models, TDI (the variables TAX, DRIVERS, and INC) and TDIP (the variables TAX, DRIVERS, INC, and POP) appear satisfactory in terms of C_P. Note that both of these models are better, in terms of total squared error, than the full model with all five predictors.

The DATA step shown below produces the annotations in the plot. The maximum value of C_P displayed is set by the macro variable, CPMAX. For values of $C_P \leq 30$, the value of the INDEX variable is written on the plot at the point (P, C_P). For $C_P > 30$, however, it uses the data percentage coordinate system (YSYS='1') to write the label in the top 10 percent of the plot. In this case, an arrow symbol (the character h in the MATH font) is drawn to indicate that these points are off-scale. To avoid overplotting, the *x* values and POSITION values for the labels are made to alternate to the left and right side of the plotted points. This is usually sufficient to keep the labels distinct.

```
%let cpmax = 30;
data labels;
   set models;
   by _P_;
   drop pop count index;
   length function color $8;
   xsys='2';
   x = _p_;
   text = index; function = 'LABEL';
   if first._p_ then do; out=0; count=0; end;
   if _cp_ > &cpmax then do;
      out+1;
      ysys='1'; y =  90 + 1.5*out;
      color='RED'; size=1.2;
      if mod(out,2)=1
         then do; x=x -.05; position='4';  end;
         else do; x=x +.05; position='6';  end;
      output;
      end;
   else do;
      count+1;
      ysys = '2'; y = _cp_;
      color='BLACK'; size=1.5;
      if mod(count,2)=1
         then do; x = x -.05; position='4'; end;
         else do; x = x +.05; position='6'; end;
      output;
      end;
   if last._p_ & out>0 then do;
      x=_p_; y=90; ysys='1'; position='5';
      function='SYMBOL'; size=3 ; color='RED';
      style='MATH'; text='h';
      output;
      end;
```

A second Annotate data set, PLINE, is constructed to draw the line corresponding to $C_P=P$ on the plot, and this is added to the end of the LABELS data set.

```
data pline;
   xsys = '2'; ysys = '2';
   color='BLUE'; line=10;
   x = 2; y = 2; function='MOVE'; output;
   x = 6; y = 6; function='DRAW'; output;
data anno;
   set labels pline;
```

In the PROC GPLOT step, the AXIS1 statement limits the vertical axis to values of $C_P \leq 30$. It is also necessary to specify OFFSET values in AXIS statements to allow room for the labels at the top of the vertical axis.

```
title h=1.5 'Fuel Consumption across the US';
proc gplot data=models  ;
   plot _cp_ * _p_ /
         anno=anno
         vaxis=axis1
         haxis=axis2
         name='GB0534' ;
   axis1 order=(0 to &cpmax by 10)
         offset=(1,12 pct)
         value=(h=1.4) label=(h=1.6 a=90 r=0);
   axis2 order=(2 to 6)
         offset=(4) minor=none
         value=(h=1.4) label=(h=1.6);
   symbol v=+ h=1.2 c=black;
```

Output 5.34
C_P Plot for Fuel
Consumption Data

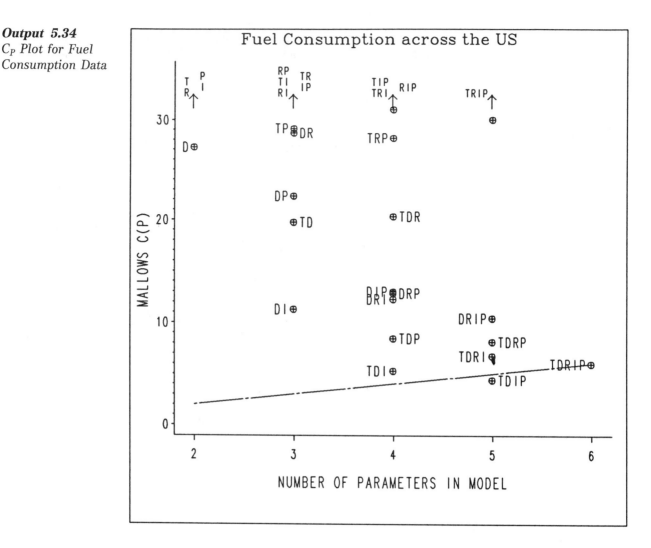

5.7.1 Related Methods

As in other situations where the graphical benchmark is a tilted line, there are advantages to re-expressing the plot so that the comparison line is horizontal. One simple alternative is to plot $(C_P - P)$ against P, so good models are represented by points close to the line $C_P - P = 0$. Or we can plot C_P/P against P and compare the points to a horizontal line at $C_P/P = 1$.

Another possibility is to plot an F statistic equivalent to C_P. Spjøtvoll (1977) shows that C_P is directly related to the incremental F statistic for testing the additional contribution of the $m - P$ variables not included in a given model,

$$F_P = 1 + (C_P - P) / (m - P)$$

where m is the total number of available predictors. F_P tests the hypothesis that the predictors omitted from a given subset have population coefficients of zero. When this hypothesis is true, then $E(F_P) \approx 1$. Hence, a plot of F_P against P has the horizontal line $F_P = 1$ as the standard for good models. The F_P plot has the additional advantage of a more familiar statistic than C_P.

However, like the C_P plot, poor models correspond to large values of the F_P statistic, so it is often necessary to exclude some of the largest values from the plot to preserve resolution among the models of interest. One way around this (Spjøtvoll 1977) is to plot the significance level of F_P, *Prob*($F > F_P$), based on the usual normality assumptions, so that poor models correspond to small probability values. Both the F_P plot and the significance level plot are used informally rather than inferentially, but the more familiar scales of these measures can help in deciding which deviations from the benchmark line are statistically important.

5.8   Regression with Geographic Data

Anscombe (1981) discusses data from the *Statistical Abstract of the United States 1970* on expenditures on public school education in each of the U.S. states and several related predictor variables (per capita income, proportion of young people, and the degree of urbanization in each state). He calls these data a "typical example of material to which simple regression methods can be applied" (p. 226).

In fact, his example is far from typical because the observation units (states) provide a geographical basis for analysis that is lacking in many regression studies. Thus, in addition to the usual sorts of scatterplots of data and of residuals available for all regression problems, a display of data or residuals on a map can help us to understand the relationships involved.

The analysis uses the data set SCHOOLS, listed in Section A2.13, "The SPENDING Data Set: School Spending Data." The quantitative variables SPENDING, INCOME, YOUTH, and URBAN have all been measured on a per capita basis, so that differences in population among the states are not a factor. The two-letter state abbreviations, ST, are mapped into the state FIPS code numbers (STATE) used as the ID variable in the MAPS.US data set. In addition, each state is classified by two regional variables, REGION and GROUP, which could be used in other analyses (such as analysis of variance or covariance) but are not used in this example. Each REGION has been subdivided into two or three groups. The DATA step and the first and last few observations are reproduced below:

```
Proc Format;
   Value $REGION
         'NE' = 'North East'   'NC' = 'North Central'
         'SO' = 'South Region' 'WE' = 'West Region';
Data Schools;
   Input ST $  SPENDING INCOME YOUTH URBAN REGION $ GROUP $;
   STATE=STFIPS(ST);
   LABEL ST = 'State'
         SPENDING='School Expenditures 1970'
         INCOME  ='Personal Income 1968'
         YOUTH   ='Young persons 1969'
         URBAN   ='Proportion Urban';
```

```
cards;
ME  189  2824  350.7  508  NE  NE
NH  169  3259  345.9  564  NE  NE
VT  230  3072  348.5  322  NE  NE
    ...
OR  233  3317  332.7  671  WE  PA
CA  273  3968  348.4  909  WE  PA
AK  372  4146  439.7  484  WE  PA
HI  212  3513  382.9  831  WE  PA
;
```

To get a geographical picture of public school spending, a choropleth map can be drawn depicting the SPENDING variable by density of shading on the U.S. map. In constructing a choropleth map of a continuous response variable such as this, the GMAP procedure ordinarily divides the range of the variable into some number of class intervals. If you use the defaults and do not specify PATTERN statements, the SAS System uses different color and pattern combinations to indicate the magnitude of the response variable, but the visual density of the fill pattern does not necessarily correspond to magnitude. In fact, if you use PATTERN statements with V= options for fill patterns, but do not specify colors with the C= option in each PATTERN statement, PROC GMAP will reuse the V= option pattern once with each available color before going on to the next PATTERN statement.*

In order to draw a map in which the visual impact of shading corresponds directly to the magnitude of a continuous response variable, it is necessary to do two things:

□ group the response variable into discrete class intervals (use a SAS format or a DATA step operation)

□ provide one PATTERN statement for each class level, specifying *both* V=fill and C=color in each.

The SPENDING variable ranges from a low of $112 per capita (not per pupil) in Alabama to a high of $372 in Alaska. The program below uses the FORMAT procedure to create a format that effectively groups spending into six $50 class intervals. The GMAP step uses the DISCRETE option in the CHORO statement and the FORMAT statement for SPENDING to make PROC GMAP respect this grouping of the SPENDING variable. The PATTERN statements specify V=Mn to select shading density and direction of the fill lines.

The map produced is shown in Output 5.35. (A color version of this output appears as Output A3.28 in Appendix 3.) In the map, Alaska stands out as the state with the highest level of school spending; California, Minnesota, and New York are also at the high end. The entire southern region of the U.S. is at the low end of the scale, with a large cluster of southern states in the lowest spending category.

---

* Because the available colors differ from one device to another, a map that looks good on a monochrome device can turn out completely differently when plotted on a multicolor device. You can override the default color selection by specifying PENMOUNTS=n or COLORS=(*colorlist*) in the GOPTIONS statement, but it is preferable to specify the V= and C= options fully in each PATTERN statement.

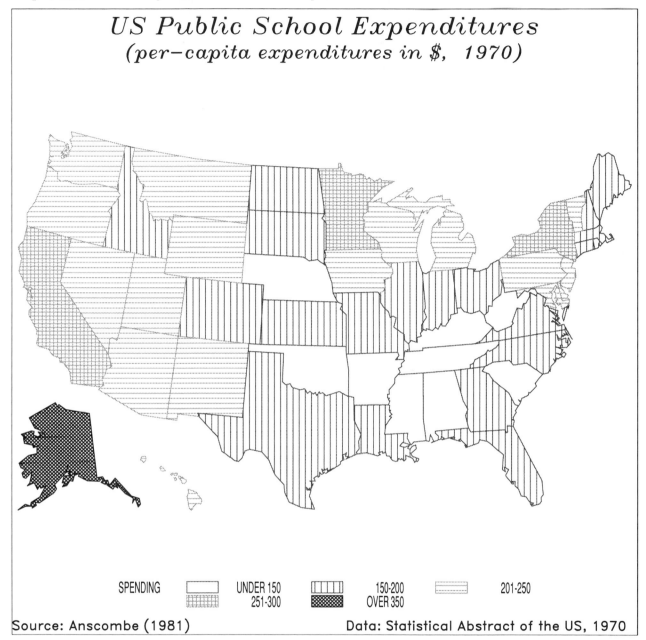

US Public School Expenditures
(per—capita expenditures in $, 1970)

SPENDING — UNDER 150   150-200   201-250
251-300   OVER 350

Source: Anscombe (1981)   Data: Statistical Abstract of the US, 1970

```
%include SPENDING;              /* On MVS, use %include ddname(file);*/
proc format;
   value spndfmt
   0-150  ='UNDER 150'
   150-200='150-200'
   201-250='201-250'
   251-300='251-300'
   301-350='301-350'
   351-999='OVER 350'
   ;
run;
title1 f=titalic      'US Public School Expenditures';
title2 f=titalic h=1.5 '(per-capita expenditures in $,  1970)';
footnote1 f=duplex j=l 'Source: Anscombe (1981)'
                   j=r 'Data: Statistical Abstract of the US, 1970  ';

proc gmap data=schools
        map=maps.us  ;
   format spending spndfmt9.;
   id state;
   choro spending / discrete name='GB0535';
   pattern1 v=e       c=red;
   pattern2 v=m2n90   c=red;
   pattern3 v=m3n180  c=green;
   pattern4 v=m4x90   c=green;
   pattern5 v=m5x45   c=black;
   pattern6 v=s       c=black;
run;
```

Next, PROC REG is used to fit a multiple regression predicting SPENDING from INCOME, YOUTH, and URBAN. The studentized residual (RSTUDENT) is output to a data set, RESIDS, to be displayed on the U.S. map.

```
proc reg;
   model spending = income youth urban ;
   id st;
   output out=resids r=residual rstudent=rstudent;
```

The studentized residual is first rounded to the nearest 0.5. The PROC FORMAT step below defines a format to group these values into class intervals centered about 0. The seven intervals correspond to the PATTERN statements below that use density of lines and crosshatching to denote the magnitude of the studentized residual, and direction and color to denote the sign of the residual. An empty fill pattern is used for residuals near zero.

```
data resids;
   set resids;
   rstudent=round(rstudent, 0.5);
```

```
proc format;
   value resfmt
   low - -2 =   'UNDER -2'
     -1.5   =  '  -1.5  '
     -1.0   =  '  -1.0  '
   -.5 - .5 =  '-.5 to .5'
      1.0   =  '  +1.0  '
      1.5   =  '  +1.5  '
    2 - high=  'OVER +2 ';

title1 f=titalic 'US Public School Expenditures';
title2 f=titalic h=1.5
                'Residuals from model SPENDING = INCOME YOUTH URBAN';
footnote;
proc gmap data=resids
        map=maps.us  ;
   id state;
   format  rstudent  resfmt9.;
   choro   rstudent  / discrete levels=7 name='GB0536';
   pattern1 v=m5x45  c=red;
   pattern2 v=m3x45  c=red;
   pattern3 v=m1n45  c=red;
   pattern4 v=e      c=green;
   pattern5 v=m1n90  c=black;
   pattern6 v=m3x90  c=black;
   pattern7 v=m5x90  c=black;
```

The residuals from the regression model indicate the extent to which actual school expenditure differs from what is predicted for states of comparable wealth, need, and urbanization. The residual map, shown in Output 5.36, indicates that two states, Connecticut and Nebraska, stand out with the largest negative residuals, while Oregon, California, and Minnesota stand out with large positive residuals. (A color version of this output appears as Output A3.29 in Appendix 3.)

In this figure, the negative residuals (less spending than predicted) appear in red, and the positive residuals in black, so it is easy to see which states have less or more spending than predicted. The use of empty patterns for residuals near zero helps to make the nonzero values stand out.

Output 5.36   *Residual Map for Public School Expenditures*

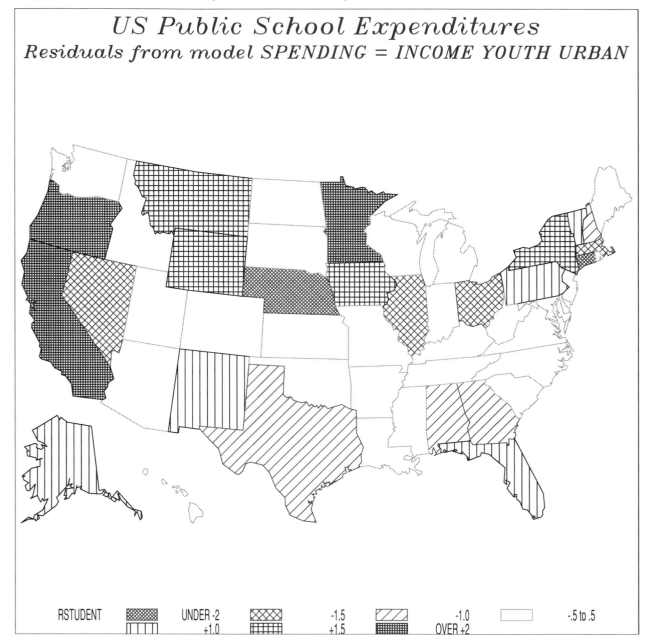

However, one difficulty with choropleth maps such as these is that the size of a region contributes to the visual impression of the variable being portrayed, so that in Output 5.36 the large negative value for Connecticut does not have as much impact as does that for Nebraska. The visual system seems to summate visual density over area. One way to do better is to dispense with shading the regions. Instead, we could annotate the map with rectangles or circles whose size is proportional to the variable being depicted, as shown in the following section. This way, the size of the state itself does not contribute to the visual impression of the response variable.

5.8.1 Bubble Map of Residuals

The program below displays the same studentized residuals in Output 5.37 with bubbles whose size (radius) is proportional to the residual for the state. (A color version of this output appears as Output A3.30 in Appendix 3.) Thus, the magnitude of the residual can be shown quantitatively, whereas in Output 5.36 the values must be grouped into class intervals. An alternative display favored by some geographers uses a framed rectangle to show magnitudes on a map. See Dunn (1987) for discussion of this technique.

Output 5.37   *Bubble Map of School Spending Residuals*

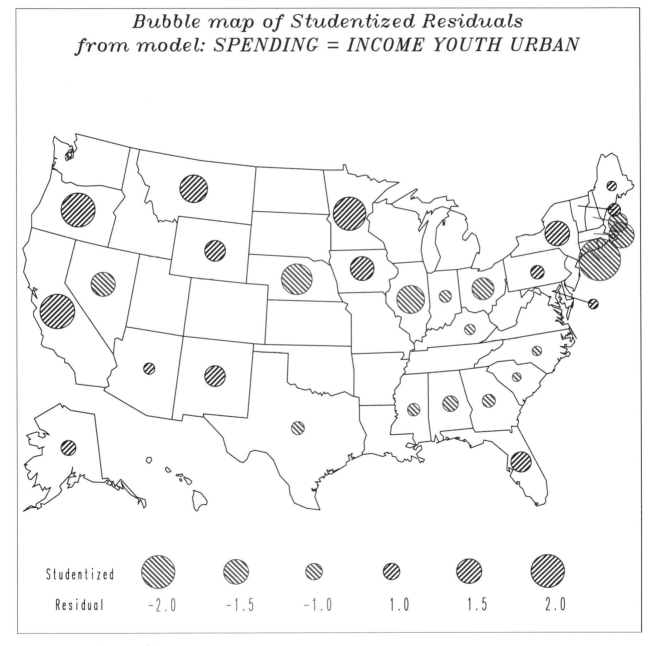

For maps of the U.S., another SAS/GRAPH software map data set, MAPS.USCENTER, gives the coordinates of the visual centers of each of the 50 states, which are used to locate the bubble symbols. In the program below, the first DATA step merges the RESIDS data set with the MAPS.USCENTER data set to get the *x, y* coordinates of the center of each state.

As in Output 5.36, only residuals greater than 0.5 in magnitude are shown. For each such value, a circle whose size is proportional to the absolute value of the residual is drawn. The color and direction of shading of the circle are determined by the sign of the residual. To avoid crowding of the symbols in the northeast and mid-Atlantic states, the MAPS.USCENTER data set contains special observations with a specification OCEAN='Y' that allow the annotation symbol to be located adjacent to the state, with a line drawn to the center.

A customized legend, showing the symbols for typical residuals, is constructed in a second DATA step and is then concatenated with the ANNOMAP data set. The DATA step reads the values of the residual for the legend from cards and uses similar logic to determine the SIZE, STYLE, and COLOR of each pie symbol. However, here we use the percentage coordinate systems (XSYS and YSYS='3') for the *x* and *y* coordinates, whereas the DATA step that produces the ANNOMAP data set uses data coordinates (XSYS and YSYS='2'). A FOOTNOTE statement is used with the PROC GMAP step to leave room at the bottom of the map for the legend.

We want all the states to be left empty, except for the bubble symbols supplied by the Annotate data set. The PROC GMAP step requests a choropleth map of STATE (any discrete variable would do), and the PATTERN statement requests an empty fill pattern for each value of the variable STATE.

```
/**----------------------------------------------------**
 | Annotate map with bubbles showing size of residual |
 **----------------------------------------------------**/
proc sort data=resids;
   by state;

data annomap;
   length color function style $8;
   merge maps.uscenter (drop=lat long)
         resids;
   by state;
   drop   flag;
   retain flag        0
          xsys ysys   '2'
          angle       0
          rotate      360
          when        'A';
```

```
   if abs(rstudent) < .5 then delete;
   function = 'PIE';
   size = .6 * abs(rstudent);
   if rstudent > 0
      then do;
         color = 'BLACK'; style = 'R3';
      end;
   else do;
         color = 'RED';   style = 'L3';
      end;
   if flag=1 then do;
      function = 'DRAW';
      size = 1;
      flag = 0;
      end;
   if ocean='Y' then do;
      output;
      function = 'MOVE';
      flag = 1;
      end;
   output;
run;
 /*-------------------------------------------*
  | Annotate map with bubble legend at bottom |
  *-------------------------------------------*/
data legend;
   length function color $8 text $12;
   retain xsys ysys '3'
          when 'B'
          x    y;
   input rstudent @@;
   if _n_=1 then do;
      x = 10;
      y = 10;
      function='LABEL';
      text='Studentized'; output;
      y =  5;
      text='Residual    '; output;
      end;
   y = 10;
   x = x + 12.5;
   angle=0; rotate=360;
   function = 'PIE';
   size = .6 * abs(rstudent);
   if rstudent > 0
      then do;
         color = 'BLACK'; style = 'R3';
      end;
   else do;
         color = 'RED';   style = 'L3';
      end;
```

```
      output;
      y =  5; text=put(rstudent,5.1);
      style=' '; size=1.2;
      function='LABEL';
      output;
   cards;
   -2.  -1.5 -1.  1.  1.5 2.
   ;
   data annomap;
      set annomap legend;
   run;
   title1 f=titalic h=1.5 'Bubble map of Studentized Residuals' ;
   title2 f=titalic h=1.5 'from model: SPENDING = INCOME YOUTH URBAN';
   footnote h=3.5 ' ';  *-- space for legend;
   proc gmap data=maps.us
             map =maps.us  ;
      id      state;
      choro   state   / annotate=annomap
                        coutline=black
                        nolegend name='GB0537';
      pattern v=empty r=66;
```

Part 4

Graphical Comparisons and Experimental Design Data

284

Chapter **6** Comparing Groups

6.1 Introduction

When data are available from two or more groups, our interest is often concentrated on comparing the mean or central value in each group. As illustrated in Section 1.2, "Advantages of Plotting Data," however, groups with the same mean value can differ widely in other properties, so it is usually useful to begin by looking at the data as a whole.

Chapter 7, "Plotting ANOVA Data," deals with graphical techniques for comparing means in designed experiments. In this chapter, the focus is on comparing the complete distributions for several groups to see how they differ in terms of central tendency, spread, and shape of distribution.

The quantile comparison plot (Section 6.2, "Quantile Comparison Plots for Pairs of Distributions") can be used to show how the distribution of two groups differs in these properties. The comparative boxplot (Section 6.3, "Comparative Boxplots") can also show these differences, but can be used for any number of groups. In a notched boxplot, the box for the central portion of each group is drawn with a notch to show graphical confidence intervals for differences in central value. A general SAS macro for constructing notched boxplots is explained and illustrated in Section 6.3. Finally, Section 6.4, "Diagnostic Plot for Equalizing Variability," describes the spread-level plot, a graphical technique for choosing a transformation of data to equalize variability.

6.2 Quantile Comparison Plots for Pairs of Distributions

In comparing groups, our interest is sometimes focused on particular aspects of the data, as when we want to compare means or measures of variability. For such cases, the boxplot provides an effective graphical display. Sometimes, however, we want to compare two (or more) data distributions as a whole. Here, empirical quantile comparison plots can be quite useful.

The quantile comparison plot (also called an empirical quantile-quantile plot; see Chambers et al. 1983, pp. 48—57) is based on the same ideas as the theoretical quantile-quantile plot (Section 3.5, "Quantile Plots"). The plot is simplest to understand and to construct when both samples have the same number (n, for example) of observations.* Let the sorted observations in two groups be denoted

$$x_{(1)}, x_{(2)}, \ldots, x_{(n)} \quad \text{and} \quad y_{(1)}, y_{(2)}, \ldots, y_{(n)} \quad .$$

Then the quantile comparison plot is a plot of $y_{(i)}$ against $x_{(i)}$. The plot compares the shape, level, and spread of the distributions as follows:

□   If the shapes are the same, the plot of $y_{(i)}$ against $x_{(i)}$ will be linear.

□   If the spreads of the distributions are the same, the plot will have a slope≈ 1.

□   If the means and spreads are the same, the plot will have a slope≈ 1 and intercept≈ 0.

In general, the number of observations in the two groups will differ. Suppose there are n_x x scores and n_y y scores, and (without loss of generality) $n_y < n_x$. Then for each $y_{(i)}$, which is the $(i-.5)/n_y$ quantile of the y data, a corresponding interpolated x quantile, $x_{(i)}^*$, is found, and we plot $y_{(i)}$ against $x_{(i)}^*$, $i=1, \ldots, n_y$. The interpolated quantiles are found in the following steps:

1.   Calculate the proportion, p_i, of observations below each $y_{(i)}$: $p_i = (i-1/2)/n_y$.

2.   The corresponding interpolated x^* has the subscript $v = p_i n_x + 1/2$. But v probably is not an integer, so let $k=$ the integer part of v and $f=$ the fractional part of v.

3.   Then the interpolated quantile is the number that falls a fraction, f, of the way from $x_{(k)}$ to $x_{(k+1)}$:

$$x_{(i)}^* = (1 - f)\, x_{(k)} + (f)\, x_{(k+1)} \quad .$$

The calculations for a set of six y values and seven x values is shown numerically in Table 6.1 and graphically in Output 6.1. The values of y on the ordinate and of x on the abscissa are spaced equally, according to the fractions, p_i. The interpolation process corresponds to finding the values of x with the same fractions on the scale of x. The value $y_{(5)} = 10$, for example, has a fraction $p = .75$ below it. The corresponding observation in x has the subscript $v = 5.75$, which is interpolated as .75 of the way from $x_{(5)}$ to $x_{(6)}$, giving $x_{(5)}^* = 11.25$. In Output 6.1, the line leaving $y = 10$ descends to $x^* = 11.25$.

*   This section borrows from Fox (1988).

Table 6.1
Calculating Interpolated Quantiles

i	y	p	v		x*	x
			k	f		
1	2	.0833	1	.0833	1.25	1
2	4	.2500	2	.2500	4.25	4
3	5	.4167	3	.4167	5.83	5
4	8	.5833	4	.5833	8.16	7
5	10	.7500	5	.7500	11.25	9
6	13	.9167	6	.9167	14.75	12
7						15

Output 6.1
Finding Interpolated Quantiles

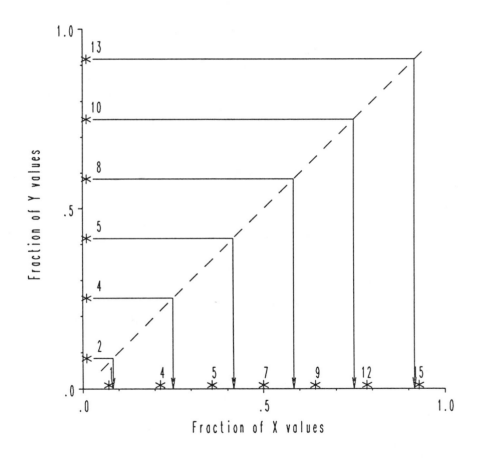

Identical distributions appear as points along a 45 degree line, $y_{(i)} = x_{(i)}{}^*$, so it is common to draw this line as a visual reference in the plot and force equal horizontal and vertical scales. Alternatively, a *sum-difference plot* makes it easier to judge departures from the line $y=x$. Here we plot the difference of the quantiles, $y_{(i)} - x_{(i)}^*$, against their sum, $y_{(i)} + x_{(i)}^*$, without forcing the scales to be the same. In effect, this rotates the plot 45 degrees and expands the vertical (difference) scale.

6.2.1   Example: Infant Mortality Data

The program below constructs a quantile comparison plot to compare (log) infant mortality rates in North and South America with those in Africa. It uses the IML procedure to calculate the interpolated quantiles (XQ) as follows:

```
p = ((1:ny)` - (.5) ) / ny;      *-- proportion below each y(i);
v = (nx * p) + (.5);             *-- subscript of ordered x(i);
k = int(v) ;                     *--    integer part;
f = v - k;                       *--    fractional part;
xq = (1-f) # x[k] + f # x[k+1];  *-- interpolated quantiles ;
```

The program reads the Nations data and calculates $\log_{10}$ (IMR) for those countries with nonmissing values. The values for the Americas (REGION=1) and Africa (REGION=2) are read into IML vectors YY and XX, and each vector is sorted. The IML module QUANTILE performs the calculations outlined above and returns the results in the vectors Y and X. An output data set QQ is created, containing the variables X, Y, SUM, and DIFF for plotting in a subsequent PROC GPLOT step.

```
Title 'Quantile comparison plots';
%include NATIONS;                /* On MVS, use %include ddname(file);*/
data nations;
     set nations;
     if imr=. then delete;
     limr = log10(imr);

proc iml;
   use nations;
   read all var (limr) where (region=1 & limr¬=.) into yy;
   read all var (limr) where (region=2 & limr¬=.) into xx;
   close nations;

   *-- Sort X, Y;
   t = xx; xx[ rank(xx),] = t;
   t = yy; yy[ rank(yy),] = t;

start quantile(x,y);           *-- x, y assumed sorted;
   nx = nrow(x);
   ny = nrow(y);
   swap=0;
   if nx < ny then do;         *-- swap x,y so ny <= nx;
     swap=1;
     t = x;  x = y;  y = t;
     nt =nx; nx =ny; ny =nt;
     end;
```

```
     p = ((1:ny) - (.5) ) / ny;          *-- proportion below each y(i);
     v = (nx * p + (.5))`;               *-- subscript of ordered x(i);
     k = int(v) ;                        *--   integer part;
     f = v - k;                          *--   fractional part;
     xq = (1-f) # x[k] + f # x[k+1];     *-- interpolated quantiles ;

     if swap=1 then do;        *-- swap back;
        t = xq; x = y; y = t;
       end;
     else x = xq;
 finish;
 x = xx;
 y = yy;
 run quantile(x,y);

 *---Output to dataset---;
   names={ x  y sum diff};
   xy = x || y || (x+y) || (y-x) ;
   print 'Interpolated Quantiles',  xy [colname = names] /;
   create qq from xy [colname = names];
   append from xy;
```

The next DATA step creates an Annotate data set to draw a dashed 45-degree reference line on the plot in the PROC GPLOT step. The *x* and *y* values for the ends of the line (1—2.5) are data dependent and correspond to the range for the axes in the AXIS statements.

```
 data annoline;
    xsys='2';
    ysys='2';
    line=3;
    input function $ x y;
 cards;
 MOVE  1.0 1.0
 DRAW  2.5 2.5
 ;
 title h=1.5 'Quantile Comparison Plot: Log10 Infant Mortality Rates';
 symbol v='+' h=2 c=black;
 proc gplot data=qq  ;
    plot y * x / vaxis=axis1
                 haxis=axis2
                 frame
                 annotate=annoline
                 name='GB0602' ;
    axis1 order=(1 to 2.5 by .5)
          label=(a=90 r=0 f=duplex h=1.5) value=(h=1.4);
    axis2 order=(1 to 2.5 by .5)
          label=(f=duplex h=1.5) value=(h=1.4);
    label y = 'Americas'
          x = 'Africa';
 run;
```

```
title h=1.5 'Sum-Difference Plot: Log10 Infant Mortality Rates';
proc gplot data=qq  ;
   plot diff * sum / frame
                 vaxis=axis1 vminor=1
                 haxis=axis2 hminor=1
                 name='GB0603' ;
   axis1 order=(-.6 to -.2 by .1)
         label=(a=90 r=0 f=duplex h=1.5) value=(h=1.4);
   axis2 order=(2 to 5 by .5)
         label=(f=duplex h=1.5) value=(h=1.4);
   label diff= 'Americas - Africa'
         sum = 'Americas + Africa';
```

The quantile comparison plot for the infant mortality data is shown in
Output 6.2. Most of the points lie along a line with slope slightly >1 and well
below the 45-degree line, indicating that rates in Africa are much higher and
slightly more spread out. The sum—difference plot shown in Output 6.3 would be
flat if the two distributions were equal in variability. The pronounced upward tilt
of the points emphasizes that the slope is greater than 1. The point in the upper
right (Libya, Brazil) stands out as quite different from the rest, indicating that
Libya is more extreme relative to the African nations than Brazil is to the
Americas.

Output 6.2
*Quantile
Comparison Plot
for Infant
Mortality Data*

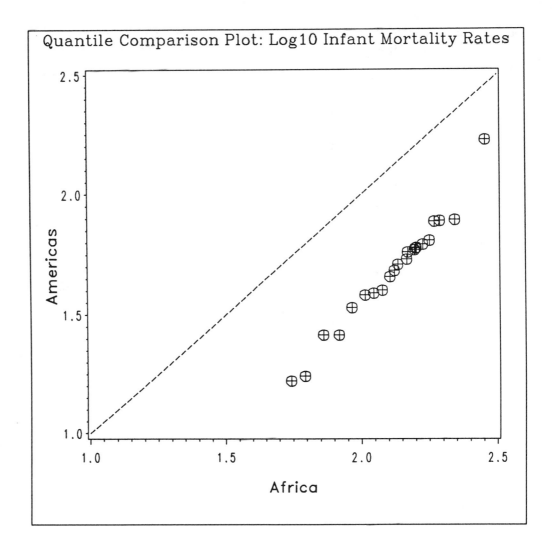

Output 6.3
Sum—Difference
Plot for Infant
Mortality Data

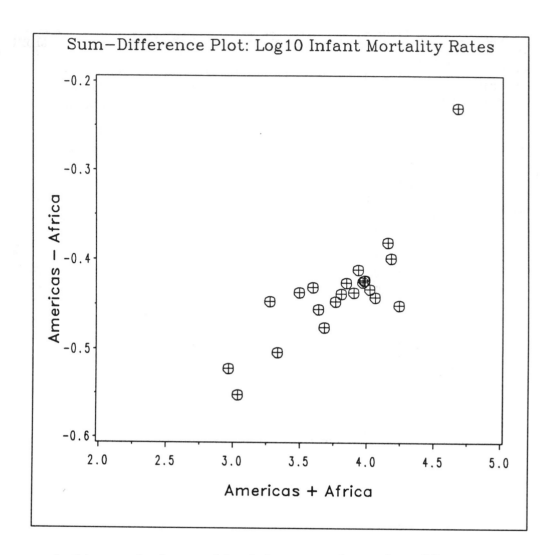

In this example, the sets of data being compared come from different observations on a single variable. The QUANTILE module can also be used to compare the distributions of observations in separate variables. For example, the NATIONS data set also contains infant mortality rates from 1980. To see how the distribution of 1970 values changed by 1980, read the 1970 data into the vector YY and the 1980 data into the vector XX, as shown below:

```
%include NATIONS;              /* On MVS, use %include ddname(file);*/
data nations;
   set nations (rename=(imr=imr70));
   if imr70=. and imr80=. then delete;

Proc IML;
   use nations;
   read all var {imr70} where (imr70¬=.) into yy;
   read all var {imr80} where (imr80¬=.) into xx;
   close nations;
   xx = log10(xx);
   yy = log10(yy);
   ...
```

The remainder of the program is the same as before. The quantile comparison plot, shown in Output 6.4, shows that the distributions are very nearly the same in 1970 and 1980, with the exception that the largest values for 1970 have become smaller in 1980. Two equal distributions would plot along the dashed line. The slope of the points (shown by the solid line) is slightly less than 1, indicating that the variance of the values is slightly smaller in 1980.

Output 6.4
1970 Plotted against 1980 Log Infant Mortality Rates

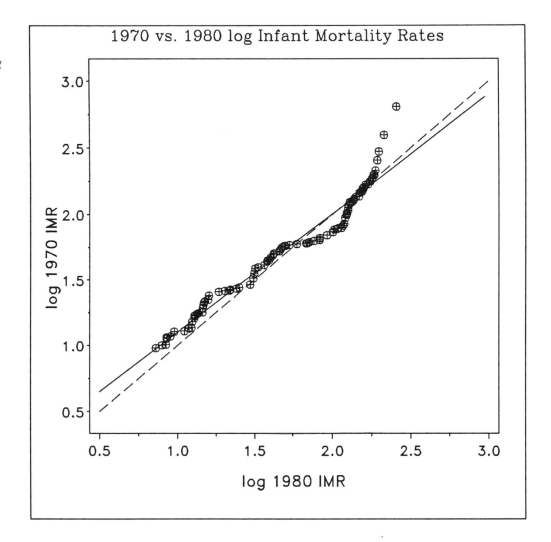

6.3 Comparative Boxplots

In many data analysis situations, we have two or more groups of observations and we want to compare the distributions of scores in the groups. Statistical procedures such as analysis of variance (see Chapter 7) are designed to determine if the means of the distributions are the same; but we would often like to know more: how do the groups compare in variability? do they have the same shape? are there unusual observations in some groups? Side-by-side boxplots help to answer these questions.

The boxplots produced by the UNIVARIATE procedure (Section 2.4, "Boxplots") are useful for some of these group comparisons, and this procedure is

certainly easy to use. For some purposes, however, a high-resolution version of
the boxplot is preferable. In Version 6 of the SAS System, you can produce
high-resolution boxplots most easily by specifying I=BOX in the SYMBOL
statement. With SAS/QC software, the BOXCHART statement in the SHEWHART
procedure draws boxplots tailored to quality control applications.

For comparing groups, a variation of the boxplot called the *notched boxplot*
(McGill, Tukey, and Larsen 1978) shows an approximate 95% confidence interval
for the location of the population median by notches in the sides of the box. Two
groups whose notches do not overlap can be considered to differ significantly in
their central values. The notches are calculated as

$$\text{Median} \pm 1.58 \left(\frac{IQR}{\sqrt{n}} \right)$$

to give approximate 95% comparison intervals for the median (for more details,
see Velleman and Hoaglin 1981). This section explains how to produce simple
boxplots with PROC GPLOT and a variety of boxplots, including notched boxplots,
with a general BOXPLOT macro.*

6.3.1 Boxplots with the GPLOT Procedure

Simple boxplots can be drawn using PROC GPLOT in Version 6 by specifying
I=BOX in the SYMBOL statement.** Output 6.5 shows boxplots of infant
mortality rates by region for the nations data. This plot is drawn by the
statements below.

Specifying I=BOXT draws horizontal lines at the tips of the whisker lines. By
default, the whisker lines extend ±1.5 IQR beyond the quartiles. Note that
MODE=INCLUDE is also specified in the SYMBOL statement so that observations
beyond the plot range, if any, are included in the boxplot calculations. (In this
example, the vertical plot range is not restricted, so specifying MODE=INCLUDE
is not actually necessary.) See Chapter 16, "The SYMBOL Statement," in
SAS/GRAPH Software: Reference, Version 6, Volume 1 for more details on
specifying I=BOX and MODE=INCLUDE in the SYMBOL statement.

```
%include NATIONS;              /* On MVS, use %include ddname(file);*/
title h=1.5 'Boxplot with I=BOX';
symbol i=boxt mode=include          /* V6.06 */
       v=+ color=black h=1.2;
proc gplot data=nations  ;
   plot imr * region
        / frame name='GB0605'
          vaxis=axis1 haxis=axis2
          vminor=1 hminor=0;
```

---

*　In Release 6.06 of SAS/QC software, the NOTCHES option in the BOXCHART statement of PROC
SHEWHART also produces notched boxplots.

**　The specification I=BOX is not available in releases prior to Release 6.06 of the SAS System.

```
axis1 label=(h=1.5 a=90 r=0)
      value=(h=1.3);
axis2 label=(h=1.5)
      value=(h=1.3)
      offset=(8 pct);
```

Output 6.5
Simple Boxplot
with I=BOX

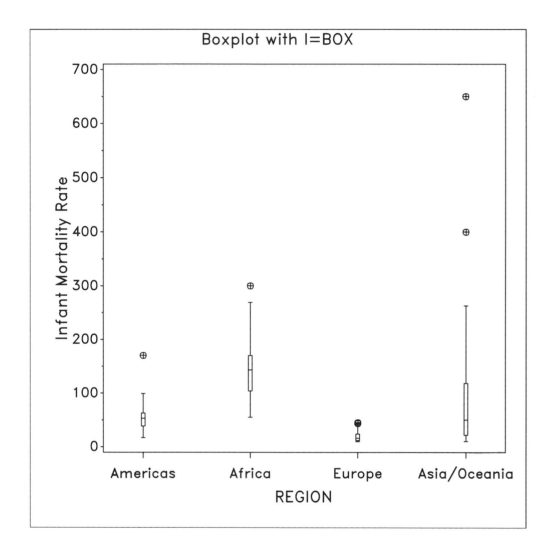

6.3.2 BOXPLOT Macro

For purposes of exploratory data analysis, a SAS macro program, BOXPLOT, described below, provides greater flexibility. The BOXPLOT macro provides the following features:

□ one or more CLASS variables, which may be either character or numeric.

□ optional lines drawn connecting the group medians.

□ optional notches showing approximate 95% confidence intervals for difference in medians.

□ outside observations labeled on the graph with the value of an ID variable.

□ controlled box width. Box widths can be constant for all groups or made proportional to a function of the sample size in each group.

□ user-supplied annotations added to the graph.

□ an output data set containing boxplot statistics that can be used for other graphs.

The main features of the BOXPLOT macro are illustrated below. See Section A1.4, "The BOXPLOT Macro," for a more complete description of parameters and the program listing.

6.3.3   Using the BOXPLOT Macro

The header for the macro below lists the parameters. Only the CLASS and VAR parameters must be supplied; if no data set is given, it uses the most recently created data set. All the other parameters have default values as shown after the equal signs. The ID parameter can supply a character variable that identifies each observation. If an ID variable is specified, outside observations (beyond the adjacent values) are labeled on the graph.

```
                        /* Description of Parameters:     */
%macro BOXPLOT(         /* ----------------------------   */
        data=_LAST_,    /* Input dataset                  */
        class=,         /* Grouping variable(s)           */
        var=,           /* Ordinate variable              */
        id=,            /* Observation ID variable        */
        width=.5,       /* Box width as proportion of maximum */
        notch=0,        /* =0|1, 1=draw notched boxes     */
        connect=0,      /* =0 or line style to connect medians*/
        f=0.5,          /* Notch depth, fraction of halfwidth */
        fn=1,           /* Box width proportional to &FN  */
        varfmt=,        /* Format for ordinate variable   */
        classfmt=,      /* Format for class variable(s)   */
        varlab=,        /* Label for ordinate variable    */
        classlab=,      /* Label for class variable(s)    */
        yorder=,        /* Tick marks, range for ordinate */
        anno=,          /* Addition to Annotate set       */
        out=boxstat,    /* Output data set: quartiles, etc. */
        name=BOXPLOT    /* Name for graphic catalog entry */
        );
```

The example below shows how the BOXPLOT macro is used. The data set TESTBOX consists of six groups that can be described by a single numeric class variable (GROUP), a single character CLASS variable (GR), or two CLASS variables (A and B). The SAS formats GRP. and AB., defined below, are used as values of the CLASSFMT parameter to provide labels for the groups on the horizontal axis. The variable NAME serves as an observation ID variable.

```
%include boxplot;              /* On MVS, use %include ddname(file);*/
 /*-----------------------*
  | Test data for BOXPLOT |
  *-----------------------*/
proc format;
              /* Format for 1 class variable, CHAR or NUM */
     value grp  1='Group A'  2='Group B'  3='Group C'
                4='Group D'  5='Group E'  6='Group F';
              /* Format for 2 class variables            */
     value ab   1='Grp A-1'  2='Grp A-2'  3='Grp A-3'
                4='Grp B-1'  5='Grp B-2'  6='Grp B-3';

data testbox;
   drop n obs;
   group=0;
   do a = 'A', 'B';
      do b = 1 to 3;
         group + 1;
         gr = put(group,grp.);
         if group <4 then n=10;
                     else n=20;
         do obs=1 to n;
            name = put(obs,2.);
            y = group + normal(156863);
            if obs <4 then y = y + 3*normal(156863);  /* add outliers */
            y = round(y, .1);
            output;
         end;
      end;
   end;
```

Output 6.6 shows a connected boxplot produced by the first %BOXPLOT statement below. The numeric CLASS variable GROUP is labeled using the GRP. format specified in the CLASSFMT parameter. Group medians are connected by a dashed line specified by CONNECT=20. The width of each box is proportional to the square root of the number of observations in that group. This choice is based on the fact that standard errors are inversely proportional to $\sqrt{n}$. (Groups D, E, and F have twice the number of observations as Groups A, B, and C.)

Output 6.6
Connected Boxplot of TESTBOX Data

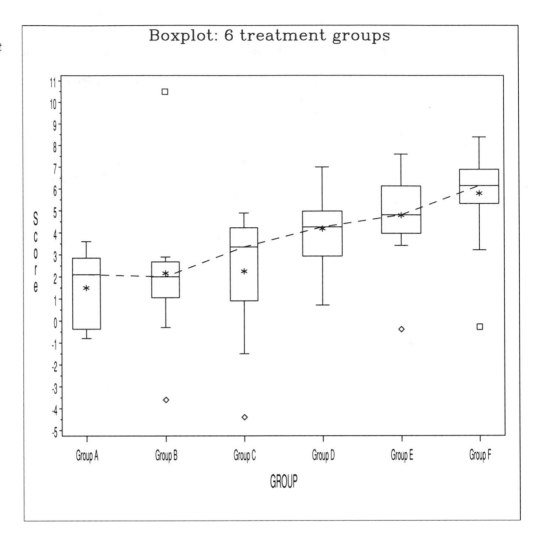

```
title h=1.5 'Boxplot: 6 treatment groups';
%boxplot(data=testbox,
      class=GROUP,                  /* Class variable (numeric)    */
      var=y,
      connect=20,
      fn=sqrt(n),                   /* Box width proportional to FN */
      classfmt=grp.,                /* Format for CLASS variable    */
      varlab=Score,
      name=GB0606);
run;
```

Output 6.7
Notched Boxplot
of TESTBOX Data

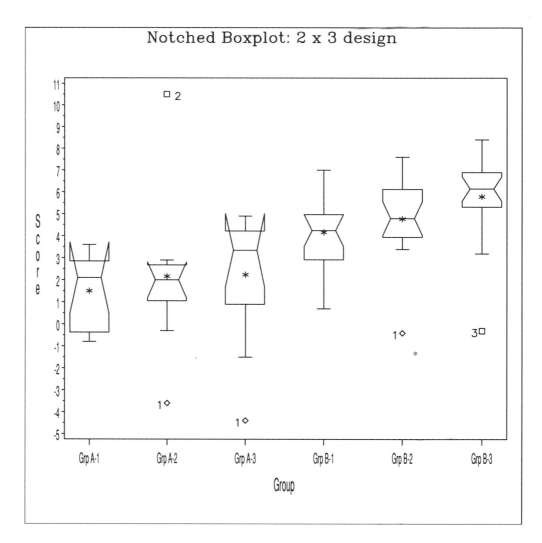

The next %BOXPLOT statement draws a notched boxplot shown in Output 6.7. Here the groups are identified by the two CLASS variables, A and B, which are labeled using the AB. format. The parameter ID=NAME causes the outside points to be labeled with the value of the NAME variable.

```
/*----------------------------------------------------*
| Test with 2 class variables, 1 CHAR, 1 NUMERIC |
*----------------------------------------------------*/
title h=1.5 'Notched Boxplot: 2 x 3 design';
%boxplot(data=testbox,
    class= a b,
    var=y,
    id=name,
    notch=1,                      /* 1=do notched boxplots        */
    f=0.4,                        /* Notch depth,  0<= F <= 1     */
    connect=0,                    /* do not connect medians       */
    classfmt=ab.,                 /* Format for class variables   */
    varlab=Score,
    classlab=Group,
    name=GB0607);
```

Note that in the notched boxplot, it is possible for the notches to extend beyond the quartiles. This is the case for groups A-1, A-2, and A-3 in Output 6.7, which makes the central boxes look as if they were folded over.

6.3.4 How the BOXPLOT Macro Works

The BOXPLOT macro is included in Section A1.4. While a complete description of the program is beyond the scope of this book, the techniques used for drawing boxplots are quite useful for those wanting to develop their own displays with SAS/GRAPH software. For other approaches to drawing boxplots, see Benoit (1985a) and Olmstead (1985).

The main idea in the program is to compose the boxplot from two parts:

1. the outside points, which are plotted individually, with a larger symbol for far-out points.

2. the inside points (between the adjacent values) that are not plotted. Instead, an Annotate data set (DOTS) is constructed to draw the outline of the box and whiskers, possibly with notched confidence intervals.

Thus, each observation is given a value of a variable, OUTSIDE, with the following values:

1 = inside observation

2 = outside observation

3 = far-out observation

If you want to draw lines connecting the medians of groups, this is done by specifying OUTSIDE=4. The plot is constructed with the following SYMBOL statements:

```
symbol1 h=1.  c=black v=none;           /* inside points */
symbol2 h=1.  c=black v=diamond;        /* outside points */
symbol3 h=1.5 c=black v=square;         /* farout points  */
symbol4 l=3   c=black v=none i=join;    /* connected mdns */
```

The following PROC GPLOT statements are also used:

```
proc gplot data=plotdat;
  plot yvar * class =outside
       / frame nolegend annotate=dots;
```

Thus, the inside points are not plotted, the outside points are plotted with different symbols, and the connected medians are drawn with joined lines but no points. (The BOXPLOT macro uses a different method to assign symbols, to allow for the fact that outside points may not occur in the data.)

The Annotate data set, DOTS, is constructed as follows:

1. For each group, the following summary values for *y* are found (using PROC UNIVARIATE and a DATA step):

   □ mean and median

   □ upper and lower quartiles (Q25, Q75) and interquartile range (IQR)

   □ upper and lower adjacent values (HI_WHISK, LO_WHISK)

   □ upper and lower notch values (HI_NOTCH, LO_NOTCH).

2. The horizontal dimensions of the box for each group are found from the following:

   □ *x* coordinate of the center of the box (CLASS value)

   □ half width of the box (WIDTH and FN)

   □ depth of the notch as a fraction of the box half width (F).

3. These values are used to form 18 *x,y* pairs around the perimeter of the box, starting at the lower whisker value. These points are numbered as shown in Figure 6.1.

Figure 6.1
Location of 18
Points Used to
Draw a Notched
Boxplot

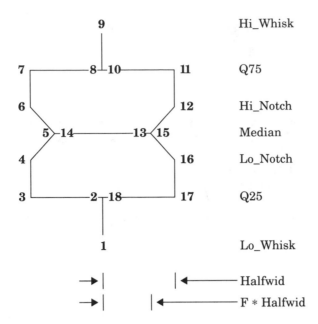

4. These 18 points are calculated by the following SAS programming statements:

```
/*     Produce connect-the-dots X, Y pairs */
X = CLASS                    ; Y= Lo_Whisk ; Dot =   1; Link Out;
X = CLASS                    ; Y= Q1       ; Dot =   2; Link Out;
X = CLASS - Halfwid          ; Y= Q1       ; Dot =   3; Link Out;
X = CLASS - Halfwid          ; Y= Lo_Notch ; Dot =   4; Link Out;
X = CLASS - (1- F)*Halfwid   ; Y= Median   ; Dot =   5; Link Out;
X = CLASS - Halfwid          ; Y= Hi_Notch ; Dot =   6; Link Out;
X = CLASS - Halfwid          ; Y= Q3       ; Dot =   7; Link Out;
X = CLASS                    ; Y= Q3       ; Dot =   8; Link Out;
X = CLASS                    ; Y= Hi_Whisk ; Dot =   9; Link Out;
X = CLASS                    ; Y= Q3       ; Dot =  10; Link Out;
X = CLASS + Halfwid          ; Y= Q3       ; Dot =  11; Link Out;
X = CLASS + Halfwid          ; Y= Hi_Notch ; Dot =  12; Link Out;
X = CLASS + (1 - F)*Halfwid  ; Y= Median   ; Dot =  13; Link Out;
X = CLASS - (1 - F)*Halfwid  ; Y= Median   ; Dot =  14; Link Out;
X = CLASS + (1 - F)*Halfwid  ; Y= Median   ; Dot =  15; Link Out;
X = CLASS + Halfwid          ; Y= Lo_Notch ; Dot =  16; Link Out;
X = CLASS + Halfwid          ; Y= Q1       ; Dot =  17; Link Out;
X = CLASS                    ; Y= Q1       ; Dot =  18; Link Out;
```

5. The program uses the POLY and POLYCONT functions of the Annotate facility to draw the polygon connecting the DOT values. The POLY function initiates drawing an arbitrary polygon defined by a set of Annotate observations with variables X and Y; the POLYCONT function draws a line from the previous point to the current *x, y* value.*

```
out:
    select;
        when ( dot=1 ) do;
            function = 'MOVE';                    output;
            function = 'POLY';                    output;
            end;
        when ( 1< dot <=18) do;
            function = 'POLYCONT';                output;
            end;
            ...
        otherwise ;
    end;
    return;
```

Example: Infant Mortality Rates

The example below uses the BOXPLOT macro to examine infant mortality rates, classified by regions of the world. Because the regions differ in size (number of countries), we make the size of the box proportional to $\sqrt{n}$ with the FN= option. REGION is a numeric variable, and a user-defined format, REGION., is used to supply descriptive labels on the abscissa. The resulting plot is shown in Output 6.10.

```
%include NATIONS;           /* On MVS, use %include ddname(file);*/
%include boxplot;           /* On MVS, use %include ddname(file);*/
title1 h=1.5 'Infant Mortality Rate by Region';
%boxplot(data=nations, class=region, var=imr,
    id=nation,
    notch=1,
    width=0.4,
    f=0.4,
    fn=sqrt(n),
    classfmt=region.,
    varlab=Infant Mortality Rate,
    classlab=Region,
    name=GB0608);
```

---

* The whiskers are usually drawn with dashed lines and the rest of the box with solid lines. One drawback to using the POLY and POLYCONT functions this way is that the line style must be the same for all segments of the polygon. The BOXAXIS macro, described in Section 4.5, "Enhanced Scatterplots," uses the DRAW Annotate function to draw the box and whisker lines with solid lines for the box and dashed lines for the whiskers.

Output 6.8
*Boxplots for Infant
Mortality Rate, by
Region*

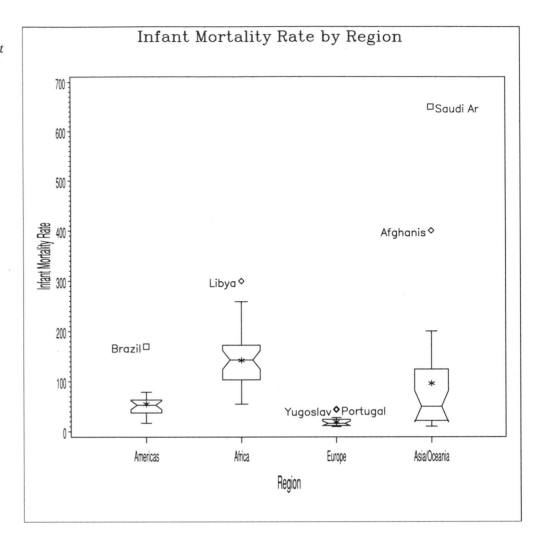

The median infant mortality rates are greatest in Africa, next in the Americas, followed by Asia/Oceania, and lowest in Europe. The notches indicate that Africa differs from all other regions and that Europe and the Americas differ from each other at an approximate 95% level of confidence. The two large values in Asia/Oceania, however, have the effect of compressing most of the data to the bottom third of the plot, making visual comparisons difficult. This problem is due, in part, to the fact that variability of each region increases with the typical value. The following section explores transformations of the data to remove this relationship between spread and level.

Output 6.9 shows the boxplot for $\log_{10}$ (IMR). The log transformation pulls in the highest values, removing some of the apparent outliers. This also makes it easier to compare the regional groups.

Output 6.9
Boxplots for Log Infant Mortality Rate, by Region

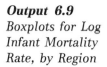

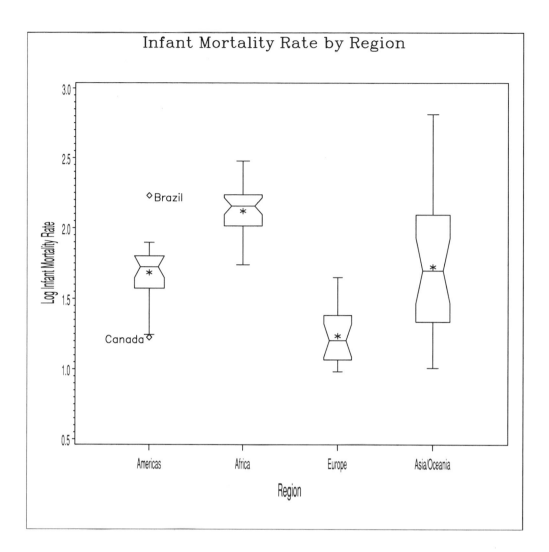

6.4 Diagnostic Plot for Equalizing Variability

In comparing groups, we are most often interested in determining whether and how they differ in central tendency—means or medians. If the groups differ in spread, this can make it harder to compare their means. Moreover, the analysis of variance assumes that each group has the same variability. For these reasons, it is often useful to try to transform the data to make the spreads more nearly equal.*

Section 3.6.4, "Finding Transformations for Symmetry," describes the use of the mid-spread plot to choose a transformation from the ladder of powers that makes a distribution approximately symmetric. The plot is constructed so that, if the points in the plot are linear with slope=b, then $p=1-b$ is the indicated power for the transformation.

In the case of equalizing spreads, there is a plot with the same slope—power relationship, called the *spread-level plot* (Emerson and Stoto 1982). Here we plot a

---

*   Alternatively, the weighted least squares method provides a way to incorporate heterogeneity of variance into the model. See Freund and Littell (1991).

measure of spread against a measure of central tendency, or level. Standard deviation against mean is one possibility, but using H-spread (IQR) against median offers resistance to outliers. Suppose that the measure of spread is proportional to a power of the measure of level:

$$\text{spread} = c(\text{level})^b \quad .$$

Then, taking logs of both sides,

$$\log(\text{spread}) = \log(c) + b \log(\text{level}) = a + b \log(\text{level})$$

where $a = \log(c)$. Thus, the plot of log(spread) against log(level) will be linear with slope b when spread varies as some power of level. The argument showing that $p = 1 - b$ is the appropriate power in the ladder of powers to remove this dependence is given by Leinhardt and Wasserman (1979) and Emerson and Stoto (1982).

Specifically, in the spread-level plot we plot

$$\log_{10}(\text{IQR}_i) \qquad \text{vs.} \qquad \log_{10}(\text{median}_i)$$

for each group, $i = 1, 2, \ldots$ and try to fit a line to the configuration of points. The plot is interpreted as follows:

- □ If the relationship is approximately linear, $p = 1 - b$ gives the power for a transformation $y \rightarrow y^p$ that equalizes spread. Again, a power $p = 0$ is taken as $\log(y)$ on the ladder of powers.

- □ If there is no systematic relation between spread and level, or if the relation is markedly nonlinear, then no power transformation will help to equate spread. Equivalently, a slope $b = 0$ in the spread-level plot corresponds to a power $p = 1$, and y^1 is the original variable.

6.4.1 Example: Infant Mortality Rates

The example below shows the calculation of log(IQR) and log(median) for the infant mortality rates in the data set NATIONS. PROC UNIVARIATE is used to obtain the values of the median and IQR for each group. The OUTPUT statement sends these results to an output data set, SUMRY, from which the log values are calculated.

```
%include NATIONS;              /* On MVS, use %include ddname(file);*/

proc sort data=nations;
   by region;
proc univariate data=nations noprint;
   by region;
   var imr;
   output out=sumry median=median qrange=iqr  n=n;
```

```
            data sumry;
               set sumry;
               logmed =  log10(median);
               logiqr = log10(iqr);
               label logmed='log Median IMR'
                     logiqr='log IQR of IMR';
            proc print data=sumry;
            proc gplot data=sumry  ;
               plot logiqr * logmed
                     / vaxis=axis1 haxis=axis2
                       hminor=1 vminor=1
                       name='GB0611' ;
               symbol1 v=+ h=1.7 i=rl c=black;
               axis1 label=(h=1.4 a=90 r=0) value=(h=1.4)
                     order=(1 to 2.2 by .2);
               axis2 label=(h=1.4)           value=(h=1.4)
                     order=(1 to 2.2 by .2);
               title h=1.5 'Spread - Level plot';

            * Use proc reg to find the slope;
            proc reg data=sumry;
               model logiqr = logmed;

            * Slope = .77 indicates that log(imr) will equalize spread;
            data nations;
               set nations;
               logimr = log10(imr);

            proc univariate plot data=nations;
               by  region;
               var logimr;
```

Output 6.10 shows the SUMRY data set with the log median and log IQR values. The plot of LOGIQR against LOGMED is used to determine the transformation to equalize variability.

Output 6.10
Output from the UNIVARIATE Procedure for Infant Mortality Rates

```
OBS   REGION        N    MEDIAN     IQR     LOGMED    LOGIQR

 1    Americas     22     53.05    25.900   1.72469   1.41330
 2    Africa       34    143.15    69.000   2.15579   1.83885
 3    Europe       18     16.05    12.525   1.20548   1.09778
 4    Asia/Oceania 27     50.00   102.600   1.69897   2.01115
```

Output 6.11 shows the spread-level plot produced by the PROC GPLOT step, with the least squares line fit by specifying I=RL in the SYMBOL statement. The configuration of points does not show a clearly linear relation between spread and level. Nevertheless, the slope of the line, $b=.777$, gives $p=1-.777=.223$, which we round to 0. This suggests that the log transformation (or square root) would help to equalize variability.

Output 6.11
Spread-Level Plot

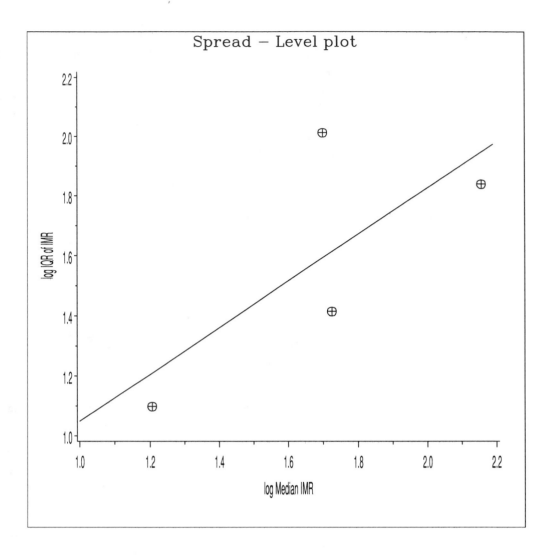

The slope value, .777, is found from the REG procedure output, shown in Output 6.12. The slope is the parameter estimate for the variable LOGMED. The boxplot of $\log_{10}$ (IMR) is shown in Output 6.9.

Output 6.12
Output from the REG Procedure for Spread-Level Plot

```
DEP VARIABLE: LOGIQR   log IQR
                                 ANALYSIS OF VARIANCE

                          SUM OF           MEAN
           SOURCE     DF  SQUARES          SQUARE      F VALUE    PROB>F

           MODEL      1   0.27368120    0.27368120      2.289     0.2694
           ERROR      2   0.23911529    0.11955764
           C TOTAL    3   0.51279649

                   ROOT MSE    0.3457711    R-SQUARE    0.5337
                   DEP MEAN    1.590268     ADJ R-SQ    0.3006
                   C.V.        21.74294

                            PARAMETER ESTIMATES

                        PARAMETER     STANDARD     T FOR H0:              VARIABLE
           VARIABLE  DF  ESTIMATE       ERROR    PARAMETER=0   PROB > |T|  LABEL

           INTERCEP   1  0.27162283   0.88853573     0.306       0.7887   INTERCEPT
           LOGMED     1  0.77739769   0.51381815     1.513       0.2694   log Median
```

6.4.2 Spread-Level Plot for Regression Data

The idea behind the spread-level plot also applies in regression applications when the variability of the response, y, varies linearly with the factor, x. Because the standard linear regression model assumes that the (residual) variance is constant, a transformation of y may be found to bring the data into line with this assumption.

In this situation, we can form groups by dividing the (x,y) points into some number of groups based on the ordered x values. Then, find the spread (IQR) of the y values and the median x for each group and plot log IQR (y) against the corresponding log median (x) for these groups. Again the slope of a line through these points indicates the power to be used in transforming the data.

Example: Baseball Data

For example, Output 6.15 shows a scatterplot of the salaries of players in the BASEBALL data set against their career batting average. The dashed vertical lines divide the 263 points into ten groups (deciles), so there are about the same number of points in each vertical strip. Because the distribution of batting average has greater density in the middle, the strips are narrower in the middle.

Note that although salary tends to increase with batting average, the variability of the points also increases from left to right. This violates the assumption of homogeneity of variance in a regression of salary on batting average. The relation also appears nonlinear, with salary increasing at a faster rate toward the high end of batting average. A transformation of salary that removes either or both of these problems would clearly be helpful.

Output 6.13   *Baseball Salaries Divided into Deciles by Batting Average*

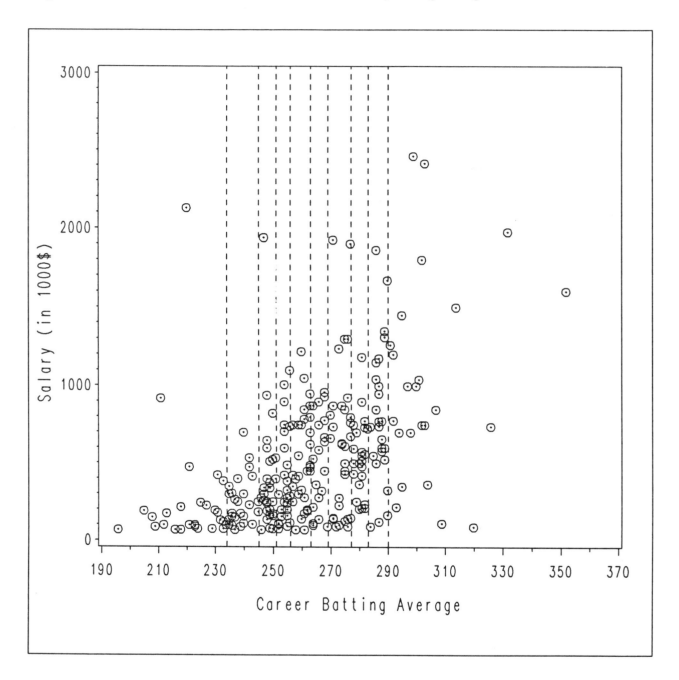

The division of the data points into ten groups is conveniently done using the RANK procedure and specifying GROUPS=10. (The choice of ten groups is arbitrary; other choices, roughly between six and twelve for a data set of this size, would also be reasonable.) The boundaries of the strips in Output 6.13 are the minimum values of batting average for each decile group, found with PROC UNIVARIATE.

The statements below read the observations in the BASEBALL data set with nonmissing SALARY values. (The variables NAME, the player's last name, and LOGSAL are used later in this example.) The RANK procedure assigns a variable DECILE, with values 0 through 9, so that there are approximately the same number of observations in each group. PROC UNIVARIATE is used to find the

median *x* values and division points. The output data set MIDDLE, shown in
Output 6.14, is used to set values for the HREF= option in the PROC GPLOT
step to produce the dashed lines in Output 6.13.

```
%include BASEBALL;              /* On MVS, use %include ddname(file);*/
%include boxplot;               /* On MVS, use %include ddname(file);*/
title h=.5 ' ';
data baseball;
   set baseball;
   if salary ¬= .;
   logsal = log10(1000 * salary);
   name = scan(name, 2, ' ');
   label logsal = 'log Salary';

   /* divide into deciles */
proc rank data=baseball out=grouped groups=10;
   var batavgc;
   ranks decile;

   /* find median batting average for each group */
proc sort data=grouped;
   by decile ;
proc univariate noprint data=grouped;
   by decile;
   var batavgc;
   output out=middle median=mdn_x n=n min=min;
proc print data=middle;
   id decile;
run;

   /* plot with reference lines at decile minima */
proc gplot data=baseball ;
   plot salary * batavgc
        / frame name='GB0613'
          vaxis=axis1 haxis=axis2
          href=234 245 251 256 263 269 277 283 290
          lhref=20 ;
   symbol v=- h=1.2 c=black;
   axis1 label=(h=1.5 a=90 r=0) value=(h=1.3);
   axis2 label=(h=1.5) value=(h=1.3);
run;
```

Output 6.14
Output Data Set
MIDDLE from the
UNIVARIATE
Procedure for
Baseball Data

DECILE	N	MIN	MDN_X
0	26	196	222.0
1	26	234	237.5
2	29	245	248.0
3	23	251	254.0
4	27	256	259.0
5	27	263	265.0
6	28	269	274.0
7	27	277	280.0
8	23	283	287.0
9	27	290	300.0

Next, the median batting average (MDN_X) is merged with the grouped data, and a boxplot is produced showing the distribution of salary values for each decile group. This plot, which appears in Output 6.15, shows more clearly the tendency for the spread of salaries (lengths of the boxes) to increase with batting average. Note that using MDN_X as the CLASS variable in the %BOXPLOT statement positions the boxplots at the median value of batting average for each group.

Output 6.15   *Boxplot of Salary by Batting Average Deciles*

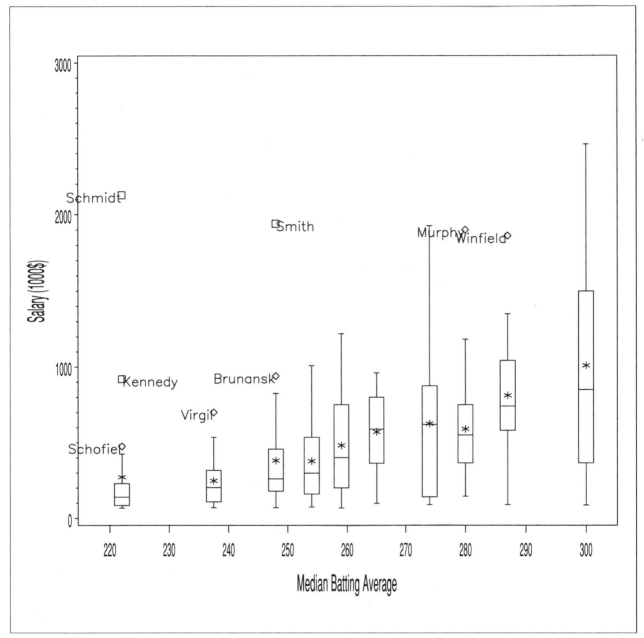

To construct the spread-level plot, we need a data set containing the median and IQR values of SALARY for each decile group. This would ordinarily be obtained with PROC UNIVARIATE, as in the infant mortality example earlier in this section (data set SUMRY). However, this information is contained in the output data set from the BOXPLOT macro, specified by OUT=SUMRY in the %BOXPLOT statement below. A subsequent DATA step is used to calculate the logs of the variables MEDIAN and IQR from the SUMRY data set. Selected variables from the SUMRY data set are shown in Output 6.16.

```
                /* merge data with median x value for each decile group */
         data grouped;
            merge grouped middle(keep=decile mdn_x);
            by decile;
         %boxplot(data=grouped, class=mdn_x,
                  var=salary,   id=name,
                  varlab=%str(Salary (1000$)),
                  classlab=Median Batting Average,
                  out=sumry,
                  name=GB0615
                  );
         run;
         data sumry;
            set sumry;
            logmed = log10(median);
            logiqr = log10(iqr);
            label logmed='log Median Salary'
                  logiqr='log IQR of Salary';
         proc print data=sumry;
            var mdn_x median iqr logmed logiqr;
```

Output 6.16
Output Data Set from the BOXPLOT Macro for Spread-Level Plot of Baseball Salaries

OBS	MDN_X	MEDIAN	IQR	LOGMED	LOGIQR
1	222.0	140.0	145.25	2.14613	2.16212
2	237.5	202.5	207.50	2.30643	2.31702
3	248.0	260.0	278.00	2.41497	2.44404
4	254.0	298.0	375.00	2.47422	2.57403
5	259.0	400.0	550.00	2.60206	2.74036
6	265.0	588.0	437.00	2.76938	2.64048
7	274.0	619.0	733.75	2.79169	2.86555
8	280.0	550.0	385.00	2.74036	2.58546
9	287.0	740.0	463.00	2.86923	2.66558
10	300.0	850.0	1135.00	2.92942	3.05500

Finally, we use PROC GPLOT to plot the LOGIQR against LOGMED, producing the plot shown in Output 6.17. To make it easier to interpret the plot, the slope of the regression line and the power for the transformation of salary are added to this plot using an Annotate data set. The slope of LOGIQR on LOGMED is calculated with PROC REG and saved in the OUTEST= data set PARMS.

Output 6.17   *Spread-Level Plot for Baseball Data*

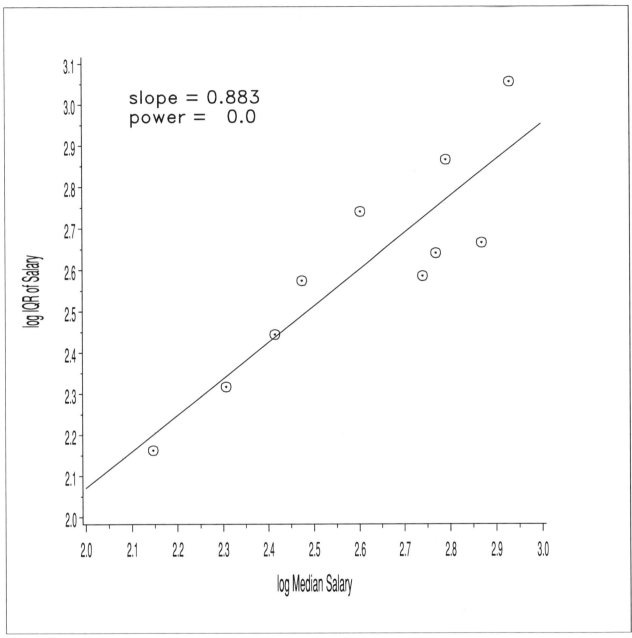

```
                  /* find slope of logiqr on logmed */
            proc reg data=sumry outest=parms;
               model logiqr = logmed;

                  /* annotate the spread-level plot with slope and power */
            data parms;
               set parms(keep=logmed);
               length text $20;
               xsys='1'; ysys='1';
               x=10;     y=90;
               function = 'LABEL';
               style    = 'DUPLEX';
               size = 1.4;
               power = round(1-logmed, .5);
               position='C'; text ='slope ='||put(logmed,f6.3);  output;
               position='F'; text ='power ='||put(power, f6.1);  output;
            proc gplot data=sumry ;
               plot logiqr * logmed
                    / name='GB0617'
                    vaxis=axis1 haxis=axis2 anno=parms;
               symbol v=- h=1.5 i=rl c=black;
               axis1 value=(h=1.3) label=(a=90 r=0 h=1.5);
               axis2 value=(h=1.3) label=(h=1.5);
            run;

                  /* boxplot of transformed Salary */
            %boxplot(data=grouped, class=decile, var=logsal, id=name,
                    yorder=4.5 to 6.5 by .5,
                    varlab=log Salary,
                    classlab=Batting Average Decile,
                    name=GB0618
                    );
```

The slope value, $b=.88$, in Output 6.17 corresponds to $p=1-.88=.12$, which is rounded to 0. This again suggests a log transformation to equalize variability. To see what effect this transformation has had, we can make a boxplot of log(SALARY) for the batting average groups. This plot is produced by the last %BOXPLOT statement above. The result is shown in Output 6.18. The spreads are now more nearly the same for most of the groups. Moreover, when expressed on a log scale, salary has more nearly a linear relation to batting average (see the trend of means in Output 6.18) than raw salary does (compare with Output 6.13).

Output 6.18   *Boxplot of Log(SALARY)*

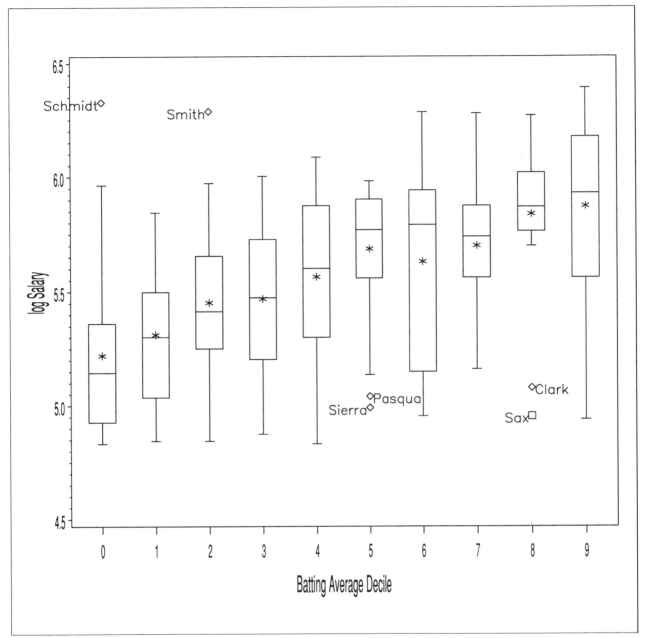

318

Chapter **7** Plotting ANOVA Data

7.1 Introduction

For designed experiments with two or more factors, the standard analysis of variance summary tables are of little help in interpreting the pattern of group differences, especially when the factors interact. That is, the ANOVA table shows which factors show significant differences among the means, but it does not indicate how the means differ.

This chapter starts with plots of means and standard errors in one-way and two-way designs. Section 7.2, "Plotting Means in a One-Way Design," illustrates simple plots of means and standard errors in a one-factor design and shows how to construct multiple comparison plots indicating which means differ. Section 7.3, "Plotting Means for Two-Factor Designs," covers plots of treatment means in factorial designs. Graphical displays designed to help understand the interaction between two or more experimental factors are discussed in Section 7.4, "Plotting Interactions," including an all effects plot that simultaneously displays all effects and residuals for any factorial design. Repeated measure designs are like factorial designs but require some special data manipulations that are described in Section 7.5, "Repeated Measure Designs."

Sometimes interactions may be reduced or eliminated by transforming the data. A simple diagnostic plot for choosing a transformation is described in Section 7.6, "Displaying Two-Way Tables for $n=1$ Designs." Finally, a SAS program for calculating the statistical power of ANOVA tests is described in Section 7.7, "Plotting Power Curves for ANOVA Designs," and a graphical display of the results is used to help determine the sample size for an experiment.

This chapter assumes that you are familiar with analysis of variance ideas and the use of the ANOVA and GLM procedures for testing ANOVA hypotheses, so the focus here is on the ideas and techniques for plotting your data. See Freund, Littell, and Spector (1991) for discussion of the use of SAS procedures for analysis of variance (ANOVA).

As well, it is assumed that you have already carried out exploratory plots to determine that the assumptions of the analysis of variance are met so that means and standard deviations are reasonable summaries of the data for each group. In particular, boxplots and the spread-level plots described in Chapter 6, "Comparing Groups," are recommended for determining whether the assumption of homogeneity of variance is met. Some potential problems arising from an uncritical use of plots of means and standard deviations are illustrated in Section 1.2.1, "Means and Standard Deviations: Potentially Misleading Measures."

7.2 Plotting Means in a One-Way Design

For data collected in a designed experiment, it is usually helpful—and sometimes necessary—to plot the response for each treatment condition in order to understand the nature of the treatment effects. The ANOVA table may tell you what effects are significant, but it sheds no light on the nature of the effect or the pattern of the means.

For an initial look at the data, boxplots may provide the most useful display, especially if you suspect that the groups may differ in variability or that some deviant scores may be present. Once you are satisfied that the assumptions of the ANOVA model are reasonably well met, a plot of treatment means with standard error bars is most commonly used for presentation purposes.

Specifying I=STD*kxxx* in the SYMBOL statement makes it quite easy to produce such plots with the GPLOT procedure from data in the same form as is used for the significance tests carried out using PROC GLM or PROC ANOVA. With the PLOT procedure, the group means and standard errors must first be obtained with the MEANS or SUMMARY procedure; these values are stored in an output data set and can then be plotted.

7.2.1 Example: Interactive Free Recall

This example displays partial data from an experiment on an interactive free recall memory task by Friendly and Franklin (1980). In this experiment, subjects were presented with a list of 40 words to memorize, with words displayed under computer control. Three groups of ten subjects each were defined by the order in which words were presented from one trial to the next. Group 1, the control group, had the words presented in a different random order on each trial (Standard Free Recall, SFR). In two other groups, the order in which words were recalled on the previous trial was used to determine the presentation order on the current trial. Group 2, BEGIN, received their previous recalled words first, in the order of recall, followed by the remaining words in random order. Group 3, MESH, received their previous recalled words in the order of recall, with the remaining words randomly interspersed. The experimental hypothesis was that both the BEGIN and MESH groups would recall more than the SFR control group.

The DATA step below reads the GROUP (group number) and SCORE variable for each subject. Descriptive labels for the treatment groups are defined with the GRP. format in a PROC FORMAT step. The printed output from the PROC ANOVA step, shown in Output 7.1, indicates that the group means differ significantly.

```
proc format;
    value grp 1='SFR' 2='BEGIN' 3='MESH';
data ifr;
    input subject group score ;
    label score='Number of Words Recalled';
    format group grp.;
cards;
 1 1 35
 2 1 25
 3 1 37
 4 1 25
 5 1 29
 6 1 36
 7 1 25
 8 1 36
 9 1 27
10 1 28
11 2 40
12 2 38
13 2 39
14 2 37
15 2 39
16 2 24
17 2 30
18 2 39
19 2 40
20 2 40
21 3 40
22 3 39
23 3 34
24 3 37
25 3 40
26 3 36
27 3 36
28 3 38
```

```
29 3 36
30 3 30
;
proc anova data=ifr;
   class group;
   model score = group;
```

Output 7.1
ANOVA Procedure
Output for IFR
Data

```
                    ANALYSIS OF VARIANCE PROCEDURE
                       CLASS LEVEL INFORMATION

                 CLASS     LEVELS     VALUES

                 GROUP        3       BEGIN MESH SFR


              NUMBER OF OBSERVATIONS IN DATA SET = 30
                    ANALYSIS OF VARIANCE PROCEDURE
DEPENDENT VARIABLE: SCORE      Number of Words Recalled

SOURCE                DF       SUM OF SQUARES       MEAN SQUARE      F VALUE

MODEL                  2        264.60000000      132.30000000       6.24

ERROR                 27        572.90000000       21.21851852      PR > F

CORRECTED TOTAL       29        837.50000000                        0.0059


R-SQUARE            C.V.           ROOT MSE          SCORE MEAN

0.315940          13.3518        4.60635632         34.50000000


SOURCE                DF            ANOVA SS     F VALUE    PR > F

GROUP                  2        264.60000000       6.24     0.0059
```

A plot of treatment means with error bars extending ± 1 standard error of the individual group means is shown in Output 7.2. The plot is produced with PROC GPLOT with these statements:

```
*-- Plot means and stderrs with PROC GPLOT ;
title ' ';
proc gplot data=ifr  ;
   plot score * group
      / frame hminor=0 vminor=4
        vaxis=axis1 haxis=axis2
        name='GB0702' ;
   symbol1 i=std1mjt v=none l=23 c=black;
   axis1 label=(a=90 r=0 h=1.5) value=(h=1.5) offset=(4)
         order=(20 to 40 by 5);
   axis2 label=(h=1.5) value=(h=1.5) offset=(4);
```

Note that the SYMBOL statement uses V=NONE to suppress plotting of the individual observations. Specifying I=STD1MJT for the interpolate option causes the individual group standard deviations to be used for error bars, so the bars may differ in length. If you specify I=STD1PJT, the length of each bar is determined by the pooled standard error, which is the $\sqrt{\text{MSE}}$ from the ANOVA procedure output.

Output 7.2
GPLOT Procedure Display of Means and Standard Errors from IFR Experiment

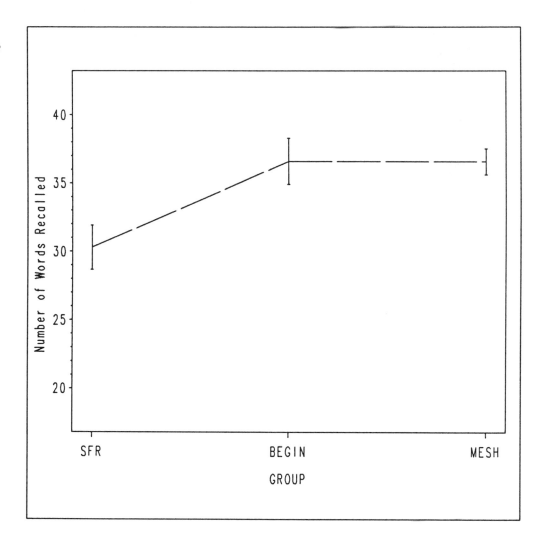

To construct a similar plot with PROC PLOT, use PROC MEANS to compute values by group and obtain an output data set containing the means and standard errors as shown below. Then the upper and lower limits can be calculated and plotted together with the means. The plot appears in Output 7.3.

```
*-- Plot means and stderrs with PROC PLOT ;
proc means data=ifr;
   by group;
   var score;
   output out=xbar mean=mean stderr=stderr n=n;
data xbar;
   set xbar;
   hi = mean + stderr;
   lo = mean - stderr;
proc plot data=xbar;
   plot mean * group = 'O'
        hi   * group = '='
        lo   * group = '='
        / overlay vaxis=20 to 40 by 5 hpos=60 vpos=40;
```

Output 7.3
PLOT Procedure
Display of Means
and Standard
Errors from IFR
Experiment

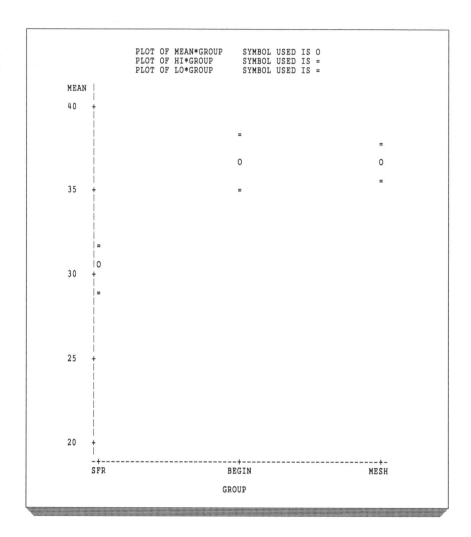

```
                    PLOT OF MEAN*GROUP    SYMBOL USED IS O
                    PLOT OF HI*GROUP      SYMBOL USED IS =
                    PLOT OF LO*GROUP      SYMBOL USED IS =

    MEAN |
         |
      40 +
         |
         |
         |
         |                            =                              =
         |
         |               O                         O
         |
         |                            =                              =
      35 +                            =
         |
         |
         |
         | =
         | O
      30 +
         |
         | =
         |
         |
         |
         |
      25 +
         |
         |
         |
         |
         |
      20 +
         |
       -+-------------------------+------------------------+-
        SFR                     BEGIN                    MESH

                                GROUP
```

7.2.2 Multiple Comparison Plots: Showing Which Means Differ

A significant F value indicates that not all means are equal, but does not tell which groups differ. When the F value is significant, you would usually use a *multiple comparison* procedure to make comparisons among the groups, controlling the overall error rate for the collection of comparisons. The MEANS statement in the PROC ANOVA step provides a variety of multiple comparison procedures to test differences among all pairs of means.

These tests are based on simultaneous $100(1-\alpha)\%$ confidence intervals for the difference between a pair of means, which have the form

$$(\bar{y}_i - \bar{y}_j) \pm \text{interval width} = (\bar{y}_i - \bar{y}_j) \pm c_{k,\nu}(\alpha) \sqrt{MSE(\frac{1}{n_i} + \frac{1}{n_j})} \qquad (7.1)$$

where $c_{k,\nu}(\alpha)$ is the critical value (or standard error multiple) for k groups and ν degrees of freedom of the MSE, which is $\nu = \Sigma(n_i - 1)$ in a one-way design. For a particular multiple comparison procedure, if the interval (equation 7.1) does not include 0, it is concluded that the means, μ_i and μ_j, differ at significance level α.

Different multiple comparison procedures are akin to different kinds of insurance policies in which you can weigh the extent of coverage (the number and kind of comparisons to be made) against the cost of the policy (risk of one or more erroneous conclusions and size of the confidence interval). For unprotected t-tests using Fisher's *least-significant difference* (LSD), the critical value c is just the ordinary two-sided critical t: $c = t_\nu(1-\alpha/2)$. These provide the shortest intervals, but the policy applies to just one comparison, and the overall risk is high (the overall error rate is a multiple of the number of comparisons). For the *Tukey HSD procedure*, the critical value c, symbolized T, is

$$T = \frac{1}{\sqrt{2}} q_{k,\nu}(\alpha) \qquad\qquad (7.2)$$

where $q_{k,\nu}(\alpha)$ is the upper αth quantile of the studentized range distribution (Tukey 1953; Kramer 1956). When all $k(k-1)/2$ pairwise comparisons are to be made, the Tukey procedure guarantees that the probability of one or more erroneous conclusions is no more than α and generally provides the shortest (protected) confidence intervals. Other procedures, such as the Bonferroni and Scheffé tests, provide other ways of balancing overall coverage against cost.

For the purpose of graphical display, however, it is more convenient to display the means in the form

$$\bar{y}_i \pm \text{decision interval}$$

where the *decision interval* is calculated to provide a simple decision rule: two means are declared significantly different if their decision intervals do not overlap.* This provides an effective way to communicate the results graphically because the significant differences may be seen directly. These plots are suggested by Gabriel (1978) and by Andrews, Snee, and Sarner (1980) and can be used with any multiple comparison procedure that produces a single confidence interval for all comparisons. The decision interval is simply one half the width of the confidence interval.

Example: IFR Data

For the IFR data, the MEANS statement in the PROC ANOVA step uses the Bonferroni procedure for multiple comparisons:

```
means group / bon;
```

The printed output from this test, in Output 7.4, indicates that groups BEGIN and MESH do not differ from each other, but that both groups score significantly more than group SFR.

---

* The term decision interval is used to avoid confusion with a confidence interval for a single mean. Andrews, Snee, and Sarner (1980) call this an uncertainty interval, but decision interval is more descriptive.

Output 7.4
Bonferroni Group Comparisons for IFR Data

```
                      ANALYSIS OF VARIANCE PROCEDURE

        BONFERRONI (DUNN) T TESTS FOR VARIABLE: SCORE
        NOTE: THIS TEST CONTROLS THE TYPE I EXPERIMENTWISE ERROR RATE
              BUT GENERALLY HAS A HIGHER TYPE II ERROR RATE THAN REGWQ

                ALPHA=0.05  DF=27  MSE=21.2185
                CRITICAL VALUE OF T=2.55246
                MINIMUM SIGNIFICANT DIFFERENCE=5.2581

        MEANS WITH THE SAME LETTER ARE NOT SIGNIFICANTLY DIFFERENT.

           BON    GROUPING          MEAN    N   GROUP

                      A            36.600   10  BEGIN
                      A
                      A            36.600   10  MESH

                      B            30.300   10  SFR
```

To show the decision intervals graphically, it is necessary to calculate the means with PROC MEANS (or PROC SUMMARY) as we did with PROC PLOT to produce Output 7.3. The program below calculates the Bonferroni decision interval around each sample mean. The error bars reflect the width of the Bonferroni decision intervals. The groups whose means do not differ (whose intervals overlap) are joined as shown in Output 7.5. (Note that the 1 standard error intervals shown by the plot using I=STD in Output 7.2 are considerably shorter than the decision intervals, so may give an overly optimistic impression.)

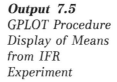

Output 7.5
GPLOT Procedure
Display of Means
from IFR
Experiment

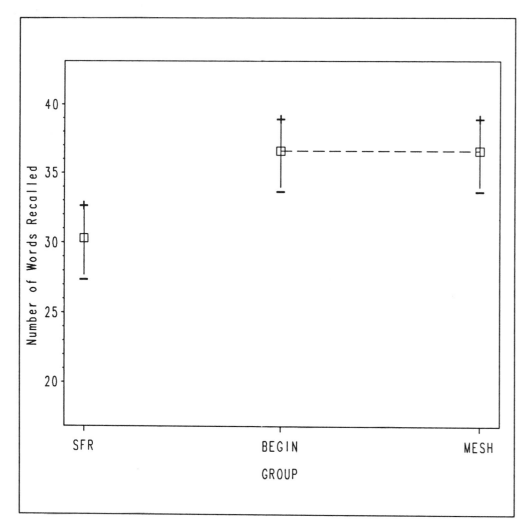

When the groups have equal sample size, n, the decision intervals for the Bonferroni procedure are given by

$$\bar{y}_i \pm B \sqrt{MSE / 2n} \qquad (7.3)$$

where the Bonferroni critical value is $B = t_\nu(1 - \alpha/2g)$, the upper $1 - \alpha/2g$ percentage point of the t distribution, and where ν is the degrees of freedom associated with the MSE and g is the number of comparisons, which is $k(k-1)/2$ for all pairwise comparisons of k groups.*

The DATA step that produces the BARS data set in the program below uses the TINV function to calculate the value B. For the IFR data, this gives the value 2.55246 labeled CRITICAL VALUE OF T in Output 7.4. The value of the MSE is

---

*   When the sample sizes are unequal, the length of the decision interval varies with sample size. In this case, an exact graphical representation of a multiple comparison procedure does not always exist. However, as long as the degree of imbalance in sample sizes is not great, the simplest approximation (Gabriel 1978) is to use n_i for each interval in equation 7.3.

taken directly from that output. For each group, the decision interval is drawn using the Annotate facility's MOVE and DRAW instructions.

```
*--Plot showing Bonferroni decision intervals;
proc means data=ifr noprint;
   by group;
   var score;
   output out=xbar mean=mean n=n;
data bars;
   set xbar;
   length function $8;
   xsys='2'; ysys='2';
   MSE = 21.2185;
   B = tinv( (1-(.05/6)), 27);          * Bonferroni critical t value;
   x = group;
   y = mean - B * sqrt( MSE / (2*n));
   function = 'MOVE'; output;
   position='5'; style='SWISS'; size=1.2;
   function ='LABEL'; text='-'; output;
   y = mean + B * sqrt( MSE / (2*n));
   function = 'DRAW'; output;
   function ='LABEL'; text='+'; output;
```

The DATA step that produces the LINES data set constructs another Annotate data set to join the means of the BEGIN and MESH groups, which did not differ, and leaves the SFR group mean unconnected. This again relies on the results of the printed output from the MEANS statement shown in Output 7.4. This data set is then appended to data set BARS.

```
data lines;
   set xbar;
   xsys='2'; ysys='2';
   x = group;
   y = mean;
   line = 21;
   select (group);
      when (1) function='COMMENT';
      when (2) function='MOVE';
      when (3) function='DRAW';
   end;
data bars;
   set bars lines;
proc gplot data=xbar  ;
   plot mean * group
      / frame hminor=0 vminor=4
        vaxis=axis1 haxis=axis2
        anno=bars
        name='GB0705' ;
   symbol1 i=none v=SQUARE h=1.5 color=black;
   label mean='Number of Words Recalled';
   axis1 label=(a=90 r=0 h=1.5) value=(h=1.5) offset=(4)
         order=(20 to 40 by 5);
   axis2 label=(h=1.5) value=(h=1.5) offset=(4);
```

A more general approach is suggested by Hochberg, Weiss, and Hart (1982). This plot, the *multiple comparison plot*, is a graphic analog of the GROUPING column of A's and B's in the printout from the MEANS statement, which is shown in Output 7.4. However, the multiple comparisons plot shows the metric spacing of the means while the printout does not.

Let $\bar{y}_{(1)}, \ldots, \bar{y}_{(k)}$ be the k treatment means sorted in increasing order. First, the ordered $\bar{y}_{(i)}$ are plotted against $i = 1, \ldots, k$, and decision intervals are drawn around each mean as before. Next, a horizontal line is drawn from the top of each interval across to the right. Output 7.6 shows an example for four groups. In this plot, any pair of means whose intervals are not connected by a horizontal line are judged to differ significantly by the multiple comparison procedure used to set the decision interval. Groups whose decision intervals are connected by a horizontal line are considered not to differ.

Output 7.6
Multiple Comparisons Plot

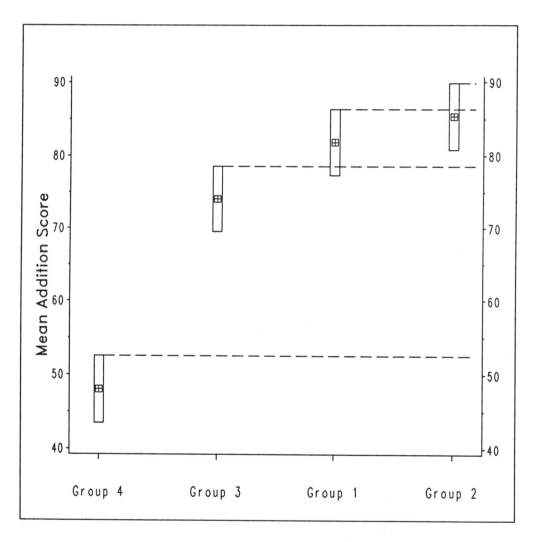

Example: Methods of Teaching Addition

The data plotted in Output 7.6 represent the scores on a test of adding four-digit numbers for groups of 20 children each, taught by four different methods (Timm 1975, p. 382). The data are shown in the DATA step that produces the

ADDITION data set, shown below. The ANOVA test results from PROC GLM (not shown) indicate that the group effect is highly significant. The output from the MEANS statement, giving decision intervals by the Tukey formula, is shown in Output 7.7. The columns labeled TUKEY GROUPING indicate that Group 4 differs from all other groups, Group 3 differs from Group 2 but not from Group 1, and so forth.

```
data addition;
    input group @ ;
    do subject = 1 to 20;
        input add @@ ;
        output;
        end;
    label add='Addition Score';
cards;
 1 100 96 90 90 87 83 85 85 78 86 67 57 83 60 89 92 86 47 90 86
 2  97 94 96 84 90 88 82 65 95 90 95 84 71 76 90 77 61 91 93 88
 3  76 60 84 86 70 70 73 85 58 65 80 75 74 84 62 71 71 75 92 70
 4  66 60 58 52 56 42 55 41 56 55 40 50 42 46 32 30 47 50 35 47
;
proc glm data=addition;
    class group;
    model add = group;
    means group / tukey ;
```

Output 7.7
Tukey Multiple Comparisons for Addition Data

```
                  GENERAL LINEAR MODELS PROCEDURE

        TUKEY'S STUDENTIZED RANGE (HSD) TEST FOR VARIABLE: ADD
        NOTE: THIS TEST CONTROLS THE TYPE I EXPERIMENTWISE ERROR RATE,
              BUT GENERALLY HAS A HIGHER TYPE II ERROR RATE THAN REGWQ

              ALPHA=0.05  DF=76  MSE=117.448
              CRITICAL VALUE OF STUDENTIZED RANGE=3.715
              MINIMUM SIGNIFICANT DIFFERENCE=9.0022

        MEANS WITH THE SAME LETTER ARE NOT SIGNIFICANTLY DIFFERENT.

            TUKEY    GROUPING        MEAN    N  GROUP

                        A          85.350   20  2
                        A
                 B      A          81.850   20  1
                 B
                 B                 74.050   20  3

                        C          48.000   20  4
```

To construct the plot in Output 7.6, we use PROC MEANS to produce an output data set and then calculate the half width (WIDTH) of the decision interval in a DATA step. Because there is no SAS function to calculate the critical value of the studentized range, this value ($Q=3.715$) is taken from the printed output in Output 7.7. The calculated value of the variable WIDTH is 4.501, half the value labeled MINIMUM SIGNIFICANT DIFFERENCE in Output 7.7.

```
proc means data=addition noprint;
    var add;
    by group;
    output out=xbar mean=mean n=n;
```

```
proc sort data=xbar;
   by mean;
data xbar;
   set xbar;
   order = _n_;
   mse = 117.45;                * (from MEANS statement output);
   q = 3.715;                   * Studentized range (ditto);
   width = sqrt(mse/n) * q / 2;
   label mean='Mean Addition Score';
proc print;
```

The data set BARS draws a rectangle centered at each mean extending up and
down by the value of the variable WIDTH. Then a dashed line is drawn
horizontally to the right side of the plot. I used rectangles for the decision
intervals and dashed horizontal lines to give more visual emphasis to the vertical
than the horizontal components of the plot.

```
data bars;
   set xbar;
   length function $8;
   xsys='2'; ysys='2';
   x = order - .04;
   y = mean - width;
       function = 'MOVE'; output;
   line=0;                         * error bar;
   x = order + .04;
   y = mean + width;
       function = 'BAR '; output;

   xsys='1';                       * horizontal dashed line;
   x = 98; line=21;
       function = 'DRAW'; output;

   xsys='2'; ysys='5';             * group label on horiz. axis;
   x = order;
   y = 4;
   size = 1.5;
   text = 'Group' || put(group, 2.);
       function = 'LABEL'; output;
```

Because the groups are to be plotted against the variable ORDER, the Annotate
function LABEL is used to label the abscissa, and the horizontal axis is suppressed
in the AXIS2 statement in the PROC GPLOT step that follows.

```
proc gplot data=xbar  ;
   plot  mean * order
          / anno=bars
            vaxis=axis1 haxis=axis2 vminor=1
            name='GB0706' ;
   plot2 mean * order
          / vaxis=axis3 vminor=1;
   symbol v=SQUARE h=1.2 i=none c=black;
```

```
axis1 label=(a=90 r=0 h=1.5 f=duplex)
       value=(h=1.3);
axis2 offset=(6) label=none
       value=none minor=none;
axis3 label=none value=(h=1.3);
```

7.2.3 Plotting a Quantitative Factor

When the factor variable is quantitative and the treatment means differ, you may want to test whether there is a regression relation (linear, quadratic, and so on) between the mean response and the value of the treatment variable. The corresponding plots can use I=RL, I=RQ, and so on, in the SYMBOL statement to draw the fitted relation through the treatment means.

Example: Running Speed and Sugar Concentration

In this example we examine data from an experiment in which rats were given a sugar solution as a reward for running a runway. The independent variable was sucrose concentration (8, 16, 32, or 64 percent), and the response variable was the rat's running speed to complete the runway. The goal is to determine how running speed depends on sucrose concentration.

The data are input to the data set SUCROSE. Notice that the data lines list the SUGAR value, followed by all scores for that group. This simply makes it easier to enter the data without duplicating the SUGAR value for each observation. The data set SUCROSE has exactly the same form as the IFR data set from the previous example: each output observation contains the value of the independent variable (SUGAR) and the score for that subject (SPEED).

```
Title 'Sucrose Data: Speed to traverse a Runway';
data sucrose;
    label SUGAR = 'Sucrose Concentration (%)'
          SPEED = 'Speed in Runway (ft/sec)';
    input SUGAR @ ;
    do SUBJECT = 1 to 8;
       input SPEED @;
       output;
       end;
cards;
 8    1.4 2.0 3.2 1.4 2.3 4.0 5.0 4.7
 16   3.2 6.8 5.0 2.5 6.1 4.8 4.6 4.2
 32   6.2 3.1 3.2 4.0 4.5 6.4 4.5 4.1
 64   5.8 6.6 6.5 5.9 5.9 3.0 5.9 5.6
 ;
```

The PROC GLM output, shown in Output 7.8, indicates that the means differ significantly. But to discover how running speed depends on sugar concentration, we must plot the means.

Output 7.8
GLM Procedure
Output for Sucrose
Data

```
                    GENERAL LINEAR MODELS PROCEDURE
DEPENDENT VARIABLE: SPEED       Speed in Runway (ft/sec)

SOURCE                 DF      SUM OF SQUARES      MEAN SQUARE       F VALUE

MODEL                   3        28.68000000        9.56000000        5.60

ERROR                  28        47.76000000        1.70571429       PR > F

CORRECTED TOTAL        31        76.44000000                         0.0039


R-SQUARE             C.V.           ROOT MSE           SPEED MEAN

0.375196           29.3490         1.30602997          4.45000000


SOURCE                 DF          TYPE I SS     F VALUE     PR > F

SUGAR                   3        28.68000000       5.60      0.0039


SOURCE                 DF         TYPE III SS    F VALUE     PR > F

SUGAR                   3        28.68000000       5.60      0.0039
```

The first plot (Output 7.9) shows the mean response and standard error in relation to sucrose concentration. Because the values of sucrose concentration increase in powers of two, we can make the horizontal axis equally spaced by plotting the variable SUCROSE on a log scale using LOGBASE=2 and LOGSTYLE=EXPAND in the AXIS statement. Also, the specification I=STD does not provide for plotting a symbol at the location of the mean. You can plot the means by obtaining the fitted values (which are the treatment means in a one-way design) from PROC ANOVA or PROC GLM and plotting these in addition to the data values.

Output 7.9
*Mean Running
Speed Plotted
against Sucrose
Concentration*

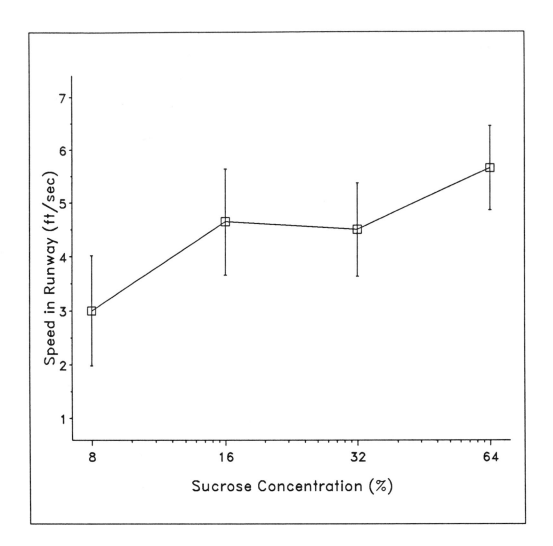

```
proc glm data=sucrose;
    classes sugar;
    model speed = sugar;
    output out=results              /* get output dataset       */
               p = mean;            /* containing fitted value  */
title ' ';
proc gplot data=results ;
    plot speed * sugar = 1
         mean  * sugar = 2
       / overlay
         vaxis=axis1 haxis=axis2
         name='GB0709' ;
    symbol1 v=none   i=stdmjt  color=black;     /* std error bars */
    symbol2 v=square i=none h=1.5;              /* means          */
    axis1 label=(a=90 r=0 h=1.4 f=duplex)
           value=(h=1.2 f=duplex) minor=(n=1)
           offset=(2);
```

```
axis2 label=(h=1.4 f=duplex)
      value=(h=1.2 f=duplex)
      logbase=2 logstyle=expand
      offset=(4);
```

More specifically, we may entertain the hypothesis that there is a definite functional relationship between mean running speed and sucrose concentration. Suppose, for example, we believe that mean running speed should increase quadratically with the sugar concentration of the reward.

To test this idea, treat SUGAR as a quantitative, regression variable and fit the model

$$\text{SPEED} = \beta_0 + \beta_1 \text{ SUGAR} + \beta_2 (\text{SUGAR})^2 + residual \quad . \tag{7.4}$$

The test is carried out with the PROC GLM step below. The printed output, shown in Output 7.10, shows that the quadratic term is not significant, though the linear term is clearly significant. Thus, a simple linear regression model provides a better account of the data than model 7.4. Nevertheless, for illustration, we proceed with the plot as if model 7.4 were appropriate.

Output 7.10
Test of Quadratic Model for Sucrose Data

```
                    GENERAL LINEAR MODELS PROCEDURE
DEPENDENT VARIABLE: SPEED        Speed in Runway (ft/sec)

SOURCE                   DF       SUM OF SQUARES       MEAN SQUARE      F VALUE

MODEL                     2         22.67838710       11.33919355         6.12

ERROR                    29         53.76161290        1.85384872       PR > F

CORRECTED TOTAL          31         76.44000000                         0.0061


R-SQUARE          C.V.            ROOT MSE          SPEED MEAN

0.296682       30.5969           1.36156113         4.45000000


SOURCE                   DF          TYPE I SS    F VALUE      PR > F

SUGAR                     1         21.30434783      11.49      0.0020
SUGAR*SUGAR               1          1.37403927       0.74      0.3963


SOURCE                   DF        TYPE III SS    F VALUE      PR > F

SUGAR                     1          4.19417287       2.26      0.1434
SUGAR*SUGAR               1          1.37403927       0.74      0.3963


                                    T FOR H0:      PR > |T|     STD ERROR OF
PARAMETER           ESTIMATE      PARAMETER=0                     ESTIMATE

INTERCEPT          2.74166667           3.52        0.0014       0.77786607
SUGAR              0.08674395           1.50        0.1434       0.05767043
SUGAR*SUGAR       -0.00065734          -0.86        0.3963       0.00076354
```

The plot of the means with the fitted quadratic curve is shown in Output 7.11. This plot uses a linear scale for the variable SUGAR. Note that the SYMBOL2 statement applies I=RQ to the mean response in the output data set RESULTS from the previous PROC GLM step. Because the sample sizes are equal, this gives the same result as for the raw data.

```
proc glm data=sucrose;
    model speed = sugar sugar*sugar;
proc gplot data=results ;
    plot speed * sugar = 1
         mean  * sugar = 2
       / overlay
         vaxis=axis1 haxis=axis2
         name='GB0711' ;
    symbol1 v=none   i=stdmjt color=black;      /* std error bars */
    symbol2 v=square i=rq h=1.5 color=red;      /* means          */
    axis1 label=(a=90 r=0 h=1.5 f=duplex)
          value=(h=1.4 f=duplex) minor=(n=1)
          offset=(2);
    axis2 label=(h=1.5 f=duplex)
          value=(h=1.4 f=duplex)    ;
```

Output 7.11
Mean Running Speed Plotted against Sucrose Concentration

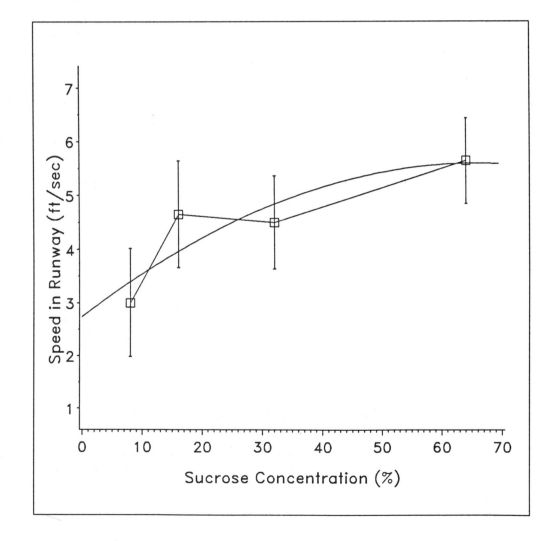

7.3 Plotting Means for Two-Factor Designs

With two or more factors in a factorial design, plotting treatment means becomes even more important in order to understand the nature of the treatment effects. For example, finding a significant interaction between two factors using PROC ANOVA by itself sheds no light on whether the interaction is substantively large or important, or whether the main effects of the factors separately can be interpreted. Plots of the mean treatment response can usually supply the answers to these questions.

For multifactor designs, specifying I=STD*kxxx* in the SYMBOL statement makes PROC GPLOT especially convenient. Consider a two-factor design with treatment factors A and B, with Y as the response variable. With PROC GPLOT, a PLOT statement of the form

```
plot Y * A = B;
```

would ordinarily plot each observation individually, using variable B to determine the plotting symbol. However, with a SYMBOL statement using I=STD*kxxx*, the mean of the multiple values of Y at each value of factor A are plotted together with vertical bars indicating $k(=1, 2,$ or $3)$ standard deviations; $k=2$ standard deviations is the default. A separate curve is drawn for each level of factor B.

The following example plots data for a design in which reaction time (RT) is the response variable measured for two treatment variables: type of item (ITEMTYPE) and size of the set of items memorized (SS). The data are taken from Loftus and Loftus (1988, Table 12-9).[*] For each treatment combination, another response variable (ERR) gives the number of errors made in this task of short term recognition.

The data are read by the DATA step below that creates the STERN data set. Although each data line lists all five observations for each condition, the OUTPUT statement inside the DO loop produces one observation per subject, in the following form:

```
ITEMTYPE SS SUBJ RT   ERR

Letters  1   1   350   4
Letters  1   2   346   6
Letters  1   3   361   7
Letters  1   4   330   7
Letters  1   5   342   3
...
```

---

[*] From *Essence of Statistics* by Geoffrey R. Loftus and Elizabeth R. Loftus. Copyright © 1988. Reprinted by permission of McGraw-Hill, Inc.

The DATA step creates a total of 30 observations. The data are required to be in this form for PROC ANOVA and for plotting.

```
title 'GPLOT display of ANOVA means with Std Error bars';
data stern;
    input itemtype $ ss  @;
    label ss= 'SetSize'
          rt= 'Reaction Time (msec)'
          err='Percent errors';
    do subj = 1 to 5;
       input rt err @;
       output;
       end;
cards;
Letters 1 350 4   346 6   361 7   330 7   342 3
Letters 3 394 5   392 8   410 8   405 10 400 9
Letters 5 456 4   471 8   464 12 460 9   450 7
Words   1 384 4   371 7   365 3   392 8   375 4
Words   3 466 6   484 9   475 11 470 8   465 7
Words   5 585 8   580 10  560 10 562 9   570 8
;
proc anova data=stern;
   class itemtype ss;
   model RT = itemtype | ss ;
```

The SYMBOL statements below define plotting symbols to give a 1 standard deviation error bar for each mean and join these with lines. The specification I=STDTMJ requests a plot with horizontal lines at the tops (T) of the bars, whose lengths are $k=2$ standard errors of the means (M) and joined (J) from group to group for each level of factor B. The line style options, L=1 and L=20, use different line styles for the levels of Item Type. It is a good idea to specify the COLOR= option in the SYMBOL statement. If no color option is given, the procedure reuses SYMBOL1 with each available color before going on to use the SYMBOL2 specifications.

```
symbol1 v=none l=1  i=stdtmj w=2 c=black;
symbol2 v=none l=20 i=stdtmj w=2 c=red  ;

title  h=1.8 'Reaction time vs. Set Size for Words and Letters' ;
title2 h=1.4 'Data from Loftus & Loftus (1988)';
proc gplot data=stern   ;
   plot rt * ss = itemtype
           / legend=legend1
             vaxis = axis1  vminor=1
             haxis = axis2
             name='GB0712' ;
   legend1 label= (h=1.5 'Item Type')
           value= (h=1.4)               /* Height of legend value */
           shape= line(4);             /* size of markers        */
   axis1   label= (a=90 r=0 h=1.5)
           value= (h=1.5)
           order= 300 to 600 by 100
           offset= (2);
```

```
axis2   label= (h=1.5)
        value= (h=1.5)
        order = 1 to 5 by 2
        offset=(3)
        minor =none;
```

The plot in Output 7.12 shows clearly that the main effects of SetSize and Item Type are interpretable even though the two factors interact significantly. (A color version of this output appears as Output A3.31 in Appendix 3, "Color Output.") And we can see that the effect of SetSize is approximately linear for both letters and words. Moreover, it is apparent that the interaction between SetSize and Item Type is accounted for by the difference in slopes between the two curves.

Output 7.12
Plot of Means and Standard Errors for Stern Data

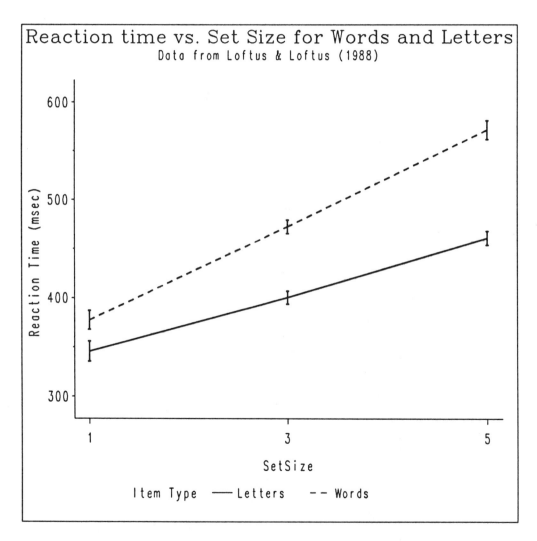

The PROC ANOVA output for these data is shown in Output 7.13.

```
DEPENDENT VARIABLE: RT          Reaction Time (msec)

SOURCE                  DF      SUM OF SQUARES        MEAN SQUARE      F VALUE

MODEL                    5      165231.36666667      33046.27333333     365.69

ERROR                   24        2168.80000000         90.36666667     PR > F

CORRECTED TOTAL         29      167400.16666667                         0.0001


R-SQUARE              C.V.           ROOT MSE             RT MEAN

0.987044            2.1712         9.50613837         437.83333333


SOURCE                  DF           ANOVA SS      F VALUE     PR > F

ITEMTYPE                 1        38377.63333333    424.69     0.0001
SS                       2       118933.26666667    658.06     0.0001
ITEMTYPE*SS              2         7920.46666667     43.82     0.0001
```

In the PROC GPLOT step, a LEGEND statement is used to define how the bottom legend for the ITEMTYPE variable is to be drawn. If this graph were being prepared for publication, you might want to draw the legend inside the frame of the graph itself. In Version 6, this can be done with the POSITION option in the LEGEND statement. For example, this LEGEND1 statement in the program above would draw the legend inside the plot frame centered just above the horizontal axis:

```
legend1 label= (h=1.5 'Item Type')
         value= (h=1.4)                /* Height of legend value */
         shape= line(4)                /* size of markers        */
         frame                         /* frame around legend    */
         mode=protect                  /* blank legend area      */
         offset=(,1 pct)               /* move up 1 %            */
         position=(bottom center inside);
```

In Version 5, legends can be created inside the plot frame with the Annotate facility. An example of a custom Annotate legend is included in Section 7.4.2, "Three-Factor Designs."

7.3.1   Plotting a Second Variable on the Graph

The data set STERN contains a second dependent variable, ERR, the proportion of errors made by a subject for each type of item. In this type of experiment, there is a well-known trade-off between reaction time and errors: a person can easily respond more quickly by allowing more errors. It is common, therefore, to display the RT (reaction time) data together with histogram bars showing the mean percentage of errors in each condition. If smaller reaction times are associated with larger errors, there is reason to suspect that a speed—accuracy trade-off has occurred. More generally, this type of plot is often used to show that a possibly confounding effect has been controlled.

The PLOT2 statement in PROC GPLOT allows a second dependent variable to be plotted against a vertical axis on the right side of the display. Adding the

statements below to the PROC GPLOT step of the program that produces Output
7.12 would plot the mean percent of errors against its own axis on the right:

```
symbol3 v=none i=stdbmj;
plot2 err * ss = itemtype ;
```

However, this would obscure the plot of reaction time, which is the main variable
of interest. The PLOT2 approach cannot, therefore, be used for this purpose in
this particular case.

The program below constructs an Annotate data set to plot small histogram
bars at the bottom of the plot showing the mean percent of errors for each
condition. After reading the data, it uses PROC MEANS to create an output data
set containing the mean of ERR for each condition.

```
proc means data=stern noprint;
    var err;
    by itemtype ss;
    output out=meanerr mean=err;
```

The DATA step below produces the Annotate data set BARS. Because the ERR
variable is a percentage and the values are typically less than 10 percent, the data
percentage coordinate system (YSYS='1') of the Annotate facility is used for the Y
variable, so that the bars occupy the bottom ten percent of the plot and will not
interfere with the RT data. Bars for the two types of items are displaced slightly to
the left and right of the value of the horizontal SS variable. After the last mean
has been processed, a small axis with a scale for error percent is drawn at the
right hand edge of the plot.

```
data bars;
    set meanerr end=eof;
    length function style text $8;
    xsys = '2';              /* use data values for X */
    ysys = '1';              /* use data percents for Y */

    line = 1;
    if itemtype= 'Words' then x= ss + .02;
                         else x= ss - .02;
    y = 0;
    function = 'MOVE';   output;
    if itemtype= 'Words' then do;
        x = ss + .15; style='R3'; color='RED  '; end;
    else do;
        x = ss - .15; style='L3'; color='BLACK'; end;
    y = err;
    function = 'BAR';     output;

    if eof then do;                  /* Draw scale for error % */
        xs = ss + .28;               /* a bit beyond last X val*/
        function = 'MOVE';
        y=0; x=xs  ;       output;    /* move to x-axis */
        function = 'DRAW';
        y=5; x=xs  ;       output;    /* 5% error      */
             x=xs - .03;   output;    /* & tic mark    */
```

```
              function = 'MOVE';
                  x=xs;                     output;
              function = 'DRAW';
              y=10;x=xs;                     output;     /* 10% error      */
                  x=xs - .03;                output;     /* tic mark       */
              function = 'LABEL';
              style='SIMPLEX'; size=1.3;
              x=xs+.10; text='10';          output;     /* label the value*/
              x=xs;
              y=14;      text='Errors';output;          /* label the axis */
              end;
```

The PROC GPLOT step is the same as used before, with the exception of the ANNOTATE option in the PLOT statement.

```
    proc gplot data=stern ;
       plot rt * ss = itemtype
              / legend=legend1
                vaxis = axis1   vminor=1
                haxis = axis2
                annotate=bars                        /* annotate with bars */
                name='GB0714' ;
       symbol1 v=none l=1  i=stdtmj w=2 c=black;
       symbol2 v=none l=21 i=stdtmj w=2 c=red;
       legend1 label=(h=1.5) value= (h=1.5) shape= line(4);
       axis1   label= (a=90 r=0 h=1.5)
               value=(h=1.4)
               order= 300 to 600 by 100
               offset= (2);
       axis2   label= (h=1.5)
               value=(h=1.4)
               order = 1 to 5 by 2
               offset=(4,8)                          /* leave room for error bars */
               minor =none;
```

If a trade-off between speed and accuracy has occurred, short reaction times tend to be accompanied by a large number of errors; the histogram bars have a pattern opposite to that of the reaction time curves. In Output 7.14, the histogram bars for percent errors demonstrate that differences in reaction time are not merely due to differences in errors. (A color version of this output appears as Output A3.32 in Appendix 3.) Thus, the reaction time data can be interpreted without concern for this possible artifact.

Output 7.14
Plot of Stern Data,
Showing Means on
a Second Variable

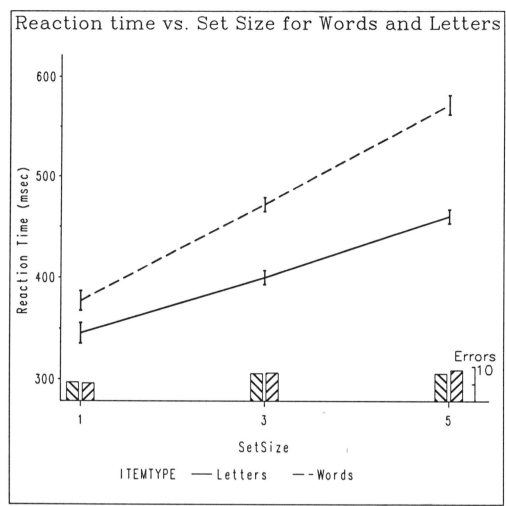

7.4 Plotting Interactions

The most challenging problems of graphical display for ANOVA data concern the plotting of interaction effects in multifactor designs. This section first describes the basic techniques for constructing these plots for two-factor designs. Plots for three-factor designs are discussed next; these plots are usually constructed with several panels that can be assembled into one figure with the GREPLAY procedure. Finally, this section presents an *all effects plot* devised by Tukey (1977) to display all effects and residuals in designs with three or more factors, and other techniques such as plots of contrasts for quantitative factors.

7.4.1 Two-Factor Designs

In a two-way design, the main effects of the factors may not be interpretable if the factors interact, depending on the size and pattern of the interaction effects. The plot of means for each treatment combination, as in Output 7.12, shows the nature of the interaction and helps you determine if the main effects are interpretable. For example, in Output 7.12, there is an apparent interaction between Item Type and SetSize because the two curves are not parallel. However, it is clear that reaction time for words is always greater than that for letters, so it is possible to interpret a main effect for Item Type. (The interaction arises because the difference between words and letters increases with set size.) Likewise, it is clear that reaction time increases with set size for both item types, so a main effect for set size can be understood.

On the other hand, if the curves cross or show a more complex pattern, you would not be able to interpret the main effects separately because the nature of the effect for one factor would differ for the levels of the other factors. Still, one plot of treatment means is usually sufficient to interpret the ANOVA procedure results.

7.4.2 Three-Factor Designs

When three or more factors are studied simultaneously, the ANOVA tests are straightforward extensions of the two-factor case. However, there are more possibilities for interaction effects, and you usually need to study several plots of the data to understand the nature of the factor effects.

In a three-factor design, with factors A, B, and C, there are three possible two-factor interactions, denoted A*B, A*C, and B*C. As in a two-factor study, an A*B interaction means that the effects of one factor, A, differ over the levels of the other factor, B. In addition, there may be a three-factor interaction, A*B*C. If present, this would indicate that a specific two-factor interaction, A*B, for example, differed over the levels of the other factor, C. To understand these effects, there are two different types of plots that can be made:

Three-way plots

First, we can plot the individual treatment means, denoted $\bar{y}_{ijk}$, where the subscripts $i=1, \ldots, a; j=1, \ldots, b; k=1, \ldots, c$ refer to the levels of factors A, B, and C, respectively. This plot is usually made up of several panels, with the data for each level of one factor shown separately. For example, we might show the effects of factors A and B (as in a two-factor design) separately for each level of factor C.

With PROC GPLOT, these panels can all be plotted in one step through the use of a BY statement. For example, separate plots for each level of factor C are produced by the following statements:

```
proc gplot;
   plot Y * A = B ;
   by C;
```

As in the two-way case, separate curves can be drawn for each level of factor B by appropriate SYMBOL statements. Then the separate panels can be combined into one figure with PROC GREPLAY, so that the pattern of A*B means can be compared visually over the levels of C.

We could, of course, choose any of the factors A, B, or C as the BY variable to define the panels, and in general these plots give different views of the data and make some visual comparisons easier than others. I generally choose a factor with the smallest number of levels such as the BY variable, so that the individual panels will not be made too small.

Two-way marginal plots

The three-way plot helps to show the pattern of the A*B*C effect and the specific two-factor effects in each panel (for example, the specific A*B effect for each level of C). But to understand any overall two-factor interactions (between factors B and C, for example), it is useful to plot the marginal means averaged over the remaining factor. For example, if the ANOVA table shows a significant B*C interaction, a plot of the means, $\bar{y}_{.jk}$, averaged over factor A should be made.

Again, there are three different such plots that we could make, and it is often useful to make more than one.

To make these two-way and three-way plots with SAS/GRAPH software, it is most convenient to obtain all the cell and marginal means (and standard errors of each) with PROC SUMMARY and plot the values from the output data set, rather than specifying I=STD in a plot of the raw data, as I did earlier. In addition, I'll illustrate a novel plot suggested by Tukey (1977) that combines all effects into one display.

Example: Effects of Linguistic Activity on Long-Term Retention

The data for this example are fictitious, but constructed for a reasonable study of human memory (Keppel 1973). Subjects learn a list of verbal materials, and their memory for these items is tested after varying retention intervals: 1, 4, or 7 hours (factor A in the example below). A classical theory of forgetting supposes that people forget because linguistic activity during the retention period interferes with the material they learned. To test this theory, subjects are exposed to either an activity that minimizes linguistic activity during the retention interval (such as a numerical task) or an activity that maximizes linguistic activity. This factor, amount of linguistic activity, is labeled factor C in the example below. A third factor, factor B in the example below, manipulates the type of material that different individuals learn: items that have a low, medium, or high correspondence to language.

Keppel (1973, Table 13-8) gives the data in terms of the sum, for $n=15$ subjects in each of the $3 \times 3 \times 2$ treatment groups. To show the analysis from raw data, the DATA step below reads each cell sum and generates 15 scores with the same (approximate) mean and standard deviation, equal to the MSE$=1.49$ given by Keppel.

```
proc format;
    value Bfmt 1='Low'     2='Med'     3='High' ; /* material */
    value Cfmt 1='Minimum' 2='Maximum';          /* activity */
data recall;
    label A='Retention Interval (hours)'
          B='Material'
          C='Activity'
          Y='Words Recalled';
```

```
      drop sum mean;
      input a b c sum;
      mean = sum / 15;
      do subj = 1 to 15;
          y = mean + sqrt(1.49)*normal(7631631);
          output;
          end;
  cards;
  1 1 1   205
  1 2 1   210
  1 3 1   208
  4 1 1   198
  4 2 1   193
  4 3 1   197
  7 1 1   182
  7 2 1   177
  7 3 1   179
  1 1 2   209
  1 2 2   203
  1 3 2   211
  4 1 2   178
  4 2 2   182
  4 3 2   197
  7 1 2   146
  7 2 2   169
  7 3 2   182
  ;
  proc glm;
      class A B C;
      model Y = A | B | C ;
```

The PROC GLM step produces the ANOVA table shown in Output 7.15. From this table, the effects of factors A (retention interval) and B (type of material) are highly significant. However, these main effects may not be interpretable due to the presence of interactions. There is a strong three-way interaction, A*B*C, and the B*C interaction between material and amount of linguistic activity is also significant.

Output 7.15   *GLM Procedure Output for Recall Data*

```
                    GENERAL LINEAR MODELS PROCEDURE
                      CLASS LEVEL INFORMATION
                   CLASS    LEVELS    VALUES
                     A         3       1 4 7
                     B         3       1 2 3
                     C         2       1 2

              NUMBER OF OBSERVATIONS IN DATA SET = 270
DEPENDENT VARIABLE: Y        Words Recalled
SOURCE               DF     SUM OF SQUARES      MEAN SQUARE      F VALUE
MODEL                17      328.06559549      19.29797621       14.95
ERROR               252      325.30721900       1.29090166      PR > F
CORRECTED TOTAL     269      653.37281450                       0.0001

R-SQUARE          C.V.          ROOT MSE            Y MEAN
0.502111        8.9264        1.13617853        12.72825655

SOURCE               DF        TYPE I SS     F VALUE    PR > F
A                     2     254.76246470       98.68    0.0001
B                     2      13.54022175        5.24    0.0059
A*B                   4       9.19084199        1.78    0.1333
C                     1       2.86265713        2.22    0.1377
A*C                   2       5.63862259        2.18    0.1147
B*C                   2       9.48171805        3.67    0.0268
A*B*C                 4      32.58906928        6.31    0.0001
```

The means and standard errors for all cells, as well as all the marginal means, are obtained and saved in the output data set MEANS with the PROC SUMMARY step below. The MEANS data set is shown in Output 7.16.

The different types of means are distinguished by the variable _TYPE_, whose values go from 0 (for the grand mean, $\bar{y}_{...}$) to 7 (for the cell means, $\bar{y}_{ijk.}$) for a three-factor design. Based on the ANOVA test results, I decided to make a three-way plot of the A*B*C means and two-way marginal plots of the A*C means (averaged over factor B) and the B*C means (averaged over factor A). To do this, each set of means must be extracted to a separate data set. This operation is carried out in one DATA step to give the data sets ABC, AC, and BC.

```
proc summary data=recall;
   class A B C;
   var   Y;
   output out=means mean=YBAR stderr=STDERR;
proc print;
   Id _TYPE_;

data ABC
     AC BC;
   set means;
   label YBAR='Mean Number Recalled'
         C   ='Activity';
```

```
If _TYPE_ = '101'B then output AC;       /* _type_ 5 */
If _TYPE_ = '011'B then output BC;       /* _type_ 3 */
If _TYPE_ = '111'B then output ABC;      /* _type_ 7 */
format B Bfmt. C Cfmt.;
```

Output 7.16
SUMMARY
Procedure Output
for Recall Data

TYPE	A	B	C	_FREQ_	YBAR	STDERR
0	.	.	.	270	12.7283	0.094847
1	.	.	1	135	12.8312	0.121992
1	.	.	2	135	12.6253	0.145181
2	.	1	.	90	12.5574	0.180833
2	.	2	.	90	12.5828	0.157904
2	.	3	.	90	13.0446	0.149148
3	.	1	1	45	12.7873	0.188919
3	.	1	2	45	12.3275	0.306913
3	.	2	1	45	12.8237	0.238119
3	.	2	2	45	12.3418	0.203784
3	.	3	1	45	12.8826	0.208499
3	.	3	2	45	13.2066	0.212902
4	1	.	.	90	13.8862	0.125407
4	4	.	.	90	12.7894	0.123897
4	7	.	.	90	11.5092	0.136699
5	1	.	1	45	13.8814	0.171700
5	1	.	2	45	13.8910	0.184771
5	4	.	1	45	12.7959	0.169145
5	4	.	2	45	12.7829	0.182997
5	7	.	1	45	11.8164	0.171404
5	7	.	2	45	11.2019	0.204759
6	1	1	.	30	13.9901	0.208800
6	1	2	.	30	13.7884	0.230800
6	1	3	.	30	13.8801	0.217316
6	4	1	.	30	12.5783	0.213157
6	4	2	.	30	12.5347	0.169966
6	4	3	.	30	13.2553	0.236112
6	7	1	.	30	11.1038	0.261497
6	7	2	.	30	11.4253	0.225837
6	7	3	.	30	11.9985	0.196206
7	1	1	1	15	13.4704	0.252046
7	1	1	2	15	14.5097	0.280214
7	1	2	1	15	14.2864	0.311321
7	1	2	2	15	13.2904	0.297209
7	1	3	1	15	13.8872	0.306157
7	1	3	2	15	13.8729	0.319237
7	4	1	1	15	12.8132	0.330758
7	4	1	2	15	12.3434	0.266367
7	4	2	1	15	12.6551	0.268337
7	4	2	2	15	12.4142	0.213548
7	4	3	1	15	12.9194	0.292953
7	4	3	2	15	13.5911	0.359196
7	7	1	1	15	12.0782	0.307058
7	7	1	2	15	10.1293	0.230988
7	7	2	1	15	11.5297	0.302039
7	7	2	2	15	11.3208	0.344251
7	7	3	1	15	11.8413	0.284048
7	7	3	2	15	12.1557	0.274360

Constructing Three-Way Plots

For the three-way plot, I used factor C as the BY variable because it has only two levels. As described in Section 7.4.2, "Three-Factor Designs," the two separate plots produced by the following PROC GPLOT step are stored in a graphics catalog (named GRAPHS) and later combined in one figure with PROC GREPLAY:

```
proc gplot;
    plot Y * A = B / gout=GRAPHS ... ;
    by C;
```

In Output 7.17, the two panels are drawn separately and put together with PROC GREPLAY. (A color version of this output appears as Output A3.33 in Appendix 3.) However, when you use Y*A = B in the PLOT statement, each panel contains the same legend beneath the horizontal axis, which is redundant.

Instead, the program uses Annotate instructions to draw a custom legend identifying the levels of factor B.

Output 7.17　　　*Three-Way Plot of A*B Means for Each Level of Factor C*

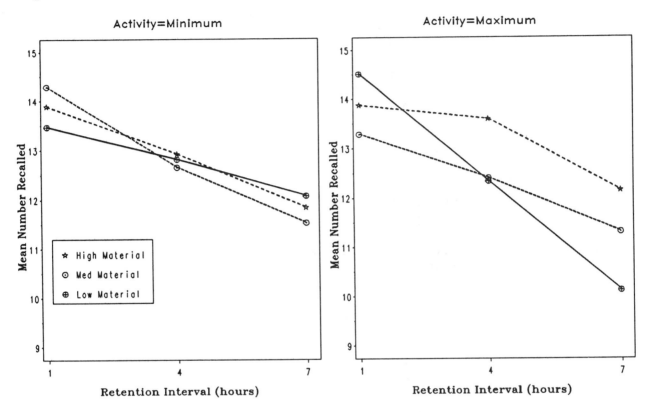

The ABC means must first be sorted by C, so that C can be used as a BY variable. The DATA step that produces the ANNOABC data set contains the instructions to draw the box with the legend for factor B in the first panel only, for C=1. In the PROC GPLOT step shown below, the Annotate data set is invoked by the following statement:

```
plot Y * A = B / ... ANNO=AnnoABC
by C;
```

In this form, the Annotate observations for C=1 are used only in the plot for C=1; Annotate observations for C=2 (if any) would be applied only to the plot for C=2. However, in order for this to work properly, all attributes (length, format, label) of the variable C must be identical in the means and Annotate data sets. One way to ensure this is to read variable C from the means data set, ABC.

```
proc sort data=ABC;
   by C;
data AnnoABC;
   length text $15 function $8;
   XSYS='2'; YSYS='2';
   set ABC (keep=C);              /* Annotate BY variable must */
   by  C;                         /*  agree in all attributes  */
   if C=1 and FIRST.C then do;    /* Legend in left panel only */
```

```
          X=1.1;
          Y=9.9;  function='MOVE';   output;
          X=3.4;  LINE=0;
          Y=11.2; function='BAR ';    output;

          X=1.4;  position='6';       size=1.7;
          Y=10.1; function='SYMBOL'; text='+';  COLOR='GREEN'; output;
                  function='LABEL';  text='  Low Material';    output;
          Y=10.5; function='SYMBOL'; text='-';  COLOR='BLUE';  output;
                  function='LABEL';  text='  Med Material';    output;
          Y=10.9; function='SYMBOL'; text='=';  COLOR='RED';   output;
                  function='LABEL';  text='  High Material';   output;
       End;
  /*-----------------------------------------------*
   |  Plot A * B * C Interaction, in two panels,  |
   |  A*B for C=1, A*B for C=2                     |
   *-----------------------------------------------*/
goptions NODISPLAY;         * device dependent;
goptions fby=duplex hby=1.8;
title ;
proc gplot data=ABC gout=GRAPHS;
   plot YBAR * A = B / vaxis=axis1 vminor=1 frame
                       haxis=axis2
                       nolegend              /* suppress bottom legend */
                       ANNO=AnnoABC;         /* add Annotate legend    */
   By C;
   axis1   label= (a=90 r=0 h=1.7 f=triplex)
           value=(h=1.7)
           order= (9 to 15) offset= (2);
   axis2   label= (h=1.7 f=triplex)
           value=(h=1.7)
           order= (1 to 7 by 3) offset=(2,4)
           minor =none;
   symbol1 v=+    h=1.7 l=1  i=join w=2 c=green;
   symbol2 v=-    h=1.7 l=3  i=join w=2 c=blue ;
   symbol3 v='='  h=1.7 l=20 i=join w=2 c=red  ;
   format B Bfmt. C Cfmt.;
run;
```

The three-way plot in Output 7.17 shows the nature of the A*B*C interaction. Under conditions of minimum linguistic activity (C=1, left panel), there is very little effect of type of material learned and little evidence of a specific A*B interaction: the curves are closely spaced and not far from being parallel. With the maximum linguistic activity (C=2, right panel), however, there is a strong A*B interaction: the curves diverge with increasing retention interval, so that the amount of forgetting is least for the high material and greatest for the low material.

Constructing Two-Way Plots

The two-way plots show the A*C means averaged over factor B and the B*C means averaged over factor A. These plots are drawn from the means contained in the data sets AC and BC, respectively, using separate PROC GPLOT steps shown below. Because these are different views of the same data, it is often useful to see them side-by-side to make visual comparisons. So in Output 7.18, these two two-way plots are combined into one display with PROC GREPLAY. (A color version of this output appears as Output A3.34 in Appendix 3.)

Output 7.18   *Side-By-Side Display of A*C and B*C Interactions*

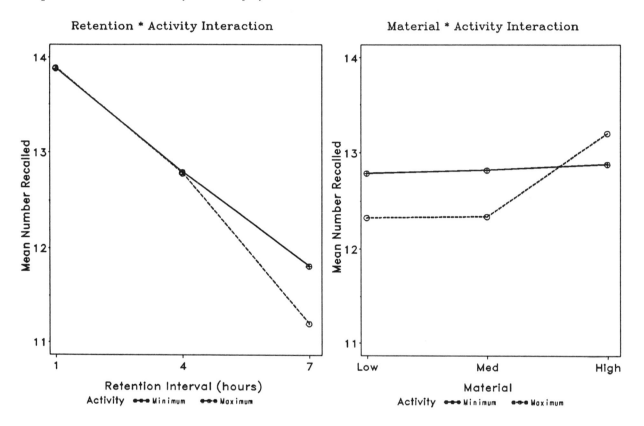

In the PROC GPLOT steps below, each of the graphs is also stored in the GRAPHS graphics catalog. The PROC GREPLAY step that follows defines one template for a two-panel display. The TREPLAY statement uses that template to plot both Output 7.17 and 7.18.

```
/*+----------------------------+*
| Plot A*C and B*C interactions |
*+----------------------------+*/
proc gplot data=AC gout=GRAPHS;
    plot YBAR * A = C / vaxis=axis1 vminor=1
                        haxis=axis2 frame
                        legend=legend1;
    legend1 label=(h=1.6 f=duplex 'Activity')
            value=(h=1.5);
    axis1   label=(a=90 r=0 h=1.8 f=duplex)
            value=(h=1.7 f=duplex)
            order=(11 to 14) offset= (2);
```

```
      axis2   label=(h=1.8 f=duplex)
              value=(h=1.7 f=duplex)
              order= (1 to 7 by 3) offset=(2,4)
              minor =none;
      Title H=1.8 'Retention * Activity Interaction';
   run;
   proc gplot data=BC gout=GRAPHS;
      plot YBAR * B = C / vaxis=axis1 vminor=1
                          haxis=axis2 frame
                          legend=legend1;
      axis2   label=(h=1.8 f=duplex)
              value=(h=1.7 f=duplex)
              offset=(2,4) minor =none;
      legend1 label=(h=1.6 f=duplex 'Activity')
              value=(h=1.5);
      title h=1.8 'Material * Activity Interaction';
   run;
   goptions DISPLAY;           * device dependent;

   proc greplay igout=graphs nofs
           template=plot2 tc=templt ;
      tdef plot2
         des="2 plots left/right"
         1/ ulx=0  uly=100   urx=50  ury=100  /*  left      */
            llx=0  lly=0     lrx=50  lry=0
         2/ copy=1 xlatex=+50;               /*        right */
      treplay 1:1 2:2 ;      /* name=GB0717 */  /* ABC plot    */
      treplay 1:3 2:4 ;      /* name=GB0718 */  /* AC, BC plot */
```

7.4.3 All Effects Plot

Tukey (1977, p. 451) illustrates a novel way to display simultaneously all effects and residuals from the analysis of a table with three or more factors. The idea is simply to display the estimates of the effects for each term in the fitted model. The result gives a visual analog of what is summarized in the ANOVA table, but also portrays large individual values for any effect. For a three-way design, the linear model for a complete factorial design expresses the observation y_{ijkm} as

$$y_{ijkm} = \mu + a_i + \beta_j + \gamma_k + (\alpha\beta)_{ij} + (\alpha\gamma)_{ik} + (\beta\gamma)_{jk} + (\alpha\beta\gamma)_{ijk} + \varepsilon_{ijkm} \qquad (7.5)$$

where a_i, $i=1, \ldots, a$ are the main effects for the levels of factor A, defined so that $\Sigma a_i = 0$; β_j and γ_k are the main effects for the levels of factors B and C respectively, with similar restrictions, and the remaining terms in equation 7.5 are the corresponding interaction effects, each defined so that they sum to zero over all subscripts.

For the recall data, the estimated (least squares) effects for the three-way design are shown in Table 7.1. In this table, each row corresponds to one cell in the design, and each column to one main effect or interaction. The sum of entries in each row (when added to the grand mean) gives the fitted value for that cell, and n times the sum of squares of each column is the ANOVA sum of squares for that effect.

Table 7.1   *Effect Values for Recall Data*

Cells			Effect Values						
A	B	C	α_i	β_j	γ_k	$(\alpha\beta)_{ij}$	$(\alpha\gamma)_{ik}$	$(\beta\gamma)_{jk}$	$(\alpha\beta\gamma)_{ijk}$
1	1	1	1.156	−0.267	0.267	0.222	−0.267	0.311	−0.444
1	2	1	1.156	−0.089	0.267	0.011	−0.267	0.022	0.211
1	3	1	1.156	0.356	0.267	−0.233	−0.267	−0.333	0.233
2	1	1	0.033	−0.267	0.267	0.078	0.078	0.311	0.011
2	2	1	0.033	−0.089	0.267	−0.133	0.078	0.022	0.000
2	3	1	0.033	0.356	0.267	0.055	0.078	−0.333	−0.011
3	1	1	−1.189	−0.267	0.267	−0.300	0.189	0.311	0.433
3	2	1	−1.189	−0.089	0.267	0.125	0.189	0.022	−0.211
3	3	1	−1.189	0.356	0.267	0.178	0.189	−0.333	−0.222
1	1	2	1.156	−0.267	−0.267	0.222	0.267	−0.311	0.444
1	2	2	1.156	−0.089	−0.267	0.011	0.267	−0.022	−0.211
1	3	2	1.156	0.356	−0.267	−0.233	0.267	0.333	−0.233
2	1	2	0.033	−0.267	−0.267	0.078	−0.078	−0.311	−0.011
2	2	2	0.033	−0.089	−0.267	−0.133	−0.078	−0.022	−0.000
2	3	2	0.033	0.356	−0.267	0.055	−0.078	0.333	0.011
3	1	2	−1.189	−0.267	−0.267	−0.300	−0.189	−0.311	−0.433
3	2	2	−1.189	−0.089	−0.267	0.125	−0.189	−0.022	0.211
3	3	2	−1.189	0.356	−0.267	0.178	−0.189	0.333	0.222

In Tukey's display, the abscissas on the plot are the various effects in the model: A, B, and C for the main effects; AB, AC, BC, and ABC for the interactions; plus the residual terms, ε_{ijkm}. Above the abscissa values A, B, and C, the estimates of the main effect values, a_i, β_j, and γ_k, are plotted individually and labeled. At the abscissa for an interaction, such as AB, the estimates of the corresponding model terms, such as $(\alpha\beta)_{ij}$, are either plotted individually or summarized by a boxplot. Finally, all the residuals are summarized by a boxplot.

Because the *F* values in the ANOVA table compare the variability of the terms for any effect with the variability of the error (residual) terms, Tukey's plot shows the information in the ANOVA table graphically.

The plot that I'll describe here, shown in Output 7.20, captures the spirit of Tukey's ingenious display, but not all its details. In particular, Output 7.19 uses I=HILO in the SYMBOL statement to show the range of effect values for interactions rather than a boxplot symbol, which would be less sensitive to extreme values.*

---

* In Version 6, you can specify I=BOX in the SYMBOL statement for interaction effects to produce a display less affected by extreme values.

To construct the display, the unique values in each column in Table 7.1 are entered in the data set EFFECTS in the format that appears in the output data set from PROC SUMMARY* (compare with Output 7.16). An additional variable, named EFFECT, is the horizontal variable for the plot, and the FORMAT procedure is used to provide labels for the values.

```
title h=1.4 'All effects plot';
proc format;
    value eff 1='A' 2='B' 3='C' 4='AB' 5='AC' 6='BC' 7='ABC' 8='Err';
data effects;
    input effect _TYPE_ A B C value ;
    factors = n(of A B C);
    label value  = 'Effect Value';
    format effect eff.;
cards;
1 4  1 . .    1.1557
1 4  4 . .    0.0333
1 4  7 . .   -1.189
2 2  . 1 .   -0.2667
2 2  . 2 .   -0.0888
2 2  . 3 .    0.3555
3 1  . . 1    0.2667
3 1  . . 2   -0.2667
4 6  1 1 .    0.2222
4 6  1 2 .    0.0108
4 6  1 3 .   -0.233
4 6  4 1 .    0.078
4 6  4 2 .   -0.1333
4 6  4 3 .    0.0553
4 6  7 1 .   -0.3002
4 6  7 2 .    0.1225
4 6  7 3 .    0.1777
5 5  1 . 1   -0.2666
5 5  1 . 2    0.0778
5 5  4 . 1    0.1888
5 5  4 . 2    0.2666
5 5  7 . 1   -0.0778
5 5  7 . 2   -0.1888
6 3  . 1 1    0.3111
6 3  . 1 2    0.0223
6 3  . 2 1   -0.3334
6 3  . 2 2   -0.3111
6 3  . 3 1   -0.0223
6 3  . 3 2    0.3334
7 7  1 1 1   -0.4443
7 7  1 1 2    0.2111
7 7  1 1 3    0.2332
```

---

* The effect values can be obtained with the SOLUTION option in the MODEL statement in PROC GLM and captured in a data set using the PRINTTO procedure. The calculations could also be done with the IML or REG procedure. However, to get these values into a data set in the required form would have required far too much programming for the point to be made here.

```
7 7  1 2 1    0.0109
7 7  1 2 2    0.0002
7 7  1 2 3   -0.0111
7 7  4 1 1    0.4334
7 7  4 1 2   -0.2113
7 7  4 1 3   -0.2221
7 7  4 2 1    0.4443
7 7  4 2 2   -0.2111
7 7  4 2 3   -0.2332
7 7  7 1 1   -0.0109
7 7  7 1 2   -0.0002
7 7  7 1 3    0.0111
7 7  7 2 1   -0.4334
7 7  7 2 2    0.2113
7 7  7 2 3    0.2221
;
```

An Annotate data set, LABELS, is used to label the points for individual effect values. The formats A., B., and C. are used to code the levels for each effect so that, for example, "1" represents a_1, and "1M" represents $(\alpha\beta)_{12}$. All main effects are labeled, as are interaction values greater than .25 in magnitude, a value chosen somewhat arbitrarily.

```
proc format;
    value a 1='1' 4='4' 7='7' .=' ';
    value b 1='L' 2='M' 3='H' .=' ';
    value c 1='c' 2='C'       .=' ';
data labels;
    set effects;
    xsys = '2'; ysys = '2';
    if factors = 1 | abs(value) > .25 then do;
        x = effect + .1;
        y = value;
        text = compress(put(A,a.)||put(B,b.)||put(C,c.));
        function = 'LABEL'; position='6'; size=1.4;
        output;
    end;
proc print;
```

Next, the residuals from the three-way model are obtained from PROC GLM and appended to the EFFECTS data set as EFFECT=8. In the PROC GPLOT step, a different SYMBOL statement is used for each EFFECT value.

```
proc glm data=recall;
    class a b c;
    model y = a | b | c;
    output out=resids r=value;
data resids;
    set resids;
    effect = 8;
data effects;
    set effects resids;
```

```
proc gplot data=effects  ;
   plot value * effect = effect
      / vaxis=axis1 haxis=axis2
        vminor=4 hminor=0
        vref=0 lvref=21
        nolegend anno=labels
        name='GB0719' ;
   axis1 label=(a=90 r=0 h=1.5 f=duplex)
         value=(h=1.5)
         order=(-2 to 2);
   axis2 label=(h=1.5 f=duplex)
         value=(h=1.5) offset=(4);
   symbol1 v=SQUARE   h=1.5 i=none color=red;    /* A   */
   symbol2 v=TRIANGLE h=1.5 i=none color=red;    /* B   */
   symbol3 v=DIAMOND  h=1.5 i=none color=red;    /* C   */
   symbol4 v=none     i=hilob  color=blue ;      /* AB  */
   symbol5 v=none     i=hilob  color=blue ;      /* AC  */
   symbol6 v=none     i=hilob  color=blue ;      /* BC  */
   symbol7 v=none     i=hilob  color=black;      /* ABC */
   symbol8 v=none     i=std3mb color=green;      /* Err */
```

Output 7.19 shows that the effect of retention interval, factor A, is the largest of all effects, with the effects of the levels approximately equally spaced. Each column on the plot shows the effects for one term in the three-way model. Terms that are large relative to the estimated standard error are labeled individually. The significant main effect of material, factor B, appears due largely to the difference between the high material and the others. The A*B interaction is not significant, but condition 7L appears low relative to expectation if these factors were completely additive. Similarly, the plot identifies treatment combinations 1Lc and 7Mc as low, and 4Lc and 4Mc as high, relative to what would be expected if there were no A*B*C interaction effect. The magnitude of all the effects can be roughly compared to the size of the bar for error.

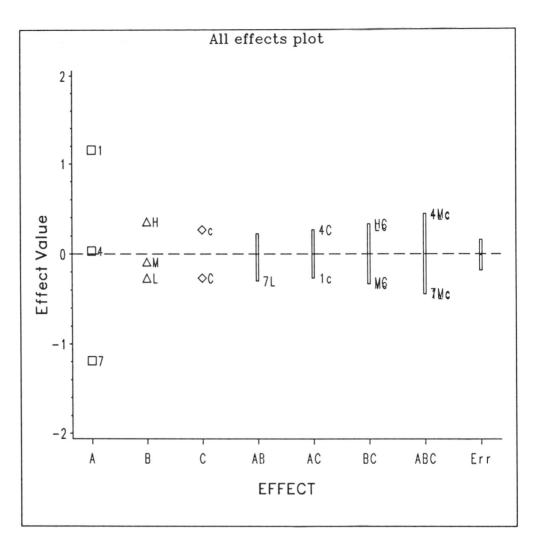

7.4.4 Related Techniques

There are several other useful graphical techniques for plotting three-way and higher-order factorial designs, though space does not permit me to explore them in detail. Fox (1987) describes a general procedure for constructing effect displays for generalized linear models, including regression models, ANOVA models, logit models, and so on. The basic idea is to identify the high-order terms in the linear model (highest order significant interactions and their low-order relatives). Then fitted (adjusted) effects are calculated and plotted by varying each independent variable in an effect over its range, while fixing the values of other independent variables.

When planned contrasts among the levels of one factor are of particular interest, one idea is to calculate the contrast values for the cell means and plot these over levels of the other factors. For example, you may want to test orthogonal polynomial contrasts (linear, quadratic, and so on) for a quantitative factor. In the recall data, factor A is the retention interval (1, 4, or 7 hours), and the linear contrast for factor A is the slope of the curves in Output 7.17, which is interpretable as rate of forgetting (number of items forgotten per hour); the quadratic contrast would reflect the curvature of these curves, or the non-linear

forgetting effect. While these contrasts are tested in PROC GLM with the
CONTRAST statement, you can also plot these values by calculating the contrasts
among the means.

The program below shows the calculation of the SLOPE and CURVE contrasts
for the ABC means in the recall data.

```
/*----------------------------------------*
 | Contrast plot for quantitative factor (A)|
 *----------------------------------------*/
proc sort data=ABC;
   by B C A;              * factor A last;
proc transpose data=ABC out=BCA;
   by B C;
   var ybar stderr;
proc print;
data BCA;
   set BCA;
   if _name_='YBAR' then do;        * compute contrasts;
      mean = mean(of col1-col3);    * B * C :  1  1  1  ;
      slope= col3 - col1;           * linear: -1  0  1  ;
      curve= 2*col2 - (col1 + col3);  * quad:   -1  2 -1  ;
   end;
   else do;                         * & their std errors;
      mean = sqrt( 2 * uss(of col1-col3) );
      slope= sqrt( 2 * uss(of col1-col3) );
      curve= sqrt( 6 * uss(of col1-col3) );
   end;
proc format;
   value cont 0='Mean'  1='Slope' 2='Curve';
data BC;
   set BCA;
   drop col1-col3;
   if _name_='YBAR';
   contrast = 0; value = mean ; output;
   contrast = 1; value = slope; output;
   contrast = 2; value = curve; output;
proc sort data=BC;
   by contrast;
proc print;
proc gplot data=BC  ;
   plot value * B = C
      / vminor=1 hminor=0 frame
        vaxis=axis1 haxis=axis2
        legend=legend1
        name='GB0720' ;
   by contrast;
   format contrast cont. B Bfmt. C Cfmt.;
   axis1 label=(a=90 r=0 h=1.5 'Contrast Value');
   axis2 label=(h=1.5) value=(h=1.5) offset=(5);
   symbol1 v=+ h=1.4 i=join color=black l=20;
   symbol2 v=$ h=1.4 i=join color=red   l=1 ;
   legend1 label=(h=1.5 'Activity') value=(h=1.5);
```

The plot of the SLOPE contrast shown in Output 7.20 shows quite dramatically how the B*C groups differ in their rate of forgetting. The value plotted is the rate of forgetting over the retention interval for each of the B*C conditions shown in Output 7.17.

Output 7.20
Contrast Plot of Slopes for Recall Data

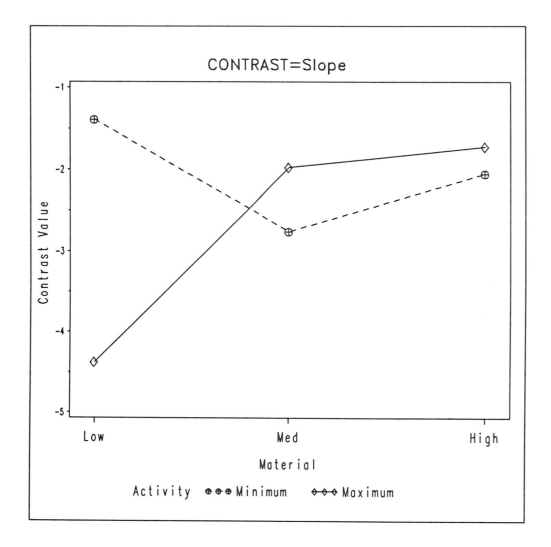

To interpret this plot, recall that the two connected curves show how rate of forgetting varies with the type of material learned. Thus, with a maximum amount of linguistic activity during the retention interval (solid lines in Output 7.20), the rate of forgetting increases as the material learned becomes more like natural language. This would be taken as support for the theory that forgetting is due, at least in part, to interference of similar activity during retention. The dashed lines in the figure show that with minimum linguistic activity during the retention interval, there is no systematic relation between rate of forgetting and how language-like the to-be-remembered material is, which would also be predicted by interference theory.

Notice that if the bulk of a higher-order effect, such as A*B*C, is accounted for by a contrast component (A-linear*B*C, here), the calculation of that contrast captures most of a three-way display on a two-dimensional plot. Some related examples for three-factor models are given by Monlezun (1979).

Another technique based on plotting contrasts is the *half-normal plot* suggested by Daniel (1959) for plotting all effects in a 2^n design. In a 2^5 design (5 factors, each with 2 levels), for example, there are 31 single degree of freedom contrasts, each of which represents one main effect or interaction in the ANOVA table. The half-normal plot is a quantile-quantile plot of the absolute values of each contrast against the quantiles of the normal distribution folded around 0 (the half normal). The idea is that the contrasts for insignificant effects all estimate the error standard deviation, σ, and should plot as a straight line whose slope is an estimate of σ. The nonnull effects appear as outliers, points with contrasts much greater than the half-normal quantiles. Johnson and Tukey (1987) extend the usefulness of half-normal plots considerably, discussing the extension to arbitrary factorial designs and the handling of a priori contrasts as opposed to those suggested by the data.

7.5 Repeated Measure Designs

In repeated measure designs, each subject is measured on one or more occasions. In a learning experiment, for example, a person may receive a series of study-test trials on a list of vocabulary words.

Statistically, repeated measure designs must be handled differently from the completely randomized (between subjects) designs described earlier, because the repeated scores for one individual are dependent, while scores for different individuals are independent. Computationally, the data for repeated measure designs must be arranged differently, and this introduces an additional step in plotting the data.

In the previous examples, each score is a separate observation in the data set, with variables Y and factor levels A, B, and so on. In a repeated measure design, analyzed using the REPEATED statement of PROC GLM, each subject is a separate observation, and PROC GLM requires the repeated scores to be variables, for example, Y1, Y2, and so on. In order to plot such data, therefore, you have to transpose the repeated variables to separate observations on one new variable and construct a second variable with values 1, 2, and so on to plot against. This puts the data in the same univariate format as in the previous examples.

7.5.1 Example: Multitrial Free Recall

The IFR data considered in Section 7.2, "Plotting Means in a One-Way Design," were actually collected in a multitrial experiment. In one experiment (Friendly and Franklin 1980, Exp. 2), subjects studied a list of words on each of five trials and attempted to recall as many as they could after each study trial. There were six groups of $n = 10$ in a 3×2 between-subjects design. One factor comprised the three presentation-order conditions described earlier (BEGIN, MESH, and IFR). The second factor represented whether the words that the subject had not recalled on the previous trial were tagged (by printing an asterisk next to the word when presented on the current trial) or not tagged.

The number of words recalled correctly on each of the five trials is recorded in the variables COR1—COR5 in the data set D8 below. The formats ORD. and TAG. are used to code the GROUP number into the levels of the ORDER and TAG factor variables.

```
title 'IFR EXPERIMENT D8';
proc format;
    value ord   1='SFR'   2='BEGIN'   3='MESH'
                4='SFR'   5='BEGIN'   6='MESH' ;
    value tag   1='NO '   2='NO '     3='NO '
                4='YES'   5='YES'     6='YES'  ;
data d8;
    length tag $3;
    input subj group cor1-cor5 ;
    tag = put(group, tag. );
    order=put(group, ord. );
cards;
 11 1     13     26     36     39     39
 98 1      6     14     19     23     25
 12 1     11     24     23     36     37
 99 1     10     12     13     18     25
 25 1      6     17     24     25     29
 ... (observations omitted)
 16 6     11     23     29     34     33
 98 6     12     21     33     35     37
 53 6      5     15     25     25     34
 96 6     13     26     36     38     39
 97 6     19     35     32     32     36
;
```

The repeated measures analysis of variance is carried out with PROC GLM using the statements below. Note that the repeated scores, COR1—COR5, are listed as separate dependent variables in the MODEL statement.

```
proc glm data=d8;
    class tag order;
    model cor1-cor5 = tag order tag*order / nouni;
    repeated trials 5 polynomial / nom;
```

For these data, we would commonly plot the learning curves, showing the mean recall for each group as a function of trials. Because there are six groups and the curves are not widely separated, the plot I'll construct has separate panels for the tagged/nontagged factor. The plot, shown in Output 7.21, also illustrates use of an Annotate data set to draw standard error bars and custom plot legends. (A color version of this output appears as Output A3.35 in Appendix 3.)

Output 7.21   *Mean Learning Curves for D8 Data*

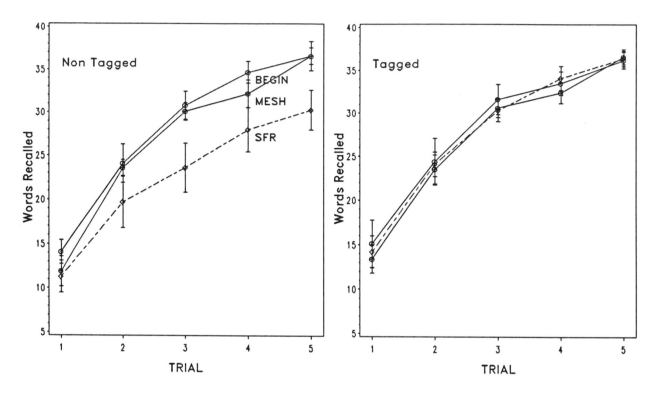

The scores, COR1—COR5, really represent a single variable, number of words recalled, measured over five trials. To plot a mean learning curve for each group, you need to rearrange the data so there is one score variable, RECALL, and a second TRIAL variable. That is, the first two data lines from data set D8 (for subjects 11 and 98) should be transposed to give the five observations for each subject, as shown below:

OBS	TAG	SUBJ	ORDER	TRIAL	RECALL
1	NO	11	SFR	1	13
2	NO	11	SFR	2	26
3	NO	11	SFR	3	36
4	NO	11	SFR	4	39
5	NO	11	SFR	5	39
6	NO	98	SFR	1	6
7	NO	98	SFR	2	14
8	NO	98	SFR	3	19
9	NO	98	SFR	4	23
10	NO	98	SFR	5	25

This rearrangement is accomplished with the DATA step that produces the data set D8PLOT, shown below.

```
data d8plot;
   set d8;
   keep subj tag order trial recall;
   trial=1; recall= cor1;  output;
   trial=2; recall= cor2;  output;
   trial=3; recall= cor3;  output;
   trial=4; recall= cor4;  output;
   trial=5; recall= cor5;  output;
```

After this step, the data are in the same form as an ordinary three-factor design, with factors TAG, ORDER, and TRIAL, so the steps in plotting the data are quite similar to what was done in Section 7.4.2. We use PROC SUMMARY to get means and standard errors for the recall score on each TRIAL factor. The _TYPE_=7 observations in the output data set contain the trial means for each combination of the TAG and ORDER factors; these means are extracted to data set GPMEAN for the plot, then sorted by the TAG factor.

```
proc summary data=d8plot;
   class tag order trial;
   var recall ;
   output out=gpvar  mean=recall stderr=sd;
data gpmean;
   set gpvar;
   if _type_=7;
proc sort data=gpmean;
   by tag;
```

The standard error bars and custom legends inside the panels are created with the Annotate data set BARS. The label for the levels of the TAG factor are drawn in the upper-left corner of the plot, using data-system coordinates. The labels for the curves for each level of the ORDER factor are drawn near the points for the mean on Trial 4 in the Non Tagged panel. These locations for the labels were determined by inspection.

```
data bars;
   set gpmean;
   by tag;
   xsys = '2'; ysys = '2';
   if first.tag then do;
      x=1; y=36;
      function = 'LABEL'; position='6';
      style = 'DUPLEX';   size=1.9;
      if tag='NO ' then text='Non Tagged';
                   else text='Tagged';
      output;
      end;
   x = trial;
   y = recall;
```

```
      select (order);
         when ('BEGIN') color='RED  ';
         when ('MESH' ) color='BLUE ';
         when ('SFR'  ) color='BLACK';
      end;
      function = 'MOVE';  output;
      y = recall - sd;
      function = 'DRAW';  output;
      link tips;
      y = recall + sd;
      function = 'DRAW';  output;
      link tips;

      if tag='NO ' & trial=4 then do;
         y = recall;
         x=trial+.1;
         position = 'F';
         size=1.7; style='DUPLEX';
         function = 'LABEL';
         text = order;
         output;
         end;
      return;

   tips:                      * draw top and bottom of bars;
     x = trial - .03; function='DRAW'; output;
     x = trial + .03; function='DRAW'; output;
     x = trial       ; function='MOVE'; output;
     return;
   run;
```

In the PROC GPLOT step, the statement

```
   BY TAG;
```

produces one plot for each value of the TAG variable. The Annotate observations
in the data set BARS are applied to the panel for the corresponding value of TAG.
The BY variable TAG must have the same attributes in both the means and
Annotate data sets.

```
   goptions NODISPLAY;           * device dependent;
   proc gplot data=gpmean gout=graphs;
      plot recall * trial = order
           / anno=bars vaxis=axis1 haxis=axis2
             hminor=0 vminor=4 frame nolegend;
      by tag;
      axis1 order=(5 to 40 by 5)
            value=(h=1.7)
            label=(a=90 r=0 h=1.9 f=duplex 'Words Recalled');
      axis2 label=(h=1.9 f=duplex)
            value=(h=1.7) offset=(4);
```

```
      symbol1 h=1.4 v=- i=join c=RED;
      symbol2 h=1.4 v=+ i=join c=BLUE;
      symbol3 h=1.4 v=$ i=join c=BLACK 1=25;
      title ' ';
   run;
```

Finally, the separate panels are joined on one plot using PROC GREPLAY, which produces Output 7.21.

```
   goptions DISPLAY;              * device dependent;

   proc greplay igout=graphs nofs
        template=plot2
        tc=templt ;
     tdef plot2
        des="2 plots left/right"
        1/ ulx=0  uly=100   urx=50  ury=100    /*     left  */
           llx=0  lly=0     lrx=50  lry=0
        2/ copy=1 xlatex=+50;                   /*    right  */
     treplay 1:1 2:2 ;     /* name=GB0721 */
```

The left panel in Output 7.21 shows that the BEGIN and MESH groups recall more than the SFR control group when the words are not presented with tags. This result is interpreted to mean that presenting the words in the order that the subject recalled them on the previous trial facilitates learning. The right panel shows that there are no differences among the groups when the words are presented with a tag indicating the words they had previously recalled.

However, it is difficult to compare the curves in the left panel with those in the right panel. If these panels were superimposed, it would be seen that the Non Tagged SFR group recalls less than the other five groups, which do not differ from each other. The question is how we can show this visually. If the two sets of curves were more widely separated, a single plot of all six curves would be satisfactory. For these data, however, such a plot would make it extremely difficult to discriminate among the groups.

A more effective solution is to reduce the display to a two-way plot, as was done in the contrast plot in Section 7.4.4, "Related Techniques." For example, the two curves for the TAG factor could be represented as a single curve showing the difference between the tagged and nontagged conditions for each presentation order. Alternatively, because there is no interaction between trials and the between-subject factors (the curves in Output 7.21 all have the same shape), we can average over trials and plot the mean recall for each of the six groups. Such a plot is shown in Output 7.22. It is now clear that the Non Tagged SFR group differs from the others, but the other five groups recall equally well. Averaging over the repeated measure factor, when it does not interact with other factors, makes between-group comparisons easier.

Output 7.22
Mean Recall for
D8 Data

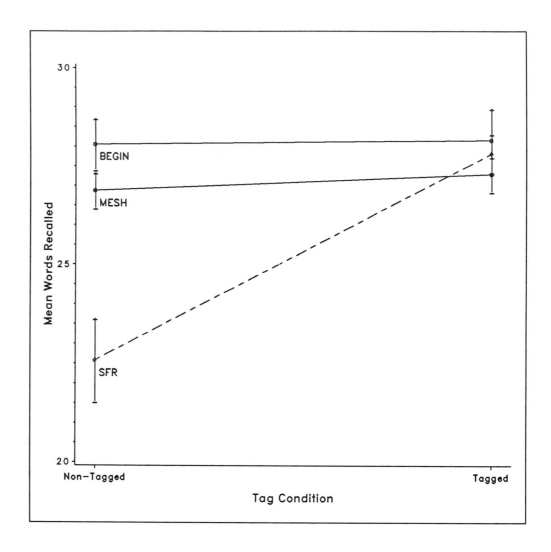

7.6 Displaying Two-Way Tables for *n* = 1 Designs

A two-way table records a single response measured for each combination of two independent variables or factors. For example, we may have the average scores on a test of reading skills for children of six different ages in each of five different school districts. Another example, shown in Table 7.2, is the reaction time, in seconds, for three subjects to decide whether each of three types of sentences is true or false.*

---

* Fictitious data.

Table 7.2
Two-Way Table
Example

DATA	SENT1	SENT2	SENT3	ROWMEAN	ROWEFF
SUBJ1	1.7	1.9	2.0	1.867	−3.1
SUBJ2	4.4	4.5	5.7	4.867	−0.1
SUBJ3	6.6	7.4	10.5	8.167	3.2
COLMEAN	4.233	4.6	6.067	4.967	0
COLEFF	−0.733	−0.367	1.1	0	0

These are situations to which we might apply a two-factor analysis of variance to study the effects of the factors, except that there is just one observation in each cell of the design. In such cases, the standard two-way ANOVA model

$$y_{ij} = \mu + a_i + \beta_j + (a\beta)_{ij} + \varepsilon_{ij} \tag{7.6}$$

cannot be used, because there is no within-cells estimate of error variance, σ^2. However, if it is reasonable to assume that the factors do not interact, then the additive model

$$y_{ij} = \mu + a_i + \beta_j + \varepsilon_{ij} \tag{7.7}$$

can be fit to test the effects of the two factors. σ^2 can be estimated by the mean square for interaction, MS(AB). (In Table 7.2, the column labeled ROWEFF gives the estimates, $\hat{a}_i$, of the row effects; the COLEFF row gives the column effect estimates, $\hat{\beta}_j$.)

There are two problems with this approach, however:

□ The validity of the additive model (7.7) depends on the assumption that the factors do not interact, and this assumption is not testable with model 7.7.

□ If the factors do interact, it might be possible to transform the data to reduce the interaction, but we need a way to determine what re-expression to try and whether it will help.

A test proposed by Tukey (1949) provides a solution to the first problem. The *Tukey test for non-additivity* tests for interactions of the form $(a\beta)_{ij} = D\, a_i\, \beta_j$ in the model

$$y_{ij} = \mu + a_i + \beta_j + D\, a_i\, \beta_j + \varepsilon_{ij} \tag{7.8}$$

where D is some constant to be fit to the data. The test for non-additivity tests the hypothesis, H_0: $D=0$. If this hypothesis is not rejected, the additive model can be assumed reasonable.

Tukey (1977) developed some graphical displays to solve the second problem. Let

$$e_{ij} = y_{ij} - \bar{y}_{i.} - \bar{y}_{.j} + \bar{y}_{..} \tag{7.9}$$

be the (interaction) residual from the additive model, and

$$c_{ij} = \frac{\hat{a}_i \hat{\beta}_j}{\bar{y}_{..}} \tag{7.10}$$

be the comparison value for that cell. Then, if a plot of e_{ij} against c_{ij} is linear with a slope b, it can be shown that the power transformation using the ladder of powers, $y{\rightarrow}y^{1-b}$, will reduce the degree of interaction. In addition, the components of the additive fit to a two-way table—the row effects, $\hat{a}_i$, column effects, $\hat{\beta}_j$, and the residuals, e_{ij}—can all be shown in a display suggested by Tukey (1977).

In exploratory data analysis, we should be particularly wary of the effects of one or two unusual observations. Robust analyses of two-way tables use row and column medians rather than means to ensure that the results are not affected by such values. Here I use analysis by means for simplicity. The same sort of plots would be made for analysis by medians. The resistant techniques for two-way and three-way tables and the associated plots are also described in Hoaglin, Mosteller, and Tukey (1983, 1985).

7.6.1 Using the TWOWAY Macro

The Tukey test for additivity and graphical displays of the additive fit and diagnostic plot for transformation have been implemented with PROC IML, as a general SAS macro called TWOWAY. (Another SAS macro for Tukey's test using PROC GLM is described by Mariam and Griffin 1986.) The TWOWAY program is included in Section A1.16, "The TWOWAY Macro." The arguments to the macro are as follows:

```
%macro TWOWAY(
          data=_LAST_,         /* Data set to be analyzed       */
          var=,                /* List of variables: cols of table*/
          id=,                 /* Row identifier: char variable   */
          response=Response,   /* Label for response on 2way plot */
          plot=FIT DIAGNOSE,   /* What plots to do?             */
          name=TWOWAY,         /* Name for graphic catalog plots */
          gout=GSEG);          /* Name for graphic catalog      */
```

The macro is used to analyze the data in Table 7.2 as shown below. Note that the observations correspond to one factor and the variables represent the other factor. (The data would be strung out as scores on a single variable for the usual analysis with PROC ANOVA.)

```
%include twoway;              /* On MVS, use %include ddname(file);*/
   /* Response times for 3 types of sentences (n=1) */
data RT;
     input Subject $ Sent1-Sent3;
cards;
SUBJ1  1.7 1.9 2.0
SUBJ2  4.4 4.5 5.7
SUBJ3  6.6 7.4 10.5
;
```

```
/*------------------------------------------------------------*
| Call TWOWAY macro for Tukey 1df test for non-additivity. |
| Note: the VAR= list must list each column variable. You |
| can't use VAR=SENT1-SENT3. |
*------------------------------------------------------------*/

%TWOWAY(data=RT, var=SENT1 SENT2 SENT3,
        id=SUBJECT,
        response=Reaction Time, name=GB0724);
```

The printed output, in Output 7.23, indicates that interaction, of the multiplicative form in equation 7.8, accounts for a significant portion of the error sum of squares, as shown by the F value for non-additivity ($F=64.79$). Thus, the usual ANOVA tests under the additive model (equation 7.7) and hence, the F values for rows and columns in Output 7.23, are not appropriate for these data as they stand. One possibility is to use the mean square for pure error (MSPE) as the error term for testing the row and column effects. This gives a much more sensitive test because the MSPE (.053) is considerably smaller than the MSE (.983) under the additive model. Another possibility, explored in Section 7.6.3, "Diagnostic Plot for Transformation to Additivity," is to transform the response variable so that the additive model applies. In order to see the pattern of the data, however, we first consider plots of the fitted values and residuals from the additive model.

Output 7.23
Output from
TWOWAY Macro

```
DATA    SENT1    SENT2    SENT3    ROWMEAN  ROWEFF

SUBJ1    1.7000   1.9000   2.0000   1.8667  -3.1000
SUBJ2    4.4000   4.5000   5.7000   4.8667  -0.1000
SUBJ3    6.6000   7.4000  10.5000   8.1667   3.2000
COLMEAN  4.2333   4.6000   6.0667   4.9667        0
COLEFF  -0.7333  -0.3667   1.1000        0        0


                 Interaction Residuals
         E     SENT1    SENT2    SENT3

SUBJ1         0.5667   0.4000  -0.9667
SUBJ2         0.2667  -2.2E-16 -0.2667
SUBJ3        -0.8333  -0.4000   1.2333


              ANALYSIS OF VARIANCE SUMMARY TABLE
            with Tukey 1 df test for Non - Additivity

SOURCE       SS          DF        MS          F

Rows       59.580        2       29.790      30.2949
Cols        5.647        2        2.823       2.8712
Error       3.933        4        0.983       1.0000
Non-Add     3.759        1        3.759      64.7935
Pure Err    0.174        3        0.058       1.0000


          D = Coefficient of alpha ( i ) * beta ( j )
         Slope of regression of Residuals on Comparison values
                  1 - slope = power for transformation

                 D        SLOPE      POWER

               0.3171    1.5750    -0.5750
```

7.6.2 Plotting Fit and Residuals from Main-Effects Model

An informative plot of the two-way fit and residuals of the additive model 7.7 for the sentence data is shown in Output 7.24. The ordinate is the response variable, Reaction Time. The intersections of the grid lines show the fitted value for the additive model, and the dashed lines show residuals that are larger than $\sqrt{\text{MSPE}}$, the pure error mean square.

Output 7.24
Two-way Display of Fit and Residuals

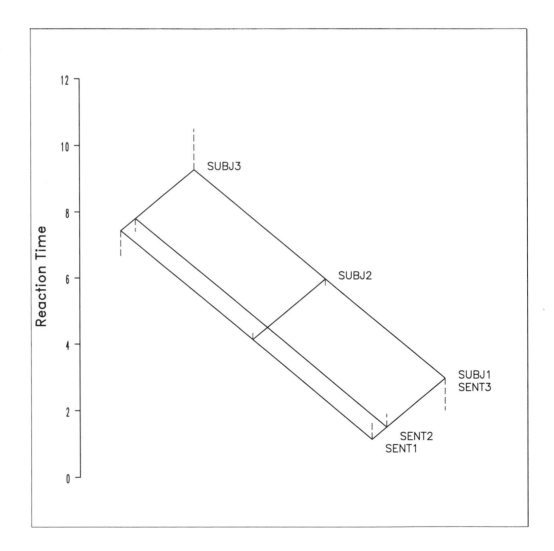

The construction of this plot is best described from a rearrangement of the data values and estimated effects, shown in Output 7.25. The DATA column shows the data values from Table 7.2, strung out rowwise. The ROWEFF values are the corresponding $\hat{a}_i$, estimated by the deviation of the row mean from the grand mean, $\bar{y}_{i.} - \bar{y}_{..}$, and the COLFIT value gives $\hat{\mu} + \hat{\beta}_j = \bar{y}_{.j}$, shown as the column means in Table 7.2.

Output 7.25
*Data and
Calculations for
the Additive and
Diagnostic Plots*

```
                      Additive & Diagnostic Plot Values

    OBS    DATA    ROWEFF    COLFIT    FIT      DIFF    RESIDUAL    COMPARE

     1     1.7     -3.1      4.233    1.133    7.333     0.567      0.4577
     2     1.9     -3.1      4.600    1.500    7.700     0.400      0.2289
     3     2.0     -3.1      6.067    2.967    9.167    -0.967     -0.6866
     4     4.4     -0.1      4.233    4.133    4.333     0.267      0.0148
     5     4.5     -0.1      4.600    4.500    4.700    -0.000      0.0074
     6     5.7     -0.1      6.067    5.967    6.167    -0.267     -0.0222
     7     6.6      3.2      4.233    7.433    1.033    -0.833     -0.4725
     8     7.4      3.2      4.600    7.800    1.400    -0.400     -0.2362
     9    10.5      3.2      6.067    9.267    2.867     1.233      0.7087
```

If the values of the row effects are plotted against the column fits, the result is a rectangular grid, shown in Output 7.26. (The dashed lines and labels are added manually to a PROC PLOT plot of ROWEFF * COLFIT = 'O'.) This plot shows the order and spacing of the effects for rows (subjects) and columns (sentences) in the data.

Notice that on this plot, lines of slope -1 correspond to lines of constant fitted value (that is, lines of $y+x=$constant). The plot in Output 7.24 is obtained by rotating the display of Output 7.26 counterclockwise 45 degrees. This is achieved by plotting

$$\text{FIT} = \text{ROWEFF} + \text{COLFIT} = \hat{\mu} + \hat{a}_i + \hat{\beta}_j$$

on the ordinate against

$$\text{DIFF} = \text{ROWEFF} - \text{COLFIT} = \hat{a}_i - \hat{\mu} - \hat{\beta}_j$$

on the abscissa. Given the values FIT and DIFF listed in Output 7.25, a PROC PLOT version of the two-way plot would be printed by the following statements:

```
proc plot;
   plot DATA* DIFF = '+'
        FIT * DIFF = '*'     / overlay;
   title 'Two-Way Fit plot';
```

The difference between the DATA value, y_{ij}, and the additive fit, $\bar{y}_{..}+\bar{y}_{i.}+\bar{y}_{.j}$, is the residual, e_{ij}. If the grid lines connecting the asterisk (*) symbols for the fitted values are drawn by hand, together with vertical lines for the residual from the asterisk (*) to the plus sign (+), the PROC PLOT version gives a reasonable two-way display.

Output 7.26
Plot of ROWEFF
and COLFIT for
Additive Model

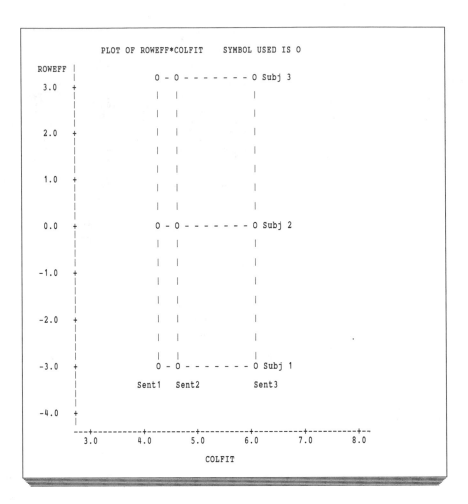

Plotting the Two-Way Display in the IML Procedure

The TWOWAY macro uses PROC IML graphics to draw the two-way display. While the complete program in Section A1.16 is left for the interested reader to study, the use of PROC IML graphics routines for this purpose is worth mention here.

The plot is drawn by the IML module TWOWAY shown below. The first section of the program collects the end points of the lines to be drawn in two arrays of (*x*,*y*) values, called FROM and TO. There is one loop for the rows and another loop for the columns, giving a total of R+C points in the FROM and TO arrays. Then residuals (called E[I, J] in the program) that are larger than

$\hat{\sigma} = \sqrt{\text{MSPE}}$ are found, and for each a line from the fitted value to the data value is added to the FROM and TO arrays.*

```
start twoway;
 /*-------------------------------------------------------------*
 | Calculate points for lines in two-way display of fitted     |
 | value. Each point is (COLFIT+ROWEFF, COLFIT-ROWEFF).        |
 *-------------------------------------------------------------*/
    do i=1 to r;
        clo  = coleff[><]+allmean;
        from = from // (clo-roweff[i] || clo+roweff[i]);
        chi  = coleff[<>]+allmean;
        to   = to   // (chi-roweff[i] || chi+roweff[i]);
        labl = labl || rl[i];
        end;
    do j=1 to c;
        rlo  = roweff[><];
        to   = to   // (coleff[j]+allmean-rlo || coleff[j]+allmean+rlo);
        rhi  = roweff[<>];
        from = from // (coleff[j]+allmean-rhi || coleff[j]+allmean+rhi);
        labl = labl || cl[j];
        end;


    /*---------------------*
    | Find large residuals |
    *---------------------*/
    do i=1 to r;
    do j=1 to c;
        if abs(e[i, j]) > sqrt(mspe) then do;
           from = from // ((cf[i,j]-re[i,j])||(cf[i,j]+re[i,j]));
           to   = to   // ((cf[i,j]-re[i,j])||(cf[i,j]+re[i,j]+e[i,j]));
           end;
        end; end;
```

The two-way plot is different from most PROC IML graphics applications because no horizontal axis (corresponding to the variable DIFF) is to be drawn. So subroutine GSCALE is used to calculate the scaling information for the *x* and *y* coordinates in the FROM and TO arrays, and the minimum and maximum (*x,y*) values are set with subroutine GWINDOW. In an ordinary scatterplot, this scaling would be handled internally by the GAXIS routines.

The lines in the display are drawn with two GDRAWL calls, in order to use a different color and line style for the grid lines than for residuals. The row and column labels are drawn using GTEXT. With the Annotate facility, you can use

---

* The two-way display might have been drawn using an Annotate data set with PROC GPLOT. However, the calculations of the end points of the lines actually require access to matrices containing the fitted and residual values, which is much more difficult in a DATA step than in PROC IML.

the POSITION variable to avoid overplotting text and other graphic elements, but in PROC IML it is necessary to calculate small offsets explicitly.

```
/**-------------------------------**
 | Find scales for the two-way plot |
 **-------------------------------**/
call gport({10 10, 90 90});
call gyaxis( {10 10}, 80, from[,2]//to[,2], 5, 0, '5.0') ;
call gscale( scale2, from[,2]//to[,2], 5);
call gscript(3, 40, "&Response",'DUPLEX',3) angle=90;

call gscale( scale1, from[,1]//to[,1], 5);
window = scale1[1:2] || scale2[1:2];
call gwindow(window);

/*-------------------------------*
 | Draw lines for fit and residuals |
 *-------------------------------*/
l = nrow(from);
call gdrawl( from[1:r+c,],   to[1:r+c,])   style=1 color={"BLACK"};
call gdrawl( from[r+c+1:l,], to[r+c+1:l,]) style=3 color={"RED"};

/*-------------------------------------*
 | Plot row and column labels at margins; |
 *-------------------------------------*/
xoffset=.04 * (to[<>,1]-to[><,1]);
yoffset=0;
do i=1 to r+c;
   if i>r then do;
      yoffset=-.04 * (to[<>,2]-to[><,2]);
      end;
   call gtext(xoffset+to[i,1],yoffset+to[i,2],labl[i]);
   end;
   call gshow;
   call gstop;
finish;
```

7.6.3 Diagnostic Plot for Transformation to Additivity

Interaction in a two-way table, of the form detected by the multiplicative term in equation 7.8, can be seen in the two-way display as a tendency for residuals in opposite corners to go in the same direction. For example, in Output 7.24, residuals in the north and south corners are positive, while those in the east and west are negative. On a conventional two-way plot, generated by the following PLOT statement, we would see a set of diverging (or converging) curves, rather than parallel ones:

```
plot Y * A = B;
```

The *diagnostic display for additivity* provides a more effective means to detect systematic departures from additivity and, more importantly, to decide what to do about them.

The diagnostic plot is based on the idea that if the residuals from the additive model show this opposite-corner pattern, then there will be a linear regression relation of these residuals on the products $\hat{a}_i\hat{\beta}_j$. It is more convenient to normalize this product by dividing by the grand mean, to give the comparison value defined in equation 7.10.

For the sentence data, the residuals and comparison values for the diagnostic plot are shown in Output 7.25 in the columns labeled RESIDUAL and COMPARE. Output 7.27 shows the plot of these columns, together with a least squares line.

If this plot shows no consistent linear pattern, we can conclude that the data do not depart from an additive model in the simple product form of Tukey's term.

If, on the other hand, the plot reveals a systematic linear pattern, we can either transform the data to remove that non-additivity or amend the model to take it into account. It can be shown that when the slope of the line in the diagnostic plot is b, the power transformation, $y \rightarrow y^p$, where p is chosen as a simple number near $p=1-b$, will reduce or remove this form of non-additivity (Emerson and Stoto 1982; Emerson 1983).

In the sentence data, the slope of the regression line, shown in Output 7.23, is $b=1.57$, so $p=-0.57$, which would lead to the transformation $y_{ij} \rightarrow -1/\sqrt{y_{ij}}$, on the ladder of powers. For response time data, the transformed measure can be understood as a measure of $\sqrt{\text{speed}}$, which is not difficult to interpret.

Output 7.27
Diagnostic Plot for
Transformation

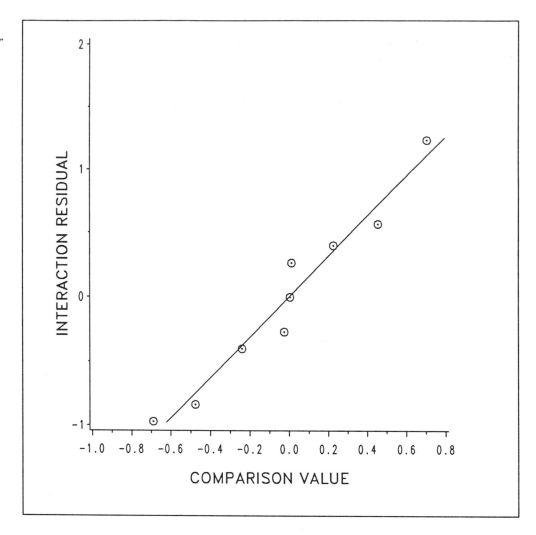

Sometimes there may be a significant *F* statistic for non-additivity shown in the ANOVA table, but the diagnostic display shows a few extreme points, with the remaining points nonsystematic. For example, Output 7.28 shows a TWOWAY analysis of reaction time scores of five subjects in a standardized learning task, after being given each of four drugs (data from Winer 1971, p. 268). From the ANOVA table, it appears that a significant portion of the error sum of squares can be accounted for by non-additivity, and the suggested power implies a log transformation.

Output 7.28
TWOWAY Output
for Drug Data

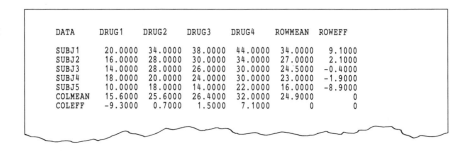

DATA	DRUG1	DRUG2	DRUG3	DRUG4	ROWMEAN	ROWEFF
SUBJ1	20.0000	34.0000	38.0000	44.0000	34.0000	9.1000
SUBJ2	16.0000	28.0000	30.0000	34.0000	27.0000	2.1000
SUBJ3	14.0000	28.0000	26.0000	30.0000	24.5000	-0.4000
SUBJ4	18.0000	20.0000	24.0000	30.0000	23.0000	-1.9000
SUBJ5	10.0000	18.0000	14.0000	22.0000	16.0000	-8.9000
COLMEAN	15.6000	25.6000	26.4000	32.0000	24.9000	0
COLEFF	-9.3000	0.7000	1.5000	7.1000	0	0

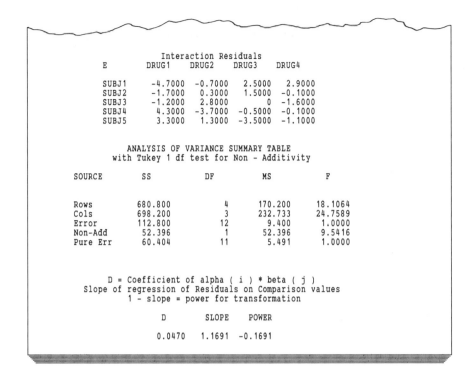

```
                          Interaction Residuals
              E        DRUG1     DRUG2     DRUG3     DRUG4

            SUBJ1     -4.7000   -0.7000    2.5000    2.9000
            SUBJ2     -1.7000    0.3000    1.5000   -0.1000
            SUBJ3     -1.2000    2.8000         0   -1.6000
            SUBJ4      4.3000   -3.7000   -0.5000   -0.1000
            SUBJ5      3.3000    1.3000   -3.5000   -1.1000


                   ANALYSIS OF VARIANCE SUMMARY TABLE
                 with Tukey 1 df test for Non - Additivity

       SOURCE        SS          DF          MS          F


       Rows       680.800         4       170.200     18.1064
       Cols       698.200         3       232.733     24.7589
       Error      112.800        12         9.400      1.0000
       Non-Add     52.396         1        52.396      9.5416
       Pure Err    60.404        11         5.491      1.0000



            D = Coefficient of alpha ( i ) * beta ( j )
         Slope of regression of Residuals on Comparison values
                  1 - slope = power for transformation

                      D       SLOPE     POWER

                   0.0470    1.1691   -0.1691
```

However, the significant non-additivity shown in the ANOVA table is misleading. The diagnostic plot for these data, shown in Output 7.29, indicates that three points are largely responsible for the linear regression relation between residual and comparison value. For these data, analysis by medians rather than means would show a clearer picture.

Output 7.29
Diagnostic Plot for
Drug Data

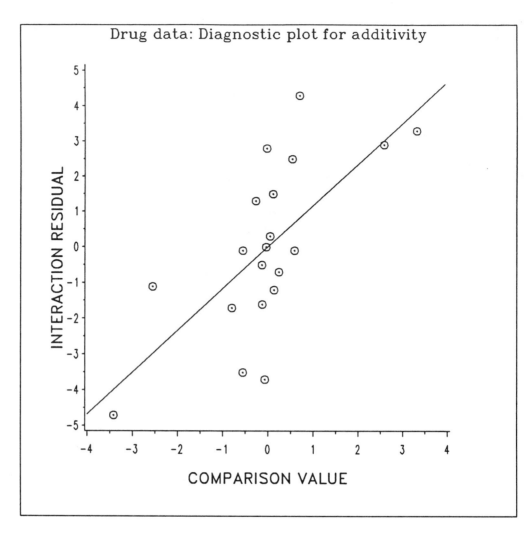

Related Methods

The model (equation 7.8) for Tukey's test for additivity is motivated by the observation that an interaction, if present, might take the simple form of a product of row and column effects, $(\alpha\beta)_{ij} = Da_i\beta_j$. Mandel (1961, 1969) extended this idea to situations where the residuals from the additive model have more general structure. In Mandel's *rows regression* model, the residuals from the additive model have the form $R_i\,\beta_j$, with a separate slope R_i estimated for each row. Tukey's model is the special case where each R_i is proportional to the row effect, a_i. The *columns regression* model has the same property for the columns, with interactions of the form $a_i\,C_j$.

Bradu & Gabriel (1978) show how these various models can be diagnosed by a *biplot* of the values $y_{ij} - y_{..}$. The biplot (see Section 8.7, "Biplot: Plotting Variables and Observations Together") displays points for both the rows and columns of any two-way table. If the rows (and columns) regression model fits, the row (and column) points will plot as a straight line. Tukey's model (equation 7.8, called the *concurrent* model) is the special case where both row and column points are collinear. If the additive model fits, the row and column points on the biplot plot as straight lines at right angles to each other.

For example, Output 7.30 shows the biplot of the sentence data from Table 7.2. This plot was constructed with the BIPLOT macro, described in Section 8.7.1, "The BIPLOT Macro." (A color version of this output appears as Output A3.36 in Appendix 3.) The points for subjects and sentences are both approximately linear, but the two lines are oblique, indicating that Tukey's model is most appropriate for these data.

Output 7.30   *Biplot of Sentence Data to Diagnose Model for Non-Additivity*

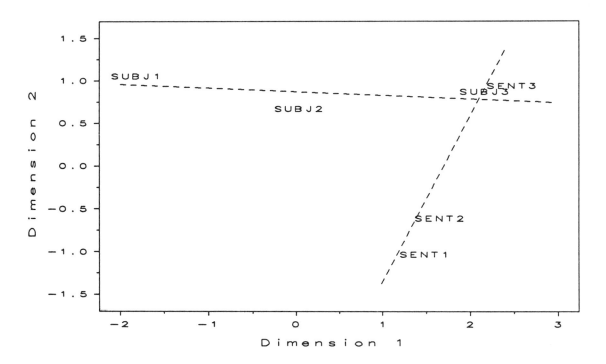

7.6.4 Two-Way Display for Quantitative Factors

In the two-way display, such as Output 7.24, levels of the row and column variables are arranged by increasing effect value, so that the graphical position of ROWEFF+COLFIT gives the fitted value for each cell. If the row and column factors are quantitative variables, the relation between the factor value and its effect may be difficult to see. In this case, it may be more effective to display the additive fit in relation to the values of the factor variables.

Example: Housing Starts

The data set HSTART below lists the number of U.S. housing starts, in thousands, by month for each of the years 1965—1973 (data from Becker, Chambers, and Wilks 1988; the original source is the U.S. Bureau of the Census, Construction Reports). While formal methods for the analysis of such data would include seasonal adjustment and time-series methods, an exploratory analysis might examine yearly and monthly effects in this two-way table.

```
proc format;
   value monfmt 1='Jan' 2='Feb' 3='Mar'  4='Apr'  5='May'  6='Jun'
                7='Jul' 8='Aug' 9='Sep' 10='Oct' 11='Nov' 12='Dec';
title h=1.75 'Monthly and Yearly Trends in Housing Starts';
data hstart;
   retain year 1965;
   input m1-m12;
   year+1;
cards;
 81.9   79   122.4 143   133.9 123.5 100   103.7  91.9  79.1  75.1  62.3
 61.7  63.2  92.9 115.9 134.2 131.6 126.1 130.2 125.8 137   120.2  83.1
 82.7  87.2 128.6 164.9 144.5 142.5 142.3 141   139.5 143.3 129.5  99.3
105.8  94.6 135.6 159.9 157.7 150.5 126.5 127.5 132.9 125.8  97.4  85.3
 69.2  77.2 117.8 130.6 127.3 141.9 143.5 131.5 133.8 143.8 128.3 124.1
114.8 104.6 169.3 203.6 203.5 196.8 197   205.9 175.6 181.7 176.4 155.3
150.9 153.6 205.8 213.2 227.9 226.2 207.5 231   204.4 218.2 187.1 152.7
147.3 139.5 201.1 205.4 234.2 203.4 203.2 199.9 148.9 149.5 134.6  90.6
 86.2 109.6 127.2 160.9 149.9 149.5 127.2 114    99.6  97.2  75.1  54.9
;
```

The two-way plot for these data is shown in Output 7.31.

Output 7.31
Two-Way Plot for
Housing Starts
Data

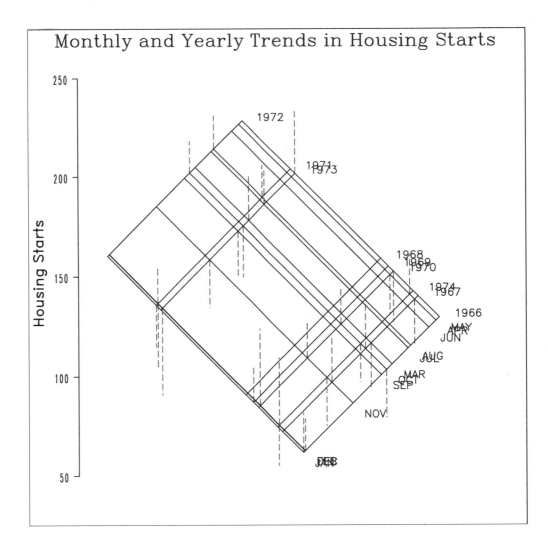

On this plot, the relation of month effect to month and year effect to year is hard to discern. An alternative display for these data plots housing starts against month, showing the year effects and residuals from an additive fit in a glyph symbol plotted for each month. The additive fit and residuals are obtained from PROC GLM. For this analysis the two-way table, with variables M1—M12, is strung out into scores on a single variable, STARTS. Notice that the MODEL statement includes only the main effects of the variables YEAR and MONTH. The output data set, RESIDS, contains the fitted value and residual for each observation.

```
data starts;
   set hstart;
   label starts='Housing Starts';
   array mon{12} m1-m12;
   drop m1-m12;
   do month=1 to 12;
      starts = mon{month};
      output;
      end;
```

```
proc glm data=starts;
   class month year;
   model starts=month year;
   output out=resids p=fit r=residual;
```

The plot, shown in Output 7.32, shows the mean number of housing starts for each month (the column fit) by a horizontal line. At each month, the year (row) effects are shown by the length of the vertical lines. Thus, the tip of the vertical line is at the fitted value for a particular month and year. The plotted points show the actual value wherever the residual is greater than $1.5 \sqrt{MSE}$.

This plot shows the trends in housing starts over time more clearly than the two-way display in Output 7.31 because the months are ordered along the abscissa and the years are ordered within the plotting symbol. The cyclic variation in housing starts over the months of the year is more apparent.

Output 7.32
*Month Fit and
Year Effects for
Housing Starts*

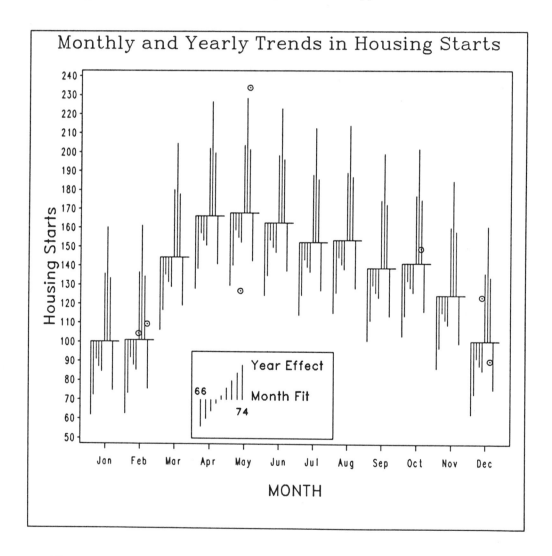

The program to construct this plot, shown below, is a detailed example of how a custom display may be made with Annotate data sets. The month fit, MFIT, is obtained with PROC MEANS as the average of the FIT values for each month. The plot glyph symbol at each month is drawn by the Annotate data set TREND.

The value of the MSE, 276.88, is taken from the PROC GLM printed output (not shown).

```
proc sort;
   by month;
 /*-------------------------*
 | Get means for each month |
 *-------------------------*/
proc means noprint;
   by month;
   var fit;
   output out=months mean=mfit;


   /*-----------------------------------------------------------*
   | Construct Annotate data set to portray month and year fits |
   *-----------------------------------------------------------*/
data trend;
   set resids; by month;
   xsys='2'; ysys='2';
   yr = year - 1965;
   drop year;
   retain mfit;
   if first.month then set months(keep=mfit);
   y =mfit;
   x =(month - .5) + yr*.8/10;          /* scale X within month */
   function='MOVE    '; output;
   y =fit;                              /* draw line from month */
   function='DRAW    '; output;         /* avg to fitted value  */

                        /* plot points with |residuals|>1.5 *s.e. */
   if abs(residual) > 1.5* sqrt(276.88) then do;
      y = starts;
      function='SYMBOL'; style=' '; text='-'; output;
   end;

   if last.month then do;            /* draw line for month fit */
      y =mfit;
      x = month - .4;
      function = 'MOVE    '; output;
      x = month + .4;
      function = 'DRAW    '; output;
      end;
```

The boxed legend at the bottom of the plot is drawn with a second Annotate data set. The LEGEND data set is then joined to the TREND data set.

```
   /*---------------------------------*
   | Annotate legend at bottom of plot |
   *---------------------------------*/
data legend;
   length text $12;
   drop yr;
   xsys='2'; ysys='2';
```

```
        do year = 1966 to 1974;
           yr = year-1965;
           y =70;
           x =3.6 + 1.5*yr/10;
           size=1;
           function='MOVE    '; output;
           y = y + 4*(yr-4.5);
           function='DRAW    '; output;
           if year=1966 | year=1974 then do;
              y=70;
              size=.9 ; style='DUPLEX';
              text=left(put((year-1900),f2.0));
              if year=1966 then position='2';
                            else position='8';
              function='LABEL'; output;
              end;
           end;
       x = x + .25;
       size=1.2; position='C';
       function='LABEL';
       text='Month Fit';   output;
       y = y +16;
       text='Year Effect'; output;
       x = 3.5; y=50;
       function='MOVE ';   output;
       x = 7.6; y=95;
       line=0;
       function='BAR  ';   output;
    data trend;
       set legend trend;
```

The plot is drawn with PROC GPLOT. The SYMBOL statement uses V=NONE to suppress all the points because everything inside the plot is drawn by the Annotate instructions. PROC GPLOT simply provides the axes.

```
proc gplot data=trend  ;
   plot starts * month
        / vaxis=axis1 haxis=axis2
          frame anno=trend
          name='GB0732' ;
   format month monfmt.;
   axis1  label=(a=90 r=0 h=1.5 f=duplex) value=(h=1.1)
          minor=none;
   axis2  label=(h=1.5 f=duplex) value=(h=1.1)
          minor=none offset=(5);
   symbol1 v=none;
```

Notice that the construction of the plot is based entirely on the fitted and residual values in the output data set from PROC GLM. If we wanted to fit and plot a different model, it would be necessary only to change the PROC GLM step.

For example, to model the year effect as a linear (regression) variable, rather than a CLASS variable, simply remove YEAR from the CLASS statement:

```
proc glm data=starts;
   class month ;
   model starts=month year;        * linear year effect ;
   output out=resids p=fit r=residual;
```

The year effects would then form a linear series, as shown in Output 7.33. This produces a worse fit for these data (the MSE is 1067), but has fewer fitted parameters and displays a simpler relation. Alternatively, we could model the month effects by a quadratic or sinusoid function, and the change in the PROC GLM step would be reflected in the plot.

Output 7.33
Month Fit and Linear Effect of Year for Housing Starts

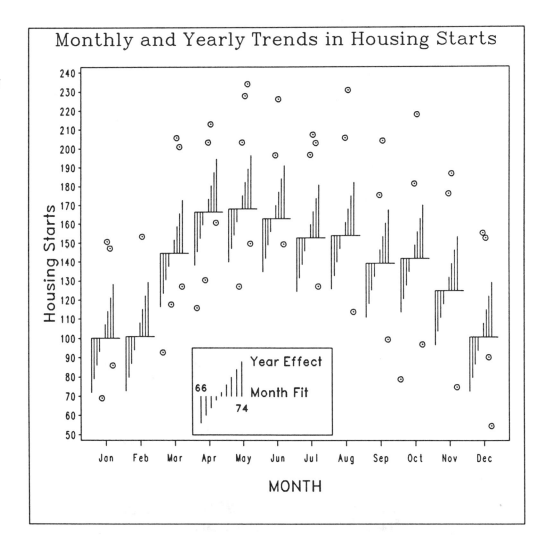

7.7   Plotting Power Curves for ANOVA Designs

For experiments using analysis of variance designs, as for other experimental situations, it is important to plan the sample size in advance to ensure that the experiment has a reasonable chance of detecting important differences among treatment groups and that estimated effects have sufficient precision to be useful.

Power analysis (Cohen 1977) provides the researcher with a way to determine the sample size required to detect an effect of given size with a specified probability. In a one-factor design with a levels of Factor A, we typically want to determine whether or not the means of the factor levels differ, as expressed by the null and alternative hypotheses,

H_0: $\mu_1 = \mu_2 = ... = \mu_a$

H_1: not all μ_i are equal   .

When the means do indeed differ, the F statistic calculated in the ANOVA procedure follows the noncentral F distribution, $F(df_h, df_e, \varphi)$ where φ is the noncentrality parameter that depends on the variance of the population μ_i; df_h and df_e are the degrees of freedom for the hypothesis and for error, respectively. For a given degree of difference among the means, power is the probability that we will reject H_0. More precisely, power is the probability that the critical value of the F statistic at a given significance level, a, is exceeded under a specified alternative hypothesis. In a two-group design the F-tests and t-tests are equivalent and the methods discussed in Section 3.2.5, "Plotting Power Curves: Student's t Distribution," can be used to determine the sample size.

For practical purposes, because we never know the true population means, the power of the F-test can be calculated by specifying the minimum range of the factor means that it is important to detect (Neter, Wasserman, and Kutner 1985). This minimum range can be specified by the quantity Δ,

$$\Delta = \frac{\max(\mu_i) - \min(\mu_i)}{\sigma}$$

in units of σ, the within-group standard deviation. Once we have specified Δ and the significance level a of the test, we can calculate sample size required to give any specified power. Cohen (1977) gives rules of thumb for specifying an *effect size* index $f = \Delta / \sqrt{2a} = \varphi / \sqrt{n}$ when data-based estimates are not available. O'Brien and Lohr (1984; Lohr and O'Brien 1984) discuss power analysis with the SAS System and show how to calculate the required sample size for any hypothesis testable with PROC GLM.

7.7.1   Example: Main Effects in a Factorial Design

The program below plots curves depicting the power of the F-test in a one-way design with $a=4$ treatment groups as a function of Δ, and sample size per group, N, when the test is carried out at the $a=.05$ significance level. The DATA step that produces the PWRTABLE data set (after Cary 1983) is general enough to calculate power for a main effect in any balanced factorial design with or without repeated measures, by setting the variables ALPHA, A, B, and so on at the beginning.* For each combination of N and DELTA, the program calculates the critical F value using the inverse F distribution function, FINV. Then it calculates power using the cumulative F distribution function, FPROB. The PROC GPLOT step is quite similar to that used for power of the t-test in Section 3.2.5. The resulting plot is shown in Output 7.34.

```
data pwrtable;
   length a b c r n trtdf 3;
   drop   fcrit ;

   alpha = .05;    *-- significance level of test of h0;
   a = 4;          *-- levels of factor a;
   b = 1;          *-- levels of factor b;
   c = 1;          *-- levels of factor c;
   r = 1;          *-- number of repeated measures;
   rho=0;          *-- intraclass correlation;

   trtdf = (a-1);  *-- treatment degrees of freedom--;
   *-- iterate through sample sizes --;
   do n = 2 to  4 by 1,
          6 to 10 by 2,
         15 to 20 by 5,
         30 to 40 by 10;

      *-- compute error degrees of freedom & f crit --;
      errdf = (a * b * c * (n-1));
      fcrit = finv (1-alpha, trtdf, errdf, 0);

      *-- iterate through values of delta --;
      do delta = .25 to 3.0 by 0.25;

         *-- Compute non-centrality parameter, & power --;
         nc = n * b * c * r * delta**2/ (2*(1+(r-1)*rho));
         power = 1 - fprob(fcrit, trtdf, errdf, nc);
         output;
         end;   *-- Do DELTA;
      end;   *-- Do N;

title1 h=1.5 'Power of F-test in 4 x 1 Design'
       h=1 a=-90 ' ';                /* allow space on right */
title2 f=complex h=1.5
   'as a function of Effect Size (' f=cgreek 'D'
   f=complex ' / ' f=cgreek 's' f=complex ') and N';
```

---

*   In a multifactor design, set the variable A equal to the number of levels of the factor for which the power calculations are to be performed. The variables B and C give the number of levels of the other between-group factors. Only the product B*C matters, so in a $4\times3\times3\times2$ design, set A=4, B=3, C=6.

```
proc gplot data=pwrtable  ;
  plot power * delta = n
       / vaxis = axis1
         haxis = axis2
         frame legend=legend1
         name='GB0734' ;
  axis1 order=(0 to 1 by .1) offset=(1) label=(h=1.5) value=(h=1.3);
  axis2 order=(0 to 3 by .5) offset=(1) label=(h=1.5) value=(h=1.3);
  symbol1  i=join v=star l=1  c=black;
  symbol2  i=join v=plus l=2  c=black;
  symbol3  i=join v='-'  l=3  c=black;
  symbol4  i=join v='+'  l=4  c=black;
  symbol5  i=join v='$'  l=5  c=black;
  symbol6  i=join v=star l=6  c=black;
  symbol7  i=join v=plus l=7  c=black;
  symbol8  i=join v='-'  l=8  c=black;
  symbol9  i=join v='+'  l=9  c=black;
  symbol10 i=join v='$'  l=10 c=black;
  legend1  across=5 down=2;
```

To find the required sample size, locate the estimated value of Δ on the abscissa and go up to the required power value. The nearest power curve gives the necessary sample size per group, or you can interpolate visually between adjacent curves to estimate the sample size. For example, if a power of 0.80 is to be achieved when $\Delta = 1.5$, Output 7.34 indicates that approximately $n = 11$ observations are required in each of the four groups.

Output 7.34
Standard Power
Curves for F-Test
in One-Way Design

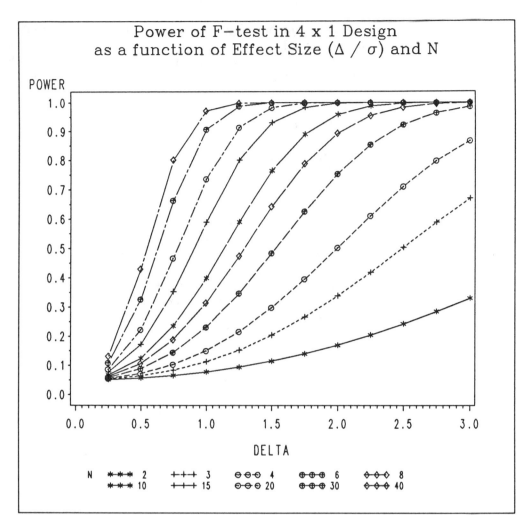

7.7.2 Linearizing Power Curves

One difficulty with plots such as Output 7.34 is that the curves become quite bunched together as POWER approaches 1.0, making it harder to find the appropriate sample size. Because all the curves are approximately normal ogives, it is possible to make the curves nearly linear and well separated in the upper region by transforming POWER to an inverse normal scale using the PROBIT function.* This is done in a second DATA step, which produces the PWRTRANS data set. The PROBIT function truncates any values outside the range -5 to $+5$, so all but the first such value for any N are set to missing.

On the plot, shown in Output 7.35, the tick marks for values -2 to 5 are labeled with the corresponding power probabilities (to 3 decimals), obtained from a table of the normal distribution. The labels for the tick marks are supplied using

---

* The correct function is the inverse of the noncentral *F* distribution, given by the FINV function. The plot would be harder to construct with the FINV function, however, and the inverse normal is sufficient for this example.

the VALUE= option in the AXIS3 statement. An Annotate data set, SCALES, is used to draw tick marks at the right side of the plot, with more conventional power values. These features of this example show, incidentally, two ways in which nonuniformly spaced tick values can be drawn for any plot.

```
   /*------------------------------------------------------*
   | Transform Power to Normal inverse to linearize plot |
   *------------------------------------------------------*/
data pwrtrans;
   set pwrtable;
   by n;
   retain firstbig;
   label zpower='POWER';
   if first.n then firstbig=0;
   zpower = probit(power);   *-- inverse normal transform;
   if zpower = 5 then do;
      if firstbig=0 then firstbig=1;
                    else zpower=.;
      end;

*-- annotate with tick marks on right side of probit plot;
data scale;
   length function $8;
   xsys = '1';     *-- data percent;
   ysys = '2';     *-- data value;
   input p @@;
   y = probit(p);
   x = 100 ; function = 'MOVE'; output;
   x = 101 ; function = 'DRAW'; output;
   text=put(p,4.3);
   position='6';
   size = 1.3;
   x = 101 ; function = 'LABEL'; output;
cards;
 .10  .25  .50  .75  .90  .95  .99  .995  .999
;
proc gplot data=pwrtrans  ;
   plot zpower* delta = n
        / vaxis = axis3
          haxis = axis2
          frame annotate=scale
          legend=legend1
          name='GB0735' ;
   axis2 order=(0 to 3 by .5) offset=(1) label=(h=1.5) value=(h=1.3);
   axis3 order=(-2 to 5 by 1)
         value=(h=1.3 '.023' '.159' '.500' '.841'
                      '.977' '.998' '.999' '1.00')
         offset=(1) label=(h=1.5);
```

Output 7.35
Power Curves
Plotted on Inverse
Normal Scale

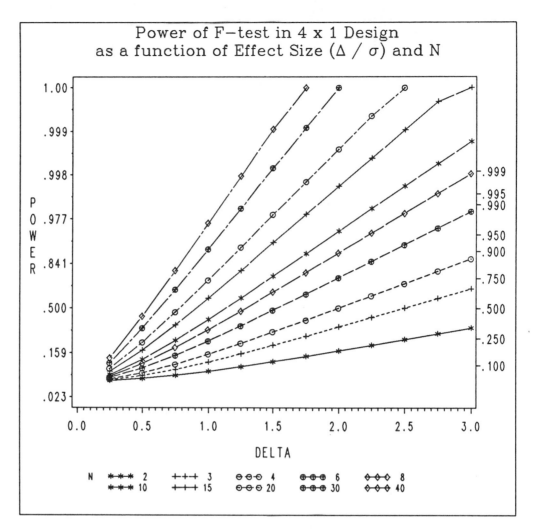

Part 5
Multivariate Data

Chapter **8** Displaying Multivariate Data

8.1   Introduction

Graphs are inherently two dimensional. Some ingenuity is therefore required to display the relationships of three or more variables on a flat piece of paper. This chapter deals mainly with informal, exploratory techniques for displaying three or more variables on one plot. Confirmatory and statistical methods for multivariate data are discussed in Chapter 9, "Multivariate Statistical Methods."

All multivariate graphics require changing or expanding the familiar visual metaphors we use for two variables, and a wide variety of methods have been developed. It is often useful to apply several of these to a given set of data. One class of techniques, described in Section 8.2, "Glyph Plots," involves choosing two primary variables for a scatterplot and representing additional variables in a glyph symbol used to plot each observation, as on the bubble plot. Another set of ideas, described in Section 8.3, "Draftsman's Display and Scatterplot Matrix," is to display the scatterplot for all pairs of variables on an organized display, called a draftsman's display or scatterplot matrix.

The following three sections discuss some other techniques that dispense with the traditional scatterplot and use some scheme for representing many variables by different attributes of graphical icons. These include star plots (Section 8.4, "Star Plots"), profile plots (Section 8.5, "Profile Plots"), and Andrews function plots (Section 8.6, "Andrews Function Plots"). Finally the biplot, a graphical display that plots any number of variables and observations together on the same plot, is described and illustrated in Section 8.7, "Biplot: Plotting Variables and Observations Together."

8.2 Glyph Plots

In Chapter 4, "Scatterplots," we examined some of the ways to portray additional variables on a basic scatterplot. Here we consider this topic again, with a view to plotting multivariate data, perhaps with many variables or observations or both.

On the bubble plot (Section 4.7.2, "Bubble Plots"), one additional variable is represented by the size of the bubble. The BUBBLE statement in the GPLOT procedure scales the size of the bubble to the absolute value of the Z variable and distinguishes positive from negative values by line style:

```
BUBBLE Y * X = Z ;
```

This plotting scheme is suitable both for single-ended Z variables (bounded below by zero, for example) and double-ended Z variables such as the residuals displayed in Section 4.7.2 (see Output 4.30). However, with large numbers of observations, the effectiveness of the bubble display decreases due to overcrowding, and if we want to show any other data characteristics in the display, we must design our own plot symbols.

Tukey and Tukey (1981) generalize this idea as follows. Assume we have partitioned the variables into two sets: two primary, or *front* variables (x, y), and the rest—one or more secondary, or *back* variables $(z_1, z_2, \ldots)$. Then a glyph plot encodes the values of each of the back variables in the symbol we draw at (x, y).

Nicholson and Littlefield (1983) define a *glyph* as "any plotting symbol which displays data by changing the appearance of the symbol" (p. 213). The most widely used glyph symbols are the weather vane symbols used by meteorologists to display cloud cover, wind direction, and wind speed simultaneously on a weather map.

There are a number of attributes of a glyph symbol we could use to code the value of a single back variable. Length and direction of lines are popular choices for single-ended quantitative variables. Color and shape are commonly used for qualitative or categorical variables. Nicholson and Littlefield (1983) and Tukey and Tukey (1981) discuss these and other possibilities.

8.2.1 Constructing a Glyph Plot with the Annotate Facility

The Annotate facility provides a relatively easy way to design your own glyph symbols. The example described in this section, inspired by Nicholson and Littlefield (1983), uses a *ray glyph*—a whisker line whose length, angle, and color vary—to display two quantitative variables and one categorical variable in addition to the two front variables of an (x, y) plot.

The plot, shown in Output 8.1, displays the relationship between sepal length and petal length in the IRIS data set. (A color version of this output appears as Output A3.37 in Appendix 3, "Color Output.") As indicated in the inset variable key, the length of the whisker line from each point is proportional to the value of the sepal width, while the angle from the horizontal axis is proportional to petal width. The species of iris is coded by both the shape of the marker and the color of the symbol.

Output 8.1
Glyph Plot of Iris Data

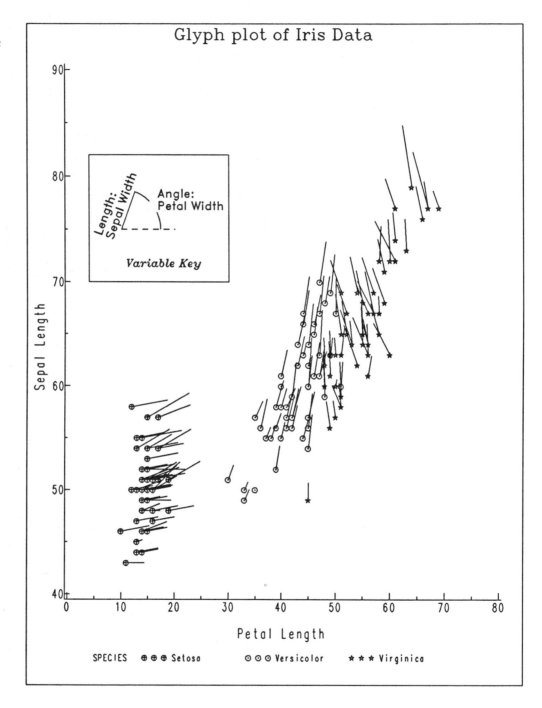

The glyph symbol is constructed in the following steps:

1. The minimum and maximum values of the glyph variables, petal width and sepal width, are found with the MEANS procedure and saved in an output data set, RANGE.

2. In the DATA step, petal width and sepal width are first scaled to (0, 1) using the minimum and maximum values from data set RANGE. The scaled petal width variable, P1, determines the variable ANGLE, which is allowed to go from 0 to 135 degrees.* The maximum ray length is set at 8 data units, a value determined by trial and error. In general, we want to use the largest values that do not cause excessive visual crowding or confusion.

3. The front variables, petal length and sepal length, determine the Annotate variables X and Y for a MOVE operation, and the conversion of ray length (8 * P2) and angle from polar to rectangular coordinates determines the X and Y values for the DRAW operation. The COLORVAL variable, which is assigned to the Annotate variable COLOR, is specified in the G3D procedure display of the IRIS data set in Section 4.8.1, "Three-Dimensional Scatterplots."

```
%include IRIS;                 /* On MVS, use %include ddname(file);*/

proc means data=iris min max ;
   var petalwid sepalwid;
   output out=range min=petalmin sepalmin
                    max=petalmax sepalmax;

data glyph;
   set iris;
   drop petalmin petalmax sepalmin sepalmax spec_no;
   length color function $8 ;
   if _n_=1 then set range;

   xsys='2'; ysys='2';

   /* Scale glyph variables to (0,1) */
   p1 = (petalwid - petalmin) / (petalmax - petalmin);
   p2 = (sepalwid - sepalmin) / (sepalmax - sepalmin);

   x = petallen; y = sepallen;
   function = 'MOVE'; output;
   angle = 135 * p1 * arcos(-1)/180;
   x = x +  8 * p2 * cos(angle);
   y = y +  8 * p2 * sin(angle);
```

---

* The maximum angle is set at 135 degrees to avoid confusion between 0 and 180. Another possibility would be to use a range of 120 degrees (30 degrees to 150 degrees from the horizontal), as in an analog voltmeter.

```
    select;
       when (spec_no = 1)   color = 'RED';
       when (spec_no = 2)   color = 'GREEN';
       when (spec_no = 3)   color = 'BLUE';
    end;
    function = 'DRAW'; output;
```

The inset variable key is constructed with a second Annotate data set, LEGEND, shown below. The arc is drawn by two observations using the PIE function to leave a small gap in the middle, and the parts of the LABEL function are made to align by choice of the POSITION values. All the X and Y values use data system coordinates, XSYS=YSYS='2'. The LEGEND data set is then added to the end of the GLYPH data set.

```
data legend;
   length function $ 8 position $ 1;
   input function $ x y color $ style $ line size angle rotate
         position $ text $ char25.;
     xsys='2'; ysys='2';
cards;
PIE     10 75 BLACK EMPTY  0 4.8  0 25 . .
PIE      .  . BLACK EMPTY  2 4.8 45 25 . .
LABEL   16 78 BLACK DUPLEX . 1.1  .  . C Angle:
LABEL   16 78 BLACK DUPLEX . 1.1  .  . F Petal Width
LABEL    6 74 BLACK DUPLEX . 1.1 70  . C Length:
LABEL    6 74 BLACK DUPLEX . 1.1 70  . F Sepal Width
LABEL   10 72 BLACK TITALIC . 1.1 0  . 6 Variable Key
MOVE    10 75  .       .    .  .   .  . . .
DRAW    20 75 BLACK    .   20  .   .  . . .
MOVE     4 70  .       .    .  .   .  . . .
BAR     30 82 BLACK EMPTY  0  .   .  . . .
;
data glyph;
   set glyph legend;

title  h=1.6 'Glyph plot of Iris Data';
proc gplot data=iris  ;
   plot sepallen * petallen = species
        / annotate=glyph
          haxis=axis2 vaxis=axis1
          hminor=1 vminor=1
          name='GB0801' ;
   axis1 order=(40 to 90 by 10)
          label=(h=1.5 a=90 r=0 'Sepal Length') value=(h=1.3)
          major=(h=-1) minor=(h=-.5);
   axis2 order=(0 to 80 by 10)
          label=(h=1.5 'Petal Length') value=(h=1.3)
          major=(h=-1) minor=(h=-.5);
   legend across=3
          value=(f=duplex) label=(h=1.2);
   symbol1 v=+ i=none color=RED;
   symbol2 v=- i=none color=GREEN;
   symbol3 v== i=none color=BLUE;
```

The ray glyph appears to work quite well in this example and in others (Bruntz et al. 1974; see Wainer and Thissen 1981). Output 8.1 shows the strong positive overall relation between the front variables, sepal length and petal length. It is easy to see the positive relation between petal width and the two length variables because the ray angle increases systematically from the lower left to upper right. The relation between sepal width (ray length) and the other variables is less easy to discern. Also, the relations among the variables are strongly associated with species, *Iris setosa* being most distinct from *versicolor* and *virginica*.

Dealing with Large Data Sets

The moderately high correlations among variables, coupled with the clear separation into groups, make the glyph plot effective for raw data with moderately large data sets ($n=150$ in the iris data). For larger data sets, or more modest relationships, it may be necessary to perform some aggregation or smoothing. That is, we may want to summarize the $(x,y,z_1,z_2, . . .)$ values across the (x,y) plane and display typical z values (such as medians or means) rather than displaying all individual ones at each (x,y) location. If the size of the data set makes the relationship of the front variables difficult to apprehend, we can combine summarization of z-level with cellulation over the viewing plane, as in the sunflower plot (Section 4.6, "Plotting Discrete Data"), but use plot symbols such as the MARK font (Section 8.2.2, "Constructing Polygon Marker Symbols with the GFONT Procedure") rather than the FLOWER font to display numerosity.

For a huge data set (over 20,000 observations), Nicholson and Littlefield (1983) use dot size to portray number of observations at each (x,y) cell (dot area $\sim$ log(count)) combined with fixed length rays whose angles display the 12.5, 50, and 87.5 percentiles of the distribution of a single z variable at each (x,y), producing a remarkably effective display.

Tukey and Tukey (1981) discuss some further ideas for smoothing z values over the viewing plane and remark that "we will ordinarily want to give up precision (or accuracy) in favor of smoothness" (p.245). The multivariate extension of the lowess smooth developed by Cleveland and Devlin seems promising for this situation (1988; also Cleveland, Devlin, and Grosse 1988).

8.2.2   Constructing Polygon Marker Symbols with the GFONT Procedure

The ability to define your own fonts for graphical display is a useful tool generally, and it is particularly handy for the kinds of multiattribute-enhanced scatterplots described here. If the default SYMBOL statement font does not serve your purpose, you can create a custom font of plotting symbols with the GFONT procedure, as we saw in Section 4.6.

This section shows how to define a generalized family of symbol characters. The symbols from this font are used in Section 8.3. The basic shapes include polygons and sunflower-type radial lines with any number of sides or lines. You can combine these basic shapes in any arbitrary way to produce a font symbol, which can be either filled or unfilled.

The symbol characters can be described parametrically by a SAS data set. Each symbol is determined by one or more observations (segments), specified by the following variables, in the data set SYMBOLS:

SEG   is a segment within a symbol (1, 2, . . .). A symbol can have any number of pieces, or segments, each defined by the remaining variables.

LP   is the variable used by PROC GFONT to determine if the current segment is empty or filled. Polygon segments can be either empty or filled, but radial lines cannot be filled. LP is the character 'L' for empty, 'P' for filled.*

N   is the number of sides or radial lines in the segment shape.

ANGLE   is the starting angle, in degrees, for the first side or ray.

TYPE   is the type for this segment, where 0=polygon; 1=radial lines (plus, star, and so on).

The DATA step below defines 18 symbol characters in this way. These symbols are shown in Output 8.2. You can think of the SYMBOLS data set as a high-level description of the characters in the symbol font.

```
data symbols;
   length lp $ 1;
   input seg lp $ n angle type desc $22-40;
   if seg=1 then chr+1;
   if lp¬='L' and lp¬='P' then error;
   if lp ='P' and type¬=0 then error;
cards;
   1  L  3    0  0    Empty triangle
   1  P  3    0  0    Filled triangle
   1  L  3  180  0    Inverted triangle
   1  P  3  180  0    Filled Inv Triangle
   1  L  3    0  0    Two triangles
   2  L  3  180  0    Two triangles
   1  P  3    0  0    Two triangles-Fill
   2  P  3  180  0    Two triangles-Fill
   1  L  4   45  0    Empty square
   1  P  4   45  0    Filled square
   1  L  4    0  0    Empty diamond
   1  P  4    0  0    Filled diamond
   1  L  4    0  1    Plus
   1  L  4   45  1    X
   1  L  4    0  1    Star
   2  L  4   45  1    Star
   1  L  4   45  0    X-ed square
   2  L  4   45  1    X-ed square
   1  L  5    0  0    Empty pentagon
   1  P  5    0  0    Filled pentagon
```

---

*   The format for the input data set to PROC GFONT is different for stroked than for filled fonts. The MARK font is defined as a filled font, and LP determines if the shape is filled or empty.

```
          1  L  8    0  0    Empty Octagon
          1  P  8    0  0    Filled Octagon
      ;
   proc print;
      id chr;
```

Now we need to expand the SYMBOL data set into one that PROC GFONT understands. The DATA step below performs this translation and assigns an alphabetic character to each symbol. This is the character you would use in the SYMBOL statement. For example, specifying V=B and FONT=MARK gives the second symbol, the filled triangle.

The POLY subroutine below calculates the X and Y values for each vertex point in the segment shape. The only difference between the radial line shapes (TYPE=1) and the polygon shapes is that we draw to the center of the shape (X=0, Y=0) at the start of each side. See Chapter 26, "The GFONT Procedure," in *SAS/GRAPH Software: Reference, Version 6, First Edition, Volume 2*, for further details.

```
   data marker;
      drop alpha desc chr a a0;
      set symbols;
      by chr ;
      alpha='ABCDEFGHIJKLMNOPQRSTUVWXYZ';
      if first.chr then segment+1;
      char = substr(alpha, segment, 1);
      a0 = (2+(angle/45))*atan(1);
      link poly;
      if ¬last.chr then do;
         x = .; y = .; output;
         end;
      return;

   poly:
      do i=0 to n ;
         if type =1 then do;        /* For radial lines, start */
            x=0; y=0; output;       /* each line at origin     */
            end;
         a= a0 + i/n*(8*atan(1));
         x=cos(a);
         y=sin(a);
         output;
        end;
       return;
   proc print;
      id char;
   proc gfont data=marker filled              /* name=GB0802 */
              name=mark
              romht=3 romfont=duplex showroman h=6
              deck  ;          /* Version 5 */
```

Version 6 of the SAS System makes it even easier to define custom fonts. PROC GFONT now understands a point-type variable (PTYPE), which makes it easy to compose a font character from circular arcs as well as lines. See Blettner

(1988) for examples. In Version 6, the GFONT procedure stores user-generated fonts in the catalog associated with the libref GFONT0. Use a LIBNAME statement to specify the location where the font is to be stored, as follows:

```
libname gfont0 'SAS-data-library';
```

Output 8.2   *MARK Font Symbols*

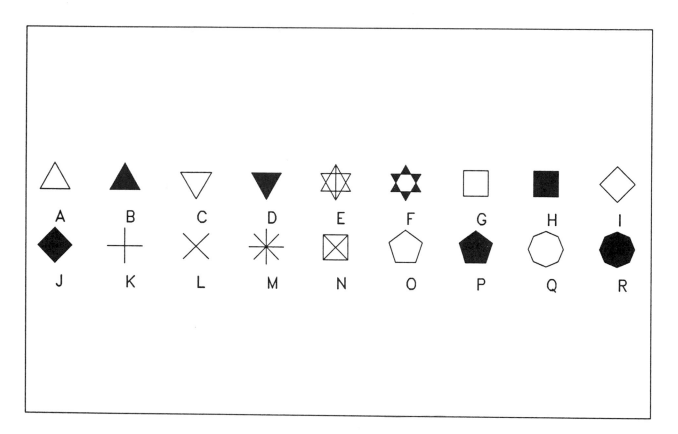

8.3 Draftsman's Display and Scatterplot Matrix

Multivariate data are often difficult to understand because it is difficult to see how the variables relate to each other. For three variables, there are 3 pairwise scatterplots to look at, for four variables there are 6 plots, and for six variables there are 15 plots. The number of scatterplots grows as the square of the number of variables.

Tukey and Tukey (1981; 1983; see also Chambers et al. 1983) proposed the *draftsman's display* as a way of plotting three or more variables together in a way that assembles all $p \times (p-1)/2$ scatterplots into a single composite display. The term draftsman's display comes from the idea that we can display three variables at once by imagining the three pairwise scatterplots pasted on the three visible faces of a cube (Cleveland, Kettenring, and McGill 1976). But instead of drawing the plot in three dimensions, we unfold the faces into a lower-triangular array of plan, elevation, and side views, so that adjacent plots share an axis in common. Each row of plots in the array has the same Y variable; each column has the same

X variable. A generalization to the plot of all pairs of *p* variables in a *p*×*p* grid is referred to as a *scatterplot matrix* (see Section 8.3.2, "Scatterplot Matrix").

In this arrangement, you can reduce the clutter by eliminating the axis labels and scales except for those on the outside margins of the plot. This helps the viewer to scan across or down the page to find where an observation on one plot appears on different plots. If observations are grouped according to some classification variable, points can be plotted with different plotting symbols.

8.3.1 Example: Automobiles Data

The example below uses the GREPLAY procedure to construct a draftsman's display of the WEIGHT, PRICE, and REPAIR record variables from the auto data.* The region of origin (U.S., Europe, and Japan) is shown by the color and shape of the plotting symbol. Thus, three quantitative variables and one qualitative variable are shown in a single display. The plots use symbols from the MARK font discussed in Section 8.2.2.

As in the interaction plots for ANOVA designs (Section 7.4.2, "Three-Factor Designs"), a GOUT= option is used to save each plot in a graphics catalog. Each of the three plots is generated with a separate PLOT statement, and each plot uses separate AXIS statements, so that the axes may be scaled individually and the labels on interior axes suppressed. The NODISPLAY option suppresses the display of the individual plots as they are generated; this option is changed to DISPLAY before the GREPLAY step.**

The GSLIDE procedure constructs a panel used as a title for the combined plot because PROC GREPLAY does not use titles other than those in the panels. The plot is shown below in Output 8.3. (A color version of this output appears as Output A3.38 in Appendix 3.)

```
%include AUTO;              /* On MVS, use %include ddname(file);*/
data auto;
    set auto;
    if rep77 ¬=. and rep78 ¬=.;
    price = price / 1000;
    weight= weight/ 1000;
    repair = sum(of rep77 rep78);
    label price = 'PRICE (1000 $)'
          weight= 'WEIGHT (1000 LBS)'
          repair= 'REPAIR RECORD';
```

---

* The repair record variables, REP77 and REP78, are measured on a 1 to 5 scale and appear as discrete horizontal bands. Summing the two variables gives REPAIR, which ranges from 2 to 10. This reduces but does not remove the discrete appearance. Jittering (see Section 4.6), by adding a small uniform random quantity to each REPAIR value, would improve the appearance and reduce clumping.

** For some devices, you may need to use other options in the GOPTIONS statement to achieve this. For example, with devices that use a graphics stream file (GSF), specify GSFMODE=NONE to suppress the display and GSFMODE=APPEND to resume display. See Chapter 5, "Graphics Options and Device Parameters Dictionary," in Volume 1 of *SAS/GRAPH Software: Reference.*

```
 /* save the display for later */
goptions NODISPLAY;           * device dependent;

title ;                                    /* suppress titles */
symbol1 v=G font=mark h=1.2 c=blue;                /* U.S.A. */
symbol2 v=B font=mark h=1.2 c=green;               /* Europe */
symbol3 v=J font=mark h=1.2 c=red;                 /* Japan */
proc gplot data=auto
    gout=graphs ;
    axis1 order=(1 to 5) offset=(2)
        label=(a=90 r=0 h=2)
        value=(a=90 r=0 h=2);
    axis2 order=(4 to 16 by 4) offset=(2)
        label=none value=none;
    plot weight * price = origin / frame nolegend name='T1'
        vaxis=axis1 haxis=axis2
        hminor=0 vminor=0;
    axis3 order=(2 to 10 by 2) offset=(2)
        label=(a=90 r=0 h=2)
        value=(a=90 r=0 h=2);
    axis4 order=(4 to 16 by 4) offset=(2)
        label=(h=2) value=(h=2);
    plot repair * price = origin /  frame nolegend name='T2'
        vaxis=axis3 haxis=axis4
        hminor=0 vminor=0;
    axis5 order=(2 to 10 by 2) offset=(2)
        label=none value=none;
    axis6 order=(1 to 5) offset=(2)
        label=(h=2) value=(h=2);
    plot repair * weight = origin /  frame nolegend name='T3'
        vaxis=axis5 haxis=axis6
        hminor=0 vminor=0;
run;
proc gslide gout=graphs name='T4' ;
title  h=3 ' ';
title2 f=triplex h=4 'Draftsman''s Display';
title3 f=triplex h=3 'Pairwise scatter plots of price,';
title4 f=triplex h=3 'weight and 77+78 repair record';
run;
```

The PROC GREPLAY step defines a template (DRAFT22) for a screen divided into four equal-sized regions in a 2 by 2 array. The four regions are assigned identifying numbers in the TDEF statement. The TREPLAY statement indicates how the four plots, named T1, T2, T3, and T4, which have been stored in the graphics catalog GRAPHS in the previous steps, are to be placed onto the four regions of the DRAFT22 template. The panels are numbered so that the three plots and the GSLIDE titles are placed on the corresponding panels of the template.

```
goptions DISPLAY;              * device dependent;

proc greplay igout=graphs nofs
            template=draft22 tc=templt ;
  tdef draft22
      des="Draftsman's display 2x2"
                1/ ulx=0  uly=100    urx=50  ury=100
                   llx=0  lly=50     lrx=50  lry=50
                4/ copy=1 xlatex= 50        /* Regions are numbered: */
                2/ copy=1 xlatey=-50               /*    1    4     */
                3/ copy=2 xlatex= 50;             /*    2    3     */
  treplay 1:T1 2:T2 3:T3 4:T4;          /* name=GB0803          */
```

Output 8.3   *Draftsman's Display of Price, Weight, and Repair Record for Automobiles Data*

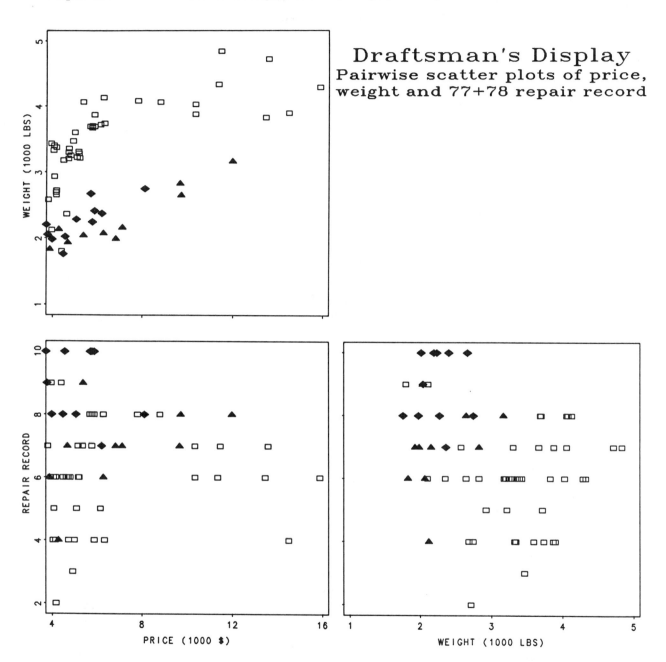

The use of plotting symbol to indicate the region of origin effectively adds another variable to the display. The upper panel depicts the same separation of the autos into two roughly parallel bands that appear on the scatterplot of weight against price (Output 4.11), but now it is clear that the upper band corresponds to the heavier North American cars. North American cars are shown by blue squares, European models by green triangles, and Japanese cars by red diamonds. On the bottom left panel, we can see that Japanese cars have among the best (highest) repair records. Similarly, the heaviest North American cars tend to be most expensive and have worse than average repair records.

A similar process can be used to display four or more variables on a draftsman's display. PROC GPLOT can be used to generate plots of all six pairs of variables. Then PROC GREPLAY will define a template based on a division of the plot into a 3 by 3 grid. The plots will be replayed in this template in the pattern in Figure 8.1 below, which uses the six cells in the lower triangle of the grid.

Figure 8.1

Template Pattern for a Draftsman's Display of Four Variables

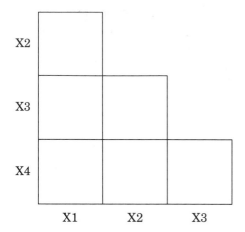

On the scatterplot matrix, the plots are displayed on a complete $p \times p$ grid (Becker and Cleveland 1987; Hartigan 1975b); each pair of variables is plotted twice, once below the diagonal and once above (with X and Y axes interchanged). In this form, it is easier to generalize the programming to apply to any set of data. A SAS macro program for such scatterplot matrices is discussed in Section 8.3.2, below.

8.3.2 Scatterplot Matrix

The draftsman's display program shows the SAS techniques for combining multiple scatterplots into one display, but it is simply too much work to write a program like this for each new set of data. Moreover, the program for the display in Output 8.3 requires one PLOT statement and two AXIS statements for each panel, so the work grows as the square of the number of variables.

Such a requirement can be met by writing a SAS macro program. The macro SCATMAT constructs a scatterplot matrix for any number of variables. The complete program is listed in Section A1.13, "The SCATMAT Macro." Some of the techniques used in the program are described below.

For p variables, $x_1, \ldots, x_p$, the scatterplot matrix is a p by p array in which the cell in row i, column j contains the plot of x_i against x_j. The diagonal cells are used for the variable names and scale markings. In the SAS macro language, this can be done with two nested %DO loops. In the code fragment below, &VAR is the list of variables to be plotted (X1 X2 X3, for example) from the data set &DATA, and &NVAR is the number of variables.

```
%do i = 1 %to &nvar;                    /* rows */
   %let vi = %scan(&var , &i );
   %do j = 1 %to &nvar;                 /* cols */
      %let vj = %scan(&var , &j );
```

```
      %if &i = &j %then %do;          /* diagonal panel */
         data title;
            length text $8;
            xsys = '1'; ysys = '1';
            x = 50; y = 50;
            text = "&vi";
            size = 2 * &nvar;
            function = 'LABEL';  output;
         proc gplot data = &data;
            plot &vi * &vi
            / frame anno=title vaxis=axis1 haxis=axis1;
         axis1 label=none value=none major=none minor=none offset=(2);
         symbol v=none i=none;
      %end;

      %else %do;                       /* off-diagonal panel */
         proc gplot data = &data;
            plot &vi * &vj
            / frame nolegend vaxis=axis1 haxis=axis1;
         axis1 label=none value=none major=none minor=none offset=(2);
         symbol v=+ i=none h=&nvar;
      %end;
   %end; /* cols */
%end;    /* rows */
```

To avoid the complexity of different AXIS statements for each panel and to
reduce the visual clutter of the display, the axis values, tick marks, and labels are
suppressed on all the plots. Instead, the variable name is written in the middle of
the diagonal panel. Note that the PROC GPLOT step for the diagonal panels plots
each variable against itself; however, specifying V=NONE in the SYMBOL
statement suppresses the points on the plot. This assures that each plot in a given
row has the vertical variable scaled uniformly and the plots align horizontally
when replayed.

In the SCATMAT macro, the Annotate data set TITLE also writes the
minimum and maximum value of that variable on the opposite corners of the
diagonal panel. Because the plots become smaller as the number of variables
increases, the size of labels and points is made proportional to the number of
variables, &NVAR.

The GREPLAY step can also be generalized with the SAS macro facility. In the
draftsman's display program, the TDEF statement defined the upper-left

corner panel and used the COPY= option to create the other panels. For a 3 by 3 scatterplot matrix, we would want TDEF and TREPLAY statements such as these:

```
TDEF scat3 DES="scatterplot matrix 3x3"
    1/ ULX=0   ULY=100   URX=34   URY=100
       LLX=0   LLY=66    LRX=34   LRY=66
    2/ copy=1 XLATEX= 33              /* panels:      */
    3/ copy=2 XLATEX= 33              /*   1   2   3 */
    4/ copy=1 XLATEY=-33              /*   4   5   6 */
    5/ copy=4 XLATEX= 33              /*   7   8   9 */
    6/ copy=5 XLATEX= 33
    7/ copy=4 XLATEY=-33
    8/ copy=7 XLATEX= 33
    9/ copy=8 XLATEX= 33 ;
TREPLAY 1:1 2:2 3:3 4:4 5:5 6:6 7:7 8:8 9:9;
```

Note that the first panel in any row after the first is copied from the row above it. The remaining panels in any row are copied from the preceding column. In the SCATMAT macro, the proper TDEF statement is constructed by an internal macro, TDEF, and the replay list in the TREPLAY statement is constructed as the plots are done.

The SCATMAT macro also enables a class or grouping variable in the data set to be used to determine the shape and color of the plotting symbol. The parameters and default values for the SCATMAT macro are shown below:

```
%macro SCATMAT(
        data =_LAST_,         /* data set to be plotted         */
        var  =_NUMERIC_,      /* variables to be plotted - can be */
                              /* a list or X1-X4 or VARA--VARB  */
        group=,               /* grouping variable (plot symbol) */
        symbols=%str(- + : $ = X _ Y),
        colors=BLACK RED GREEN BLUE BROWN YELLOW ORANGE PURPLE,
        gout=GSEG);           /* graphic catalog for plot matrix */
```

Example: Automobiles Data

The SCATMAT macro is used to plot a scatterplot matrix for the PRICE, WEIGHT, REPAIR, and MPG variables from the auto data as shown below. Country of origin is used to define the plotting symbol.

```
data AUTO;
    Set AUTO;
    if REP77 ¬=. and REP78 ¬=.;
    PRICE = PRICE / 1000;
    WEIGHT= WEIGHT/ 1000;
    REPAIR = sum(of REP77 REP78);
    Label PRICE = 'PRICE (1000 $)'
          WEIGHT= 'WEIGHT (1000 LBS)'
          REPAIR= 'REPAIR RECORD';
```

```
%scatmat(data=auto, var=price weight repair mpg,        /* name=GB0804 */
         symbols=+ SQUARE STAR,colors=RED GREEN BLUE,
         group=origin );
```

The plot produced is shown in Output 8.4. (A color version of this output appears as Output A3.39 in Appendix 3.) U.S. models are shown in black, European models in red, and Japanese models in green. Note that it is somewhat easier to see how one variable, PRICE, for example, relates to the others in the set by scanning across a given row (or down a column) than it is in the draftsman's display (Output 8.3).

On this plot, we can see the following features:

□ Correlations between the variable MPG and both the variables PRICE and WEIGHT are moderately strong (negative).

□ High mileage cars also tend to have better repair records (REPAIR) and are mostly Japanese.

□ There is a positive relation between the variables PRICE and WEIGHT for all three regions of origin, with U.S. models generally heavier.

□ The relationship between PRICE and REPAIR is complex and possibly nonlinear.

Output 8.4   *Scatterplot Matrix for Automobiles Data*

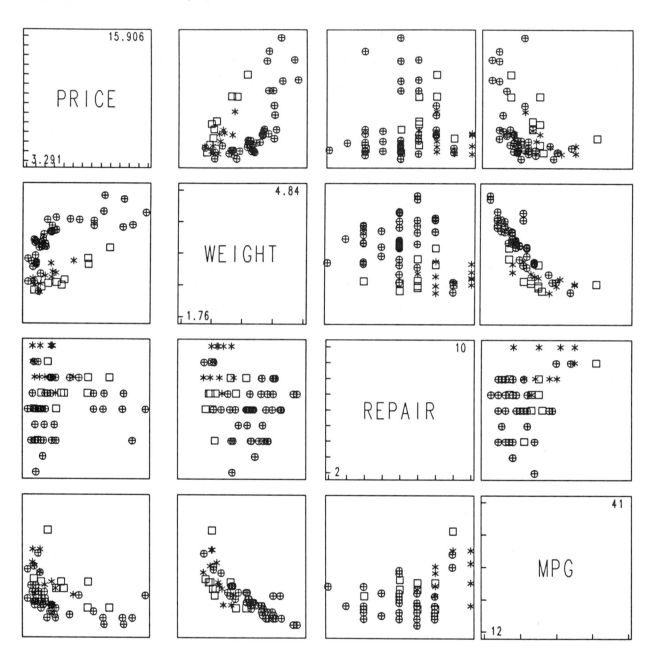

Example: Iris Data

The scatterplot matrix for the iris data, shown in Output 8.5, is produced by the following statements. (A color version of this output appears as Output A3.40 in Appendix 3.)

```
%include scatmat;            /* On MVS, use %include ddname(file);*/
%scatmat(data=iris,                           /* name=GB0805 */
         var=SEPALLEN SEPALWID PETALLEN PETALWID,
         group=spec_no );
```

This plot may be compared with the glyph plot of these data in Output 8.1, which corresponds to the plot in row 1, column 3 in Output 8.5. *Iris setosa* is shown in black, versicolor in red, and virginica in green. In the scatterplot matrix, it is easier to see the strong relation between the two petal variables, as well as the fact that the *Iris setosa* sample is most different from the other two species on all variables. Moreover, there is one runt *setosa* flower that stands apart from the rest of its group in all but the top row of the scatterplot matrix.

Output 8.5   *Scatterplot Matrix for Iris Data*

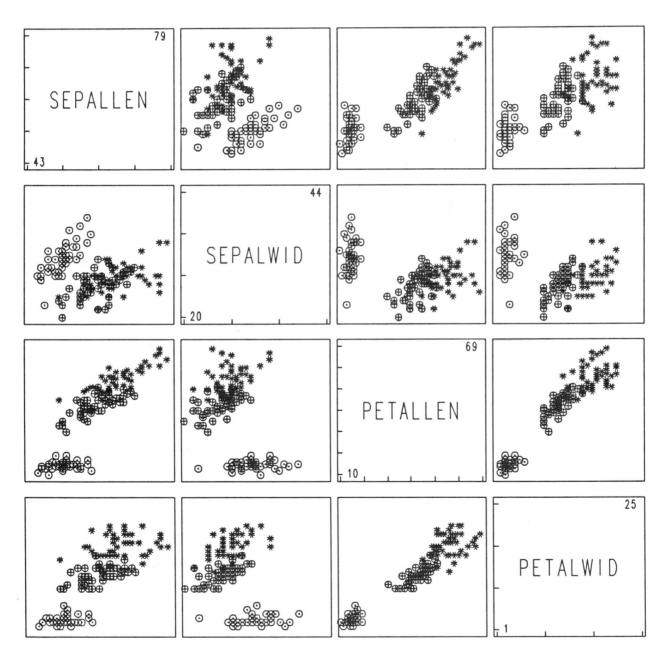

8.4 Star Plots

Star plots (Friedman et al. 1972; Chambers et al. 1983, pp. 158—162) are a useful way to display multivariate observations with an arbitrary number of variables. Each observation is represented as a star-shaped figure with one ray for each variable. For a given observation, the length of each ray is made proportional to the size of that variable. Star plots differ from glyph plots (Section 8.2) in that all variables are used to construct the plotted star figure; there is no separation into foreground and background variables. Instead, the star-shaped figures are usually arranged in a rectangular array on the page. It is somewhat easier to see patterns in the data if the observations are arranged in some nonarbitrary order, and if the variables are assigned to the rays of the star in some meaningful order.

Star plots can be drawn using the STAR statement in the GCHART procedure. In Release 6.06 of SAS/QC software, star symbols can be used to portray additional multivariate data on charts from the SHEWHART procedure.* Both of these procedures work best if there are only a few observations or if the stars are used to portray subgroup summaries. For the greatest flexibility, a custom display is constructed with the Annotate facility in the example below.

8.4.1 Example: Automobiles Data

Output 8.6 shows a star plot of the 12 numeric variables in the automobiles data from Chambers et al. (1983). Each star represents one car model; each ray in the star is proportional to one variable. Only the 15 lightest car models (top three rows) and 15 heaviest models (bottom three rows) are shown. These 12 variables are arranged around the perimeter as shown in the variable assignment key in Output 8.7. In the key, the variables toward the sides and bottom are related to size, and the others are related to price and performance.

---

*   This information is taken from a paper entitled "Selected SAS/QC Software Examples, Release 6.06," prepared by R. N. Rodriguez for the fifteenth annual SAS Users Group International Conference (April 1990). For additional discussion, see SAS Technical Report P-188, *SAS/QC Software Examples, Version 6.*

Output 8.6   *Star Plot of Automobiles Data*

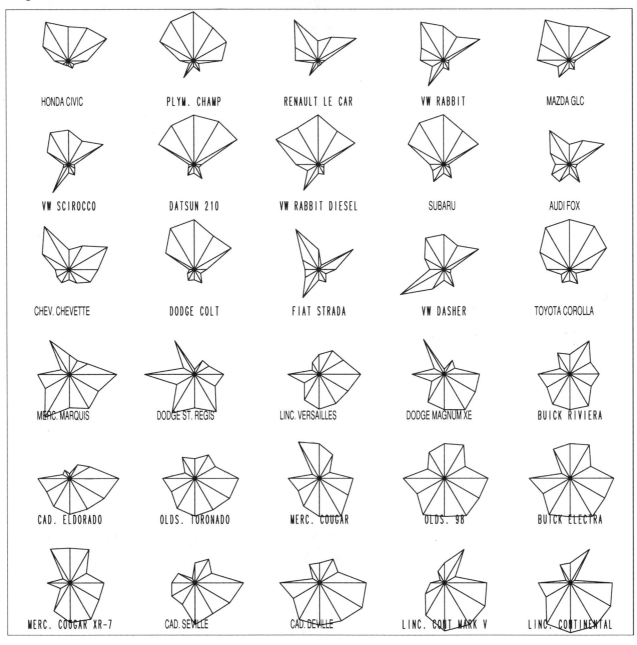

Output 8.7
Variable Assignment Key for Star Plot

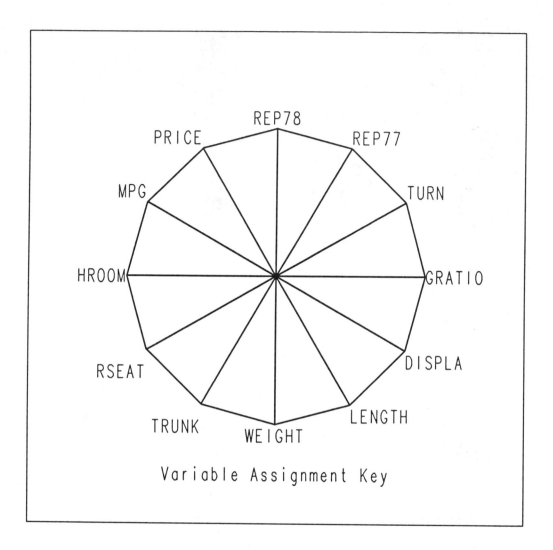

The star plot is constructed using the Annotate facility with PROC GSLIDE. To make this procedure general, it has been written as a SAS macro program, STARS, which takes the following parameters (and default values, shown after the equal signs). The complete STARS program is given in Section A1.14, "The STARS Macro."

```
%macro STARS(
            data=_LAST_,       /* Data set to be displayed        */
            var=_NUMERIC_,     /* Variables, as ordered around the */
                               /* star from angle=0 (horizontal)  */
            id=,               /* Observation identifier  (char)  */
            minray=.1,         /* Minimum ray length, 0<=MINRAY<1 */
            across=5,          /* Number of stars across a page   */
            down=6             /* Number of stars down a page     */
            );
```

The star plot is most useful when all of the variables have their scales aligned in the same direction so that increasing values have a similar meaning for all variables. For the auto data, this means that large values of a variable should reflect a better car and appear as long rays. To do this, the signs of the variables

PRICE, TURN, and GRATIO are changed by this DATA step before the STARS program is used:

```
proc sort;
     by weight;

data autot;
     set auto;
     price = -price;                /* change signs so that large  */
     turn  = -turn;                 /* values represent 'good' cars*/
     gratio= -gratio;
     if _n_ <= 15 or _n_ > 66-15 ;  /* keep 15 lightest & heaviest */
```

After observations with missing repair records are deleted, the observations are sorted by the variable WEIGHT, so that the first 15 and last 15 observations will contain the lightest and heaviest cars, respectively. With the default values for the variables DOWN and ACROSS, these 30 observations all fit on one plot. Then the STARS program is invoked by the following program lines to produce Output 8.6 and Output 8.7.

```
%stars(data=autot,
            /* Variables arranged so horizontal and down are
                size-linked & others are price/performance */
        vars= gratio turn rep77 rep78 price    mpg
           hroom rseat  trunk weight length displa,
        id=model,
        minray=.1);
```

The dominant pattern in Output 8.6 is that the star symbols in the top three rows have long rays on the top (good price and performance) and short rays on the bottom (small in size variables), but the reverse is generally true for the 15 heaviest models in the bottom three rows.

Deviations from this general pattern can also be seen. Among the 15 lightest cars, three Volkswagon models stand out with larger than average trunks, and the VW Dasher has an unusually large rear seat. Among the larger cars, the two Dodge models and the three Mercuries stand out with low prices, and the two Lincoln Continentals have unusually large turning diameters for cars in this weight range.

Note that on the star plot we tend to see the configural properties of the collection of variables represented for each observation, and that this perception is affected by the ordering of variables around the perimeter and by the arrangement of stars on the page. Other arrangements might lead to noticing other features of the data, so it might be useful to try several alternatives.

Another idea is to use a principal components analysis to arrange the variables. Use the variable with the largest loading on the first component to arrange the star figures on the page. The order of variables around the star can be determined using the loadings on the first two components. This analysis is most usefully carried out with the FACTOR procedure using the REORDER option, which causes the variables to be reordered in the printout according to the pattern of loadings. Variables with their highest absolute loading on the first factor are printed first, ordered by their loading, followed by the variables with their highest absolute loading on the second factor, and so on. The sign of loadings on the first factor can be used to determine which variables to reflect, in the absence of other considerations. For the auto data, PROC FACTOR would be used as follows to give the (rotated) two-component solution shown in Output 8.8.

```
proc factor data=auto
     method=prin nfactor=2 rotate=varimax reorder;
```

The WEIGHT variable has the largest loading on the first component, which consists largely of the size measures. The order of the variables in this table would be suitable for the star plot.

Output 8.8
Rotated
Two-Component
Solution for
Automobiles Data

```
                    ROTATED FACTOR PATTERN

              FACTOR1   FACTOR2

    WEIGHT    0.93664   -0.23876    WEIGHT (LBS)
    LENGTH    0.93023   -0.19509    LENGTH (IN.)
    DISPLA    0.89201   -0.25923    DISPLACEMENT (CU IN)
    TRUNK     0.82550    0.03736    TRUNK SPACE (CU FT)
    TURN      0.81984   -0.42173    TURN CIRCLE (FT)
    RSEAT     0.72453   -0.04316    REAR SEAT (IN.)
    HROOM     0.64219    0.04244    HEADROOM (IN.)
    PRICE     0.60278    0.15430    PRICE
    GRATIO   -0.75990    0.32385    GEAR RATIO
    MPG      -0.77644    0.33283    MILEAGE
    REP78    -0.20105    0.90874    REPAIR RECORD 1978
    REP77     0.03244    0.88569    REPAIR RECORD 1977
```

The star plot and scatterplot matrix give complementary views of a data set. The scatterplot matrix shows the relation between each pair of variables, but it is harder to see clusters of observations that have similar values on more than two variables. The star plot displays the configuration of all variables for each observation, but it is more difficult to see the joint relation between pairs of variables on the star plot.

8.4.2   How the STARS Program Works

The star plot is constructed in the following steps:

1. Each variable is rescaled to range from c=MINRAY to 1, so that the smallest and largest values of all variables are represented by rays of the same length. This is done with the IML procedure using the following formula:

$$x_{ij}^{*} = c + (1 - c)\, \frac{x_{ij} - \min_i x_{ij}}{\max_i x_{ij} - \min_i x_{ij}} \quad . \tag{8.1}$$

2. A DATA step constructs the Annotate data set, named STARS, to draw the star figure for each observation. A good deal of code is devoted to drawing the variable assignment key and positioning the stars properly on one or more pages as determined by the DOWN and ACROSS macro parameters. However, the process is made simpler than it would be otherwise by using the *percentage coordinate system* of the Annotate facility that treats X and Y values

as if they referred to percentage values of the screen area. In essence, drawing the star involves two steps:

a. Calculate the ray angles equally spaced around a circle according to $Ang_k = 2\pi(k-1)/\&NV$, where &NV is the number of variables. The sine and cosine of each angle are stored in SAS arrays S and C, respectively:

```
array s(k)  s1-s&nv;              /* sines of angle    */
array c(k)  c1-c&nv;              /* cosines of angle  */

   do k= 1 to &nv;
       ang = 2 * 3.1415926 * (k-1)/&nv;
       s = sin( ang );
       c = cos( ang );
   end;
```

b. Draw the outline of the star using the Annotate functions POLY and POLYCONT. In the lines below, P is an array containing the scaled values of the variables, R is the maximum radius of the star, and X0, Y0 is the origin of the star for the current observation.

```
DrawStar:
    *-- Draw star outline;
    do k = 1 to &nv;
       x = x0 + p * r * c;
       y = y0 + p * r * s;
       if k=1 then function = 'POLY';
               else function = 'POLYCONT';
       output;
    end;
```

Then draw a set of rays from the center (X0, Y0) to each vertex of the star using the MOVE and DRAW functions.

```
    *-- draw rays from center to each point;
    do k = 1 to &nv;
       x=x0; y=y0;
       function = 'MOVE';    output;
       x = x0 + p * r * c;
       y = y0 + p * r * s;
       function = 'DRAW'; output;
    end;
 return;
```

3. Finally, the Annotate data set STARS is drawn with PROC GSLIDE. The variable PAGE in the data set is used to separate the star figures into as many pages as are needed (each page contains DOWN×ACROSS figures), with PAGE=0 used for the variable assignment key. Unfortunately, PROC GSLIDE does not support BY variables, so the multiple pages are produced by calling GSLIDE within a SAS macro %DO loop.

```
/*----------------------------------------*
|  Plot each page with GSLIDE:            |
|   - Copy observations for current page  |
|   - Draw plot                           |
|   - Delete page data set                |
*----------------------------------------*/
%do pg = 0 %to &pages;
   data slide&pg;                    /* Select current page to plot */
      set stars;
      if page = &pg;
   run;

   proc gslide annotate=slide&pg;    /* Plot current page           */
   title;
   run;
   proc delete data=slide&pg;        /* Delete temporary data set   */
%end;
```

8.5   Profile Plots

A *profile plot* displays a set of variables simultaneously as a set of line segments connecting points, one for each variable. Profile plots are also called *parallel coordinate plots* because *p* variables are represented by a set of *p* coordinate axes drawn parallel to one another rather than orthogonally as in the scatterplot. Bolorforoush and Wegman (1988) discuss the statistical interpretation of parallel coordinate plots, whose geometry was developed by Inselberg (1985; Inselberg and Dinsmore 1988).

Output 8.9 is one example of a profile plot, showing the rates of seven types of crimes (number of incidents per 100,000 population) in 16 U.S. cities. (A color version of this output appears as Output A3.41 in Appendix 3.) The data plotted here are shown in Table 8.1 and are taken from Hartigan (1975a, p. 28), whose source was the United States Statistical Abstracts, 1970.*

---

* From *Clustering Algorithms* by John A. Hartigan. Copyright © 1975 by John Wiley & Sons, Inc. Reprinted by permission of John Wiley & Sons, Inc.

Output 8.9   *Naive Profile Plot of Raw Crime Data*

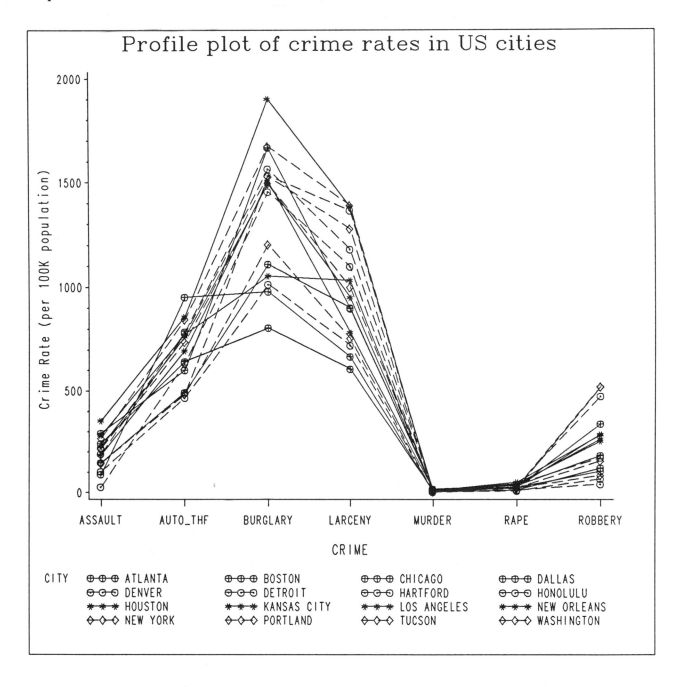

Table 8.1   *Data on Crime Rates in 16 U.S. Cities*

CITY	MURDER	RAPE	ROBBERY	ASSAULT	BURGLARY	LARCENY	AUTO_THF
ATLANTA	16.5	24.8	106	147	1112	905	494
BOSTON	4.2	13.3	122	90	982	669	954
CHICAGO	11.6	24.7	340	242	808	609	645
DALLAS	18.1	34.2	184	293	1668	901	602
DENVER	6.9	41.5	173	191	1534	1368	780
DETROIT	13.0	35.7	477	220	1566	1183	788
HARTFORD	2.5	8.8	68	103	1017	724	468
HONOLULU	3.6	12.7	42	28	1457	1102	637
HOUSTON	16.8	26.6	289	186	1509	787	697
KANSAS CITY	10.8	43.2	255	226	1494	955	765
LOS ANGELES	9.7	51.8	286	355	1902	1386	862
NEW ORLEANS	10.3	39.7	266	283	1056	1036	776
NEW YORK	9.4	19.4	522	267	1674	1392	848
PORTLAND	5.0	23.0	157	144	1530	1281	488
TUCSON	5.1	22.9	85	148	1206	757	483
WASHINGTON	12.5	27.6	524	217	1496	1003	739

The plot in Output 8.9 is a very poor display for these data, however. In plotting the raw data, the high frequency crimes such as burglary completely dominate the display, and the low frequency crimes (murder, rape) are compressed to the bottom of the plot. Because of this, the possibility that the cities may differ in their patterns of crime rate is effectively hidden.

When the variables differ widely in range or are measured on different scales, it is usually preferable to standardize them in some way. Output 8.10 shows the same data after the crime variables have each been scaled to have a mean of 0 and standard deviation of 1 with the STANDARD procedure. (A color version of this output appears as Output A3.42 in Appendix 3.)

Output 8.10   *Profile Plot of Standardized Crime Data*

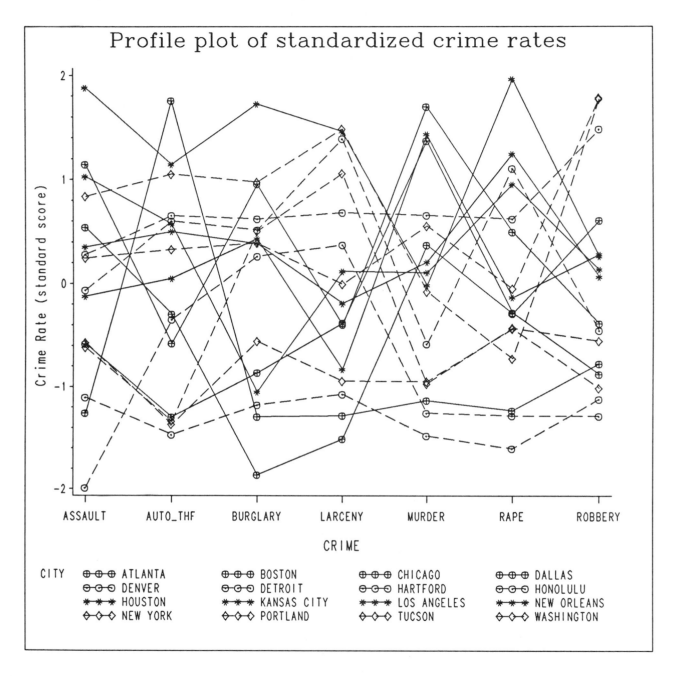

While profile plots of standardized scores are often effective, for these data Output 8.10 is equally dismal. The large number of crossings and the peaks and troughs in the profiles make it hard to see any similarities among the cities in their patterns of crime. (The star plot is, in effect, a profile plot wrapped around a circle; plotting the star figures separately is one way to solve the problem of crossings.) However, a close look at Output 8.10 reveals some consistencies: Los Angeles, Detroit, and New York are typically high on all crimes, while Tucson and Hartford are typically low. We can also see that Bostonians are generally law abiding, but can't resist an unattended automobile (Hartigan 1975a). Bolorforoush and Wegman (1988) explain how the correlations between adjacent variables on the profile plot can be understood visually in terms of the number of crossings of the line segments, but we do not exploit this here because our interest is in understanding the patterns of crime rates across cities.

Part of the problem in this example is that the crimes are arranged alphabetically and are spaced equally along the horizontal axis. The plot in Output 8.10 is produced with these statements:

```
title h=1.6 'Profile plot of standardized crime rates';
proc standard data=crime mean=0 std=1 out=stdcrime;
proc transpose out=crime2 name=crime;
   by city;
proc sort data=crime2 (rename=(col1=rate));
   by crime city;

goptions colors=(black red blue green);
proc gplot data=crime2  ;
   plot rate * crime = city
        / vaxis=axis1 haxis=axis2
          vminor=4
          legend=legend1
          name='GB0810' ;
   axis1 label=(a=90 h=1.2 'Crime Rate (standard score)' );
   axis2 offset=(2)
         label=(h=1.2 'CRIME');
   symbol1 v=+ i=join l=1;         /* each symbol statement */
   symbol2 v=- i=join l=21;        /* is used 4 times, once */
   symbol3 v=: i=join l=1;         /* with each color       */
   symbol4 v=$ i=join l=21;
   legend1 across=4;
```

With a character variable such as CRIME, PROC GPLOT arranges the values alphabetically on the axis. For this plot, note that the data values in Table 8.1 are transposed to observations on the single variable RATE to give the data set that is plotted, CRIME2. This allows the profiles of all variables to be plotted with the following statement:

```
PLOT RATE * CRIME = CITY;
```

8.5.1 Optimal Linear Profiles

A far more effective display of profiles is suggested by Hartigan (1975a, Section 1.6). The goal of this technique is to make the profiles as smooth as possible. This is done by choosing the position of the variables along the horizontal scale in an optimal way. A sensible definition of smoothness requires that each profile be as nearly linear as possible.

If the data values are denoted y_{ij}, we first allow each variable to be transformed linearly to

$$y_{ij}^* = (y_{ij} - o_j) / s_j \tag{8.2}$$

where o_j and s_j are scale parameters defining the origin and scale for the variable j. This allows us to equate variables with very different ranges, such as the crime variables. Usually, the origin is defined as the mean $\bar{y}_j$, and the scale is defined by the standard deviation for variable *j*, as is done for the profile plot of standardized data (Output 8.10).

Suppose the variables are positioned on the horizontal scale at positions x_j, the values we want to find. Then if each case has a linear profile, that means that

$$y_{ij}^* = a_i + b_i x_j \tag{8.3}$$

with slope b_i and intercept a_i. Hartigan (1975a) shows that the solution can be expressed in terms of the first two eigenvectors of the correlation matrix **C** of the y_{ij} (or the covariance matrix if the scales of the variables are not to be equated). Let $\mathbf{e}_1$ and $\mathbf{e}_2$ be the eigenvectors corresponding to the largest two eigenvalues of **C**. Then the optimal position for variable *j* is

$$x_j = \frac{e_{2j}}{e_{1j}} \tag{8.4}$$

and intercepts and slopes that give the fitted linear profile for each case are defined by

$$a = (\mathbf{D}\ \mathbf{e}_1) / \Sigma e_{1j}^2 \tag{8.5}$$

$$b = (\mathbf{D}\ \mathbf{e}_2) / \Sigma e_{2j}^2 \tag{8.6}$$

where **D** is the matrix whose elements are $(y_{ij} - \bar{y}_j)/e_{1j}$.

The calculations for the linear profiles display are most easily done in PROC IML. The program below, LINPRO SAS, calculates the scale values for all numeric variables in any input data set. The case label is represented by the macro variable ID. The output data to produce the plot are stored in data set LINPLOT. A second output data set, LINFIT, contains the residuals bordered by the fitted scale values. This data set also contains the case and variable names used to label the plot. Following the PROC IML program, an example shows how these data sets are used to plot the linear profiles for the city crime data, giving the display shown in Output 8.11. (A color version of this output appears as Output A3.43 in Appendix 3.)

```
proc iml;
  start linpro;
    *--- Get input matrix and dimensions;
    use _last_;
    read all into x[ rowname=&id  colname=vars ];
    n = nrow(x);
    p = ncol(x);

    *--- Compute covariance matrix of x;
    m = x[ : ,];                    *- means;
    d = x - j( n , 1) * m;          *- deviations from means;
    c = d` * d / n;                 *- variance-covariance matrix;
    sd = sqrt( vecdiag( c ))`;      *- standard deviations;
```

```
*--- Prestandardize if necessary;
ratio = sd[<>] / sd[><];
if ratio > 2.5 then do;
   s = sd;
   c = diag( 1 / sd) * c * diag( 1 / sd);
   print "Analyzing the correlation matrix";
end;
else do;
   s = j( 1 , p);
   print "Analyzing the covariance matrix";
end;
print c [ colname=vars rowname=vars ];

*--- Eigenvalues & vectors of C;
call eigen ( val , vec , c );
tr = val[+];
pc = val / tr ;
cum = pc[1];
do i = 2 to p;
   cum = cum // ( cum[ i-1 ] + pc[ i ]);
end;
val = val || (100 * pc) || (100 * cum ) ;
cl = { 'EIGVALUE' '%_TRACE' 'CUM_%'};
print "Eigenvalues and Variance Accounted for"
      , val [ colname=cl ];

*--- Scale values;
e1 = vec[ , 1] # sign( vec[<> , 1]);
e2 = vec[ , 2];
d = d / ( j( n , 1) * ( s # e1` ) );
pos = e2 / e1;

*--- For case i, fitted line is  Y = F1(I) + F2(I) * X   ;
f1 = ( d * e1 ) / e1[## ,];
f2 = ( d * e2 ) / e2[## ,];
f = f1 || f2;

*--- Output the results;
scfac = round( ( sd` # e1 ) , .1);
table = ( m` ) || ( sd` ) || e1 || e2 || scfac || pos;
ct = { 'MEAN' 'STD_DEV' 'EIGV1' 'EIGV2' 'SCALE' 'POSITION'};
print , table [ rowname=vars colname=ct ];

*--- Rearrange columns;
s = rank( pos);
zz = table;  table [ s , ] = zz;
zz = vars;   vars [ , s ] = zz;
zz = pos;    pos [ s , ] = zz;
zz = scfac;  scfac [ s , ] = zz;
zz = d;      d [ , s ] = zz;
```

```
            print "TABLE, with variables reordered by position"
                   , table [ rowname=vars colname=ct ];
            lt = { 'INTERCPT' 'SLOPE'};
            print "Case lines"      , f [ rowname=&id colname=lt ];

            *--- Fitted values, residuals;
            fit = f1 * j( 1 , p) + f2 * pos`;
            resid = d - fit;
            print "Fitted values" , fit [ rowname=&id colname=vars format=8.3 ];
            print "Residuals"      , resid [ rowname=&id colname=vars format=8.3 ];

            *--- Construct output array -
                  residuals bordered by fitted scale values;
            v1 = val[ 1 ] || {0};
            v2 = {0} || val[ 2 ];
            xout = ( resid || f ) //
                   ( pos` || v1 ) //
                   ( scfac` || v2 );
            rl = { 'VARIABLE','SCALE'};
            rl = shape( &id ,0 , 1) // rl;
            cl = { 'INTERCPT','SLOPE'};
            cl = shape( vars , p) // cl;
            create linfit from xout[ rowname=rl colname=cl ];
            append from xout[ rowname= rl ];
            free rl cl xout;

            *--- Output the array to be plotted;
            do col = 1 to p;
                rows = j(n,1,pos[col,]) ||  fit[,col] || resid[,col];
                pout = pout // rows;
                rl = rl // shape(&id,1);
                end;
            cl = { 'POSITION' 'FIT' 'RESIDUAL'};
            create linplot from pout[ rowname=rl colname=cl ];
            append from pout[ rowname=rl ];
         finish;
         run linpro;
      quit;
      data linplot;
        set linplot(rename=rl=&id);
```

The program below applies the LINPRO algorithm to the city crime data and plots the linear profiles. This program produces the plot shown in Output 8.11.

```
      data crime;
         input city &$16. murder rape robbery assault burglary larceny auto_thf;
      cards;
      ATLANTA          16.5  24.8  106  147  1112   905  494
      BOSTON            4.2  13.3  122   90   982   669  954
      CHICAGO          11.6  24.7  340  242   808   609  645
      DALLAS           18.1  34.2  184  293  1668   901  602
      DENVER            6.9  41.5  173  191  1534  1368  780
      DETROIT          13.0  35.7  477  220  1566  1183  788
```

```
HARTFORD            2.5    8.8    68   103   1017    724   468
HONOLULU            3.6   12.7    42    28   1457   1102   637
HOUSTON            16.8   26.6   289   186   1509    787   697
KANSAS CITY        10.8   43.2   255   226   1494    955   765
LOS ANGELES         9.7   51.8   286   355   1902   1386   862
NEW ORLEANS        10.3   39.7   266   283   1056   1036   776
NEW YORK            9.4   19.4   522   267   1674   1392   848
PORTLAND            5.0   23.0   157   144   1530   1281   488
TUCSON              5.1   22.9    85   148   1206    757   483
WASHINGTON         12.5   27.6   524   217   1496   1003   739
;
title 'LINEARLY OPTIMAL PROFILES ALGORITHM';
title2 'Data: Crime in U.S. cities';

%let ID=city;
%include linpro;                        /* On MVS, use %include ddname(file);*/
```

The information in the LINFIT data set is used to create an Annotate data set VLABEL to label the scaled positions of the variables on the horizontal axis and the names of the cities at the right end of each profile line. The variable RL is the row label, which is the city name except in the last two observations. The observation with RL='VARIABLE' contains the position of each variable on the horizontal axis that is used to draw the variable name.

```
data vlabel;
   set linfit;
   length text $12 function $8;
   array vars MURDER RAPE ROBBERY ASSAULT BURGLARY LARCENY AUTO_THF;
   Select;
      when (RL='VARIABLE') do;
         xsys='2'; ysys='1';
         angle = 45;
         position='6';
         do over vars;
            x = vars;
            y = 0; function='MOVE';  output;
            y = 1; function='DRAW';  output;
            call vname(vars,text);
            y = 2; function='LABEL'; output;
            end;
      end;
      when (RL='SCALE') ;          /* nothing */
      otherwise do;               /* observations */
         xsys='2'; ysys='2';
         angle = 0;
         position='C';
         function='LABEL';
```

```
        text = RL;                /* name of city */
        x = 2.15;                 /* data dependent: largest position */
        y = INTERCPT + SLOPE * x;
        x = x + .1;
        output;
      end;
   end; /* select */
title h=1.5 'Optimal Linear Profiles for City Crime Rates';
proc gplot data=linplot ;
   plot fit * position = city
       / vaxis=axis1 haxis=axis2
         anno=vlabel nolegend
         name='GB0811' ;
   symbol1 i=join v=- h=.5 l=1 c=black r=4;
   symbol2 i=join v=- h=.5 l=3 c=red   r=4;
   symbol3 i=join v=- h=.5 l=5 c=blue  r=4;
   symbol4 i=join v=- h=.5 l=9 c=green r=4;
   axis1 label=(h=1.4 a=90 r=0);
   axis2 minor=none style=0 ;
   label fit ='FITTED CRIME INDEX';
```

Output 8.11 *Optimal Linear Profiles for Crime Data*

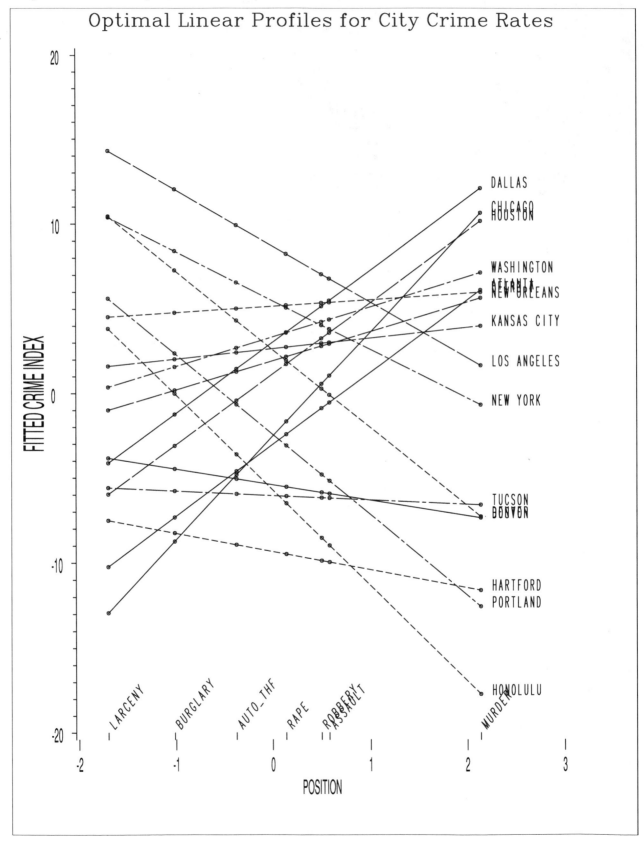

Optimal Linear Profiles for City Crime Rates

In Output 8.11, the variables have been ordered roughly in terms of property crimes (to the left) versus personal and violent crimes (to the right). Murder is widely separated from the other crimes, while robbery and assault are nearly coincident. The cities are ordered roughly in overall crime rate by their intercepts, the vertical positions where $x=0$. Each city is represented by a line, each variable by a horizontal position. The vertical scale corresponds to increasing crime rate. Thus Honolulu, Hartford, Tucson, Boston, and Portland are relatively low in crime, while Los Angeles, New York, Washington, and New Orleans are relatively high. Cities whose profiles have a negative slope, such as Honolulu, Boston, and Los Angeles, have a greater proportion of property than personal crime, while cities with a positive slope, such as Dallas, Chicago, and Atlanta, have a greater proportion of personal than property crime.

8.6 Andrews Function Plots

Another method for displaying data on many variables simultaneously is the *Andrews function plot* (Andrews 1972). This description follows Gnanadesikan (1977, Section 6.2).

The technique consists of representing an observation on p variables, $\mathbf{x}=(x_1, x_2, x_3, \ldots, x_p)'$ by a smooth function, $Z_x(t)$, of a single variable t over some range. Andrews defines $Z_x(t)$ as a finite Fourier series,

$$Z_x(t) = x_1 / \sqrt{2} + x_2 \sin t + x_3 \cos t + x_4 \sin 2t + x_5 \cos 2t + \ldots \quad .(8.7)$$

For each observation, the function is calculated over a range of t from $-\pi$ to $+\pi$ and plotted against t. Thus the initial multivariate observations, which are n points in a p-dimensional space, appear as n two-dimensional curves in a display of the function value, $Z_x(t)$ versus t.

This function has the attractive property that it preserves the mean and distance between observations. The curve representing the mean vector is the average of the individual curves. Observations that are far apart in p-dimensional space will plot as curves with large distances between them; conversely, points that are close together appear as close curves on the function plot.

Another statistical property is that these curves have the same variance over t, if the variables are uncorrelated and have equal variances. For this reason, the method is often applied to scores on principal components (see Section 9.4, "Plotting Principal Components and Component Scores") or discriminant functions (Section 9.5.1, "Plotting Discriminant Functions"). This property enables one to calculate the standard error of the Andrews function, to decide whether two observations differ significantly in multivariate space. When applied to raw data, the standard error of the function varies with t, but the plot can still be used to detect clustering or outliers.

8.6.1 Example: Ashton Fossil Data

The program below uses the data presented by Andrews (1972) and was modified from FUNPLOT SAS in the SAS Sample Library (Version 5). The data consist of $p=8$ measurements on premolar teeth of different groups of humans and apes, together with the same data on a set of fossil teeth. The data have been analyzed by Ashton, Healy, and Lipton (1957) by discriminant function analysis in order to classify the fossil teeth in relation to the known species of men and apes.

The observations are actually the mean scores for each species on a set of eight discriminant functions, linear combinations of the original variables designed to separate the groups maximally (see Section 9.5.1). The discriminant function scores are uncorrelated and have equal variance, so the standard error calculations are appropriate. The fossil specimens are the last four observations in the data set FOSSIL:

```
title 'ASHTON FOSSIL DATA FOR ANDREWS FUNCTION PLOTS';
data fossil;
    input id $1 name :&$char18. x1-x8;
cards;
A WEST AFRICAN         -8.09   .49   .18   .75 -.06  -.04  .04   .03
B BRITISH              -9.37  -.68  -.44  -.37  .37   .02 -.01   .05
C AUSTRAIL ABORIG      -8.87  1.44   .36  -.34 -.29  -.02 -.01  -.05
D GORILLA MALE          6.28  2.89   .43  -.03  .10  -.14  .07   .08
E GORILLA FEMALE        4.82  1.52   .71  -.06  .25   .15 -.07  -.10
F ORANGOUTANG MALE      5.11  1.61  -.72   .04 -.17   .13  .03   .05
G ORANGOUTANG FEMALE    3.60   .28 -1.05   .01 -.03  -.11 -.11  -.08
H CHIMPANZEE MALE       3.46 -3.37   .33  -.32 -.19  -.04  .09   .09
I CHIMPANZEE FEMALE     3.05 -4.21   .17   .28  .04   .02 -.06  -.06
J PITHICANTHROPUS      -6.73  3.63  1.14  2.11 -1.9   .24 1.23  -.55
M PARANTHROPUS CRAS    -7.79  4.33  1.42   .01 -1.8  -.25  .04  -.87
N MEGANTHROPUS         -8.23  5.03  1.13  -.02 -1.41 -.13 -.28  -.13
O PROCONSUL AFRICAN     1.86 -4.28 -2.14 -1.73  2.06 1.80 2.61  2.48
```

The first step in the analysis is to calculate the Andrews function score, $z = Z(t)$ over $(-\pi, +\pi)$, for each observation:

```
data andrews;
  set fossil;
  keep z t id legend;
  length legend $20;
  legend = id || ' ' || name;      /* GPLOT associates SYMBOLn with */
                                    /* sorted values of the legend   */
                                    /* variable.                     */
  *--apply the orthonormal functions of t from -pi to pi;
  pi=3.14159265;
  s=1/sqrt(2);
  inc=2*pi/100;
  do t=-pi to pi by inc;
     z=s*x1+sin( t )*x2 + cos( t )*x3 + sin(2*t)*x4 + cos(2*t)*x5 +
             sin(3*t)*x6 + cos(3*t)*x7 + sin(4*t)*x8;
     output;
     end;
  run;
```

The LEGEND variable will be used in PROC GPLOT with the following PLOT statement:

```
PLOT Z * T = LEGEND ... ;
```

In order for this to work with the SYMBOL*n* statements, the LEGEND variable must be sorted, so in the DATA step above the one-character ID variable is concatenated with the species NAME to form the LEGEND variable.

The plotting part of the program is shown below. It begins with a step to produce an Annotate data set to draw a small scale showing ±1 standard error (STD) of the Andrews function. The variance of $Z(t)$ for uncorrelated standardized measures is $p/2$ when p is odd, and between $(p-1)/2$ and $(p+1)/2$ when p is even. The program is followed by the plot it produces, Output 8.12. (A color version of this output appears as Output A3.44 in Appendix 3.)

```
data conf;
   drop p varx varz;
   length function $8 text $10;
   p   = 8;                        /* number of variables */
   varx= 1;                        /* var of any X        */

   if mod(p,2) = 1                 /* Calculate variance of Z(t) */
      then varz = varx * p/2;
      else varz = varx * mean ( (p-1)/2, (p+1)/2 );
   std = sqrt(varz);
   y0=   10;
   x     =-2.8; y=y0;
   xsys = '2';
   ysys = '2';
   function = 'SYMBOL'; text='PLUS';  output;
   function = 'MOVE ';                output;
   y =  y0 + std;
   function = 'DRAW ';                output;
   function = 'SYMBOL'; text='PLUS';  output;
   y =  y0 - std;
   function = 'DRAW ';                output;
   function = 'SYMBOL'; text='PLUS';  output;
   x = -2.7;  position = '8';
   function = 'LABEL';  text='2 Std Err'; output;
proc print;
```

The SYMBOL statements to plot the curves were constructed so that each of the three groups of observations—humans, apes, and fossils—are plotted in different colors, and the four fossil samples are plotted with thicker lines. SYMBOL*n* is assigned to the *n*th sorted value of the LEGEND variable.

```
symbol1  v=none i=join c=red   l=1;
symbol2  v=none i=join c=red   l=2;
symbol3  v=none i=join c=red   l=3;
symbol4  v=none i=join c=green l=4;
symbol5  v=none i=join c=green l=5;
symbol6  v=none i=join c=green l=6;
symbol7  v=none i=join c=green l=7;
symbol8  v=none i=join c=green l=8;
symbol9  v=none i=join c=green l=9;
```

```
symbol10 v=none i=join c=blue  l=10 w=3;
symbol11 v=none i=join c=blue  l=11 w=3;
symbol12 v=none i=join c=blue  l=12 w=3;
symbol13 v=none i=join c=blue  l=13 w=3;

title  f=duplex 'ANDREWS FUNCTION PLOTS FOR ASHTON FOSSIL DATA';
title2 h=1.5 f=duplex 'Premolar teeth of apes, humans, and fossils';
proc gplot data=andrews annotate=conf ;
    plot z * t = legend
        / vaxis=axis1 vminor=0 frame
          haxis=axis2
          name='GB0812' ;
    label  legend='SPECIES';
    axis1 order=(-12 to 12 by 2)
          label=(a=90 r=0 h=1.5 f=duplex 'Z(t)' )
          offset=(2,2);
    axis2 order=(-3.14, 0, 3.14) label=(h=1.5 f=duplex 't')
          offset=(2);
```

Output 8.12   *Andrews Function Plot for Multivariate Data*

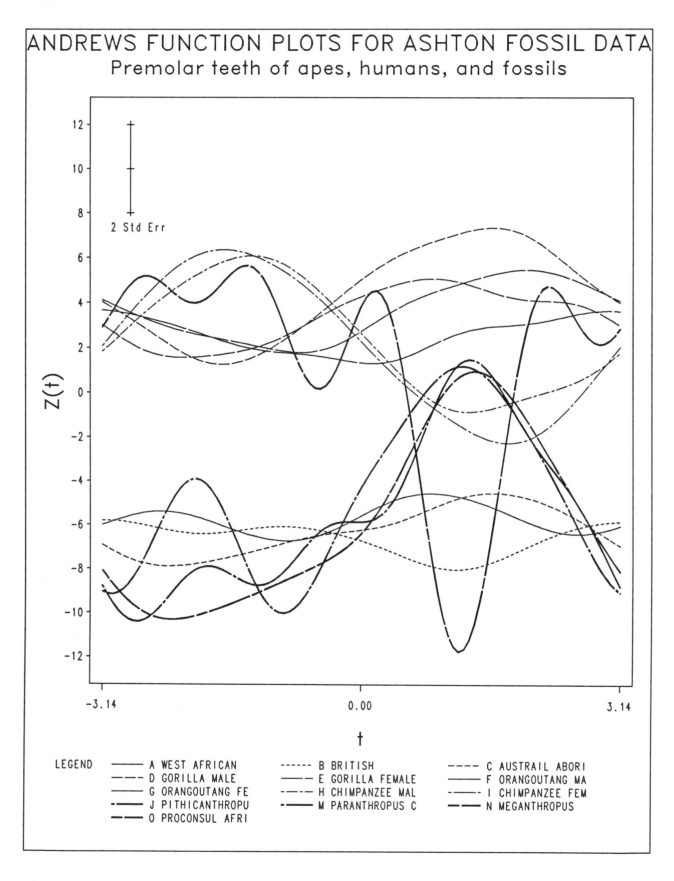

ANDREWS FUNCTION PLOTS FOR ASHTON FOSSIL DATA
Premolar teeth of apes, humans, and fossils

In Output 8.12, each observation is represented by a sum of sine and cosine functions, each weighted by a variable value. It is clear that the human groups (plotted in red) are well distinguished from the apes (in green) and that within the apes, the males and females of each species are more similar to each other than they are to the same sex of other species. So for these data, the plot appears to group the known species in the expected way.

Of greater interest is the classification of the fossil species (in blue), and it is apparent that one species, *Proconsul africanus*, stands out as being quite different from the other three and generally more apelike than human except at one region of *t*, where it dips far down. The other fossil species generally appear more similar to the humans, except at the same part of the range where they rise to the band of the ape data. Andrews (1972) supports these conclusions with formal tests based on the standard error calculations.

8.6.2 Strengths, Weaknesses, and Related Techniques

A strong advantage of the Andrews function plot is that it can display a relatively large number of variables without visual crowding. As additional variables are added, they are mapped into higher frequency components of the function and so remain visually more distinct than they do with methods such as the star plot (Section 8.4). As a result, the method may be better at detecting outliers or high-dimensional clustering of the observations (similar profiles on three or more measures jointly) than some of the methods considered earlier in this chapter.

On the other hand, the individual variables tend to lose their individual identity and are replaced by a configural gestalt. Thus, in Output 8.12 it is hard to see on which variables the three more human fossils differ from the human species, although we can see that *Proconsul* differs from the apes in variables X4 and X5, which have frequency 2*t* in the sine and cosine functions.

A disadvantage, compared with the star plot, is that as the number of observations increases beyond 20 or 30, for example, it becomes more difficult to distinguish the individuals. A novel solution suggested by Gnanadesikan (1977, p. 210ff) is to plot only selected quantiles or percentage points (for example, the 10th, 25th, 50th, 75th, and 90th percentiles) of the function values at each *t*, perhaps with some individual outside observations plotted as well. This plot, called a *quantile contour plot*, is akin to a continuous boxplot, with the middlemost observations replaced by their quantiles.

Another potential disadvantage, which the Andrews plot shares with the star plot, is that patterns in the display differ when the variables are ordered differently. The lower-numbered variables in the ordering $x_1, x_2, x_3, \ldots, x_p$ are mapped into low-frequency components of the Fourier function, which are easier to see on the plot than the high-frequency components. It is useful, therefore, to order the variables in terms of importance, with x_1 most important. If the raw data are first transformed to principal component scores, the transformed variables will be ordered by variance accounted for.

A related technique, called the *constellation graphical method* (Wakimoto and Taguri 1978), maps each multivariate observation into a (segmented) curve in the plane in a different way. First, all variables are rescaled to angular measure from 0 to 180 degrees. Each observation is portrayed by a set of rays of fixed length, but varying in angle, one for each variable. Starting from the origin, the rays are drawn head-to-tail to trace a path from the first variable to the last. All observations on such a plot lie in the positive half circle above the horizontal axis.

Observations with a relatively flat profile plot appear as relatively straight paths, while those that vary from one variable to the next tend to wiggle and do not extend as far. Wakimoto and Taguri (1978) claim that this scheme preserves the structure of each observation while it keeps the variables visually distinct.

8.7 Biplot: Plotting Variables and Observations Together

Glyphs, stars, and function plots focus all attention on the observations and portray the relations among the variables only implicitly. The *biplot*, proposed by Gabriel (1971; 1980; 1981), displays the observations and variables on the same plot in a way that depicts their joint relationships.

The biplot is based on the idea that any data matrix, $\mathbf{Y}$ ($n \times p$), can be represented approximately in d dimensions (d is usually 2 or 3) as the product of two matrices, $\mathbf{A}$ ($n \times d$) and $\mathbf{B}$ ($p \times d$),

$$\mathbf{Y} \approx \mathbf{A} \, \mathbf{B}' \tag{8.8}$$

so that any element y_{ij} of $\mathbf{Y}$ is approximated by the inner product of row i of $\mathbf{A}$, and of column j of $\mathbf{B}'$,

$$y_{ij} \approx \mathbf{a}_i' \, \mathbf{b}_j = a_{i1}b_{1j} + a_{i2}b_{2j} \tag{8.9}$$

Thus, the rows of $\mathbf{A}$ represent the observations in a two- or three-dimensional space, and the columns of $\mathbf{B}'$ represent the variables in the same space. The prefix *bi* in the name biplot stems from the fact that both the observations and variables are represented on the same plot, rather than from the fact that a two-dimensional representation is usually used.

The approximation used in the biplot is like that in *principal components analysis* (see Section 9.4): the biplot dimensions account for the greatest possible variance of the original data matrix. A biplot display has the following features:

□ The observations are usually plotted as points. The configuration of points is essentially the same as scores on the first two principal components.

□ The variables are plotted as vectors from the origin. The angles between the vectors represent the correlations among the variables.

For a least-squares approximation, the biplot vectors, $\mathbf{a}_i$, and $\mathbf{b}_j$, are obtained by the singular value decomposition (SVD) (Householder and Young 1938). By the SVD, any matrix $\mathbf{Y}$ may be decomposed exactly as

$$\mathbf{Y} = \mathbf{U} \, \Lambda \, \mathbf{V}' = \sum_{k=1}^{p} \lambda_k \, \mathbf{u}_k \, \mathbf{v}'_k \tag{8.10}$$

where Λ is a diagonal matrix, whose diagonal entries, $\lambda_1 \geq \lambda_2 \geq \ldots \geq \lambda_p$ are the *singular values*, which correspond to the eigenvalues of a square matrix and which reflect the proportion of variance accounted for by that dimension. Thus, a two-dimensional biplot results from taking the first two singular values and the corresponding columns of $\mathbf{U}$ and $\mathbf{V}$,

$$\mathbf{Y} \approx \hat{\mathbf{Y}} = \mathbf{U}_{[2]} \, \Lambda_{[2]} \, \mathbf{V}'_{[2]} = \sum_{k=1}^{2} \lambda_k \, \mathbf{u}_k \, \mathbf{v}'_k$$

and the measure of goodness of fit of this approximation is the ratio of variance accounted for to total variance,

$$\text{Goodness of fit} = (\lambda_1^2 + \lambda_2^2) / \sum_{k=1}^{p} \lambda_k^2 \quad .$$

For a three-dimensional biplot, we simply take the first three singular values and the corresponding columns of $\mathbf{U}$ and $\mathbf{V}$.

In practice, the SVD (equation 8.10) is applied not to the raw data, but to the matrix $\mathbf{Y} - \overline{\mathbf{Y}}$, with column means removed ($\overline{\mathbf{Y}}$ is an $n \times p$ matrix with all entries in each column equal to the mean of the corresponding column of $\mathbf{Y}$). If the variables are measured on different scales, it is common to standardize the data matrix to unit variance as well, in the same way that PCA is applied to the correlation matrix in this situation.

The factoring of this matrix according to equation 8.8 is not unique. There are three common ways to represent the (standardized) data matrix as the product $\mathbf{AB}'$ in terms of the SVD, shown in the following table. While they all have the property that the product of row $\mathbf{a}_i$ and column $\mathbf{b}_j$ approximates y_{ij}, they differ in how the lengths and angles of the vectors and distances between points approximate other properties of the data.

Factorization	$\mathbf{A}$	$\mathbf{B}'$
GH′	$\mathbf{U}$	$\mathbf{\Lambda V}'$
Symmetric	$\mathbf{U \Lambda}^{1/2}$	$\mathbf{\Lambda}^{1/2}\mathbf{V}'$
JK′	$\mathbf{U \Lambda}$	$\mathbf{V}'$

Specifically, the three methods of factorization work as follows:

GH′ biplot   represents the relations among the variables most closely, in that

     □   when the variables are not standardized, the length of the variable vector $\mathbf{b}_j$ reflects the standard deviation of column $\mathbf{y}_j$.

     □   the cosine of the angle between the vectors $\mathbf{b}_j$ and $\mathbf{b}_{j'}$ approximates the correlation between variables y_j and $y_{j'}$.

JK′ biplot   represents the relations among the observations most closely, in that

     □   the distance between points $\mathbf{a}_i$ and $\mathbf{a}_{i'}$ approximates the Euclidean distance between rows i and i' of the data matrix.

Symmetric biplot   scales the columns of **A** and **B** in the same way. That is,

$$\mathbf{A} = (\sqrt{\lambda_1}\, \boldsymbol{u}_1,\ \sqrt{\lambda_2}\, \boldsymbol{u}_2) \quad \text{and} \quad \mathbf{B} = (\sqrt{\lambda_1}\, \boldsymbol{v}_1,\ \sqrt{\lambda_2}\, \boldsymbol{v}_2) \quad .$$

This is often most convenient for plotting because the ranges of the row points and variable vectors will tend to fill the same region. Specifically, this factorization equates the lengths of the corresponding observation and variable vectors for each of the biplot dimensions.

Note that these factorizations differ only in how the columns of the row points, **A**, and variable vectors, **B**, are scaled relative to one another. The configurations are stretched or shrunk along the coordinate dimensions differently, but are otherwise the same.

8.7.1 The BIPLOT Macro

For the symmetric factorization, the biplot coordinates can be calculated using the FACTOR or PRINCOMP procedure.* With PROC PRINCOMP, the coordinates for the observations are the principal component scores; the coordinates for the variables are the component weights, which are obtained as the _TYPE_='SCORE' observations in the OUTSTAT= data set from PROC PRINCOMP.

It is more convenient, however, to calculate the biplot coordinates using PROC IML directly because this allows the GH' and JK' factorizations as well as the symmetric PCA factorization. In addition, it is often useful to scale the lengths of the variable vectors, multiplying them all by some constant, so that the row and column points have similar ranges in the biplot.

The row and column values for the biplot are calculated with PROC IML using the SVD function for the singular value decomposition. For generality, the program was written as a SAS macro program, BIPLOT, listed in Section A1.2. The main steps in the program are shown below:

```
* Singular value decomposition:
      Y is expressed as U * diag(Q) * V`
      Q contains singular values, in descending order;
  Call SVD(U,Q,V,Y);

* Extract first  d  columns of U & V, and first  d  elements of Q;
  U = U[,1:d];
  V = V[,1:d];
  Q = Q[1:d];
```

---

* If the variables are measured on ordinal scales and a nonmetric analysis is desired, the PRINQUAL procedure in Version 6 of SAS/STAT software can be used to transform the variables to maximize the variance accounted for by the first two components of the transformed variables. See Chapter 34, "The PRINQUAL Procedure," in the *SAS/STAT User's Guide, Version 6, Fourth Edition, Volume 2.* A PRINQUAL macro program is available in the Version 5 SAS Sample Library.

```
* Scale the vectors by QL, QR;
* Scale factor 'scale' allows expanding or contracting the variable
  vectors to plot in the same space as the observations;
QL= diag(Q ## power );
QR= diag(Q ## (1-power));
A = U * QL;
B = V * QR # scale;
OUT=A // B;
```

The BIPLOT macro takes the following parameters:

```
%macro BIPLOT(
        data=_LAST_,    /* Data set for biplot                  */
        var =_NUMERIC_, /* Variables for biplot                 */
        id  =ID,        /* Observation ID variable              */
        dim =2,         /* Number of biplot dimensions          */
        factype=SYM,    /* Biplot factor type: GH, SYM, or JK   */
        scale=1,        /* Scale factor for variable vectors    */
        out =BIPLOT,    /* Output dataset: biplot coordinates   */
        anno=BIANNO,    /* Output dataset: Annotate labels      */
        std=MEAN,       /* How to standardize columns: NONE|MEAN|STD*/
        pplot=YES);     /* Produce printer plot?                */
```

The number of biplot dimensions is specified by the DIM= parameter and the type of biplot factorization by the FACTYPE= value. The BIPLOT macro constructs two output data sets, identified by the parameters OUT= and ANNO=. The macro will produce a printed plot (if PPLOT=YES), but leaves it to the user to construct a PROC GPLOT plot because the scaling of the axes should be specified to achieve an appropriate geometry in the plot. This is illustrated in the example below.

Example: Crime Rates in the U.S.

As an example of the biplot technique, we consider data on the rates of crimes (per 100,000 population) in the U.S. contained in the data set CRIME (see Section A2.5, "The CRIME Data Set: State Crime Data"). The first and last few observations are shown below:

```
data crime;
   input state $1-15 murder rape robbery assault burglary larceny
         auto st $;
   cards;
ALABAMA         14.2 25.2  96.8 278.3 1135.5 1881.9 280.7   AL
ALASKA          10.8 51.6  96.8 284.0 1331.7 3369.8 753.3   AK
ARIZONA          9.5 34.2 138.2 312.3 2346.1 4467.4 439.5   AZ
ARKANSAS         8.8 27.6  83.2 203.4  972.6 1862.1 183.4   AR
```

```
CALIFORNIA      11.5 49.4 287.0 358.0 2139.4 3499.8 663.5   CA
... (observations omitted)
WASHINGTON       4.3 39.6 106.2 224.8 1605.6 3386.9 360.3   WA
WEST VIRGINIA    6.0 13.2  42.2  90.9  597.4 1341.7 163.3   WV
WISCONSIN        2.8 12.9  52.2  63.7  846.9 2614.2 220.7   WI
WYOMING          5.4 21.9  39.7 173.9  811.6 2772.2 282.0   WY
;
```

Output 8.13 shows the biplot of these data. Because the variables have widely different scales (standard deviations), the biplot is applied to the standardized variables.

Output 8.13   *Biplot of the Crime Data*

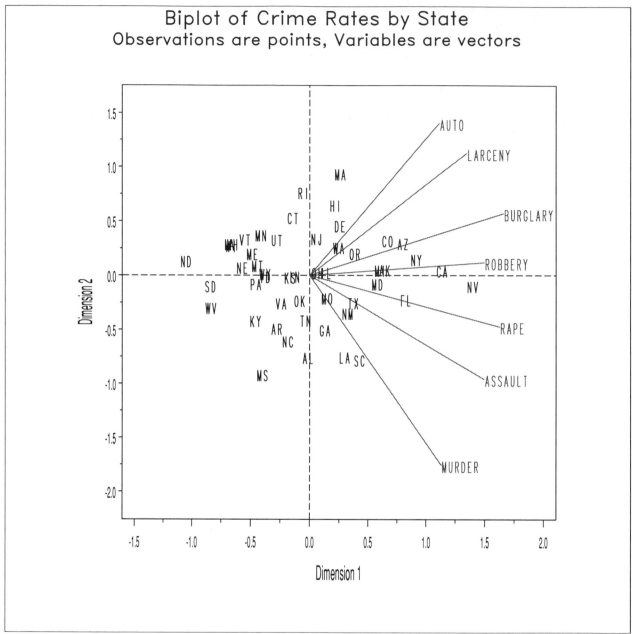

The statements below invoke the BIPLOT macro and plot the results using PROC GPLOT to produce Output 8.13. The %BIPLOT statement specifies the symmetric factorization (FACTYPE=SYM), which gives results equivalent to a principal components analysis; see Section 9.4 where this data set is analyzed by PROC PRINCOMP. The output data set BIPLOT contains the coordinates in the variables DIM1 and DIM2, which are plotted in the PROC GPLOT step. These coordinates are shown in the printed output from the BIPLOT macro in Output 8.14. The output data set BIANNO contains the Annotate instructions to label the points and plot the variable vectors.

```
%include CRIME ;                /* On MVS, use %include ddname(file);*/
%include biplot;                /* On MVS, use %include ddname(file);*/
title h=1.5 f=duplex 'Biplot of Crime Rates by State';
%biplot( data= crime,
          var = MURDER RAPE ROBBERY ASSAULT BURGLARY LARCENY AUTO,
          id  = ST, factype=SYM, std = STD );

proc gplot data=biplot ;
   plot dim2 * dim1 / anno=bianno frame
                      href=0 vref=0 lvref=3 lhref=3
                      vaxis=axis2 haxis=axis1
                      vminor=1 hminor=1
                      name='GB0813' ;
   axis1 length=4.5 in  order=(-1.5 to 2 by .5)
         offset=(2)
         label=(h=1.3 'Dimension 1');
   axis2 length=4.5 in  order=(-2 to 1.5 by .5)
         offset=(2)
         label=(h=1.3 a=90 r=0 'Dimension 2');
   symbol v=none;
   title2 h=1.3 f=duplex
          'Observations are points, Variables are vectors';
```

Output 8.14
Printed Output from BIPLOT Macro

```
              Biplot of Crime Rates by State
            Singular values and variance accounted for

        Singular     Percent      Cum %
         Values

         14.1998       58.79       58.79
          7.7909       17.70       76.48
          5.9636       10.37       86.85
          3.9377        4.52       91.37
          3.5554        3.69       95.06
          3.2985        3.17       98.23
          2.4655        1.77      100.00

                 Biplot Factor Type
                  Symmetric (PCA)

              Biplot coordinates
                          DIM1      DIM2

        OBS AL        -0.0132   -0.7510
        OBS AK         0.6426    0.0597
        OBS AZ         0.7999    0.3027
        OBS AR        -0.2798   -0.4820
        OBS CA         1.1368    0.0513
        OBS CO         0.6659    0.3284
        OBS CT        -0.1437    0.5378
        OBS DE         0.2560    0.4646
        OBS FL         0.8258   -0.2164
        OBS GA         0.1301   -0.4947
        OBS HI         0.2184    0.6535
        OBS ID        -0.3801   -.002868
        OBS IL         0.1361    0.0338
        OBS IN        -0.1327   9.3E-06
```

(continued on next page)

```
(continued from previous page)
            OBS  IA        -0.6851    0.2955
            OBS  KS        -0.1683   -0.0100
            OBS  KY        -0.4583   -0.4108
            OBS  LA         0.2973   -0.7464
            OBS  ME        -0.4847    0.2074
            OBS  MD         0.5793   -0.0698
            OBS  MA         0.2597    0.9426
            OBS  MI         0.6033    0.0555
            OBS  MN        -0.4125    0.3785
            OBS  MS        -0.4000   -0.9124
            OBS  MO         0.1476   -0.2001
            OBS  MT        -0.4426    0.0971
            OBS  NE        -0.5707    0.0809
            OBS  NV         1.3977   -0.0905
            OBS  NH        -0.6543    0.2956
            OBS  NJ         0.0578    0.3454
            OBS  NM         0.3222   -0.3406
            OBS  NY         0.9162    0.1551
            OBS  NC        -0.1856   -0.5984
            OBS  ND        -1.0520    0.1389
            OBS  OH         0.0636    0.0324
            OBS  OK        -0.0853   -0.2237
            OBS  OR         0.3845    0.2100
            OBS  PA        -0.4565   -0.0702
            OBS  RI        -0.0535    0.7690
            OBS  SC         0.4255   -0.7746
            OBS  SD        -0.8418   -0.0912
            OBS  TN        -0.0362   -0.4066
            OBS  TX         0.3707   -0.2441
            OBS  UT        -0.2786    0.3355
            OBS  VT        -0.5478    0.3386
            OBS  VA        -0.2431   -0.2482
            OBS  WA         0.2470    0.2643
            OBS  WV        -0.8353   -0.2917
            OBS  WI        -0.6642    0.2797
            OBS  WY        -0.3781    0.0225
            VAR  MURDER     1.1315   -1.7562
            VAR  RAPE       1.6270   -0.4729
            VAR  ROBBERY    1.4955    0.1179
            VAR  ASSAULT    1.4947   -0.9589
            VAR  BURGLARY   1.6586    0.5676
            VAR  LARCENY    1.3466    1.1230
            VAR  AUTO       1.1123    1.4024
```

On a biplot, it is usually important to ensure that equal distances in the vertical and horizontal directions correspond to equal data ranges. This makes it possible to interpret distances on the plot between points and vectors. To equate the data units per inch, the AXIS1 and AXIS2 statements both specify the same LENGTH and have the same range in the ORDER= parameter.

The horizontal dimension in Output 8.13 is interpreted as overall crime rate, just as in the principal component analysis of these data (see Section 9.4). The states at the right (Nevada, California, New York) are high in crime; those at the left (North Dakota, South Dakota) are low. The vertical dimension is a contrast between property crime and violent crime. These two dimensions account for over 76 percent of the variance of the standardized data.

The angles between the variable vectors reflect the correlations of the variables, so that variables with small angles, like AUTO (auto theft) and LARCENY, are highly correlated, and variables at right angles (AUTO and MURDER) are nearly uncorrelated.* In this data set there are no negative correlations, but if there were, they would appear as vectors with angles greater than 90 degrees.

---

* The representation of correlations as angles between vectors is correct only in the **GH′** factorization. It is distorted in the symmetric factorization, but we can still interpret small and large angles.

In this biplot, the variables have been standardized to unit variance, so the lengths of the vectors reflect the relative proportion of the variance of each variable that is captured by the two-dimensional biplot. All the vectors are approximately the same length, except for ROBBERY, indicating that this variable is slightly less well represented in two dimensions.

More importantly, the biplot shows how the clusters of observations relate to the variables. Visually, this is judged by seeing where an observation point projects onto a variable vector. The direction of a variable vector is interpreted as high on that variable, and the opposite direction is interpreted as low. Thus, the southern states (SC, LA, AL, and MS) are relatively high on the vectors for violent crimes of murder and assault, but relatively low on the property crimes of auto theft and larceny. The New England states, particularly MA, RI, and CT, tend to have the opposite pattern.

Chapter 9 Multivariate Statistical Methods

9.1 Introduction

This chapter surveys a number of plotting techniques related to standard statistical methods for multivariate data.

Common to most of these methods is the assumption that the data or residuals have a multivariate normal distribution. Section 9.2, "Assessing Multivariate Normality," describes a χ^2 probability plot for determining whether this assumption is reasonable. A robust version of this plot, described in Section 9.3, "Detecting Multivariate Outliers," is used to detect multivariate outliers.

Many multivariate statistical methods involve finding linear combinations of observed variables that optimize some criterion. Graphical techniques related to principal components analysis are discussed in Section 9.4, "Plotting Principal Components and Component Scores." Methods for discriminating among groups, including discriminant function analysis, canonical discriminant analysis, and multivariate analysis of variance (MANOVA), are described in Section 9.5, "Discriminating among Groups."

9.2 Assessing Multivariate Normality

A graphical technique for assessing multivariate normality is an application using the ideas of Q-Q plots described in Section 3.5, "Quantile Plots." The basic principle is to calculate a quantity from each multivariate observation, such that this quantity follows a known probability distribution when the data follow the multivariate normal distribution. Then a Q-Q plot of this quantity against the quantiles of the reference distribution will plot as a straight line. Gnanadesikan (1977) discusses the assessment of multivariate normality in much detail and presents a variety of graphical and statistical tests that may be more sensitive to some forms of departure from multivariate normality than is the simple χ^2 probability plot described here.

For testing multivariate normality, the quantity we compute is the generalized (Mahalanobis) squared distance between the ith observation vector $\mathbf{x}_i' = (x_{i1}, x_{i2}, \ldots, x_{ip})$ and the mean vector $\overline{\mathbf{x}}$ for the total sample. This squared distance for observation i is defined by

$$D_i^2 = (\mathbf{x}_i - \overline{\mathbf{x}})'\, \mathbf{S}^{-1}\, (\mathbf{x}_i - \overline{\mathbf{x}}) \tag{9.1}$$

where $\mathbf{S}$ is the $p \times p$ sample variance-covariance matrix. D^2 is the multivariate analog of the square of the standard score for a single variable, $z_i = (x_i - \overline{x})/s$, which measures distance from the mean in units of the standard deviation, and z_i^2 is distributed as $\chi^2(1)$.

Analogously, D_i^2 measures distance from the mean vector in relation to the variance-covariance matrix, which takes into account the precision of the variables and their intercorrelations. With p variables, D_i^2 is distributed approximately as χ^2 with p degrees of freedom for large samples from the multivariate normal distribution. Therefore, a Q-Q plot of the ordered distance values, $D_{(i)}^2$, against the corresponding quantiles of the $\chi^2(p)$ distribution should yield a straight line through the origin for multivariate normal data. The χ^2 quantiles can be calculated with SAS software using the GAMINV function or the CINV function.*

9.2.1 Example: Automobiles Data

The program below uses the IML procedure to calculate the Mahalanobis squared-distance values and the corresponding χ^2 quantiles for the AUTO data set. These values are produced in an output data set that is then plotted with the GPLOT procedure. (A method for calculating the D^2 values using the PRINCOMP procedure instead of the IML procedure is shown in the next section.)

```
%include auto;                 /* On MVS, use %include ddname(file);*/
Title 'Chi-square probability plot for multivariate data';
%let id=model;
%let var=PRICE MPG REP78 REP77 HROOM RSEAT TRUNK WEIGHT LENGTH
         TURN DISPLA GRATIO;
proc iml;
reset;
start dsquare;
   use _LAST_ ;
   read all var {&var} into  X[ colname=vars rowname=&id ];
```

---

* In Version 5 of the SAS System, the CINV function is available only in the SUGI Supplemental Library.

```
   n = nrow(X);
   p = ncol(X);
   rl= &id;

   *---- Compute covariance matrix of X;
   M = X[ : , ];              *-- means ;
   D = X - J( n , 1) * M;
   S = D` * D / ( n - 1 );    *-- var-cov matrix ;

   *---- Calculate Mahanalobis distances;
   dsq = vecdiag( D * inv(S) * D`);

   *---- Rank them in ascending order;
   r = rank( dsq);
   val = dsq;  dsq [ r , ] = val;
   val = rl;    &id [ r ] = val;

   *---- Calculate Chi-square percentiles;
   z = ( (1:n)` - .5 ) / n ;
   chisq = 2 * gaminv( z, p/2);

   *---- Join & output;
   result = dsq || chisq;
   cl = { 'DSQ' 'CHISQ'};
   create dsquare from result[ colname=cl rowname=&id];
   append from result [ rowname=&id];
finish;

run dsquare;
quit;
proc print data=dsquare;
   var model dsq chisq;

   /*---------------------------------------------------------------*
    | If the sample comes from a multivariate normal distribution, |
    | the plot of D-square vs. Chi-square percentiles should plot  |
    | as a straight line.                                          |
    *---------------------------------------------------------------*/

title  f=complex H=1.5 ' Chi-square probability plot';
title2 f=complex H=1.5 ' for multivariate normality';
proc gplot data=dsquare  ;
    plot dsq * chisq / vaxis=axis1 vminor=4
                       haxis=axis2 hminor=4
                       name='GB0901' ;
    symbol f=special v=K h=1.7 l=8 i=rl;
    axis1  label=(a=90 r=0 h=1.5 f=duplex) value=(h=1.4);
    axis2  label=(h=1.5 f=duplex) value=(h=1.4)
           offset=(1);
    label chisq = 'Chi-square quantile'
          DSQ   = 'Mahalanobis D-square';
```

The χ^2 probability plot for these data is shown in Output 9.1. Note that the reference line is produced by specifying I=RL in the SYMBOL statement and is not constrained to go through the origin. If the data are multivariate normal, the reference line should have a slope of 1 and intercept of 0. Specifying I=RL0 would force the reference line to have a 0 intercept. There is some indication of departure from the reference line in the upper right, but it is probably not great enough to reject the assumption of multivariate normality for practical purposes.

Output 9.1
Chi-Square
Probability Plot
for Automobiles
Data

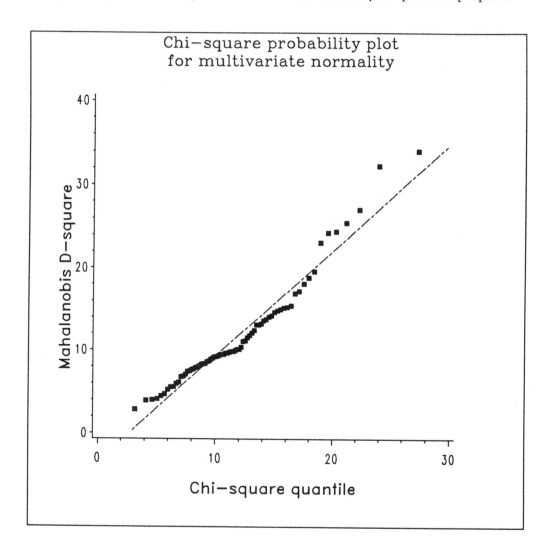

9.3 Detecting Multivariate Outliers

For single variables, outliers are cases with extreme standard scores, $|z_i| > 4$ or more, for example. Analogously, we could define a multivariate outlier as a case with a large Mahalanobis distance. On the χ^2 probability plot, these would appear as points in the upper right that are substantially above the line for the expected χ^2 quantiles. Unfortunately, like all classical (least-squares) techniques, the χ^2 plot for multivariate normality is not resistant to the effects of outliers. A few discrepant observations not only affect the mean vector, but also inflate the variance-covariance matrix. Thus, the effect of the few wild observations is spread through all the D^2 values. Moreover, this tends to decrease the range of the D^2 values, making it harder to detect extreme ones.

One reasonably general solution is to use a *multivariate trimming* procedure to calculate squared distances that are not affected by potential outliers (Gnanadesikan and Kettenring 1972; Gnanadesikan 1977). This is an iterative process where, on each iteration, some proportion of the observations with the largest D^2 values are temporarily set aside, and the trimmed mean, $\bar{\mathbf{x}}_{(-)}$, and trimmed variance-covariance matrix, $\mathbf{S}_{(-)}$, are computed from the remaining observations. Then new D^2 values are computed using the robust mean and covariance matrix,

$$D_i^2 = (\mathbf{x}_i - \bar{\mathbf{x}}_{(-)})' \, \mathbf{S}_{(-)}^{-1} \, (\mathbf{x}_i - \bar{\mathbf{x}}_{(-)}) \quad . \tag{9.2}$$

The effect of trimming is that observations with large distances do not contribute to the calculations for the remaining observations.

One way to carry out this process in the SAS System is suggested by F. W. Young and W. S. Sarle.* First, their idea avoids the necessity to calculate the inverse of the variance-covariance matrix by transforming the data to standardized principal component scores. For if $\mathbf{z}_i$ is the vector of standardized principal component scores corresponding to $\mathbf{x}_i$, then the squared distance is just the sum of squares of the elements in $\mathbf{z}_i$,

$$D_i^2 = \mathbf{z}_i' \, \mathbf{z}_i = \sum_{j=1}^{p} z_{ij}^2 \quad . \tag{9.3}$$

Thus, the squared distances for variables X1—X5, for example, can be calculated with PROC PRINCOMP and a DATA step as follows:

```
proc princomp std out=pc;
   var x1-x5;
data pc;
   set pc;
   dsq = uss(of prin1-prin5);
```

This method avoids the necessity to use PROC IML as in the previous section, though PROC IML is probably more efficient. The STD option here is necessary to get principal component scores standardized to unit variance.

The second part of Young and Sarle's idea is to use the WEIGHT statement in PROC PRINCOMP to carry out the multivariate trimming. Initially, all observations are given weights of 1, and the distance values are calculated as above. Then an observation whose D_i^2 represents a significantly large χ^2 is given a weight of 0, and the process is repeated, typically until no new observations are trimmed or a given number of iterations has been carried out.

9.3.1 OUTLIER Macro

This scheme for outlier detection has been implemented in a general SAS macro, OUTLIER, listed in Section A1.11, "The OUTLIER Macro." The arguments to the

---

* The original source is *SAS Views: Multivariate Data Analysis, 1982 Edition*, Cary, NC: SAS Institute Inc. It was replaced by *Multivariate Statistical Methods: Practical Applications Course Notes* (1989), Cary, NC: SAS Institute Inc.

macro are shown below. PVALUE is the probability, such that an observation is trimmed when its D^2 has a probability less than PVALUE. The macro produces an output data set (the OUT= parameter) containing the variables DSQ (the squared distance) and EXPECTED (the χ^2 quantile) in addition to the input variables.

```
%macro OUTLIER(
        data=_LAST_,     /* Data set to analyze         */
        var=_NUMERIC_,   /* input variables             */
        id=,             /* ID variable for observations */
        out=CHIPLOT,     /* Output dataset for plotting  */
        pvalue=.1,       /* Prob < pvalue -> weight=0    */
        passes=2,        /* Number of passes             */
        print=YES);      /* Print OUT= data set?         */
```

By default, two passes (PASSES=) are made through the iterative procedure. A printer plot of DSQ*EXPECTED is produced automatically, and you can use PROC GPLOT afterward with the OUT= data set to give a high-resolution plot with labeled outliers, as shown in the example below.

Example: Mammals' Teeth

The data set TEETH from the SAS Sample Library is often used to demonstrate multivariate procedures. The data set contains counts of eight kinds of teeth in the variables V1—V8 found in 32 species of mammals. The complete data set is shown in Section A2.14, "The TEETH Data Set: Mammals' Teeth Data." The first few and last few observations are shown below:

```
data teeth;
    title "Mammals' Teeth Data";
    input mammal $ 1-16 a21 (v1-v8) (1.);
    length id $2;
    id=put(_n_,z2.);
    format v1-v8 1.;
    label v1='Top incisors'
          v2='Bottom incisors'
          v3='Top canines'
          v4='Bottom canines'
          v5='Top premolars'
          v6='Bottom premolars'
          v7='Top molars'
          v8='Bottom molars';
    cards;
BROWN BAT          23113333
MOLE               32103333
SILVER HAIR BAT    23112333
PIGMY BAT          23112233
HOUSE BAT          23111233
RED BAT            13112233
    ... (observations omitted)
```

```
REINDEER              04103333
ELK                   04103333
DEER                  04003333
MOOSE                 04003333
;
```

The OUTLIER macro is applied to these data with the statements below:

```
title 'Multivariate outlier detection - Mammals' Teeth data';
%include TEETH;              /* On MVS, use %include ddname(file);*/
%include outlier;           /* On MVS, use %include ddname(file);*/

%outlier(data=teeth,
         var=v1-v8, id=mammal, out=chiplot);
```

Output 9.2 and 9.3 show the printed output from the OUTLIER program for the mammals' teeth data. The output in Output 9.2 shows that three observations have significantly large DSQ values in pass 1. When these are given a weight of 0 in the next PROC PRINCOMP step, their DSQ values increase, while the DSQ values for other observations tend to decrease or remain the same, and no new observations are trimmed.

Output 9.2
Observations Trimmed by the Iterative Procedure

```
                              MAMMALS' TEETH
            Observations trimmed in calculating Mahalanobis distance

   OBS    MAMMAL            PASS    CASE     DSQ          PROB

    1     MOLE               1       2     18.7217     0.0164212
    2     RACCOON            1      16     14.9452     0.0602202
    3     ELEPHANT SEAL      1      28     17.0421     0.0296739
    4     MOLE               2       2     67.9701     0.0000000
    5     RACCOON            2      16     35.0354     0.0000263
    6     ELEPHANT SEAL      2      28     48.1781     0.0000001
```

The output data set CHIPLOT, shown in Output 9.3, can be used to plot the DSQ values against their expected χ^2 values under the hypothesis of multivariate normality. In constructing this plot, it is helpful to draw the reference line and label potential outliers.

Output 9.3
Output Data Set CHIPLOT from OUTLIER Macro

```
                              MAMMALS' TEETH
                Possible multivariate outliers have _WEIGHT_=0

MAMMAL          V1 V2 V3 V4 V5 V6 V7 V8 ID _WEIGHT_  DSQ      PROB      EXPECTED

MARTEN           3  3  1  1  4  4  1  2 17     1    3.2789  0.915656   2.3175
WOLVERINE        3  3  1  1  4  4  1  2 19     1    3.2789  0.915656   2.9181
GREY SEAL        3  2  1  1  3  3  2  2 27     1    3.7076  0.882493   3.3693
WEASEL           3  3  1  1  3  3  1  2 18     1    4.3401  0.825212   3.7519
BADGER           3  3  1  1  3  3  1  2 20     1    4.3401  0.825212   4.0946
GRAY SQUIRREL    1  1  0  0  1  1  3  3 11     1    4.3802  0.821298   4.4119
PORCUPINE        1  1  0  0  1  1  3  3 13     1    4.3802  0.821298   4.7119
PIGMY BAT        2  3  1  1  2  2  3  3 04     1    5.3651  0.717930   5.0000
BROWN BAT        2  3  1  1  3  3  3  3 01     1    5.6382  0.687681   5.2800
```

(continued on next page)

(continued from previous page)

SEA OTTER	3	2	1	1	3	3	1	2	22	1	5.9486	0.652991	5.5547
WOLF	3	3	1	1	4	4	2	3	14	1	5.9890	0.648469	5.8264
BEAR	3	3	1	1	4	4	2	3	15	1	5.9890	0.648469	6.0970
BEAVER	1	1	0	0	2	1	3	3	09	1	6.2647	0.617612	6.3682
GROUNDHOG	1	1	0	0	2	1	3	3	10	1	6.2647	0.617612	6.6415
RIVER OTTER	3	3	1	1	4	3	1	2	21	1	6.3267	0.610693	6.9185
SILVER HAIR BAT	2	3	1	1	2	3	3	3	03	1	6.4221	0.600065	7.2006
HOUSE MOUSE	1	1	0	0	0	0	3	3	12	1	8.6408	0.373499	7.4895
HOUSE BAT	2	3	1	1	1	2	3	3	05	1	9.1562	0.329291	7.7870
PIKA	2	1	0	0	2	2	3	3	07	1	9.1949	0.326118	8.0950
FUR SEAL	3	2	1	1	4	4	1	1	25	1	10.1738	0.253040	8.4158
SEA LION	3	2	1	1	4	4	1	1	26	1	10.1738	0.253040	8.7521
RABBIT	2	1	0	0	3	2	3	3	08	1	10.1759	0.252899	9.1072
JAGUAR	3	3	1	1	3	2	1	1	23	1	10.4162	0.237019	9.4852
COUGAR	3	3	1	1	3	2	1	1	24	1	10.4162	0.237019	9.8915
DEER	0	4	0	0	3	3	3	3	31	1	12.2376	0.140913	10.3330
MOOSE	0	4	0	0	3	3	3	3	32	1	12.2376	0.140913	10.8198
REINDEER	0	4	1	0	3	3	3	3	29	1	13.0345	0.110669	11.3659
ELK	0	4	1	0	3	3	3	3	30	1	13.0345	0.110669	11.9930
RED BAT	1	3	1	1	2	2	3	3	06	1	13.1942	0.105341	12.7373
RACCOON	3	3	1	1	4	4	3	2	16	0	35.0354	0.000026	13.6656
ELEPHANT SEAL	2	1	1	1	4	4	1	1	28	0	48.1781	0.000000	14.9257
MOLE	3	2	1	0	3	3	3	3	02	0	67.9701	0.000000	16.9814

Output 9.4 shows the plot, which is produced by the statements below. Note that the labels for outliers alternate from left to right of the plotted point to avoid possible overplotting (which does not occur here). The correct reference line is drawn by plotting the EXPECTED value against itself.

```
data labels;
   set chiplot;
   if prob < .05;
   xsys='2'; ysys='2';
   y = dsq;
   n + 1;
   if mod( n ,2) = 0 then do;       * alternate label position;
      x = 0.98*expected; position = '4';
      end;
   else do;
      x = 1.02*expected; position = '6';
      end;
   function = 'LABEL';
   text = mammal;
   size = 1.4;
proc gplot data=chiplot  ;
   plot dsq      * expected = 1
        expected * expected = 2
        / overlay anno=labels
          vaxis=axis1 haxis=axis2
          vminor=1 hminor=4
          name='GB0904'  ;
   symbol1 f=special v=K h=1.5 i=none c=black;
   symbol2 v=none i=join c=red;
   label dsq      ='Squared Distance'
         expected='Chi-square quantile';
   axis1 label=(a=90 r=0 h=1.5 f=duplex) label=(h=1.3);
   axis2 order=(0 to 20 by  5) label=(h=1.5 f=duplex) label=(h=1.3);
   title h=1.5 'Outlier plot for Mammals teeth data';
```

Output 9.4
*Chi-Square
Probability Plot
for Teeth Data*

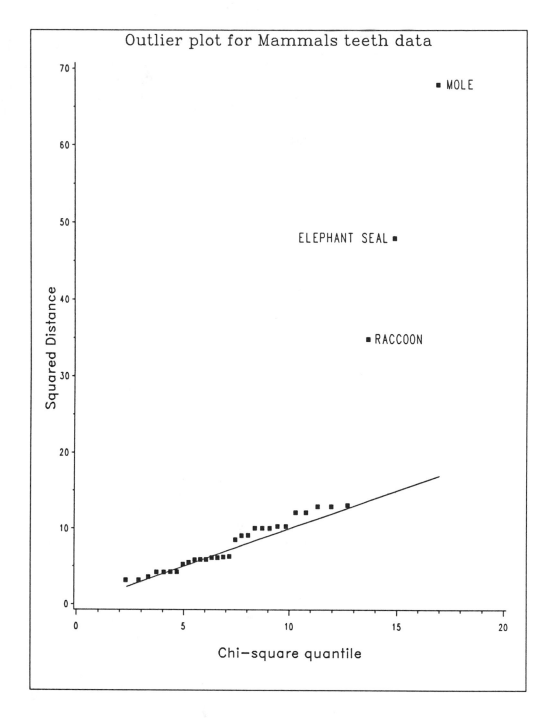

It should be noted that detecting multivariate outliers is a complex problem, and the results of any automatic procedure such as this must be tempered by substantive judgment. In the mammals' teeth data, the mole, raccoon, and elephant seal appear to differ from the other mammals, but deciding whether they are truly outliers requires knowledge of biology. Also, perhaps a more conservative probability value, on the order of $p < .001$, should be used to declare a case an outlier.

Example: Automobiles Data

The original χ^2 plot for the automobiles data (Output 9.1), which did not use trimming, showed only a slight tendency for points to drift away from the reference line. However, the OUTLIER macro indicates that six observations are suspiciously deviant from the rest. These do correspond to the points in the upper right of Output 9.1, but they do not appear discrepant there.

The OUTLIER macro was run on the automobiles data with the following SAS program:

```
%include outlier;           /* On MVS, use %include ddname(file);*/
%include AUTO;              /* On MVS, use %include ddname(file);*/
data AUTO;
   set AUTO;
   repair = mean(of REP77 REP78);
   if repair=. then delete;
%OUTLIER(data=AUTO,
   var=PRICE MPG REPAIR HROOM RSEAT TRUNK WEIGHT LENGTH TURN DISPLA GRATIO,
   pvalue=.05, id=MODEL, out=CHIPLOT, print=NO);
```

The principal components procedure used by the OUTLIER program does not allow for missing data (the component scores are all missing if any variable is missing). In the automobiles data, eight observations have missing values for either REP77 or REP78. To keep the maximum amount of data, the two repair record variables have been averaged, and only observations with missing values for both have been deleted.

The printed output, shown in Output 9.5, indicates that six observations had probability values more extreme than the .05 cutoff used. The plot of DSQ against EXPECTED values (Output 9.6) uses a similar Annotate data set LABELS and PROC GPLOT step to that used for the teeth data. The observations singled out by the procedure apparently differ from the rest, particularly the one for the Volkswagen Dasher. However, whether these automobiles should be excluded from other analyses or not is once more a substantive question, not a statistical one.

Output 9.5
OUTLIER Macro
Output for
Automobiles Data

```
          Observations trimmed in calculating Mahalanobis distance

    OBS    MODEL              PASS    CASE     DSQ        PROB

     1     AMC PACER            1       2    22.7827    0.0189645
     2     CAD. SEVILLE         1      14    23.8780    0.0132575
     3     CHEV. CHEVETTE       1      15    23.5344    0.0148453
     4     VW RABBIT DIESEL     1      66    25.3503    0.0080987
     5     VW DASHER            1      68    36.3782    0.0001464
     6     AMC PACER            2       2    34.4366    0.0003068
     7     CAD. SEVILLE         2      14    42.1712    0.0000151
     8     CHEV. CHEVETTE       2      15    36.7623    0.0001263
     9     PLYM. CHAMP          2      52    20.9623    0.0337638
    10     VW RABBIT DIESEL     2      66    44.2961    0.0000064
    11     VW DASHER            2      68    78.5944    0.0000000
```

Output 9.6
Outlier Chi-Square Plot for Automobiles Data

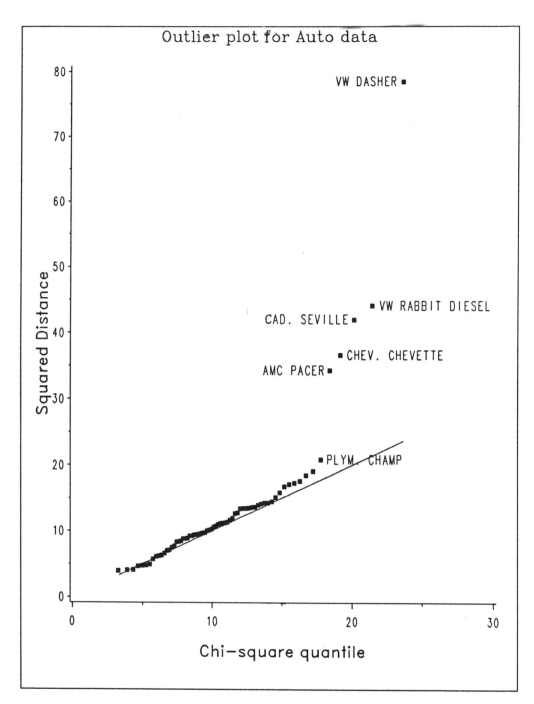

9.4 Plotting Principal Components and Component Scores

One of the central goals of multivariate methods is to find linear combinations of observed variables that optimize some goodness-of-fit criterion. In multiple regression, for example, the regression weights are found so that the weighted sum of predictors has the highest possible correlation with the criterion.

Principal components analysis is a data reduction technique, designed to explain a set of variables in terms of a small number of linear combinations of those variables. These linear combinations, called *principal components*, are determined so that each one in succession accounts for the maximum variance of the observed scores, and are all mutually uncorrelated. If there are p variables, then p principal components will reproduce the data exactly; however, often a substantial portion of the data variation can be accounted for by a smaller number, $k < p$ of the principal components.

Geometrically, principal components can be interpreted in terms of the locus of points of constant generalized distance from the centroid of the space of the variables. The equation

$$(x - \bar{x})' \, S^{-1} \, (x - \bar{x}) = c$$

defines the data ellipsoid (see Section 4.5, "Enhanced Scatterplots") in the p-dimensional space of x, on whose surface all points have squared distance from the centroid $\bar{x}$ equal to c. The principal component vectors, the directions of maximum variation, are the principal axes of this ellipsoid. The principal component scores are the projections of the observation points onto these principal axes. The reduction of dimensionality occurs when k of the p axes of the ellipse are large relative to the remaining ones, so, to a reasonable approximation, the data are concentrated in a k-dimensional subspace.

Algebraically, the principal component vectors are the eigenvectors of the sample covariance matrix S; the variance accounted for by each component is the corresponding eigenvalue. This is because the covariance matrix may always be resolved as the product

$$S = V \, D_\lambda \, V' \tag{9.4}$$

where the columns of V are the eigenvectors of S, and D_λ is a diagonal matrix of the eigenvalues, $\lambda_1 \geq \lambda_2 \geq \ldots \geq \lambda_p \geq 0$. Then the scores on the jth principal component use the jth eigenvector as weights in the linear combination

$$z_j = (x - \bar{x})' v_j \quad . \tag{9.5}$$

In the geometric interpretation, the principal axes of the ellipse are the eigenvectors, v_j, and the length of each semi-major axis is $\sqrt{\lambda_j}$.

This description of components analysis assumes that the variables are measured on the same scale, so that it is sensible to measure the importance of a variable by its variance. If the variables are not commensurable, the analysis should be carried out with the correlation matrix, or equivalently, with the data matrix converted to standard scores.

Principal components analysis is most easily carried out in the SAS System with PROC PRINCOMP. The examples below show how to plot principal component scores and the raw data with the principal component vectors and ellipse.

9.4.1 Example: January and July Temperatures

This example uses data on mean monthly temperature in January and July from selected cities to illustrate the plotting of principal components. Because there are only two variables, there is no point in using principal components analysis as a data reduction technique, but this simple example illustrates the geometry of principal components analysis and the plotting techniques involved. The data are contained in the data set CITYTEMP listed in Section A2.4, "The CITYTEMP Data Set: City Temperatures Data." The temperature variables are named JANUARY and JULY. The steps below perform a principal components analysis and construct initial plots of the data and component scores:

```
%include CITYTEMP;              /* On MVS, use %include ddname(file);*/


proc princomp cov data=citytemp
     out=prin outstat=stats;
   var january july;


proc plot data=prin;
   plot july * january = city /        /* plot the data    */
       hspace=10 vspace=5              /* aspect ratio=.5  */
       haxis= 0 to  70 by 10
       vaxis=60 to 100 by 10;
   plot prin2 * prin1 = city /         /* plot the scores  */
       hspace=10 vspace=5
       haxis=-30 to 40 by 10
       vaxis=-20 to 20 by 10;
```

Because the two variables are measured on the same scale, the components analysis is done on the variance-covariance matrix using the COV option in PROC PRINCOMP. In the printed output from the procedure (shown in Output 9.7), the weights for the optimal linear combinations are given by the columns of the EIGENVECTORS matrix, which are applied to the deviations of the observed variables from their means. So the two principal components for these data are

$$\text{PRIN1} = 0.9391 \, (\text{JANUARY} - 32.10) + 0.3435 \, (\text{JULY} - 75.61)$$

$$\text{PRIN2} = -0.3435 \, (\text{JANUARY} - 32.10) + 0.9391 \, (\text{JULY} - 75.61) \quad .$$

The JANUARY and JULY variables have variances of 137.18 and 26.29 respectively, for a total of 163.47. Of this, PRIN1 accounts for 154.31 (94.4 percent), and PRIN2 accounts for the remaining 9.16 (5.6 percent).

Output 9.7
Printed Output from the PRINCOMP Procedure for the CITYTEMP Data Set

```
          Mean temperature in January and July for selected cities

                      Principal Component Analysis

64 Observations
 2 Variables

                          Simple Statistics

                              JANUARY                 JULY

              Mean        32.09531250          75.60781250
              StD         11.71243309           5.12761910

                          Covariance Matrix

                               JANUARY                JULY

            JANUARY        137.1810888          46.8282912
            JULY            46.8282912          26.2924777

                    Total Variance = 163.47356647

                 Eigenvalues of the Covariance Matrix

           Eigenvalue       Difference       Proportion       Cumulative

PRIN1        154.311          145.148          0.943948          0.94395
PRIN2          9.163             .             0.056052          1.00000

                             Eigenvectors

                                PRIN1          PRIN2

                 JANUARY      0.939141       -.343532
                 JULY         0.343532       0.939141
```

In addition to the printed output from the procedure, PROC PRINCOMP can produce an output data set containing component scores (OUT= option) and a second output data set (OUTSTAT= option) containing some of the statistics from the printout. These two data sets are used in plotting the results. The OUT= data set, PRIN, contains all the original variables plus the principal component scores, which are named PRIN1 and PRIN2.

For two variables, the principal component scores are simply a rigid rotation of the data values after a change of origin from the centroid to zero. To see this, the PROC PLOT step plots the data values and the component scores, giving the plots shown in Output 9.8 and 9.9. Note that the PLOT statements shown above specify the same number of data units for the axes on both plots and use the HSPACE= and VSPACE= options to equate the horizontal and vertical axes on each plot. (The values used, HSPACE=10 and VSPACE=5, assume an output device that prints at ten characters/inch and five lines/inch.) Thus, both plots are scaled identically and can be compared directly.

Comparing the data in Output 9.8 with the component scores in Output 9.9 shows that the configuration of points is the same on both plots. The PRIN1 axis is the direction of greatest variation in the data space; this direction is given by the components of the PRIN1 eigenvector (0.9391, 0.3435). PRIN2 is the direction orthogonal to PRIN1, given by the second eigenvector. These two axes represent a clockwise rotation of the data points by 20 degrees (the angle whose cosine is 0.9391).

On a geometrically correct plot, we can also see the relative amounts of variation accounted for by the principal components. The eigenvalues shown in Output 9.7 have a ratio of 154.3/9.16 or about 16 to 1, so the standard deviations of PRIN1 and PRIN2 have a ratio of about 4 to 1. In Output 9.9, we can see that the variation of the points is about four times greater in the PRIN1 direction than in the PRIN2 direction. The January and July temperatures in these cities is thus largely one dimensional.

Output 9.8
Data Values for Principal Components Example

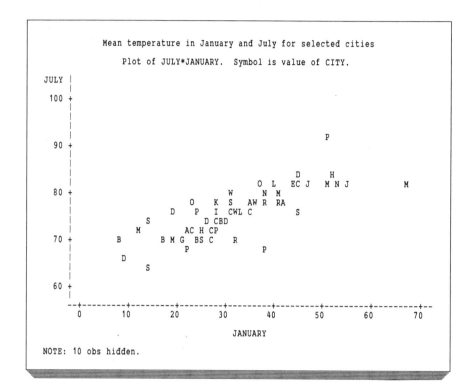

NOTE: 10 obs hidden.

Output 9.9
Component Scores
for City
Temperature Data

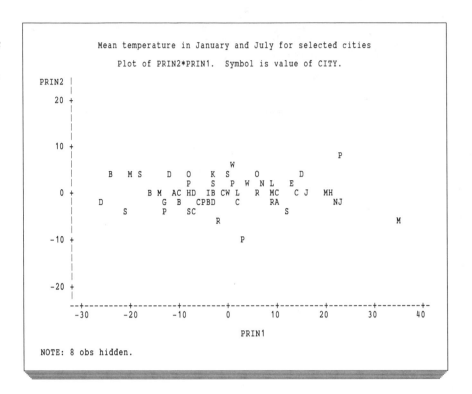

To understand what the components represent, we also plot the principal component scores, but it is necessary to label the points with the city name. However, the goal of presenting the geometry of the points usually conflicts with the goal of presenting the label information. For example, equating the axes as in Output 9.9 means that the vertical space for labels is only about 25 percent of the horizontal space, so equating the axes produces many more label collisions.

As in Section 4.3, "Labeling Observations," you can reduce this problem by abbreviating the labels or by labeling only some of the points. In principal components analysis, you would label only the extreme points on each component. Alternatively, you may decide to leave the axes unequated for the purpose of interpreting the components, but bear in mind that the relative magnitudes of the components are not portrayed accurately.

These techniques are illustrated first for printer plots constructed with the IDPLOT procedure and then for high-resolution plots drawn with PROC GPLOT.*

Two plots of the component scores constructed with PROC IDPLOT are shown in Outputs 9.10 and 9.11. Both plots use only the first three letters of the city name as the plotting symbol. Output 9.11 shows the PROC IDPLOT results. The axes are equated in Output 9.10, preserving the geometry but obscuring the observation labels. The axes are scaled independently in Output 9.11, making more labels visible at the expense of distorting information about the relative importance of PRIN1 and PRIN2.

---

* PROC IDPLOT is part of the Version 5 SUGI Supplemental Library. It is not available in Version 6 of the SAS System.

Output 9.10   *IDPLOT Procedure Plot of Component Scores*

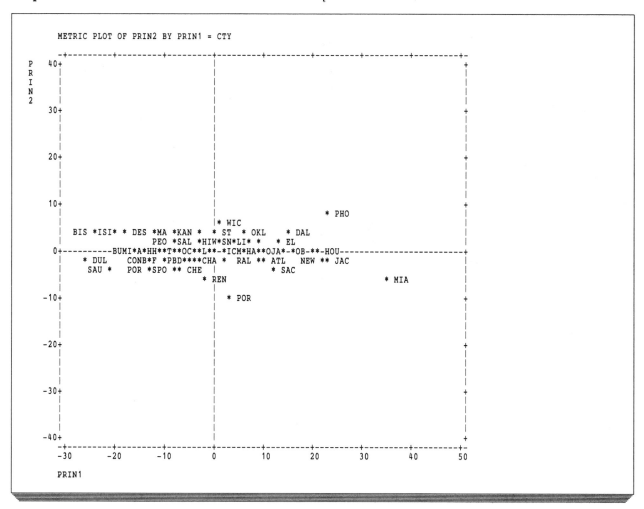

Output 9.11   *IDPLOT Procedure Plot of Component Scores*

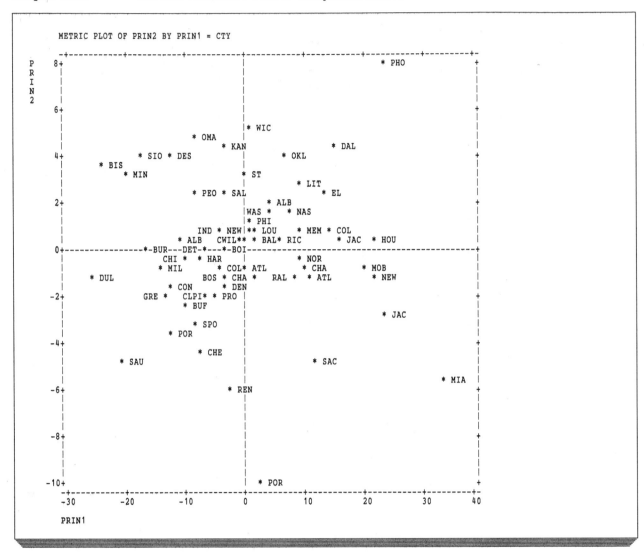

Output 9.10 and 9.11 are produced by the following statements:

```
options ls=100;                    * metric plots, using 3-char abbrev (CTY);
proc idplot data=prin metric voverh=.5;
   plot prin2 * prin1 = cty /
        href=0 vref=0 hinc=10 vinc=10
        htickpos=10 vtickpos=5;                * equated, for 15cpi, 7.5lpi;
   plot prin2 * prin1 = cty /
        href=0 vref=0 hinc=10 ;                 * not equated;
   format prin1 prin2 best4.1;
   title 'Principal component scores';
```

In the first PLOT statement, the HTICKPOS= and VTICKPOS= options have the same role in equating the axes as the HSPACE= and VSPACE= options in the PLOT procedure.

Note that we can interpret either plot by considering the extreme points on both components. The first component, PRIN1, can be interpreted as reflecting the overall warmth of a place or as a north/south dimension. The warmest (southern) cities—Miami, Jacksonville, and New Orleans—are at the right; the coldest (northern) cities—Bismark, Duluth, and Sault Ste. Marie—are at the left. PRIN2, plotted vertically, appears to reflect seasonal variation, with cities at the top having a large difference between the temperatures in January and July and cities at the bottom having more consistent winter and summer temperature.

Because the display is usually too crowded when there are more than 30 or so points, you may choose to plot the component scores with labels only for the extreme points on each principal component. This is illustrated with PROC GPLOT, using the Annotate data set LABEL shown below. To reduce overplotting, the Annotate variable POSITION is calculated from the signs of the PRIN1 and PRIN2 scores to position the labels away from the origin. The axes are not equated on this plot. The plot is shown in Output 9.12.

```
data label;
   set prin;
   retain xsys '2' ysys '2' color 'RED';
   length position $1 text $12;
   if abs(prin1)>20 | abs(prin2)>10 then do;
      x = prin1;
      y = prin2;
      function = 'LABEL';
      pos = 1 + 2*(prin1>0) + (prin2>0);
      position = substr('DAFC',pos,1);
      text = city;
      output;
   end;

proc gplot data=prin  ;
   plot prin2 * prin1
      / frame anno=label
        vminor=1 hminor=1
        vaxis=axis1 haxis=axis2
        href=0 vref=0 lhref=20 lvref=20
        cvref=BLUE chref=BLUE
        name='GB0912'  ;
   label prin1 = 'PRIN1: Warmth'
         prin2 = 'PRIN2: Variation';
   axis1 order=(-10 to 10 by 5)  offset=(4)
         label=(a=90 r=0 h=1.3 f=duplex);
   axis2 order=(-40 to 40 by 10) offset=(6)
         label=(h=1.3 f=duplex);
   symbol1 h=1.2 v=plus c=black;
```

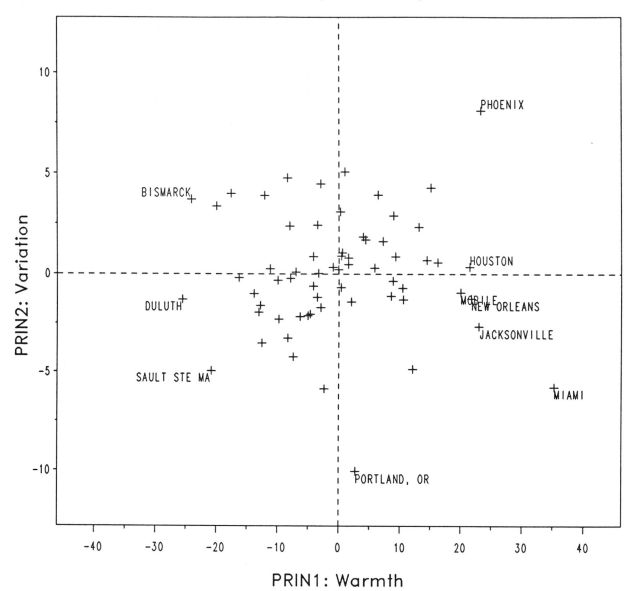

Mean temperature in January and July for selected cities

9.4.2 Plotting Observed Variables with Principal Component Vectors

It may also be helpful to plot the principal component vectors in the space of the observed variables. This plot is shown in Output 9.13 for the city temperature data. The vectors PRIN1 and PRIN2 are located at the mean point, and each has a length proportional to the square root of the corresponding eigenvalue. The direction of each vector is given by the entries in the eigenvector. The plot also shows the concentration ellipse of the variance-covariance matrix, whose major and minor axes are the principal component vectors. Output 9.13 shows that

PRIN1 is the direction in the space of the JANUARY and JULY variables that has the greatest variation, and that PRIN1 and PRIN2 are orthogonal vectors in this space.[*]

Output 9.13   *Raw Data with Principal Component Vectors and Concentration Ellipse*

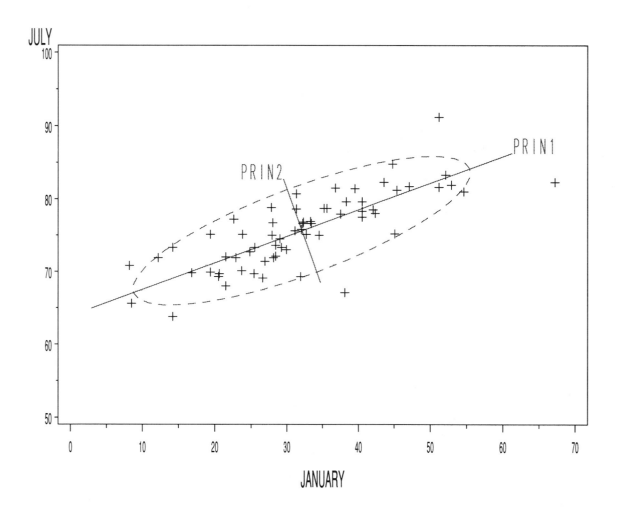

CityTemp data with principal component vectors

The plot in Output 9.13 is constructed from the information in the OUTSTAT= data set STATS produced by PROC PRINCOMP. This data set contains the statistical measures from the printed output in the form shown in Output 9.14. The various summary statistics are identified by the _TYPE_ variable. The _TYPE_='SCORE' observations give the component weights, and the individual components are labeled by the _NAME_ variable.

Output 9.14
OUTSTAT= Data
Set from the
PRINCOMP
Procedure for City
Temperature Data

```
              OUTSTAT dataset from PROC PRINCOMP

    OBS   _TYPE_    _NAME_      JANUARY     JULY

     1    MEAN                   32.095    75.6078
     2    N                      64.000    64.0000
     3    COV       JANUARY     137.181    46.8283
     4    COV       JULY         46.828    26.2925
     5    EIGENVAL              154.311     9.1630
     6    SCORE     PRIN1         0.939     0.3435
     7    SCORE     PRIN2        -0.344     0.9391
```

The component vectors are drawn from the _TYPE_='SCORE' observations. To draw them on the plot of JULY against JANUARY temperature, it is convenient to rearrange the STATS data set so that the means of the variables and lengths of the vectors are additional variables rather than separate observations. This is done with the DATA step below, whose result is shown in Output 9.15. Note that L1 in this data set is the length of PRIN1, and L2 is the length of PRIN2.

```
data weights;
   set stats;
   retain mx my l1 l2;
   drop _TYPE_;
   if _TYPE_='MEAN' then do;
      mx= JANUARY; my= JULY;
      end;
   if _TYPE_='EIGENVAL' then do;        * get lengths of PC vectors;
      l1= sqrt(JANUARY); l2= sqrt(JULY);
      end;
   if _TYPE_='SCORE' then output;
proc print data=weights;
   id _name_;
   title2 'WEIGHTS dataset';
```

Output 9.15
OUTSTAT= Data
Set Transformed
for Plotting

```
                         WEIGHTS dataset

_NAME_    JANUARY      JULY        MX         MY        L1        L2

PRIN1     0.93914    0.34353    32.0953    75.6078    12.4222    3.02704
PRIN2    -0.34353    0.93914    32.0953    75.6078    12.4222    3.02704
```

Then the vectors can be drawn with Annotate instructions by moving to the mean (MX, MY) and drawing a line whose (x,y) increments are proportional to the JANUARY and JULY components of the score components. This is carried out in the Annotate data set below.

```
data vectors;
   set weights;
   length function text $8;
   xsys='2'; ysys='2';
   drop january july mx my l1 l2;

   x = mx; y = my;
   function = 'MOVE  ';   output;
   color = 'RED';
   if _NAME_= 'PRIN1'                    /* lengths of vectors*/
      then do; l = 2.5 * l1; position='C'; end;
      else do; l = 2.5 * l2; position='A'; end;

   x = mx - l * JANUARY;                 /* draw & label them*/
   y = my - l * JULY;  function = 'DRAW';  output;
   x = mx + l * JANUARY;
   y = my + l * JULY;  function = 'DRAW';  output;
   size = 1.4;
   text = _NAME_;       function = 'LABEL'; output;

proc print data=vectors ;
title 'ANNOTATE= data set';
```

The ellipse is also drawn from the WEIGHTS data set. The first DATA step below generates *x,y* points around a circle at the origin and expands the coordinates by factors of L1 and L2 to give an ellipse. These coordinates are rotated by the angle, TILT, that the PRIN1 vector makes with the *x* axis and are finally translated to the mean, (MX, MY).

```
data ellipse;
   set weights;
   length function text $8;
   xsys='2'; ysys='2';
   drop january july mx my l1 l2 xp yp ang;
   if _NAME_='PRIN1' then do;
      tilt = atan(july/january);      * angle of major axis;
      siz=2;                          * size of ellipse;
      color='RED';
      line = 20;
      do a = 0 to 360 by 10;
         ang = a*arcos(-1)/180;                 * convert to radians;
         x = siz * l1  * cos(ang);
         y = siz * l2  * sin(ang);
         xp= (x*cos(tilt)) - (y*sin(tilt));   * rotate;
         yp= (x*sin(tilt)) + (y*cos(tilt));
         x = xp+ mx ;                          * translate;
         y = yp+ my ;
         If a=0 then FUNCTION = 'MOVE    ';
                 else FUNCTION = 'DRAW    ';
         output;
         end;
   end;
```

```
data vectors;
   set vectors ellipse;
proc gplot data=prin  ;
   plot july * january
        / frame anno=vectors
          vminor=1 hminor=1
          vaxis=axis1 haxis=axis2
          name='GB0913'  ;
   axis1 order=(50 to 100 by 10) length=3.93 in
         label=(h=1.5);
   axis2 order=( 0 to  70 by 10) length=5.50 in
         label=(h=1.5) offset=(2);
   title h=1.5 'CityTemp data with principal component vectors';
   symbol1 h=1.2 v=plus c=black;
```

The LENGTH= options in the AXIS statements are used to equate the axes in Output 9.13. The range for AXIS1 is 50, and the range for AXIS2 is 70. If the lengths for AXIS1 and AXIS2 are in the same proportion, the number of data units per inch will be the same for both axes.

Example: U.S. Crime Rates

As a more typical example of data that may profitably be studied by principal components analysis, we consider data on the rates of occurrence (per 100,000 population) of seven types of crime in each of the 50 U.S. states. The data set CRIME, described in Section A2.5, "The CRIME Data Set: State Crime Data," is also used as an example of component analysis in the *SAS User's Guide: Statistics, Version 5 Edition*.

Although the variables are all measured in the same units, they differ widely in variability (see the standard deviations in Output 9.16), so it is better to analyze the correlation matrix. This is the default for PROC PRINCOMP when the COV option is not used. As in the previous example, the OUT= and OUTSTAT= options are used to obtain the component scores and other statistics for plotting.

```
%include CRIME ;
proc princomp data=crime
              out=crimcomp
              outstat=crimstat;
```

The printed output from PROC PRINCOMP is shown in Output 9.16. The first two components account for 76 percent of the total variance; the first three account for nearly 87 percent. Values after the third eigenvalue are generally small and trail off uniformly, which is an indication that three components should be retained.

Output 9.16
Printed Output
from the
PRINCOMP
Procedure for
Crime Data

```
                    PRINCIPAL COMPONENT ANALYSIS

        50 OBSERVATIONS
         7 VARIABLES


                         SIMPLE STATISTICS

             MURDER     RAPE   ROBBERY  ASSAULT  BURGLARY  LARCENY     AUTO

  MEAN      7.44400  25.7340   124.092  211.300   1291.90  2671.29  377.526
  ST DEV    3.86677  10.7596    88.349  100.253    432.46   725.91  193.394


                            CORRELATIONS

             MURDER     RAPE   ROBBERY  ASSAULT  BURGLARY  LARCENY     AUTO

  MURDER     1.0000   0.6012    0.4837   0.6486    0.3858   0.1019   0.0688
  RAPE       0.6012   1.0000    0.5919   0.7403    0.7121   0.6140   0.3489
  ROBBERY    0.4837   0.5919    1.0000   0.5571    0.6372   0.4467   0.5907
  ASSAULT    0.6486   0.7403    0.5571   1.0000    0.6229   0.4044   0.2758
  BURGLARY   0.3858   0.7121    0.6372   0.6229    1.0000   0.7921   0.5580
  LARCENY    0.1019   0.6140    0.4467   0.4044    0.7921   1.0000   0.4442
  AUTO       0.0688   0.3489    0.5907   0.2758    0.5580   0.4442   1.0000


                 EIGENVALUE   DIFFERENCE    PROPORTION      CUMULATIVE

       PRIN1     4.11496        2.87624       0.587851       0.58785
       PRIN2     1.23872        0.51291       0.176960       0.76481
       PRIN3     0.72582        0.40938       0.103688       0.86850
       PRIN4     0.31643        0.05846       0.045205       0.91370
       PRIN5     0.25797        0.03593       0.036853       0.95056
       PRIN6     0.22204        0.09798       0.031720       0.98228
       PRIN7     0.12406           .          0.017722       1.00000


                            EIGENVECTORS

             PRIN1     PRIN2     PRIN3     PRIN4     PRIN5     PRIN6     PRIN7

  MURDER   0.300279  -.629174  0.178245  -.232114  0.538123  0.259117  0.267593
  RAPE     0.431759  -.169435  -.244198  0.062216  0.188471  -.773271  -.296485
  ROBBERY  0.396875  0.042247  0.495861  -.557989  -.519977  -.114385  -.003903
  ASSAULT  0.396652  -.343528  -.069510  0.629804  -.506651  0.172363  0.191745
  BURGLARY 0.440157  0.203341  -.209895  -.057555  0.101033  0.535987  -.648117
  LARCENY  0.357360  0.402319  -.539231  -.234890  0.030099  0.039406  0.601690
  AUTO     0.295177  0.502421  0.568384  0.419238  0.369753  -.057298  0.147046
```

The interpretation of the first two components is fairly clear from the component weights in Output 9.16. The PRIN1 vector has positive, approximately equal weights for all the variables, so it would be interpreted as an index of overall crime rate. This pattern is typical for variables that have all positive correlations. The PRIN2 vector represents a contrast of crimes against property (with positive weights) versus crimes against persons (with negative weights). The third component appears to represent a contrast of robbery and auto theft with other forms of larceny, though this interpretation is not compelling.

Plots of the component scores with observation labels can be made with PROC IDPLOT or PROC GPLOT as in the previous example.* Using the full state name (variable STATE) results in too much overlap of the labels, but the two-letter state abbreviation (ST) gives useful plots with either PROC IDPLOT or PROC GPLOT. (See also Section 4.3 for techniques for labeled scatterplots.)

The plots of the vectors PRIN2 against PRIN1 and PRIN3 against PRIN1 are shown in Output 9.17 and 9.18. (Color versions of Output 9.17 and 9.18 appear

---

* You can also use PROC PLOT, but only the first character of the state name will be used as the plotting symbol.

as Output A3.45 and A3.46 in Appendix 3, "Color Output.") On the plot of the first two components, several regional groupings are apparent. The southern states generally have negative scores on the second component, reflecting a greater prevalence of violent crimes than property crimes. States in the northeast tend to be located at the upper end of the second component, with more property crimes than violent crimes. To highlight these regional differences, each state label is drawn in a color associated with its region through a format, $REGCLR.. The Annotate data set LABEL and the PROC GPLOT step used to produce Output 9.17 are shown below, followed by the output.

```
proc format;
   value $regclr
      "MT","ID","WY","CO","AZ","NM","UT","NV",          /* West      */
      "CA","OR","WA","AK","HI"              ='BLACK'

      "DE","MD","VA","WV","NC","SC","OK","TX",          /* South     */
      "AR","LA","KY","TN","MS","AL","FL","GA"  ='RED'

      "NY","PA","NJ","ME","NH","VT","MA","RI",          /* North East*/
      "CT"                         ='BLUE'
      "ND","SD","NE","KS","WI","MI","IL","IN",          /* North     */
      "OH","MN","IA","MO"                  ='GREEN'; /*  Central  */

data label;
   set crimcomp;
   retain xsys '2' ysys '2' ;
   x = prin1;
   y = prin2;
   color = put(ST, $regclr.);
   function = 'LABEL';
   text = st;
proc gplot data=crimcomp ;
   plot prin2 * prin1 /
        anno=label vm=1 hm=1 frame name='GB0917'
        vaxis=axis1 haxis=axis2;
   label prin1 = 'PRIN1: Overall Crime Rate'
         prin2 = 'PRIN2: Property vs Personal Crime';
   axis1 order=-3 to 3  length=3.85 in
         label=(a=90 r=0 h=1.2 f=duplex)
         offset=(1);
   axis2 order=-4 to 6  length=5.50 in
         label=(h=1.2 f=duplex) offset=(2);
   format prin1-prin2 6.;
   symbol v=none;
   title ' ';
```

Output 9.17   *Plot of Principal Component Scores on PRIN1 and PRIN2*

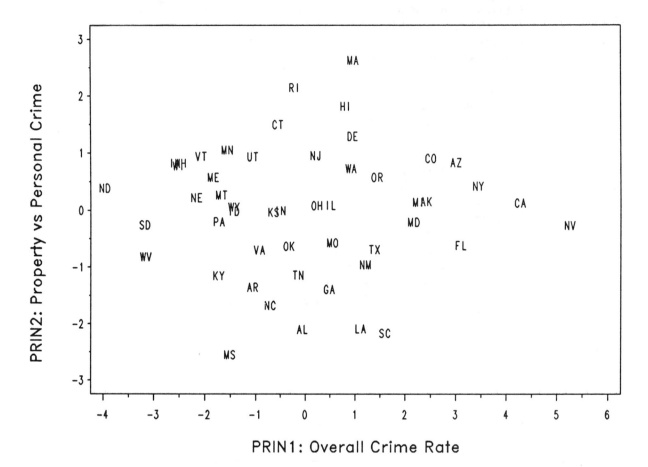

Because all points are labeled, the SYMBOL statement specifies V=NONE to suppress plotting of the points. The states appear to cluster by geographic region, coded by the color of the plotted label.

The plot of PRIN1 against PRIN3 shown in Output 9.18 is produced by the DATA step and PROC GPLOT step below. These steps are parallel to those shown above for the plot of PRIN1 and PRIN2. Equating the axes in Output 9.18 produces more overlap among the state labels than we like, but shows the relative importance of these two components correctly.

```
data label;
    set crimcomp;
    retain xsys '2' ysys '2' ;
    x = prin1;
    y = prin3;
    color = put(ST, $regclr.);
    function = 'LABEL';
    text = st;
proc gplot data=crimcomp ;
    plot prin3 * prin1 /
        anno=label vm=1 hm=1 frame name='GB0918'
        vaxis=axis1 haxis=axis2;
    label prin1 = 'PRIN1: Overall Crime Rate'
        prin3 = 'PRIN3: Larceny vs Other Theft';
    axis1 order=-2 to 3  length=2.75 in
        label=(a=90 r=0 h=1.2 f=duplex);
    axis2 order=-4 to 6  length=5.50 in
        label=(h=1.2 f=duplex) offset=(2);
    format prin1-prin2 6.;
```

Output 9.18   *Plot of Principal Component Scores on PRIN1 and PRIN3*

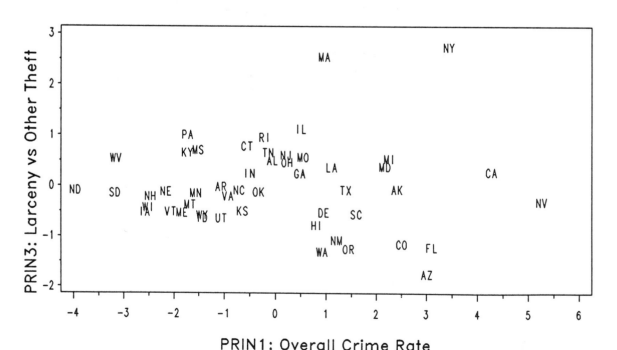

9.4.3 Plotting Component Weights

The components for the crime data were interpreted by visually inspecting the pattern of weights in each column of the eigenvectors in Output 9.16. A labeled plot of these values can make it easier to see the relations between the variables and the components.

Again the component weights can be obtained from the _TYPE_='SCORE' observations in the OUTSTAT= data set from PROC PRINCOMP. For the crime data, this data set (CRIMSTAT) is shown in Output 9.19.

Output 9.19
OUTSTAT= Data
Set from the
PRINCOMP
Procedure for
Crime Data

```
                                    OUTSTAT= data set

    OBS _TYPE_    _NAME_    MURDER   RAPE  ROBBERY ASSAULT BURGLARY LARCENY  AUTO

     1 MEAN                 7.4440 25.7340 124.092 211.300 1291.90 2671.29 377.526
     2 STD                  3.8668 10.7596  88.349 100.253  432.46  725.91 193.394
     3 N                   50.0000 50.0000  50.000  50.000   50.00   50.00  50.000
     4 CORR      MURDER     1.0000  0.6012   0.484   0.649    0.39    0.10   0.069
     5 CORR      RAPE       0.6012  1.0000   0.592   0.740    0.71    0.61   0.349
     6 CORR      ROBBERY    0.4837  0.5919   1.000   0.557    0.64    0.45   0.591
     7 CORR      ASSAULT    0.6486  0.7403   0.557   1.000    0.62    0.40   0.276
     8 CORR      BURGLARY   0.3858  0.7121   0.637   0.623    1.00    0.79   0.558
     9 CORR      LARCENY    0.1019  0.6140   0.447   0.404    0.79    1.00   0.444
    10 CORR      AUTO       0.0688  0.3489   0.591   0.276    0.56    0.44   1.000
    11 EIGENVAL             4.1150  1.2387   0.726   0.316    0.26    0.22   0.124
    12 SCORE     PRIN1      0.3003  0.4318   0.397   0.397    0.44    0.36   0.295
    13 SCORE     PRIN2     -0.6292 -0.1694   0.042  -0.344    0.20    0.40   0.502
    14 SCORE     PRIN3      0.1782 -0.2442   0.496  -0.070   -0.21   -0.54   0.568
    15 SCORE     PRIN4     -0.2321  0.0622  -0.558   0.630   -0.06   -0.23   0.419
    16 SCORE     PRIN5      0.5381  0.1885  -0.520  -0.507    0.10    0.03   0.370
    17 SCORE     PRIN6      0.2591 -0.7733  -0.114   0.172    0.54    0.04  -0.057
    18 SCORE     PRIN7      0.2676 -0.2965  -0.004   0.192   -0.65    0.60   0.147
```

The plot of the variables in the space of PRIN1 and PRIN2 is shown in Output 9.20; the corresponding plot for PRIN2 and PRIN3 appears in Output 9.21. These plots show each variable as a vector from the origin to the points whose coordinates are the weights on the two principal components of the plot. Output 9.20 shows quite clearly that the second component separates crimes against persons from crimes against property, and that all variables have roughly equal weights on the first component. As in the biplot (Section 8.7, "Biplot: Plotting Variables and Observations Together"), the correlation between variables is reflected approximately in the angle between the variable vectors in the first two dimensions.

Output 9.20   *Component Weights on PRIN1 and PRIN2*

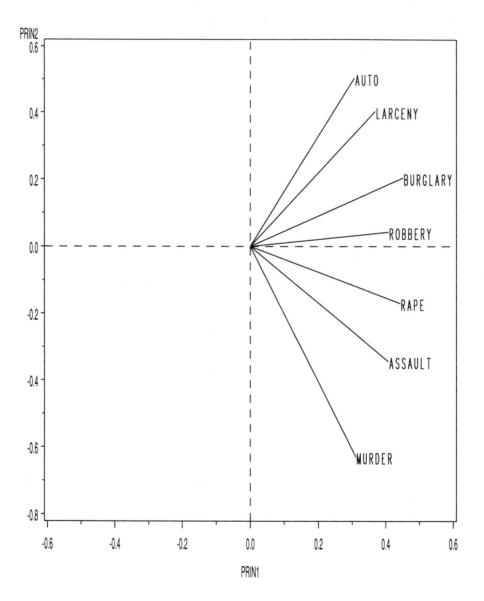

Output 9.21   *Component Weights on PRIN2 and PRIN3*

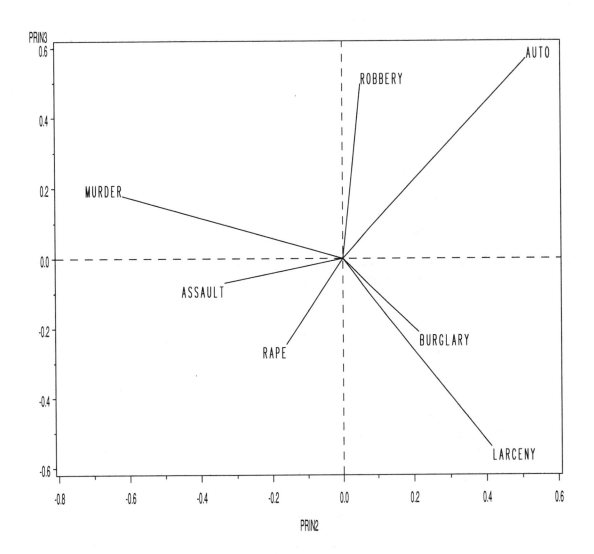

Component pattern

To produce these plots, the _TYPE_='SCORE' observations are selected from the CRIMSTAT data set and transposed so that each crime variable becomes an observation in a new data set, PATTERN. The statements below are used to draw Output 9.20:

```
data vectors;
   set crimstat;
   drop _TYPE_;
   if _TYPE_='SCORE';
proc transpose out=pattern;
proc print;
```

```
data labels;
   set pattern;
   retain xsys '2' ysys '2' position '6';
   x = 0 ; y = 0;
   function = 'MOVE ' ; output;
   x = prin1 + .01;
   y = prin2;
   function = 'DRAW ' ; output;
   text = _NAME_;
   function = 'LABEL' ; output;

proc gplot data=pattern ;
   plot prin2 * prin1
      / anno=labels frame vaxis=axis1 haxis=axis2
        vm=1 hm=1 href=0 vref=0 lhref=20 lvref=20 name='GB0920';
   axis1 order=(-.8 to .6 by .2) length=5.00 in;
   axis2 order=(-.6 to .6 by .2) length=4.28 in;
   title h=1.5 'Component pattern';
   symbol v=none;
```

9.5 Discriminating among Groups

When you have multivariate data for two or more groups, you may want to determine how the groups differ on the quantitative variables. To test whether differences exist, you can perform an ANOVA test on each variable separately, but this will not take the correlations among the variables into account. Instead, a multivariate analysis of variance (MANOVA) can be used to test mean differences on the collection of variables.

Alternatively, methods of discriminant analysis can be used to describe the differences among the groups on the quantitative variables. The term *discriminant analysis* refers to a collection of methods with the following goals:

classificatory discriminant analysis
> finds a mathematical classification rule, or *discriminant function*, for classifying observations into the groups, based solely on the quantitative variables. The DISCRIM procedure provides both parametric and nonparametric classification methods. When the data are approximately multivariate normal within each group, parametric methods can be used to find linear or quadratic discriminant functions. Nonparametric methods do not require assumptions about the form of the distributions.

canonical discriminant analysis
> is a dimension-reduction technique that finds a small set of linear combinations of the observed variables that best discriminate among the groups. The CANDISC procedure finds one or more linear combinations of the quantitative variables on which the means of the groups differ most.

stepwise discriminant analysis
> finds a reduced set of the original quantitative variables that best discriminate among the groups. The STEPDISC procedure provides stepwise variable selection techniques analogous to those in stepwise regression.

This section describes graphical methods related to the first two of these goals. Methods available in Version 6 of PROC DISCRIM for plotting the discriminant functions are described in Section 9.5.1, "Plotting Discriminant Functions." These plots show the observed variables directly with the discriminant functions, but are limited to two-variable views. Plots of the canonical discriminant functions from PROC CANDISC are illustrated in Section 9.5.2, "Canonical Discriminant Analysis and MANOVA." These plots show the data in a reduced space of two dimensions designed to maximize discrimination among the groups. Either of these types of plots may also be useful when the goal is to test for differences in means in a one-way MANOVA test.

9.5.1 Plotting Discriminant Functions

Methods for determining a rule for classifying observations into groups differ according to whether the distribution of the variables within each group is assumed to be multivariate normal or not. In the former case, the classification function is based on a measure of generalized squared distance, analogous to Mahalanobis D^2 (see Section 9.2, "Assessing Multivariate Normality"). If the covariance matrices are assumed equal in the groups, the pooled within-group covariance matrix is used, and the discriminant functions are linear functions of the quantitative variables. Otherwise, the individual covariance matrices are used, and the discriminant functions are quadratic. Each observation is classified into the group from which it has the smallest generalized squared distance; equivalently, the observation is assigned to the group for which the discriminant function gives the largest value.

When the data are not assumed to follow a multivariate normal distribution, nonparametric methods are used to estimate a nonparametric probability density for each group, as in the kernel method described in Section 3.4, "Histogram Smoothing and Density Estimation." The probability density, together with estimates of prior probabilities for each group, is used to calculate the posterior probability of membership in each group, for a given set of scores on the quantitative variables. Observations are then classified into the group that gives the largest posterior probability.

See Chapter 20, "The DISCRIM Procedure," in the *SAS/STAT User's Guide, Version 6, Fourth Edition, Volume 1* for further details on the methods available in PROC DISCRIM. Seber (1984, Chapter 6) gives a detailed mathematical description of these techniques; Tabachnick and Fidell (1989) provide a practical and nontechnical introduction to the parametric methods.

Two methods for plotting the discriminant functions in the plane of any two variables are described below. The first (*contour method*) classifies each point of a grid of points and plots the contours of the boundaries between adjacent groups. This method gives an approximation to the discriminant functions, but can be used with any method performed by PROC DISCRIM. The second (*coefficient method*) derives the discriminant functions from the coefficients output by PROC DISCRIM and therefore gives an exact representation, but the program is more complex and can be used only with the linear discriminant functions. Both methods apply to Version 6 of PROC DISCRIM.

When more than two variables are necessary to discriminate among groups, the discriminant function boundaries (linear, quadratic, or nonparametric) cannot be well represented in a bivariate plot of just two observed variables. In this case, it is more useful to plot the data in the reduced space of the canonical discriminant variables as shown in Section 9.5.2.

Contour Method

For any of these methods, discriminant analysis is often used with one sample of data, called a *calibration data set*, to derive the classification function. This function can be applied to a second sample, called a *test data set*, to assign each new observation into one of the groups. For example, we may want to plot the discriminant functions on a bivariate plot of two quantitative variables. Then if a rectangular grid of points in the plane of these two variables is used as the test data set, PROC DISCRIM will classify each point into one of the groups. A contour plot of the classification variable will then display the boundaries of the classification regions. This method for plotting the discriminant functions depends on the TESTOUT= data set, which is only available in Version 6 of the DISCRIM procedure. Plots of the normal-theory, linear, and quadratic discriminant functions are illustrated below. The same technique used to construct the plots can be used with the nonparametric methods as well.

Example: iris data

The iris data (Section A2.10, "The IRIS Data Set: Iris Data") were used by Fisher (1936) to classify iris species on the basis of four measurements: sepal length, sepal width, petal length, and petal width. Not all four variables are necessary for adequate discrimination, however. The scatterplot matrix for these data (Output 8.5), for example, shows that the iris species are well separated in each of the pairwise scatterplots.

This example produces plots of the petal width and petal length variables in the iris data showing the boundaries of the parametric classification regions, assuming multivariate normality. PROC DISCRIM is first used with the pooled within-group covariance matrix (POOL=YES) to produce linear discriminant functions. A second analysis is run using POOL=TEST to test the assumption of homogeneity of covariance matrices. The test rejects this assumption, and PROC DISCRIM therefore uses the individual covariance matrices to produce quadratic discriminant functions.

To produce a plot of the discriminant functions, a grid of points is created in a DATA step. The PETALWID and PETALLEN variables, which are used in the plot, are varied throughout their range. The SEPALWID and SEPALLEN variables are not used in the PROC DISCRIM step or on the plot.

The GRID data set is used as the test data set in the PROC DISCRIM step that follows. It contains a large number of points, which makes the PROC DISCRIM step slow, but the discriminant functions are approximated well. Using a smaller number of points would make the PROC DISCRIM step faster, at the expense of producing a coarser approximation of the discriminant functions.

The PROC DISCRIM step uses the IRIS data set as the calibration data and classifies each point in the test data set GRID. Specifying METHOD=NORMAL and POOL=YES requests the normal-theory, linear discriminant functions. The result of the classification is produced in the TESTOUT= data set PLOTC, as the value of a variable named _INTO_. The _INTO_ variable uses the same set of values as the CLASS variable, so the CLASS variable must be a numeric variable in this application. In the iris data, the SPEC_NO variable has the values 1, 2, and 3, corresponding to *Iris setosa*, *versicolor*, and *virginica*.

Then a contour plot of the PLOTC data set can be used to show the classification functions. The PROC PLOT step below produces the contour plot of the _INTO_ variable shown in Output 9.22. The plot, shown for demonstration purposes, displays the linear discriminant functions but does not show the data.

```
%include IRIS;                    /* On MVS, use %include ddname(file);*/
data grid;
   do petalwid = 1 to 30;
      do petallen = 1 to 70;
         output;
      end;
   end;

*-- Get discriminant classification of grid --  Requires Version 6 ;
title h=1.5 'Linear Discriminant Functions: Iris Data';
proc discrim data=iris
             method  =normal
             pool    =yes       /* linear function?   */
             testdata=grid      /* points to classify */
             testout =plotc     /* grid classification */
             short   noclassify;
      classes spec_no;          /* numeric class variable */
      var petallen petalwid;
run;
proc plot data=plotc;
   plot petallen * petalwid = _into_
      / contour=3 s1='S' S2='.' S3='V';
```

Output 9.22
Contour Plot of
Classification
Results for Iris
Data

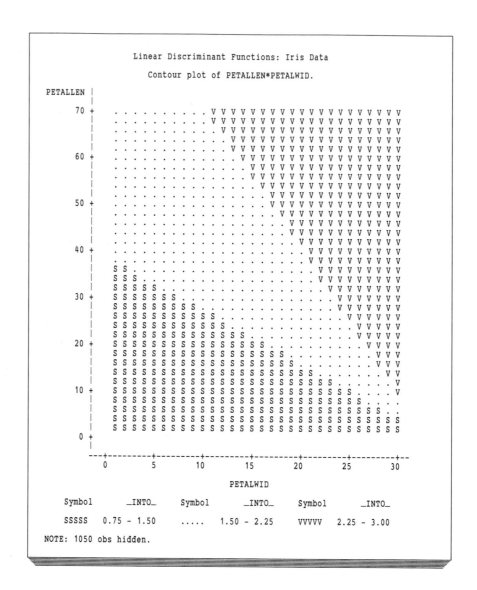

To show the data as well, we use the GCONTOUR procedure to draw the discriminant function boundaries and the Annotate facility to add the data. In this example, the CONTOUR macro (Section A1.5, "The Contour Macro") is used to draw a 75 percent data ellipse representing the bivariate normal distribution (assumed) for each iris species. The OUT= data set ELLIPSE is used as the Annotate data set in the PROC GCONTOUR step. The actual data points could be plotted on the same display by appending observations to the data set ELLIPSE.

```
*-- Get data ellipse for each species to annotate on plot;
%include contour;              /* On MVS, use %include ddname(file);*/
%contour(data=iris,
         x=petalwid, y=petallen,
         group=species,
         out=ellipse,
         pvalue=.75, plot=NO,
         colors=red green blue);
```

```
*-- Plot contours of classification variable with data ellipses;
proc gcontour data=plotc  ;
   plot petallen * petalwid = _into_
      / levels=1 to 3
        llevels=2 2 2
        clevels=black black black
        anno=ellipse
        nolegend
        caxis=black
        vaxis=axis1 haxis=axis2
        hminor=4 vminor=4
        name='GB0923' ;
   axis1 order=(0 to 70 by 10)
        label=(h=1.3 a=90 r=0 'Petal Length')
        value=(h=1.2) offset=(2);
   axis2 order=(0 to 25 by 5)
        label=(h=1.3 'Petal Width')
        value=(h=1.2) offset=(2);
   title2 ;
```

The plot produced by this program is shown in Output 9.23. The linear discriminant functions are slightly jagged because the contour is approximated on a discrete grid. Note that the data ellipses do not all have the same size and shape, indicating that the assumption of equal covariance matrices may be violated.

Output 9.23
Linear Discriminant Functions for Iris Data

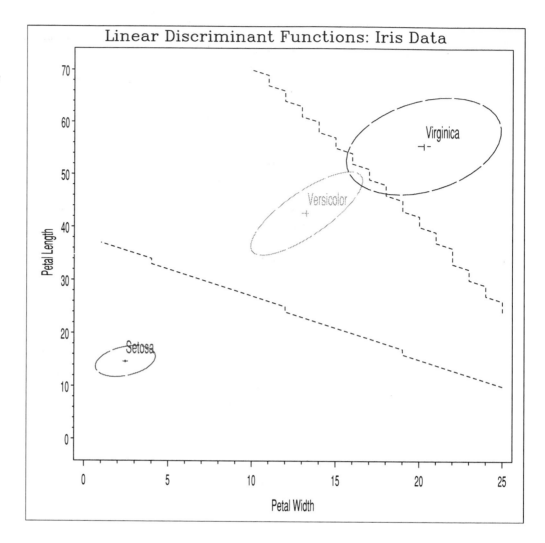

A second discriminant analysis is run with the specification POOL=TEST. This causes PROC DISCRIM to carry out a test of the hypothesis that the covariance matrices are equal in all groups. Except for the POOL=TEST specification, the PROC DISCRIM step and the rest of the program are identical to that given above. The printed output for the test of homogeneity of covariance matrices is shown in Output 9.24. The χ^2 value for the test is highly significant, indicating that the variances and covariances differ across the iris species. (The plot in Output 9.23 suggests that within-group variances increase systematically with both petal width and petal length.)

Output 9.24
Test of Homogeneity of Covariance Matrices for Iris Data

```
                    Quadratic Discriminant Functions: Iris Data
Discriminant Analysis      Test of Homogeneity of Within Covariance Matrices

        Notation: K      = Number of Groups

                  P      = Number of Variables

                  N      = Total Number of Observations - Number of Groups

                  N(i) = Number of Observations in the i'th Group - 1

                          _                        N(i)/2
                          || |Within SS Matrix(i)|
                  V     = ----------------------------------
                                                    N/2
                             |Pooled SS Matrix|


                          _           _   2
                          |   1       1  | 2P + 3P - 1
                  RHO  = 1.0 - | SUM -----  - ---  | -------------
                          |_    N(i)      N _|  6(P+1)(K-1)


                  DF   = .5(K-1)P(P+1)

                                        _                   _
                                        |        PN/2        |
                                        |   N          V     |
Under null hypothesis:   -2 RHO ln |   ------------------  |
                                        |    _         PN(i)/2  |
                                        |_   || N(i)          _|

is distributed approximately as chi-square(DF)


Test Chi-Square Value =    112.262384
with        6 DF     Prob > Chi-Sq = 0.0001


Since the chi-square value is significant at the  0.1000 level,
the within covariance matrices will be used in the discriminant function.

Reference: Morrison, D.F. (1976)    Multivariate Statistical Methods p252.
```

When the specification POOL=TEST rejects homogeneity, PROC DISCRIM does not pool the covariance matrices. The discriminant functions are then quadratic functions of the quantitative variables. The contour plot of the classification function using the same grid of points is shown in Output 9.25. The boundaries between adjacent regions are now sections of an ellipse. The boundaries could be made smoother by using more points in the GRID data set. Note that the data points in the IRIS data set are classified approximately the same way by both the linear and quadratic functions. They differ in how new points, outside the observed clusters (in the upper-left and lower-right corners of the plot), would be classified.

Output 9.25
Quadratic
Discriminant
Functions for Iris
Data

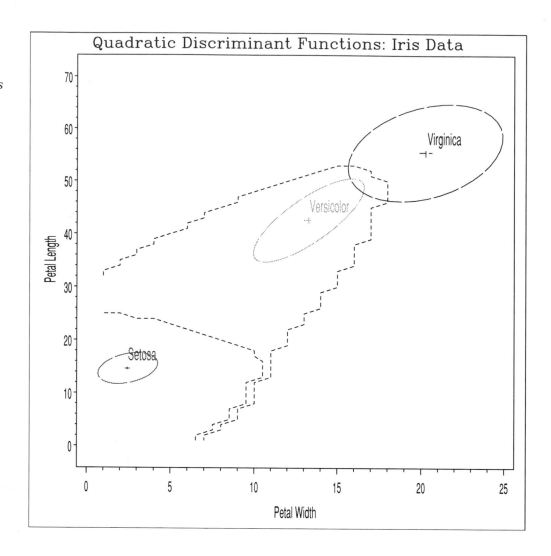

Coefficient Method

The second method for plotting discriminant functions calculates the boundaries of the discriminant regions explicitly, using the coefficients for the discriminant functions that can be output to an OUTSTAT= data set in Version 6 of PROC DISCRIM. This method calculates the discriminant functions exactly and is more efficient than the contour technique. The method is illustrated with the linear discriminant functions for the iris data. This method could, in principle, be used to plot quadratic functions, but the programming would be more complex.

Example: iris data

The program below plots the same linear discriminant functions shown in Output 9.23. Specifying OUTSTAT=COEFFS in the PROC DISCRIM step causes the linear coefficients to be included in the output data set COEFFS. These coefficients appear in the portion of the printed output shown in Output 9.26. The output data set COEFFS contains a wide variety of statistics from the discriminant analysis. Two DATA steps are used to extract the observations with _TYPE_='LINEAR' and to put the constant term into each observation.

```
%include IRIS;                    /* On MVS, use %include ddname(file);*/
title h=1.5 'Linear Discriminant Functions: Iris Data';

*-- Get discriminant function coefficients --  Requires Version 6 ;
proc discrim data=iris
             pool=yes            /* linear function */
             outstat=coeffs      /* statistics  V6  */
             noclassify;
     classes species;
     var petallen petalwid;

*-- select coefficients for LDF;
data coeffs;
   set coeffs;
   if _type_='LINEAR';
proc sort;
   by species _name_;

*-- get constant term as an additional variable;
data coeffs;
   set coeffs;
   by species;
   retain _const_;
   drop _name_;
   if _name_='_CONST_'
      then do;
         _const_=petallen;
      end;
   else output;
proc print data=coeffs;
   title2 'Linear coefficients';
```

Output 9.26
Linear Discriminant Function Coefficients for Iris Data

```
                   Linear Discriminant Functions: Iris Data

            Discriminant Analysis     Linear Discriminant Function

                         -1  _                              -1  _
          Constant = -.5 X' COV   X      Coefficient Vector = COV   X
                         j        j                              j

                                SPECIES

                 Setosa     Versicolor     Virginica     Label

     CONSTANT    -5.89972    -50.84965     -91.92744
     PETALLEN     0.85476      2.05271       2.46118    Petal length in mm.
     PETALWID    -0.28338      1.07494       2.33021    Petal width  in mm.
```

For classification, the discriminant function coefficients are applied to the variables as follows. For each observation, calculate the linear functions as follows, where x refers to the values of the two iris variables:

$$L_1(x) = -5.90 + 0.855 \text{ PETALLEN} - 0.283 \text{ PETALWID}$$

$$L_2(x) = -50.85 + 2.053 \text{ PETALLEN} + 1.075 \text{ PETALWID}$$

$$L_3(x) = -91.93 + 2.461 \text{ PETALLEN} + 2.330 \text{ PETALWID}$$

Then a new observation x is assigned to the species with the largest value of $L_i(x)$. It is more convenient to calculate the difference between each pair of functions:

$$d_{ij} = L_i(x) - L_j(x) \quad .$$

The boundaries between adjacent classification regions are then found by solving $d_{ij}(x)=0$ for one variable in terms of the others.

This method is carried out by the following program steps. The data set COEFFT, containing the differences in coefficients D12 and D13, is shown in Output 9.27. Note that each coefficient for the difference function is just the difference in coefficients for the two species. The DATA step that produces the LINDISC data set varies the value of petal width (X) throughout its range and calculates the value of petal length (Y) by solving $d_{ij}(x)=0$.

```
*-- transpose to calculate pairwise difference in coeffs;
proc transpose data=coeffs
               out=coefft;
   id species;

*-- get difference function for each pair of groups;
data coefft;
   set coefft;
   drop _label_;
   d12 = setosa   - versicol;
   d23 = versicol - virginic;
proc print;
   title2 'Coefficients for LDFs';

data coefft;
   set coefft;
   drop setosa versicol virginic;
proc transpose out=coeffd;
proc print data=coeffd;

*-- Get min, max of plotting variables;
proc means noprint data=iris;
   var petalwid petallen;
   output out=minmax min=minx miny max=maxx maxy;
```

```
*-- Calculate linear discriminant function lines for petalwid and
    petallen;
data lindisc;
  set coeffd;
  keep _name_ x y;
  do x = 1 to 25;                        * range of petalwid;
    y = -( _const_ + petalwid * x ) / petallen;
    output;
  end;
```

Output 9.27
Difference
Coefficients for Iris
Data

```
                  Linear Discriminant Functions: Iris Data
                            Coefficients for LDFs

 OBS    _NAME_     SETOSA    VERSICOL    VIRGINIC      D12        D23

  1    PETALLEN    0.8548     2.0527      2.4612     -1.1980    -0.4085
  2    PETALWID   -0.2834     1.0749      2.3302     -1.3583    -1.2553
  3    _CONST_    -5.8997   -50.8497    -91.9274     44.9499    41.0778
```

Finally, the data set LINDISC is transformed into an Annotate data set for the plot of petal length and petal width. The values calculated for petal length (Y) are in data coordinates and must be clipped to conform with the range of Y chosen for the plot. The PROC GPLOT step plots the data points together with the discriminant lines; the result is shown in Output 9.28.

```
*-- Clip to range of y and generate labels;
data lindisc;
  set lindisc;
  by _name_;
  xsys='2'; ysys='2';
  length function $ 8;
  retain function flag;
  if first._name_ then flag=1;
  if y>= 0 and y<=70;                    *-- clip;
  if flag=1 then do;
    position='8';
    function='LABEL'; text=_name_; output;
    function='MOVE';                output;
    flag=0;
  end;
  else do;
    function='DRAW'; output;
  end;
run;
proc gplot data=iris  ;
  plot petallen * petalwid = species
    / frame anno=lindisc
      vaxis=axis1 haxis=axis2
      vminor=4 hminor=4
      name='GB0928' ;
  symbol1 h=1.1 v=star   c=red;
  symbol2 h=1.1 v=square c=green;
  symbol3 h=1.1 v=-      c=blue;
```

```
axis1 order=(0 to 70 by 10)
      label=(h=1.3 a=90 r=0 'Petal Length')
      value=(h=1.2) offset=(2);
axis2 order=(0 to 25 by 5)
      label=(h=1.3 'Petal Width')
      value=(h=1.2) offset=(2);
title2 ;
```

Output 9.28
Linear Discriminant Functions for Iris Data

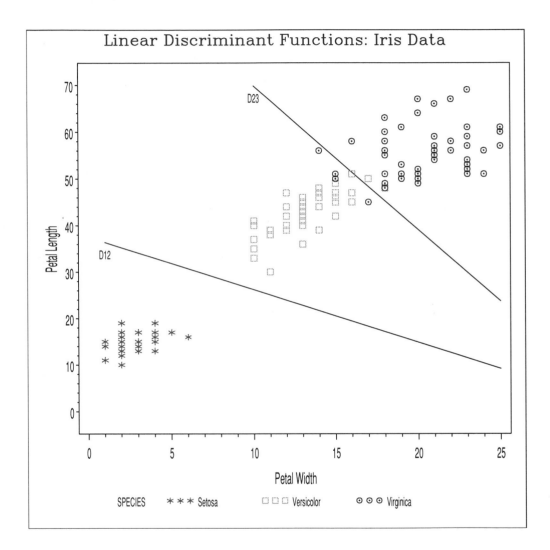

9.5.2 Canonical Discriminant Analysis and MANOVA

Canonical discriminant analysis is a dimension-reduction technique for analyzing differences in means of groups on multiple measures in terms of a small number of dimensions. The mathematics of canonical discriminant analysis and a one-way multivariate analysis of variance (MANOVA) are the same. In MANOVA, the emphasis is on testing the hypothesis that the mean vector is the same for all groups, taking the pattern of intercorrelations into account. Canonical discriminant analysis is concerned with finding the smallest number of linear combinations of the variables that maximally discriminate among the group mean vectors, in the sense of giving the largest possible univariate F statistic.

That is, for p variables, $x' = (x_1, \ldots, x_p)$, with data for n_i subjects in each of g groups, the analog of the univariate SSE is the $p \times p$ matrix, $\mathbf{E}$, of within-group sum of squares and cross-products (SSCP),

$$\mathbf{E} = \sum_{i=1}^{g} \sum_{j=1}^{n_i} (x_{ij} - \bar{x}_{i.})(x_{ij} - \bar{x}_{i.})' \tag{9.6}$$

and the analog of the univariate sum of squares SSH for group differences is the between-groups SSCP matrix, $\mathbf{H}(p \times p)$,

$$\mathbf{H} = \sum_{i=1}^{g} n_i(\bar{x}_{i.} - \bar{x}_{..})(\bar{x}_{i.} - \bar{x}_{..})' \quad . \tag{9.7}$$

Then if we were to combine the variables into a linear combination z with weights given by the vector c, so that $z_{ij} = c'x_{ij}$, the univariate F statistic of a one-way ANOVA would be

$$F(c) = \frac{SSH / (g - 1)}{SSE / (n - g)} = \frac{c'\mathbf{H}c / (g - 1)}{c'\mathbf{E}c / (n - g)} \quad . \tag{9.8}$$

In canonical discriminant analysis, the weights c are found so as to maximize this F statistic. That is, the canonical weights indicate the linear combination of the variables that best discriminates among groups, in the sense of having the greatest univariate F. These weights turn out to be the eigenvector c_1 corresponding to the largest eigenvalue λ_1 of the hypothesis matrix $\mathbf{H}$ relative to (in the metric of) the error matrix $\mathbf{E}$ or, equivalently, of the matrix $\mathbf{H} \mathbf{E}^{-1}$.

As in components analysis, there are additional eigenvectors that maximally separate the groups in orthogonal directions, but with decreasing effectiveness. The eigenvector c_2 associated with the second largest eigenvalue λ^2 provides a second, orthogonal dimension that best separates the groups. The maximum number of canonical dimensions needed to represent all differences in group means is s, the smaller of the number of quantitative variables and one less than the number of groups. In MANOVA, the various multivariate test statistics (such as Wilks' lambda and Pillai's trace criterion) are different functions of the set of eigenvalues, $\lambda_1, \lambda_2, \ldots, \lambda_s$, $s = \min(p, g-1)$. These statistics differ in how the strength of discrimination reflected in the eigenvalues is combined across the dimensions.

In the SAS System, PROC CANDISC carries out a canonical discriminant analysis and can produce an output data set containing the scores on the linear combinations z_1, z_2, and so on that discriminate maximally. These variables are usually named CAN1, CAN2, and so on. Plotting the first pair, CAN1 and CAN2 (or pairs among CAN1—CAN3), is often helpful for studying differences among the groups in terms of a reduced number of variables and sometimes for detecting outliers. If the number of quantitative variables is large, but the first two eigenvalues account for a large total proportion of the between-group variance, a two-dimensional display will capture the bulk of differences among the groups.

Typically, there are two aspects to interpreting how the canonical discriminant scores separate the groups along these orthogonal dimensions. First, the order and separation of the groups on each canonical variable indicate how that linear combination of the predictors discriminates among the groups. Second, the correlations between the predictors and the discriminant functions indicate how each predictor contributes to that discriminant dimension. Information relevant to each of these questions can be added to the basic plot of the canonical variables.

In order to see how the groups are distinguished on each dimension, it is useful to display the group means on the canonical variates on the plot. Groups widely separated on a given dimension are discriminated well by that linear combination of the variables. Moreover, because the canonical variates are uncorrelated, an approximate $100(1-a)\%$ confidence region (Seber 1984) for the population mean, μ_i, of group i is the circular region given by

$$n_i(\bar{z}_{i.} - \mu_i)(\bar{z}_{i.} - \mu_i)' \leq \chi_2^2 (1-a) \quad . \tag{9.9}$$

Equation 9.9 assumes that the discriminant scores are standardized to unit variance. The CANn scores from PROC CANDISC are standardized so that the average within-group variance of each CANn is unity, so a circle of radius $\sqrt{\chi_2^2(1-a)/n_i}$ provides the required confidence region.

To interpret the meaning of the canonical variables, it is common to examine the correlations (called *structure coefficients*) or the standardized *canonical coefficients* between the predictors and each canonical variable. The larger the value of the correlation or the canonical coefficient, the greater is the contribution of a given predictor to discrimination along a particular dimension. We can aid this aspect of interpretation by drawing vectors on the plot of the canonical scores, CAN1 and CAN2, for example, representing the contribution of each variable to each of the canonical dimensions. For example, if a variable has correlations of 0.7 and 0.2 with CAN1 and CAN2, a vector from the origin in the direction (0.7, 0.2) will indicate the relation of this variable to each canonical dimension.

Example: Fisher's Iris Data

On the glyph plot of the iris data (Section 8.2.1, "Constructing a Glyph Plot with the Annotate Facility"), there is a clear separation of the three iris species in the sepal length and petal length variables; on the plot of the petal length and petal width variables for the linear discriminant functions (Output 9.27), there is a similar pattern. However, the relation of all four variables to the groups is hard to see in these two-dimensional plots. Here we see what light canonical discriminant analysis can shed on these data.

PROC CANDISC is used to analyze these data as shown below. The OUT= option produces an output data set containing the discriminant scores, and the OUTSTAT= option gives an output data set containing statistics, as with PROC PRINCOMP.

```
%include IRIS;              /* On MVS, use %include ddname(file);*/
title h=1.5 f=duplex 'Canonical Discriminant Analysis: Iris Data';

proc candisc data=iris
           out=disc          /* scores     */
           outstat=discstat; /* statistics */
      classes species;
      var sepallen sepalwid petallen petalwid;
```

```
proc sort data=disc;
   by spec_no;
proc means noprint ;
   var can1 can2;
   by spec_no;
   output out=means mean=can1 can2 n=n;
```

The PROC MEANS step gives a data set containing the class means on the (standardized) discriminant scores.

The printed output from PROC CANDISC (Output 9.29) shows that both CAN1 and CAN2 significantly discriminate among the iris groups, but the size of the canonical correlations and eigenvalues indicates that CAN1 discriminates much more than CAN2.

Output 9.29
CANDISC
Procedure Printed
Output for Iris
Data

```
                  Canonical Discriminant Analysis: Iris Data

                     CANONICAL DISCRIMINANT ANALYSIS

        150 OBSERVATIONS       149 DF TOTAL
          4 VARIABLES          147 DF WITHIN CLASSES
          3 CLASSES              2 DF BETWEEN CLASSES

                          ADJUSTED        APPROX         SQUARED
            CANONICAL     CANONICAL       STANDARD       CANONICAL
            CORRELATION   CORRELATION     ERROR          CORRELATION

        1   0.984675      0.984359        0.002492       0.969584
        2   0.470483      0.460705        0.063789       0.221354

                       EIGENVALUES OF INV(E)*H
                         = CANRSQ/(1-CANRSQ)

            EIGENVALUE    DIFFERENCE      PROPORTION     CUMULATIVE

        1   31.8773       31.5931         0.9912         0.9912
        2    0.2843         .             0.0088         1.0000

        TESTS OF H0: THE CANONICAL CORRELATION IN THE CURRENT ROW
                    AND ALL THAT FOLLOW ARE ZERO

            LIKELIHOOD
            RATIO           F           NUM DF    DEN DF    PR > F

        1   0.02368335    197.9273        8         288      0.0
        2   0.77864562     13.7403        3         145      0.0001


             MULTIVARIATE TEST STATISTICS AND F APPROXIMATIONS
                      S=2     M=0.5     N=71

    STATISTIC                 VALUE         F        NUM DF   DEN DF   PR > F

    WILKS' LAMBDA           0.02368335    197.927       8      288     0.0
    PILLAI'S TRACE          1.190938       53.360       8      290     0.0001
    HOTELLING-LAWLEY TRACE  32.16162      574.889       8      286     0.0
    ROY'S GREATEST ROOT     31.87734     1155.554       4      145     0.0

        NOTE: F STATISTIC FOR ROY'S GREATEST ROOT IS AN UPPER BOUND
              F STATISTIC FOR WILKS' LAMBDA IS EXACT

                       TOTAL CANONICAL STRUCTURE

                                    CAN1          CAN2

                    SEPALLEN       0.7919        0.2176
                    SEPALWID      -0.5308        0.7580
                    PETALLEN       0.9850        0.0460
                    PETALWID       0.9728        0.2229
```

The plot of canonical discriminant scores with confidence regions for the mean of each group is shown in Output 9.30.* (A color version of this output appears as Output A3.47 in Appendix 3.) Note that although there are four variables, there are only three groups, so $s=2$ dimensions provide an exact representation of the differences in mean vectors. The plot also displays the structure correlations for each variable whose values appear in the printed output in Output 9.29 (TOTAL CANONICAL STRUCTURE). On the plot, these values are drawn as vectors from the origin. On this plot, the length of each variable vector is proportional to its contribution to separating the iris groups, and the direction of the vector indicates its relative contribution to the CAN1 and CAN2 linear combinations. The plus signs mark the means on the canonical variates, and the circles give a 99% confidence region for these means.

From Output 9.30, it can be seen that the three iris species are separated quite well along CAN1, with *setosa* differing most from the other two. There is little overlap between the positions of the three groups of flowers along the CAN1 dimension. The petal width and petal length variables contribute most to separating the groups in this direction. These vectors point in the direction of increasing variable values, so the *virginica* species is the largest in petal length and width. The CAN2 dimension has a much smaller separation among the groups, with the *versicolor* species lower than the other two. This dimension reflects mainly the sepal width variable.

---

*   Although the confidence regions are circular, they will appear elliptical if the scales for CAN1 and CAN2 differ on the plot. It is preferable to equate the vertical and horizontal scales; however, for these data, there would be too great a loss of resolution.

Output 9.30   *Plot of CAN1 and CAN2 for Iris Data*

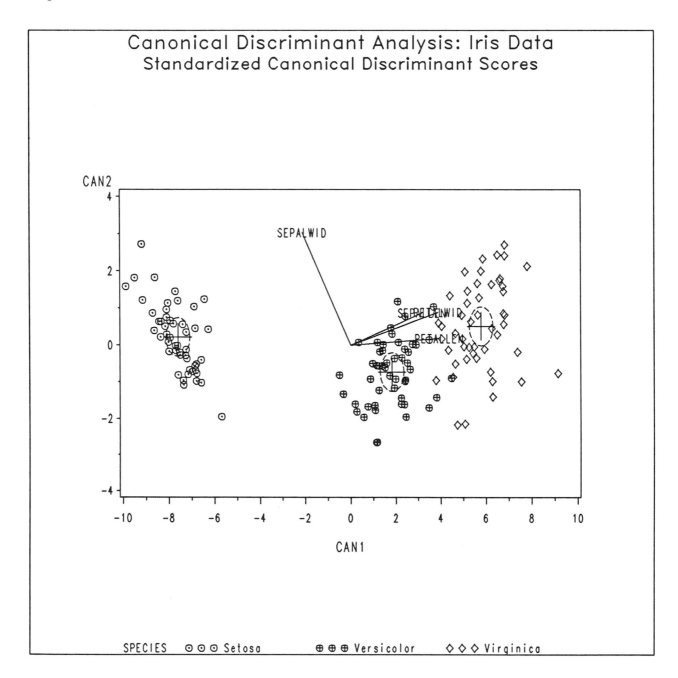

The class means on CAN1 and CAN2 and the confidence regions are drawn by the Annotate data set REGIONS. The radius of the 99.9% confidence circle, determined by the CINV function, is $\sqrt{\chi^2_2(.999)/n_i} = \sqrt{13.81/50}$ for these data.

```
proc format;
    value gpclr 1='RED' 2='GREEN' 3='BLUE';
data regions;
    set means;
    retain xsys '2' ysys '2';
    length text $8;
```

```
drop n can1 can2 a ang;
x = can1;
y = can2;
color=put(spec_no, gpclr.);

/* mark the class mean        */
text = 'PLUS';
size = 4;
function = 'SYMBOL';       output;

/* draw 99.9% confidence region */
size = sqrt( cinv(.999, 2, 0) / n ) ;      * radius ;
line = 3;
do a = 0 to 360 by 10;                      * draw a "circle" ;
   ang = a*arcos(-1)/180;                    * convert to radians;
   xp=  size * cos(ang);
   yp=  size * sin(ang);
   x = xp+ can1;                             * translate;
   y = yp+ can2;
   If a=0 then FUNCTION = 'MOVE     ';
           else FUNCTION = 'DRAW     ';
   output;
end;
```

The structure correlations for the plot are obtained from the OUTSTAT= data set from PROC CANDISC by selecting the observations with _TYPE_='STRUCTURE' (in Version 6) or _TYPE_='TSTRUCT' (in Version 5). This data set is transposed to make CAN1 and CAN2 variables, with observations for each of the iris predictor variables. Then an Annotate data set named VECTORS is constructed to draw and label a vector from the origin proportional to the coefficients of CAN1 and CAN2 for each predictor. Finally, the REGIONS and VECTORS data sets are combined for the PROC GPLOT step.

```
data coeffs;
   set discstat;            * get standardized coefficients ;
   drop _TYPE_;
   if _type_ = 'STRUCTUR' |   /* Version 6 */
      _type_ = 'TSTRUCT'  ;   /* Version 5 */
proc transpose out=coeffs;
proc print ;
data vectors;
   set coeffs;
   retain xsys '2' ysys '2' position '5';
   x = 0 ; y = 0;
   function = 'MOVE ' ; output;
   x = 4*can1;
   y = 4*can2 ;
   function = 'DRAW ' ; output;
   text = _NAME_;
   function = 'LABEL' ; output;
data regions;
   set regions vectors;
```

```
proc gplot data=disc ;
   plot can2 * can1 = species
        / anno=regions frame
          vminor=1 hminor=1
          vaxis=axis1 haxis=axis2
          name='GB0930' ;
   symbol1 v='-' i=none color=red;
   symbol2 v='+' i=none color=green;
   symbol3 v='$' i=none color=blue;
   axis1 label=(h=1.2)
         order=(-4 to 4 by 2) length=3.2 in;  * length is device-specific;
   axis2 label=(h=1.2)
         order=(-10 to 10 by 2) length=4.8in;  * length is device-specific;
   title2 h=1.3 f=duplex 'Standardized Canonical Discriminant Scores';
```

Example: Diabetes Data

Reaven and Miller (1979) examined the relationship between measures of blood plasma glucose and insulin in 145 nonobese adult patients at the Stanford Clinical Research Center in order to examine ways of classifying people as normal, overt diabetic, or chemical diabetic. Each patient underwent a glucose tolerance test, and the following variables were measured: relative weight (RELWT), fasting plasma glucose (GLUFAST), test plasma glucose (GLUTEST, a measure of intolerance to insulin), steady state plasma glucose (SSPG, a measure of insulin resistance), and plasma insulin during test (INSTEST).

Because there are again three groups, there are at most two canonical discriminant dimensions. PROC CANDISC (output not shown) indicates that both dimensions are highly significant. The plot of canonical discriminant scores with confidence regions for the mean of each group and with structure coefficient vectors for the variables is shown in Output 9.31. (A color version of this output appears as Output A3.48 in Appendix 3.) The program for this plot is the same as that for the iris data.

We can see from Output 9.31 that the three groups are ordered on CAN1 from normal to overt diabetic with chemical diabetics in the middle. The two glucose measures (GLUTEST and GLUFAST) and the measure SSPG of insulin resistance are most influential in separating the groups along the first canonical dimension. The chemical diabetics are apparently higher than the other groups on these three variables. Among the remaining variables, INSTEST is most influential and contributes mainly to the second canonical dimension.

Output 9.31 *Plot of CAN1 and CAN2 for Diabetes Data*

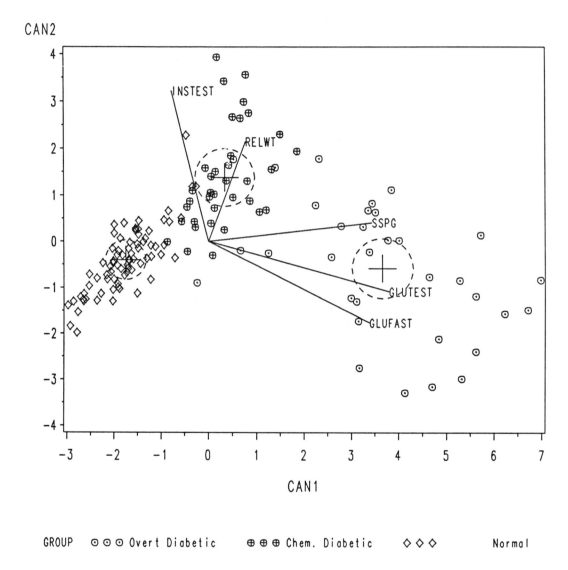

For these data, the two-dimensional plot in Output 9.31 again accounts completely for the differences in group mean vectors because there are three groups. The wide separation of the 99% confidence circles portrays the strength of group differences on the canonical dimensions, and the variable vectors help to interpret the meaning of these dimensions.

Chapter 10 Displaying Categorical Data

10.1   Introduction

The methods discussed previously in this book deal mainly with plotting continuous, quantitative response variables. The analysis of *categorical data*, or *contingency-table data*, on the other hand, deals with discrete, qualitative variables, such as the number (frequency) of individuals cross-classified by hair color and eye color. Statistical methods for categorical data, such as log-linear models and logistic regression, represent discrete analogs of the analysis of variance and regression methods for continuous response variables. However, while graphical display techniques are common adjuncts to analysis of variance and regression, methods for plotting contingency table data are not as widely used.

This chapter deals with two types of exploratory graphical techniques for such data, designed to help show how categorical variables are related. Section 10.2, "Rectangular Displays," describes methods to portray the pattern of association between the variables in a contingency table as a whole. The *association plot* and *mosaic display* both display each cell in the table as a rectangle whose area represents the cell frequency or deviation from independence. Section 10.3, "Correspondence Analysis," describes the method of *correspondence analysis*, which summarizes the pattern of association between the row and column variables in a two-dimensional display.

Both types of techniques are illustrated principally with two-way tables, though several of them can be applied to multi-way contingency tables as well. For higher-order tables, log-linear and logistic methods have become widely used for determining which of several categorical variables and their interactions need to be included in a model to account for the data. These confirmatory methods are beyond the scope of this book, however.

10.2 Rectangular Displays

A contingency table gives the joint distribution of two or more discrete, categorical variables. In a two-way table, one typically uses the χ^2 test of association to determine if the row and column variables can be considered independent.

Several schemes for representing such data graphically are based on the fact that when the row and column variables are independent, the expected frequencies are products of the row and column totals (divided by the grand total). Then values in each cell can be represented by a rectangle whose area shows the cell frequency or deviation from independence.

To establish notation, let $\mathbf{F} = \{f_{ij}\}$ be the observed frequency table with I rows and J columns. In what follows, an index is replaced by a plus sign when summed over the corresponding variable, so $f_{i+} = \Sigma_j f_{ij}$ gives the total frequency in row i, $f_{+j} = \Sigma_i f_{ij}$ gives the total frequency in column j, and $f_{++} = \Sigma\Sigma_{ij} f_{ij}$ is the grand total, which is also symbolized by n. The same table expressed as proportions of the grand total is denoted $\mathbf{P} = \{p_{ij}\} = \mathbf{F}/n$.

The usual (Pearson) χ^2 statistic for testing independence is

$$\chi^2 = \sum_{ij} \frac{(f_{ij} - e_{ij})^2}{e_{ij}} \tag{10.1}$$

with $(I-1)(J-1)$ degrees of freedom; e_{ij} is the expected frequency assuming independence, $e_{ij} = (f_{i+} f_{+j})/n$.

10.2.1 Association Plots

One type of rectangular display (suggested by a figure in Becker, Chambers, and Wilks 1988, p. 512) shows the deviations from independence by the area of rectangular bars, one for each cell. The signed contribution to χ^2 for cell i, j is

$$d_{ij} = \frac{f_{ij} - e_{ij}}{\sqrt{e_{ij}}} \tag{10.2}$$

which can be thought of as a standardized residual because $\chi^2 = \Sigma\Sigma_{ij} d_{ij}^2$ and χ^2 is a sum of squares of standard normal deviates.

In the association plot, each cell is shown by a rectangle whose (signed) height is proportional to the standardized residual d_{ij} and whose width is $\sqrt{e_{ij}}$. Then the area of each bar is

$$d_{ij} \cdot \sqrt{e_{ij}} = \left(\frac{f_{ij} - e_{ij}}{\sqrt{e_{ij}}}\right) \cdot \sqrt{e_{ij}} = f_{ij} - e_{ij} \quad .$$

The example below shows the construction of the association plot using data on the relation between hair color and eye color among 592 subjects (students in a statistics course) collected by Snee (1974).* The Pearson χ^2 for these data is 138.3 with 9 degrees of freedom, indicating substantial departure from independence. The question is how to understand the nature of the correlation between hair and eye color, and this is precisely what the association plot is designed to show.

Constructing the Association Plot

The calculations for the plot are most conveniently carried out using the matrix operations of the IML procedure because it is necessary to have access to several complete arrays at once. The plot is drawn with the Annotate facility and the GPLOT procedure. The module ASSOC calculates the matrices of expected frequencies, **E**, and standardized deviates, **D**, from the input frequencies in **F**. These results for the sample data are shown in Output 10.1.

Output 10.1
Expected
Frequencies and
Standardized
Deviates for Hair
and Eye Color

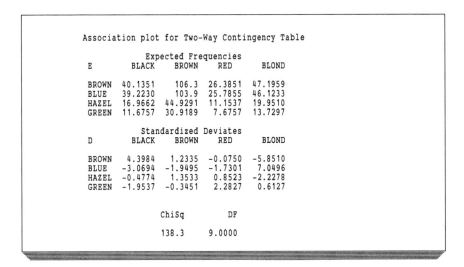

```
        Association plot for Two-Way Contingency Table

                    Expected Frequencies
        E        BLACK      BROWN      RED       BLOND

        BROWN   40.1351    106.3     26.3851    47.1959
        BLUE    39.2230    103.9     25.7855    46.1233
        HAZEL   16.9662     44.9291  11.1537    19.9510
        GREEN   11.6757     30.9189   7.6757    13.7297

                    Standardized Deviates
        D        BLACK      BROWN      RED       BLOND

        BROWN    4.3984     1.2335   -0.0750    -5.8510
        BLUE    -3.0694    -1.9495   -1.7301     7.0496
        HAZEL   -0.4774     1.3533    0.8523    -2.2278
        GREEN   -1.9537    -0.3451    2.2827     0.6127


                    ChiSq         DF

                    138.3        9.0000
```

To locate the (x, y) positions of the boxes, the maximum height for each row is found from the columns of **D**, and the maximum width of each column is found from the rows of **E**. These values, plus suitable spacings between rows and columns, are used to determine the center of each box. The results necessary for the plot are output to an output data set named ASSOC. For the sample data, this data set is shown in Output 10.2. CHI is the standardized deviate, d_{ij}; EXP$= \sqrt{e_{ij}}$. These determine the height and width of the bars. X and Y locate the box centers.

---

* From "Graphical Display of Two-Way Contingency Tables" by Ronald D. Snee, *The American Statistician*, Volume 28. Copyright © 1974 by the American Statistical Association. Reprinted by permission of the American Statistical Association.

Output 10.2
Output Data Set
ASSOC from the
IML Procedure for
Association Plot

```
                Association plot for Two-Way Contingency Table

  OBS     CHI      EXP       X         Y          PAIRS

    1    4.3984   6.3352   12.0655   84.1134    BROWN BLACK
    2    1.2335  10.3094   28.1051   84.1134    BROWN BROWN
    3   -0.0750   5.1366   38.9720   84.1134    BROWN RED
    4   -5.8510   6.8699   51.5722   84.1134    BROWN BLOND
    5   -3.0694   6.2628   12.0655   68.3573    BLUE  BLACK
    6   -1.9495  10.1916   28.1051   68.3573    BLUE  BROWN
    7   -1.7301   5.0779   38.9720   68.3573    BLUE  RED
    8    7.0496   6.7914   51.5722   68.3573    BLUE  BLOND
    9   -0.4774   4.1190   12.0655   59.1390    HAZEL BLACK
   10    1.3533   6.7029   28.1051   59.1390    HAZEL BROWN
   11    0.8523   3.3397   38.9720   59.1390    HAZEL RED
   12   -2.2278   4.4667   51.5722   59.1390    HAZEL BLOND
   13   -1.9537   3.4170   12.0655   49.2654    GREEN BLACK
   14   -0.3451   5.5605   28.1051   49.2654    GREEN BROWN
   15    2.2827   2.7705   38.9720   49.2654    GREEN RED
   16    0.6127   3.7054   51.5722   49.2654    GREEN BLOND
```

The PROC IML program to calculate the results in Output 10.1 and produce the output data set shown in Output 10.2 is listed below:

```
title h=1.5 'Association plot for Two-Way Contingency Table';
proc iml;
start assoc(f,row,col);
   /*------------------------------------------------------------*
    | module to calculate expected frequencies and std. deviates |
    |   F  = frequency table                                     |
    |   ROW= vector of row labels                                |
    |   COL= vector of col labels                                |
    *------------------------------------------------------------*/
   nr= nrow(f);
   nc= ncol(f);
   nrc=nr#nc;
   r = f[,+];              * row totals;
   c = f[+,];              * col totals;
   n = c[+];               * grand total;

   e = r * c / n;          * expected frequencies;
   d = (f - e) / sqrt(e);  * standard deviates;
   print "Expected Frequencies", e [rowname=row colname=col];
   print "Standardized Deviates", d [rowname=row colname=col];

   chisq = ssq( d );
   df = (nr-1) * (nc-1);
   reset noname;
   print chisq[colname={'ChiSq'}]
         df[colname={DF}];
   e = sqrt(e);

   *-- find maximum height for each row;
   rowht = d[ ,<> ] - d[ ,>< ];
   totht = rowht[+];
```

```
     *-- find maximum width for each col;
     colht = e[ <>, ] ;
     totwd = colht[+];

     *-- Find x, y locations of box centers;
     x0 = 0; y0 =100;
     do i = 1 to nr;
        y0 = y0 - rowht[i] - .20 * totht;
        y  = y // repeat ( y0, nc, 1 );
        rl = rl // repeat( row[i], nc, 1 );
        end;
     do j = 1 to nc;
        x0 = x0 + colht[j] + .20 * totwd;
        x  = x // x0 ;
        end;
     x = repeat(x, nr, 1);

     *-- Construct character vector of cell names;
     pairs = concat (rl, repeat(' ', nrc, 1)) ||
             shape( col, nrc, 1);
     pairs = rowcat(pairs);

     out = shape( d, nrc, 1) ||
           shape( e, nrc, 1) || x || y;
     cols = {CHI EXP X Y};
     create assoc from out [colname=cols rowname=pairs];
     append from out [rowname=pairs];
  finish;
```

The following IML statements create the frequency table (**F**) and labels (ROW and COL) and call the ASSOC module. You can run this program with different sets of data by changing these assignments.

```
  /*  blk  brn red  blond :hair / eye color */
  f = { 68  119 26   7 ,       /* brown */
        20   84 17  94 ,       /* blue  */
        15   54 14  10 ,       /* hazel */
         5   29 14  16 };      /* green */

   col = {black brown red blond};
   row = {brown blue hazel green};
   run assoc( f, row, col );
  quit;
  proc print data=assoc;
```

The data set ASSOC is then used to construct the bars and labels on the plot using the Annotate functions. The bars are constructed in the data set BOXES with the BAR function of the Annotate facility and filled in the cross-hatched STYLE, 'X1.' The data sets LABELROW and LABELCOL add the category labels to the plot margins, and these are joined with the BOXES data set to give the Annotate data set ANNO. The LABELROW step also plots a row of dots at the baseline for each row of the table, which represents zero deviation. The program is shown below, followed by the plot it produces, shown on Output 10.3. (A color version of this output appears as Output A3.49 in Appendix 3, "Color Output.")

```
data boxes;
   set assoc;
   length function $8;
   xsys = '2'; ysys = '2';
   x = x - exp/2;                  * center the bar;
   function = 'MOVE';  output;
   x = x + exp;                    * width is sqrt(e ij);
   y = y + chi;                    * height is chi;
   if chi > 0 then color='BLACK';
              else color='RED';
   line=0; style='X1';
   function = 'BAR';   output;
proc print data=boxes;

data labelrow;
   set assoc;
   by descending y;
   ysys='2';
   if first.y then do;
      xsys='1'; x=100;   position='6';
      function='LABEL';  size=1.3;
      text=scan(pairs ,1);
      output;
      do x = 4 to 96 by 4;
         function='POINT';
         output;
         end;
      end;
proc sort data=assoc;
   by x;
data labelcol;
   set assoc;
   by x;
   xsys='2';
   if first.x then do;
      ysys='1'; y=0;    position='5';
      function='LABEL'; size=1.3;
      text=scan(pairs ,2);
      end;
data anno;
   set boxes labelrow labelcol;
```

In the PROC GPLOT step, note that it is the data set BOXES that is plotted, rather than ASSOC. The former includes the X, Y values for each corner of each box, while the latter just includes their centers. The AXIS statements suppress the axes entirely, and the SYMBOL statement suppresses the points. In essence, PROC GPLOT is used here simply to scale the X, Y values to the plot page.

```
proc gplot data=boxes ;
    plot y * x / anno=anno name='GB1003'
                 vminor=0 hminor=0 vaxis=axis1 haxis=axis2;
    symbol1 v=none;
    axis1 value=none major=none style=0
          label=(h=1.5 a=90 'Eye Color');
    axis2 value=none major=none style=0
          label=(h=1.5 'Hair Color');
```

Output 10.3
Association Plot for Hair and Eye Color Data

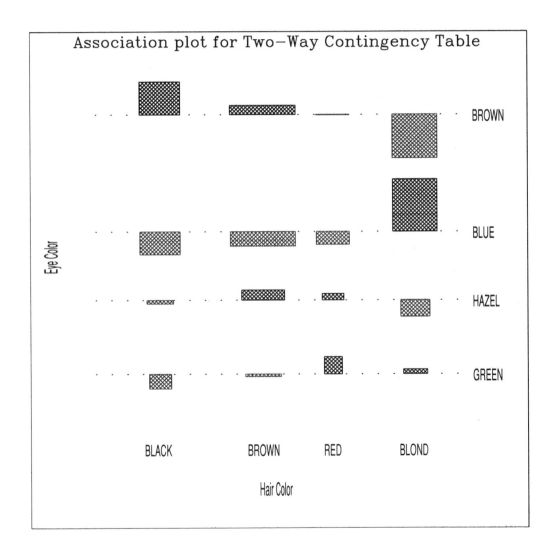

To interpret the plot, note that the (signed) height of each bar indicates deviation from a model in which hair color and eye color are independent. Boxes plotted in black thus represent combinations that occur more often than independence predicts, while boxes drawn in red are combinations that occur less often. The area of each box is the difference between observed and expected frequencies.

We see that, in this sample, two cells make the largest contribution to χ^2: people with blond hair are considerably more likely to have blue eyes and less likely to have brown eyes than independence predicts. Also, people with brown eyes are somewhat more likely to have black or brown hair, and red hair and green eyes is a more likely combination than one would expect if hair and eye color were independent. You can see the same things in the d_{ij} values (CHI) in Output 10.2 if you look hard enough, but the graphic display makes the pattern more apparent.

10.2.2 Mosaic Displays

A similar idea is the *mosaic display*, proposed by Hartigan and Kleiner (1981), which represents the observed and expected frequencies in a contingency table directly, by areas proportional to the numbers. The position and shape of the rectangles are chosen to depict the deviation from independence represented by the row-column frequency value. Similar displays for contingency tables are described by Bertin (1983, p. 224).

For the data on hair color and eye color, the expected frequencies (shown in Output 10.1) can be represented by rectangles whose widths are proportional to the total frequency in each column, f_{+j}, and whose heights are proportional to the total frequency in each row, f_{i+}, so that the area of each rectangle is proportional to the expected frequency, e_{ij}. Output 10.4 shows such rectangles. The column totals are shown at the bottom, and the row totals are shown at the right.

Output 10.4
*Expected
Frequencies under
Independence
Depicted by Areas
of Rectangles*

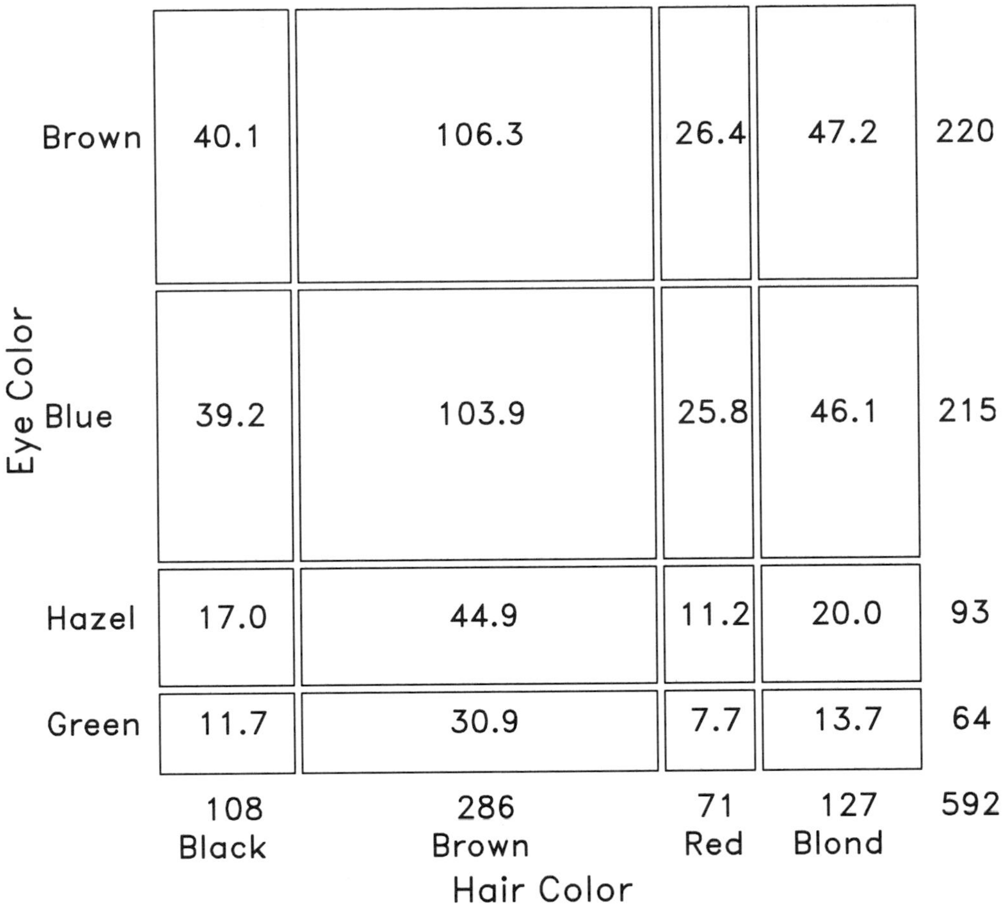

	Black	Brown	Red	Blond	
Brown	40.1	106.3	26.4	47.2	220
Blue	39.2	103.9	25.8	46.1	215
Hazel	17.0	44.9	11.2	20.0	93
Green	11.7	30.9	7.7	13.7	64
	108	286	71	127	592

Eye Color (vertical axis label) — Hair Color (horizontal axis label)

In the mosaic display, shown in Output 10.5, we draw the boxes in proportion to the actual frequency in each cell, with a dotted line showing the expected frequencies. The width of each box is still proportional to the overall frequency in each column of the table. The height of each box, however, is proportional to the cell frequency, and the dotted line in each row shows what the frequencies would be under independence.

Thus, the deviations from independence, $f_{ij} - e_{ij}$, are shown by the areas between the rectangles and the dotted lines for each cell. In Output 10.5, positive deviations appear as rectangles that descend below the dotted line, such as the excess number of blue-eyed blondes. Negative deviations, such as the smaller-than-expected number of brown-eyed blondes, appear as rectangles that fall short of the dotted line.

Output 10.5
Mosaic Display for
Hair and Eye
Color Data

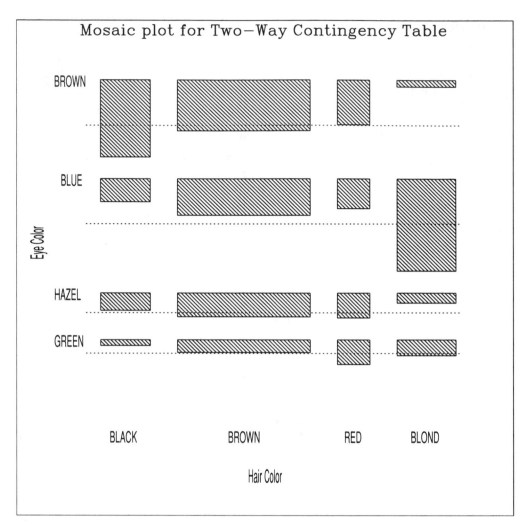

Constructing the Mosaic Display

The program for the mosaic display (shown below) is very similar to that for the association plot. The expected frequencies and (*x*, *y*) locations of the boxes are calculated with PROC IML in the module MOSAIC. The results required for the plot are the column proportions of the total frequency (CP), the observed cell proportions of each row total (P), and the expected cell proportions of each row total (EXP). These values are shown in Output 10.6.

```
title h=1.5 'Mosaic plot for Two-Way Contingency Table';
proc iml;
start mosaic(f,row,col);
   /*----------------------------------------------------------*
    |  module to calculate coordinates for mosaic display      |
    |    F  = frequency table                                  |
    |    ROW= vector of row labels                             |
    |    COL= vector of col labels                             |
    *----------------------------------------------------------*/
   nr= nrow(f);
   nc= ncol(f);
   nrc=nr#nc;
   r = f[,+];              * row totals;
   c = f[+,];              * col totals;
   n = c[+];               * grand total;

   e = r * c / n;          * expected frequencies;
   rp= r / n ;             * row proportions  ;
   cp= c / n ;             * col proportions ;
   p = f / shape (c, nr,nc);
   print "Expected Frequencies",    e [rowname=row colname=col]
                                    r [rowname=row format=6.0];
   print "Col totals",          c [colname=col] n;
   print "Proportion of col total", p [rowname=row colname=col]
                                    rp [rowname=row] ;
   print "Column Proportions",      cp [colname=col];
   e = e / shape(c, nr,nc);      * Scale expected to conform with p;
   print "Expected proportion of col total", e[rowname=row colname=col];

   *-- find maximum height for each row;
   rowht = 0 // p[ ,<> ] ;
   totht = rowht[+];
   *-- find maximum width for each col;
   colwd = cp      ;
   totwd = colwd[+];

   *-- Find x, y locations of box UR corners;
   x0 = 0; y0 =1;
   do i = 1 to nr;
      y0 = y0 - rowht[i] - ( .10 * totht);
      y  = y // repeat ( y0, nc, 1 );
      rl = rl // repeat( row[i], nc, 1);
      end;
   do j = 1 to nc;
      x0 = x0 + colwd[j] + ( .10 * totwd);
      x  = x // x0 ;
      end;
   x =  repeat(x, nr, 1);
```

```
                    *-- Construct character vector of cell names;
                    pairs = concat (rl, repeat(' ', nrc, 1)) ||
                              shape( col, nrc, 1);
                    pairs = rowcat(pairs);

                    out = shape( p, nrc, 1) || shape( e, nrc, 1) ||
                              shape(cp, nrc, 1) || x || y;
                    cols = {p exp cp x y};
                    create mosaic from out [ colname=cols rowname=pairs];
                    append from out [rowname=pairs];
                 finish;
                  /*   blk  brn  red  blond :hair / eye color */
                 f = { 68  119   26    7 ,        /* brown */
                       20   84   17   94 ,        /* blue  */
                       15   54   14   10 ,        /* hazel */
                        5   29   14   16 };       /* green */

                 col = {black brown red blond};
                 row = {brown blue hazel green};
                 run mosaic( f, row, col );
              quit;
              proc print data=mosaic;
                 format _numeric_ 7.4;
```

Output 10.6
IML Procedure
Results for Mosaic
Plot

```
               Mosaic plot for Two-Way Contingency Table
                        Expected Frequencies
      E       BLACK      BROWN     RED       BLOND      R        COL1

    BROWN    40.1351    106.3    26.3851    47.1959    BROWN     220
    BLUE     39.2230    103.9    25.7855    46.1233    BLUE      215
    HAZEL    16.9662    44.9291  11.1537    19.9510    HAZEL      93
    GREEN    11.6757    30.9189   7.6757    13.7297    GREEN      64


                             Col totals
      C       BLACK      BROWN     RED       BLOND      N        COL1

    ROW1     108.0      286.0    71.0000    127.0      ROW1     592.0


                       Proportion of col total
      P       BLACK      BROWN     RED       BLOND      RP       COL1

    BROWN    0.6296     0.4161   0.3662     0.0551     BROWN    0.3716
    BLUE     0.1852     0.2937   0.2394     0.7402     BLUE     0.3632
    HAZEL    0.1389     0.1888   0.1972     0.0787     HAZEL    0.1571
    GREEN    0.0463     0.1014   0.1972     0.1260     GREEN    0.1081


                          Column Proportions
             CP       BLACK     BROWN     RED       BLOND

            ROW1     0.1824    0.4831    0.1199    0.2145


                  Expected proportion of col total
             E        BLACK     BROWN     RED       BLOND

            BROWN    0.3716    0.3716    0.3716    0.3716
            BLUE     0.3632    0.3632    0.3632    0.3632
            HAZEL    0.1571    0.1571    0.1571    0.1571
            GREEN    0.1081    0.1081    0.1081    0.1081
```

As before, these matrices are assembled into columns of an output data set, named MOSAIC (Output 10.7), which is used to draw the plot with PROC GPLOT. In the data set MOSAIC, P and EXP represent the observed and expected frequencies, expressed as proportions of the total frequency in each row. The variables X and Y locate the upper-right corner of each rectangle.

Output 10.7
Output Data Set
MOSAIC for
Mosaic Plot

```
                    Mosaic plot for Two-Way Contingency Table

  OBS      P        EXP       CP        X          Y        PAIRS

   1     0.6296    0.3716   0.1824    0.2824     0.8236    BROWN BLACK
   2     0.4161    0.3716   0.4831    0.8655     0.8236    BROWN BROWN
   3     0.3662    0.3716   0.1199    1.0855     0.8236    BROWN RED
   4     0.0551    0.3716   0.2145    1.4000     0.8236    BROWN BLOND
   5     0.1852    0.3632   0.1824    0.2824     0.0175    BLUE  BLACK
   6     0.2937    0.3632   0.4831    0.8655     0.0175    BLUE  BROWN
   7     0.2394    0.3632   0.1199    1.0855     0.0175    BLUE  RED
   8     0.7402    0.3632   0.2145    1.4000     0.0175    BLUE  BLOND
   9     0.1389    0.1571   0.1824    0.2824    -0.8990    HAZEL BLACK
  10     0.1888    0.1571   0.4831    0.8655    -0.8990    HAZEL BROWN
  11     0.1972    0.1571   0.1199    1.0855    -0.8990    HAZEL RED
  12     0.0787    0.1571   0.2145    1.4000    -0.8990    HAZEL BLOND
  13     0.0463    0.1081   0.1824    0.2824    -1.2726    GREEN BLACK
  14     0.1014    0.1081   0.4831    0.8655    -1.2726    GREEN BROWN
  15     0.1972    0.1081   0.1199    1.0855    -1.2726    GREEN RED
  16     0.1260    0.1081   0.2145    1.4000    -1.2726    GREEN BLOND
```

Several DATA steps are used to create a data set BOXES and Annotate data sets to label the rows and columns on the plot drawn by PROC GPLOT. These steps are almost the same as those used to create the association plot.

```
data boxes;
   set mosaic;
   length function $8;
   xsys = '2'; ysys = '2';
   function = 'MOVE';  output;
   x = x - cp ;                    * width is col proportion;
   y = y - p  ;                    * height is actual freq;
   line=0; style='L1';
   function = 'BAR';  output;
proc print data=boxes;
data labelrow;
   set mosaic;
   by descending y;
   ysys='2';
   if first.y then do;
      xsys='1'; x=4;       position='4';
      function='LABEL';  size=1.3;
      text=scan(pairs ,1);
      output;
```

```
            y = y - exp;                * line at expected freq;
            do x = 4 to 100 by 1;
                function='POINT';
                output;
                end;
            end;
proc sort data=mosaic;
    by x;
data labelcol;
    set mosaic;
    by x;
    xsys='2';
    x = x - cp/2;
    if first.x then do;
        ysys='1'; y=0;     position='5';
        function='LABEL'; size=1.3;
        text=scan(pairs ,2);
        end;
data anno;
    set boxes labelrow labelcol;
proc gplot data=boxes  ;
    plot y * x
            / anno=anno
              hminor=0 vminor=0
              vaxis=axis1 haxis=axis2
              name='GB1005'  ;
    symbol1 v=none;
    axis1 value=none major=none style=0
            label=(h=1.5 a=90 'Eye Color');
    axis2 value=none major=none style=0
            label=(h=1.5 'Hair Color');
```

The mosaic plot in Output 10.5 highlights the same major deviations from independence as are found on the association plot (Output 10.3): an excess of blue-eyed blonds and dark-haired people with brown eyes. The smaller deviations from independence are not as apparent on the mosaic plot, however, because these appear as small departures from the dotted lines of expected frequencies. On the other hand, the mosaic plot shows the data (cell frequencies) directly and may therefore be easier to interpret, especially for less sophisticated viewers.

For large tables, the amount of empty space inside the mosaic plot may make it harder to see patterns, especially when there are large deviations from independence. In these cases, it may be more useful to separate the rectangles in each column by a small constant space, rather than force them to align in each row. This is done in Output 10.8. (A color version of this output appears as Output A3.50 in Appendix 3.) It is actually this condensed form of the display that suggests the name *mosaic*. Again, the area of each box is proportional to the cell frequency, and complete independence is shown when the boxes in each row all have the same height, as in Output 10.4.

The size of each box in Output 10.8 is the same as in Output 10.5: proportional to the conditional frequency of each row (eye color) for a given column (hair color). The spacing between columns is somewhat larger than that within a column. This makes it easier to make comparisons within hair-color groups (columns), but harder to make comparisons within eye-color groups because the rows are no longer aligned, except for the first and last. A similar plot could be made with the first division proportional to the row totals, which would facilitate comparisons among eye-color groups.

Output 10.8
Condensed Column Proportion Mosaic, Constant Row and Column Spacing

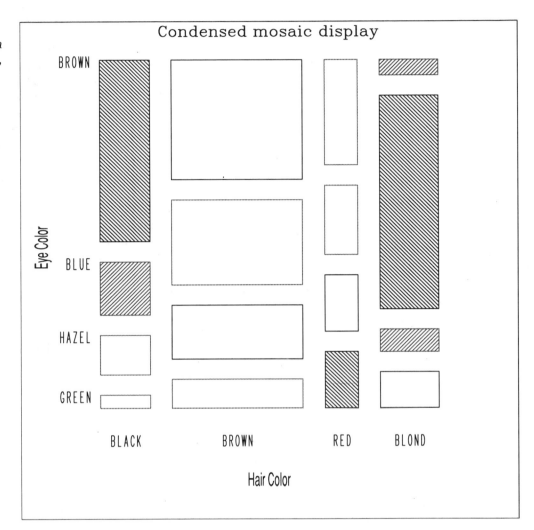

In Hartigan and Kleiner's original version (1981), all the boxes are unshaded and drawn in one color, so only the relative sizes of the rectangles indicate deviations from independence. Output 10.8 extends the mosaic plot, showing the standardized deviation from independence, d_{ij}, by the color and shading of each rectangle: cells with positive deviations are drawn black; negative deviations are drawn red. The absolute value of the deviation is portrayed by shading: cells with $|d_{ij}| \geq 2$ are filled; cells with absolute values less than 2 are empty. This use of color and shading gives the mosaic display most of the advantages of the association plot.

Multi-Way Tables

In this condensed form, the mosaic plot also generalizes readily to the display of multidimensional contingency tables in ways that the association plot does not. Imagine that each cell of the two-way table for hair and eye color is further classified by one or more additional variables—sex and level of education, for example. Then each rectangle on the mosaic plot can be subdivided horizontally to show the proportion of males and females in that cell, and each of those horizontal portions can be subdivided vertically to show the proportions of people at each educational level in the hair-eye-sex group. See Hartigan and Kleiner (1981; 1984) for revealing examples of mosaic displays of three-way and four-way tables.

For a three-way table with frequencies f_{ijk}, the scheme is this. First, the total area is divided into vertical strips proportional to the marginal totals of one of the variables—the first, for example; so the widths are proportional to f_{i++}. Then each vertical strip is subdivided in horizontal proportion to joint frequencies with the second variable, f_{ij+}. Each box therefore has an area proportional to the conditional frequency of the second variable given the first, f_{ij+}/f_{i++}. Finally, each such rectangle is divided in vertical proportion to the joint frequencies with the third variable, f_{ijk}.

These final rectangles are then set slightly apart, with the greatest space for the first division and the least space for the last. This process can be carried out for any number of variables. The addition of color and shading to show deviations from independence, as in Output 10.8, makes it much easier to see patterns, even with four or more variables. An algorithm for constructing n-way mosaic displays is given by Wang (1985).

10.3   Correspondence Analysis

Correspondence analysis is a technique related to the biplot (Section 8.7, "Biplot: Plotting Variables and Observations Together") and to principal components analysis (Section 9.4, "Plotting Principal Components and Component Scores") that finds a multidimensional representation of the association between the row and column categories of a two-way contingency table. Like the biplot, the technique uses the singular value decomposition of a matrix to find scores for the row and column categories on a small number of dimensions. Like principal components analysis, these dimensions account for the greatest proportion of the χ^2 statistic for association between the row and column categories, just as components account for maximum variance. For graphical display, two (or at most three) dimensions are typically used to give a reduced rank approximation to the data.

Correspondence analysis has a very large literature and is also discussed under the names *dual scaling*, *reciprocal averaging*, and *canonical analysis of categorical data*. See Greenacre (1984), Nishisato (1980), or Lebart, Morineau, and Tabard (1977) for a detailed treatment of the method and other applications. Greenacre and Hastie (1987) provide an excellent discussion of the geometric interpretation, and van der Heijden and de Leeuw (1985) develop some of the relations between correspondence analysis and log-linear methods.

The matrix that is decomposed is a matrix of deviations from independence, **D**, expressed in terms of the proportions, p_{ij}:

$$\mathbf{D} = \{d_{ij}\} = \frac{(p_{ij} - p_{i+}p_{+j})}{\sqrt{p_{i+}p_{+j}}} = \frac{(f_{ij} - e_{ij})}{\sqrt{e_{ij}}\sqrt{n}} \quad . \tag{10.3}$$

Thus, each d_{ij} is $(1/\sqrt{n})$ times that cell's contribution to the χ^2 statistic, and the sum of squares of the d_{ij} over all cells in the contingency table is χ^2/n. Thus, correspondence analysis is designed to show how the data deviate from expectation when the row and column variables are independent, as in the association plot. However, the association plot depicts every cell in the table, and for large tables it may be difficult to see patterns. Correspondence analysis shows only row and column categories in the two (or three) dimensions that account for the greatest proportion of deviation from independence.

The points for each row and column are found from the singular value decomposition of the residual matrix **D**,

$$\mathbf{D} = \mathbf{U} \, \Lambda \, \mathbf{V}' \quad , \tag{10.4}$$

where $\mathbf{U}' \, \mathbf{U} = \mathbf{I}$, $\mathbf{V}' \, \mathbf{V} = \mathbf{I}$, and Λ is a diagonal matrix with singular values λ_i in descending order. **U** and **V** contain the scores for the row and column categories. These scores are standardized as follows to give the principal coordinates for rows and columns:

$$\mathbf{X}_r = \mathbf{R}^{-1/2} \mathbf{U} \, \Lambda \tag{10.5}$$

$$\mathbf{X}_c = \mathbf{C}^{-1/2} \mathbf{V} \, \Lambda \tag{10.6}$$

where $\mathbf{R} = \mathbf{diag}\{p_{i+}\}$ and $\mathbf{C} = \mathbf{diag}\{p_{+j}\}$ are diagonal matrices containing the row and column totals of **P**, respectively. Other scalings of the coordinates are possible; however, the principal coordinates are most useful for graphical display because the distances among the row points and the distances among column points can be interpreted in the same way. Distances between the row and column points, however, are not directly meaningful in any scalings of the coordinates.

In correspondence analysis, this solution has the following properties:

□ The distances between row points or between column points in (Euclidean) space are equal to chi-square distances. That is, row (or column) points that are close together in the space correspond to rows (or columns) with similar profiles in the frequency table. The profiles of the marginal row and column frequencies are both projected to the origin in this space. Thus, when the profile of a given row (column) is similar to the marginal profile, the corresponding distance of the row (column) point to the origin is small.

□ The squared singular values, λ_i^2 (called *principal inertias*), are proportional to the contribution of the corresponding dimension to the overall χ^2 statistic for association between row and column variables. In the usual two-dimensional solution, the proportion of the χ^2 statistic explained is therefore $n(\lambda_1^2 + \lambda_2^2)/\chi^2$.

□ Assigning the scores in the first column of $\mathbf{X}_r$ to the row categories and in the first column of $\mathbf{X}_c$ to the column categories provides the maximum (Pearson) correlation, equal to λ_1, between these optimally scaled variables. As in canonical correlation analysis, the second and subsequent columns of scores have the greatest correlation (equal to the corresponding λ_i), subject to being uncorrelated with the optimally scaled variables of the previous columns.

Thus, correspondence analysis can be seen either as a technique for decomposing the overall χ^2 statistic into a small number of independent dimensions or as a method for quantifying or scaling the two categorical variables to produce the maximal correlations. In Version 6 of the SAS System, correspondence analysis is carried out by the CORRESP procedure. A SAS macro program, CORRESP, is also provided here for Version 5 users.

10.3.1   The CORRESP Procedure

In Version 6, correspondence analysis is performed using PROC CORRESP in SAS/STAT software. The procedure can accept data either in the form of a contingency table or in the form of raw responses on two or more categorical variables. An OUT= data set from PROC CORRESP contains the row and column coordinates, which can be plotted with the PLOT or GPLOT procedure. The procedure has many options for scaling row and column coordinates and for printing various statistics that aid interpretation. Only the basic use of the procedure is illustrated here. For further details, see Chapter 19, "The CORRESP Procedure," in the *SAS/STAT User's Guide, Version 6, Fourth Edition, Volume 1*.

Example: Hair and Eye Color

The program below reads the hair and eye color data into the data set COLORS and calls the CORRESP procedure. This example illustrates the use of the Annotate facility with PROC GPLOT to produce a labeled display of the correspondence analysis solution. To input a contingency table in the CORRESP step, the hair colors (columns) are specified as the variables in the VAR statement, and the eye colors (rows) are indicated as the ID variable.

```
title ' ';
data colors;
   input BLACK BROWN RED BLOND    EYE $;
   cards;
        68   119   26    7      Brown
        20    84   17   94      Blue
        15    54   14   10      Hazel
         5    29   14   16      Green
;
proc corresp data=colors out=coord short;
   var black brown red blond;
   id eye;
proc print data=coord;
   var _type_ eye dim1 dim2 quality;
```

The printed output from the CORRESP procedure is shown in Output 10.9. (Additional printed output, giving other statistics for the points, was suppressed in this example with the SHORT option.) The section labeled "Inertia and Chi-Square Decomposition" indicates that over 98 percent of the χ^2 statistic for association is accounted for by two dimensions, with most of that attributed to the first dimension. The coordinates (using the default scaling PROFILE=BOTH) are computed according to equations 10.5 and 10.6.

Output 10.9
CORRESP
Procedure Results
for Hair and Eye
Color Data

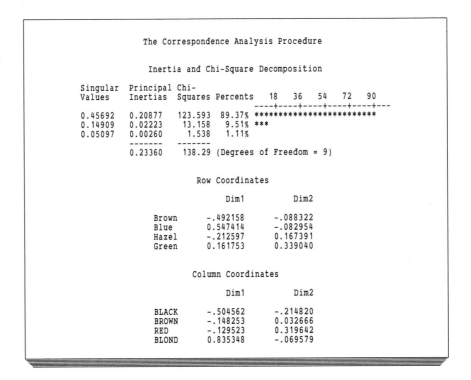

```
                  The Correspondence Analysis Procedure

                  Inertia and Chi-Square Decomposition

        Singular  Principal Chi-
        Values    Inertias  Squares Percents   18   36   54   72   90
                                             ----+----+----+----+----+---
        0.45692   0.20877   123.593  89.37%  **************************
        0.14909   0.02223    13.158   9.51%  ***
        0.05097   0.00260     1.538   1.11%
                  -------   -------
                  0.23360   138.29 (Degrees of Freedom = 9)


                          Row Coordinates

                            Dim1        Dim2

            Brown        -.492158    -.088322
            Blue        0.547414     -.082954
            Hazel        -.212597     0.167391
            Green       0.161753     0.339040


                        Column Coordinates

                            Dim1        Dim2

        BLACK            -.504562    -.214820
        BROWN            -.148253     0.032666
        RED              -.129523     0.319642
        BLOND           0.835348     -.069579
```

The singular values, λ_i, in equation 10.4, are also the canonical correlations between the optimally scaled categories. Thus, if the DIM1 scores for hair color and eye color are assigned to the 592 observations in the table, the correlation of these variables would be 0.4569. This interpretation provides an additional way to understand the strength of the association between hair and eye color in these data. The DIM2 scores give a second, orthogonal scaling of these two categorical variables, whose correlation would be 0.1491.

A plot of the row and column points can be constructed from the OUT= data set COORD requested in the PROC CORRESP step. The variables of interest in this example are shown in Output 10.10. Note that row and column points are distinguished by the variable _TYPE_. (The observation with _TYPE_='INERTIA' is ignored in this application.) The QUALITY variable gives a measure of the proportion of a point's distance from the origin, which is accounted for by the two-dimensional solution. This measure indicates that all the points are represented extremely well in two dimensions; the smallest value, 0.88 for hazel eyes, is quite adequate.

Output 10.10
Output Data Set
COORD from the
CORRESP
Procedure

```
     OBS   _TYPE_    EYE      DIM1      DIM2     QUALITY

      1   INERTIA              .         .         .
      2   OBS      Brown    -0.49216  -0.08832   0.99814
      3   OBS      Blue      0.54741  -0.08295   0.99993
      4   OBS      Hazel    -0.21260   0.16739   0.87874
      5   OBS      Green     0.16175   0.33904   0.94843
      6   VAR      BLACK    -0.50456  -0.21482   0.98986
      7   VAR      BROWN    -0.14825   0.03267   0.90633
      8   VAR      RED      -0.12952   0.31964   0.94507
      9   VAR      BLOND     0.83535  -0.06958   0.99963
```

A labeled PROC GPLOT display of the correspondence analysis solution is constructed with a DATA step to produce an Annotate data set LABELS from the COORD data set and a PROC GPLOT step shown below. In the PROC GPLOT step, it is crucial to scale the plot so that the number of data units per inch is the same for both dimensions. Otherwise, the distances on this plot would not be represented accurately. This is done with the AXIS statements: AXIS1 specifies a length and range that are both twice that in the AXIS2 statement, so that the ratio of data units to plot units is the same in both dimensions. Note that it is not necessary to make the axes the same length; rather, a one-unit distance must be the same physical length on both axes.

```
data label;
    set coord;
    xsys='2'; ysys='2';
    x = dim1; y = dim2;
    text = eye;
    size = 1.2;
    function='LABEL';

proc gplot data=coord;
    plot dim2 * dim1
        / anno=label frame
          href=0 vref=0 lvref=3 lhref=3
          vaxis=axis2 haxis=axis1
          vminor=1 hminor=1;
    axis1 length=6 in  order=(-1. to 1. by .5)
         label=(h=1.5          'Dimension 1');
    axis2 length=3 in  order=(-.5 to .5 by .5)
         label=(h=1.5 a=90 r=0 'Dimension 2');
    symbol v=none;
```

Interpreting the Correspondence Analysis Plot

The two-dimensional plot shown in Output 10.11 is interpreted in terms of the distances among the points representing the row categories and the distances among the points for the column categories. (Recall that the origin on the plot represents both the marginal row and column profiles.) Dimension 1 is a light—dark contrast between blond and black hair: these two columns differ most in the profiles of eye color frequency. This dimension also contrasts blue and brown eyes. The vertical axis reflects the difference between the profiles for red hair and the other hair colors, and a corresponding difference of green and hazel eyes from the profiles for brown and blue eyes. These differences represent a second, orthogonal dimension of association between hair color and eye color.

Notice that in the description above, we do not interpret the distance between the hair and eye color points. However, the relative positions of the row and column points on the coordinate axes can be interpreted as follows. From equation 10.4, it can be shown that each deviation from independence, d_{ij}, is proportional to the weighted sum, across all dimensions, of products of the coordinates, $x_r(i)$ for row i and $x_c(j)$ for column j (Greenacre 1984). Thus, pairs of row-column points with similar coordinates on both axes, such as blue-blond and brown-black, represent cells with large positive deviations from independence, that is $f_{ij} \!>\! > \! e_{ij}$; pairs of points with large coordinates of opposite sign, such as blue-black, reflect cells whose frequency is far less than independence predicts, $f_{ij} \ll e_{ij}$. Thus, we can say that the correspondence analysis plot shows that blue eyes go along with blond hair, while brown eyes are associated with black hair; but to be precise, this interpretation is based on comparing their signed distances from the origin rather than the distance between the row and column points.

Output 10.11   *Correspondence Analysis Plot for Hair-Eye Color Data*

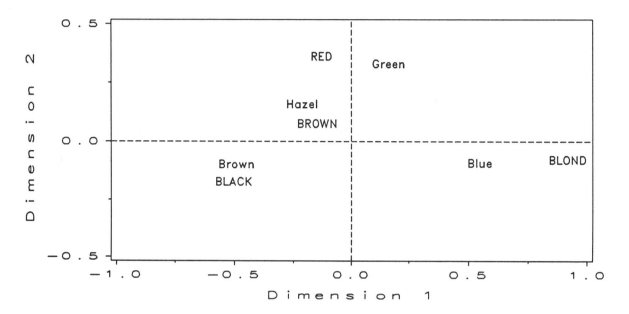

Comparing the correspondence analysis plot in Output 10.11 to the association plot in Output 10.3, we see they show similar patterns of association between hair and eye color. For example, both plots show the positive association of blue eyes with blond hair and brown eyes with black hair. Correspondence analysis, however, resolves the pattern of association and the overall χ^2 into orthogonal dimensions and summarizes the data in two dimensions, whereas the association mosaic plots display the whole pattern. Experience with these graphical displays seems to indicate that correspondence analysis is particularly useful for large tables, while the association plot and mosaic display, being more direct, are suitable for smaller tables.

10.3.2 The CORRESP Macro

For Version 5 of the SAS System, correspondence analysis can be carried out using the SAS macro program, CORRESP, provided in Section A1.6, "The CORRESP Macro." The macro is designed to be similar in use to PROC CORRESP and to produce results similar to those produced by the basic analysis illustrated in the previous section. As with the association plot, the calculations for correspondence analysis are done most easily with PROC IML. The parameters of the CORRESP macro are shown below:

```
%macro CORRESP(
      data=_LAST_,        /* Name of input data set                */
      var=,               /* Column variables                      */
      id=,                /* ID variable: row labels               */
      out=COORD,          /* output data set for coordinates       */
      anno=LABEL,         /* name of annotate data set for labels  */
      rowht=1,            /* height for row labels                 */
      colht=1             /* height for col labels                 */
      );
```

The input to the CORRESP macro must be a two-way table of frequencies. The columns of the table are specified by the variables listed in the VAR= parameter; the rows of the table are identified by the ID= parameter.

The CORRESP macro computes the row and column coordinates in two dimensions, which are returned in the variables DIM1 and DIM2 in the output data set specified by the OUT= parameter. The macro also constructs an Annotate data set to label the row and column points in the output data set specified by the ANNO= parameter. The OUT= data set is plotted in a separate step with PROC GPLOT, as illustrated in the examples below.

Example: Hair and Eye Color

The program below carries out the same analysis of the hair and eye color data using the CORRESP macro. The %CORRESP statement is very similar to the PROC CORRESP step in the previous example. The hair colors (columns) are specified as the variables (VAR=), and the eye colors (rows) are indicated as the ID= parameter.

```
%include corresp;                 /* On MVS, use %include ddname(file);*/
data colors;
   input BLACK BROWN RED BLOND    EYE $;
   cards;
         68   119   26    7       Brown
         20    84   17   94       Blue
         15    54   14   10       Hazel
          5    29   14   16       Green
;
%corresp (
    data=COLORS,
    var=BLACK BROWN RED BLOND,id=EYE,
    out=COORD,anno=LABEL,rowht=1.4,colht=1.4 );
```

Because the CORRESP macro also produces the Annotate data set LABEL, only a PROC GPLOT step is needed to produce the plot. The axes are equated in the same way as in the PROC CORRESP example.

```
proc gplot data=COORD  ;
   plot dim2 * dim1
        / anno=LABEL frame name='GB1011'
          href=0 vref=0 lvref=3 lhref=3
          vaxis=axis2 haxis=axis1
          vminor=1 hminor=1;
   axis1 length=5 in   order=(-1. to 1. by .5)
        label=(h=1.5          'Dimension 1') value=(h=1.3);
   axis2 length=2.5 in  order=(-.5 to .5 by .5)
        label=(h=1.5 a=90 r=0 'Dimension 2') value=(h=1.3);
   symbol v=none;
```

Output 10.12 shows the printed output from the CORRESP macro. It contains essentially the same information as the output from PROC CORRESP (Output 10.9).

Output 10.12
CORRESP Macro
Results for Hair
and Eye Color
Data

```
                          Hair & Eye color

                       Overall Association
                         ChiSq        DF

                         138.3         9


          Singular values, Inertia, and Chi-Square Decomposition

      Singular    Principal     Chi-        Percent      Cum %
       Values     Inertias     Squares

       0.4569      0.2088      123.593       89.37        89.37
       0.1491      0.0222       13.158        9.51        98.89
       0.0510      0.0026        1.538        1.11       100.00



                          Row Coordinates
                              DIM1      DIM2

                  Brown    -0.4922   -0.0883
                  Blue      0.5474   -0.0830
                  Hazel    -0.2126    0.1674
                  Green     0.1618    0.3390


                        Column Coordinates
                              DIM1      DIM2

                  BLACK    -0.5046   -0.2148
                  BROWN    -0.1483    0.0327
                  RED      -0.1295    0.3196
                  BLOND     0.8353   -0.0696
```

10.3.3 Multi-Way Tables

A three- or higher-way table can be analyzed by correspondence analysis in several ways. One approach, called *multiple correspondence analysis*, starts with a matrix, **Z**, of indicator (dummy) variables, one for each category of each variable, with one row for each respondent. For the hair-eye color data, this matrix would have $n = 592$ rows and 8 columns (four hair color categories and four eye color categories). The matrix $\mathbf{Z'} \mathbf{Z}$, called the *Burt matrix*, is then an 8×8 matrix that is analyzed in multiple correspondence analysis. This scheme generalizes to three or more variables as follows: For a three-way table, with $I+J+K$ categories, the indicator matrix has $(I+J+K)$ columns and the Burt matrix is of order $(I+J+K) \times (I+J+K)$. This analysis is provided by the MCA option of PROC CORRESP. See Example 2 in Chapter 19 of the *SAS/STAT User's Guide* for an example of multiple correspondence analysis of a seven-way table. However, for more than two variables, the geometric representation in multiple correspondence analysis is not a straightforward generalization of simple correspondence analysis. See Greenacre and Hastie (1987) for a more detailed discussion.

A second approach, which does maintain the same geometric interpretation, is called stacking. A three-way table of size $I \times J \times K$ can be sliced into I two-way tables, each $J \times K$. If the slices are concatenated vertically, the result is one two-way table, of size $(I \times J) \times K$. In effect, the first two variables are treated as a single composite variable, which represents the main effects and interaction between the original variables that were combined. Van der Heijden and de Leeuw (1985) discuss this use of correspondence analysis for multi-way tables and show how each way of slicing and stacking a contingency table corresponds to the analysis of a particular log-linear model.

Example: Suicide Rates

To illustrate this second method of analysis for three-way tables, we use data on suicide rates in West Germany, classified by age, sex, and method of suicide used. The data, from Heuer (1979, Table 1), have been discussed by van der Heijden and de Leeuw (1985) and others.* The table is a three-way contingency table, $2 \times 17 \times 9$. In this example, the table is structured as (2×17) rows by 9 columns, so the rows represent the joint effects of sex and age.

The data set SUICIDE is created in the following DATA step. Here we have defined a new variable, SEXAGE, to represent the age-sex combinations. Output 10.13 shows the data set created in this step.

```
title 'Suicide Rates by Age, Sex and Method';
data suicide;
   input sex $1 age poison cookgas toxicgas hang drown gun knife
                    jump other;
   length sexage $ 4;
   sexage=trim(sex)||trim(left(put(age,2.)));
cards;
M 10     4    0    0  247    1  17    1    6    0
M 15   348    7   67  578   22 179   11   74  175
M 20   808   32  229  699   44 316   35  109  289
M 25   789   26  243  648   52 268   38  109  226
   ... (observations omitted)
F 75   495    8    1  420  161   2   29  129   35
F 80   292    3    2  223   78   0   10   84   23
```

* From "Correspondence Analysis Used Complementary to Loglinear Analysis" by Peter G. M. van der Heijden, *Psychometrika*, Volume 50, Number 4, pp. 429—447. Copyright © 1985 by The Psychometric Society. Reprinted by permission of The Psychometric Society.

```
          F 85   113   4   0   83   14   0   6   34   2
          F 90    24   1   0   19    4   0   2    7   0
          ;
          proc print;
```

Output 10.13
SUICIDE Data Set

```
                     Suicide Rates by Age, Sex and Method

  SEXAGE SEX AGE POISON COOKGAS TOXICGAS HANG DROWN GUN KNIFE JUMP OTHER

   M10   M   10     4      0       0     247    1   17    1     6     0
   M15   M   15   348      7      67     578   22  179   11    74   175
   M20   M   20   808     32     229     699   44  316   35   109   289
   M25   M   25   789     26     243     648   52  268   38   109   226
   M30   M   30   916     17     257     825   74  291   52   123   281
   M35   M   35  1118     27     313    1278   87  293   49   134   268
   M40   M   40   926     13     250    1273   89  299   53    78   198
   M45   M   45   855      9     203    1381   71  347   68   103   190
   M50   M   50   684     14     136    1282   87  229   62    63   146
   M55   M   55   502      6      77     972   49  151   46    66    77
   M60   M   60   516      5      74    1249   83  162   52    92   122
   M65   M   65   513      8      31    1360   75  164   56   115    95
   M70   M   70   425      5      21    1268   90  121   44   119    82
   M75   M   75   266      4       9     866   63   78   30    79    34
   M80   M   80   159      2       2     479   39   18   18    46    19
   M85   M   85    70      1       0     259   16   10    9    18    10
   M90   M   90    18      0       1      76    4    2    4     6     2
   F10   F   10    28      0       3      20    0    1    0    10     6
   F15   F   15   353      2      11      81    6   15    2    43    47
   F20   F   20   540      4      20     111   24    9    9    78    47
   F25   F   25   454      6      27     125   33   26    7    86    75
   F30   F   30   530      2      29     178   42   14   20    92    78
   F35   F   35   688      5      44     272   64   24   14    98   110
   F40   F   40   566      4      24     343   76   18   22   103    86
   F45   F   45   716      6      24     447   94   13   21    95    88
   F50   F   50   942      7      26     691  184   21   37   129   131
   F55   F   55   723      3      14     527  163   14   30    92    92
   F60   F   60   820      8       8     702  245   11   35   140   114
   F65   F   65   740      8       4     785  271    4   38   156    90
   F70   F   70   624      6       4     610  244    1   27   129    46
   F75   F   75   495      8       1     420  161    2   29   129    35
   F80   F   80   292      3       2     223   78    0   10    84    23
   F85   F   85   113      4       0      83   14    0    6    34     2
   F90   F   90    24      1       0      19    4    0    2     7     0
```

The program below applies the CORRESP macro to the suicide data and uses a PROC GPLOT step to plot the OUT= data set COORD with the labels given in the Annotate data set LABELS. The axes are again equated by the use of LENGTH options in the AXIS statements. The two-dimensional plot is shown in Output 10.14. (A color version of this output appears as Output A3.51 in Appendix 3.)

```
%include SUICIDE;            /* On MVS, use %include ddname(file);*/
%include corresp;            /* On MVS, use %include ddname(file);*/
title h=1.5 'Suicide Rates by Age, Sex and Method';
%corresp (
    data=SUICIDE,
    var =POISON COOKGAS TOXICGAS HANG DROWN GUN KNIFE JUMP OTHER,
    id=SexAge,
    out=coord,anno=LABEL,
    rowht=.85, colht=1.1);
```

```
proc gplot data=coord ;
   plot Dim2 * Dim1 / anno=LABEL frame
                      href=0 vref=0 lvref=3 lhref=3
                      vaxis=axis2 haxis=axis1
                      vminor=1 hminor=1
                      name='GB1014' ;
   axis1 order=(-.75 to .75 by .25)  length=5.2 in   /* device dependent */
         label=(h=1.5 'Dimension 1: Sex');
   axis2 order=(-.75 to .75 by .25)  length=5.2 in   /* device dependent */
         label=(h=1.5 a=90 r=0 'Dimension 2: Age');
   symbol v=none;
```

Output 10.14   *Two-Dimensional Correspondence Analysis of Suicide Data*

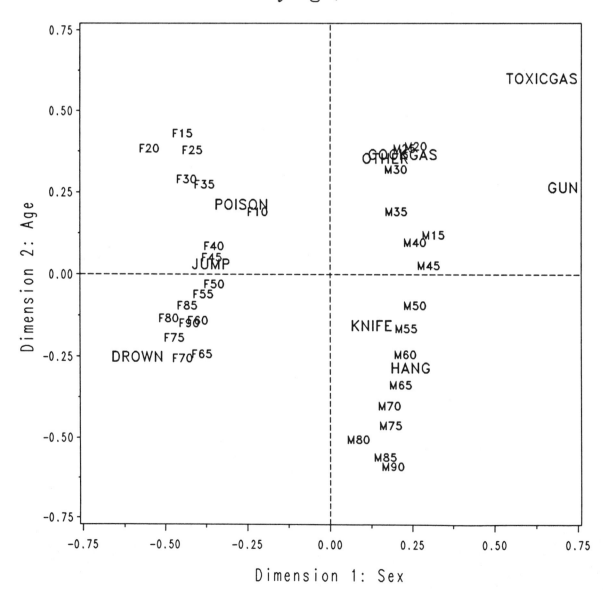

Output 10.15 gives the printed output from the analysis (excluding the row and column coordinates). The χ^2 statistic for "Overall Association" indicates a highly significant relation between method and the age-sex combinations. The second table in Output 10.15 lists the singular values of the deviations matrix and the contribution of dimension each to the overall χ^2 value. For these data, the two dimensions shown in Output 10.14 account for almost 90 percent of the overall association.

Output 10.15
Printed Output
from CORRESP
Macro for Suicide
Data

```
                     Suicide Rates by Age, Sex and Method

                          Overall Association
                            ChiSq      DF

                           10037.3     264


          Singular values, Inertia, and Chi-Square Decomposition

        Singular    Principal     Chi-      Percent     Cum %
         Values     Inertias    Squares

         0.3128      0.0978     5202.142     51.83       51.83
         0.2681      0.0719     3822.273     38.08       89.91
         0.1006      0.0101      537.879      5.36       95.27
         0.0628      0.0039      209.792      2.09       97.36
         0.0512      0.0026      139.249      1.39       98.75
         0.0344      0.0012       63.085      0.63       99.37
         0.0247      0.0006       32.415      0.32       99.70
         0.0239      0.0006       30.465      0.30      100.00
```

Dimension 1 in the plot (Output 10.14) separates males and females. This dimension, which by itself accounts for over 51 percent of the χ^2 value, indicates a strong difference between suicide profiles of males and females. The second dimension, accounting for 38 percent of the association, is mostly ordered by age with younger groups at the top and older groups at the bottom. Note also that the positions of the age groups are approximately parallel for the two sexes. Such a pattern indicates that sex and age do not interact in this analysis. The relation between the age-sex groups and methods of suicide can be interpreted in terms of similar distance and direction from the origin, which represents the marginal row and column profiles. Young males are more likely to commit suicide by gas or a gun, older males by knife or hanging; young females are more likely to ingest some toxic agent, older females to jump or drown.

The suicide data in Output 10.13 are actually a set of profiles for each age-sex combination in an eight-dimensional space; there are nine variables, but the row percentages must add up to 100 percent. Correspondence analysis gives a two-dimensional summary, which in this case accounts for 90 percent of the association between the age-sex groups and methods of suicide.

It should be emphasized that correspondence analysis, like the association and mosaic plots, may suggest patterns of association among the variables in contingency tables, but it does not establish whether those patterns are significant. Log-linear methods provide χ^2 tests of independence and generalized independence for multi-way tables, but do not reveal the pattern of nonindependence for significant effects in the model. These confirmatory methods should therefore be used to determine which variables are related; the exploratory methods of this chapter can then be used to understand how those variables are related.

Part 6

Appendices

Appendix 1 Macro Programs

A1.1   Introduction

Several of the plotting programs in the book were developed as general-purpose SAS macro programs that could be used for any set of data. Their generality, however, makes them more complex than is useful to describe completely in the body of the text. In most cases, therefore, the text discusses the application and use of the program. The programs are provided here to help make the techniques of statistical graphics more widely accessible.

The macro code for the programs listed below is contained in this appendix. All of these programs require SAS/GRAPH software. Additional requirements are indicated in parentheses after the program descriptions:

BIPLOT (A1.2)	biplot technique (SAS/IML software)
BOXANNO (A1.3)	marginal boxplot annotations for scatterplot
BOXPLOT (A1.4)	boxplots and notched boxplots
CONTOUR (A1.5)	elliptical contours for scatterplots
CORRESP (A1.6)	correspondence analysis (SAS/IML software)
DENSITY (A1.7)	histogram smoothing (SAS/IML software)
DOTPLOT (A1.8)	dot charts
LOWESS (A1.9)	lowess scatterplot smoothing (SAS/IML software)
NQPLOT (A1.10)	normal Q-Q plots
OUTLIER (A1.11)	multivariate outlier detection
PARTIAL (A1.12)	partial regression residual plots (SAS/IML software)
SCATMAT (A1.13)	scatterplot matrices
STARS (A1.14)	star plots (SAS/IML software)
SYMPLOT (A1.15)	symmetry transformation plots
TWOWAY (A1.16)	analysis of two-way tables (SAS/IML software)

All of the macro programs use keywords for the required and optional parameters. Default values (if any) are given after the equal sign in the parameter list below. Thus, it is only necessary to specify parameters that differ from the default value, and these parameters may be specified in any order in the macro call. The following conventions (which generally follow SAS software conventions in PROC steps) are used for naming parameters and default values:

DATA=   name of the input data set to be analyzed or plotted. The default is usually DATA=_LAST_, which means that the most recently created data set is the default if no data set is specified.

VAR=   names of the input variables in the DATA= data set. VAR=_NUMERIC_ means that all numeric variables in the data set are analyzed if no variables list is specified. Some of the macros understand a variable list specified as a range of variables, such as VAR=X1-X5 or VAR=EDUC--INCOME, as in the VAR statement. Others, especially those using the IML procedure, require the variables to be listed individually, for example VAR=X1 X2 X3 X4 X5.

ID=   name of an input variable used to label observations. There is usually no default ID variable.

CLASS=   name of an input variable used to classify observations into
GROUP=   groups.

OUT=   name of the output data set created by the macro. OUT=_DATA_ means that the output data set is named automatically according to the DATA*n* convention: the first such data set created is called DATA1, the second is called DATA2, and so on. Typically, this contains the data that are plotted. In some cases, the macro leaves it to the user to plot the OUT= data set, so that axis labels, values, and ranges can be controlled.

ANNO=   name of an input or output data set used for annotating the plot.

NAME=   name assigned to the graphs in the graphic catalog. The default is usually the name of the macro.

GOUT=   name of the graphics catalog used to save the output for later replay. The default is WORK.GSEG, which is erased at the end of your session. To save graphs in a permanent catalog, use a two-part name.

In the section "Program Listing" in each of the macro descriptions in this appendix, each of the macro parameters is briefly described in the comment at the right of the program line in the %MACRO statement. Where further description is necessary, it is given in the section "Parameters."

Note: Since the initial publication of *SAS System for Statistical Graphics,* many of the macro programs have been revised and extended. Documentation and source code for the current versions are available on the World Wide Web at http://www.math.yorku.ca/SCS/sssg/

A1.2   The BIPLOT Macro

The BIPLOT macro uses PROC IML to carry out the calculations for the biplot display described in Section 8.7, "Biplot: Plotting Variables and Observations Together." The program produces a printer plot of the observations and variables by default, but it does not produce a GPLOT procedure graph because a proper graph should equate the axes. Instead, the coordinates to be plotted and the labels for observations are returned in two data sets, specified by the parameters OUT= and ANNO=, respectively. A typical plotting step, using the default specifications OUT=BIPLOT and ANNO=BIANNO, is shown below:

```
proc gplot data=BIPLOT;
   plot dim2 * dim1 / anno=BIANNO frame href=0 vref=0
                      vaxis=axis2 haxis=axis1 vminor=1 hminor=1;
   axis1 length=5 in offset=(2) label=(h=1.5 'Dimension 1');
   axis2 length=5 in offset=(2) label=(h=1.5 a=90 r=0 'Dimension 2');
   symbol v=none;
```

The axes on the plot should be equated. See the examples in Section 8.7.

A1.2.1   Parameters

DATA=_LAST_	name of the input data set for the biplot.
VAR =_NUMERIC_	variables for the biplot. The list of variables must be given explicitly; the range notation X1-Xn cannot be used.
ID =ID	name of a character variable used to label the rows (observations) in the biplot display.
DIM =2	number of biplot dimensions.
FACTYPE=SYM	biplot factor type: GH, SYM, or JK.
SCALE=1	scale factor for variable vectors. The coordinates for the variables are multiplied by this value.
OUT =BIPLOT	output data set containing biplot coordinates.
ANNO=BIANNO	output data set containing Annotate labels.
STD=MEAN	specification for standardizing the data matrix before the singular value decomposition is computed. If STD=NONE, only the grand mean is subtracted from each value in the data matrix. This option is typically used when row and column means are to be represented in the plot, as in the diagnosis of two-way tables (Section 7.6.3, "Diagnostic Plot for Transformation to Additivity"). If STD=MEAN, the mean of each column is subtracted. This is the default,

and assumes that the variables are measured on commensurable scales. If STD=STD, the column means are subtracted and each column is standardized to unit variance.

PPLOT=YES   specification that indicates the first two dimensions should be plotted.

A1.2.2 The OUT= Data Set

The results from the analysis are saved in the OUT= data set. This data set contains two character variables (_TYPE_ and _NAME_) that identify the observations and numeric variables (DIM1, DIM2, . . .) that give the coordinates of each point.

The value of the _TYPE_ variable is 'OBS' for the observations that contain the coordinates for the rows of the data set and 'VAR' for the observations that contain the coordinates for the columns. The _NAME_ variable contains the value of the variable specified by the ID= parameter for the row observations and the variable name for the column observations in the output data set.

A1.2.3 Missing Data

The program makes no provision for missing values on any of the variables to be analyzed.

A1.2.4 Program Listing

```
%macro BIPLOT(
         data=_LAST_,      /* Data set for biplot                    */
         var =_NUMERIC_,   /* Variables for biplot                   */
         id =ID,           /* Observation ID variable                */
         dim =2,           /* Number of biplot dimensions            */
         factype=SYM,      /* Biplot factor type: GH, SYM, or JK     */
         scale=1,          /* Scale factor for variable vectors      */
         out =BIPLOT,      /* Output dataset: biplot coordinates     */
         anno=BIANNO,      /* Output dataset: annotate labels        */
         std=MEAN,         /* How to standardize columns: NONE|MEAN|STD*/
         pplot=YES);       /* Produce printer plot?                  */

%let factype=%upcase(&factype);
     %if &factype=GH  %then %let p=0;
%else %if &factype=SYM %then %let p=.5;
%else %if &factype=JK  %then %let p=1;
%else %do;
   %put BIPLOT: FACTYPE must be GH, SYM, or JK. "&factype" is not valid.;
   %goto done;
   %end;
```

```
Proc IML;
Start BIPLOT(Y,ID,VARS,OUT, power, scale);
   N = nrow(Y);
   P = ncol(Y);
   %if &std = NONE
       %then Y = Y - Y[:] %str(;);              /* remove grand mean */
       %else Y = Y - J(N,1,1)*Y[:,] %str(;);    /* remove column means */
   %if &std = STD %then %do;
     S = sqrt(Y[##,] / (N-1));
     Y = Y * diag (1 / S );
   %end;

   *-- Singular value decomposition:
        Y is expressed as U diag(Q) V prime
        Q contains singular values, in descending order;
   call svd(u,q,v,y);

   reset fw=8 noname;
   percent = 100*q##2 / q[##];
     *-- cumulate by multiplying by lower triangular matrix of 1s;
   j = nrow(q);
   tri= (1:j)`*repeat(1,1,j)  >= repeat(1,j,1)*(1:j) ;
   cum = tri*percent;
   Print "Singular values and variance accounted for",,
         q        [colname={' Singular   Values'} format=9.4]
         percent [colname={'Percent'} format=8.2]
         cum      [colname={'Cum %  '} format=8.2];

   d = &dim ;
   *-- Extract first  d  columns of U & V, and first  d  elements of Q;
   U = U[,1:d];
   V = V[,1:d];
   Q = Q[1:d];

   *-- Scale the vectors by QL, QR;
   * Scale factor 'scale' allows expanding or contracting the variable
     vectors to plot in the same space as the observations;
   QL= diag(Q ## power );
   QR= diag(Q ## (1-power));
   A = U * QL;
   B = V * QR # scale;
   OUT=A // B;

   *-- Create observation labels;
   id = id // vars`;
   type = repeat({"OBS "},n,1) // repeat({"VAR "},p,1);
   id  = concat(type, id);

   factype = {"GH" "Symmetric" "JK"}[1 + 2#power];
   print "Biplot Factor Type", factype;
```

```
      cvar = concat(shape(("DIM"),1,d), char(1:d,1.));
      print "Biplot coordinates",
            out[rowname=id colname=cvar];
      %if &pplot = YES %then
      call pgraf(out,substr(id,5),'Dimension 1', 'Dimension 2', 'Biplot');
         ;
      create &out  from out[rowname=id colname=cvar];
      append from out[rowname=id];
   finish;

      use &data;
      read all var{&var} into y[colname=vars rowname=&id];
      power = &p;
      scale = &scale;
      run biplot(y, &id,vars,out, power, scale );
      quit;

    /*----------------------------------*
    |  Split ID into _TYPE_ and _NAME_ |
     *---------------------------------*/
   data &out;
      set &out;
      drop id;
      length _type_ $3 _name_ $16;
      _type_ = scan(id,1);
      _name_ = scan(id,2);
    /*-------------------------------------------------*
    | Annotate observation labels and variable vectors |
     *------------------------------------------------*/
   data &anno;
      set &out;
      length function text $8;
      xsys='2'; ysys='2';
      text = _name_;

      if _type_ = 'OBS' then do;          /* Label the observation   */
         color='BLACK';
         x = dim1; y = dim2;
         position='5';
         function='LABEL   '; output;
         end;
```

```
            if _type_ = 'VAR' then do;        /* Draw line from    */
               color='RED  ';
               x = 0; y = 0;                   /* the origin to     */
               function='MOVE'    ; output;
               x = dim1; y = dim2;             /* the variable point */
               function='DRAW'    ; output;
               if dim1 >=0
                  then position='6';           /* left justify      */
                  else position='2';           /* right justify     */
               function='LABEL   '; output;    /* variable name     */
               end;
         %done:
         %mend BIPLOT;
```

A1.3   The BOXANNO Macro

The BOXANNO program contains two SAS macros to annotate a scatterplot with marginal boxplots of one or more of the variables plotted with either the GPLOT or the G3D procedure.

BOXAXIS   creates an Annotate data set to draw a boxplot for one axis on a two- or three-dimensional scatterplot.

BOXANNO   uses two calls to the BOXAXIS macro to create an Annotate data set for boxplots on both axes.

Use the BOXANNO macro to draw the boxplots for both variables on a scatterplot. For a PROC G3D scatterplot, use one call to BOXANNO for two of the variables and BOXAXIS for the third. See the examples in Section 4.5, "Enhanced Scatterplots."

A1.3.1   Parameters for BOXAXIS

DATA=_LAST_   name of the input data set.

OUT=_DATA_   name of the output Annotate data set.

VAR=   variable for which a boxplot is constructed.

BAXIS=X   axis on which the boxplot goes. It must be X, Y, or Z.

OAXIS=Y   the other axis in the plot.

PAXIS=Z   the third axis (ignored in PROC GPLOT).

BOXWIDTH=4   width of the box in percent of the data range.

POS=98   position of the center of the box on OAXIS in data percent. POS−BOXWIDTH/2 and POS+BOXWIDTH/2 must both be between 0 and 100.

A1.3.2 Parameters for BOXANNO

DATA=_LAST_   data set to be plotted

XVAR=   horizontal variable

YVAR=   vertical variable

OUT=BOXANNO   output Annotate data set

A1.3.3 Program Listing

```
%macro BOXAXIS(
        data=_LAST_,        /* Input dataset                    */
        out=_DATA_,         /* Output Annotate dataset          */
        var=,               /* Variable to be plotted           */
        baxis=x,            /* Axis on which it goes- X, Y, or Z */
        oaxis=y,            /* The other axis in the plot        */
        paxis=z,            /* The 3rd axis (ignored in GPLOT)   */
        boxwidth=4,         /* width of box in data percent      */
        pos=98);            /* position of box on OAXIS 0<POS<100*/

%local bx by bz;            /* macro symbols for annotate x,y,z  */
%if ( &baxis = %str() or &oaxis = %str() ) %then %do;
    %put ERROR: Box (BAXIS) and other (OAXIS) axes must be specified;
%end;
%else %do;
   %if (&baxis=&oaxis) %then %do;
        %put ERROR: BAXIS (&BAXIS) cannot be the same as OAXIS (&OAXIS);
   %end;
%end;
 /*-----------------------------------*
 | Find median & quartiles          |
 *-----------------------------------*/
proc univariate data=&data noprint;
    var &var;
    output out=quartile
          n=n q1=q1 q3=q3 median=median qrange=iqr mean=mean;
run;
 /*-----------------------------------------------*
 | Find outside & farout points                 |
 *-----------------------------------------------*/
data plotdat;
    set &data;
    if _n_=1 then set quartile;
    retain q1 q3 iqr;
    keep &var outside;
    outside=1;
    if &var < (q1-1.5*iqr) or &var > (q3+1.5*iqr)
       then outside=2;
```

```
           if &var < (q1-3.0*iqr) or &var > (q3+3.0*iqr)
              then outside=3;
   run;
    /*--------------------------------------------------*
     |  Whiskers go from quartiles to most extreme values |
     |  which are *NOT* outside.                           |
     *--------------------------------------------------*/
   data whis;
       set plotdat;
       if outside = 1;
   proc univariate data=whis noprint;
       var &var;
       output out=whisk min=lo_whisk max=hi_whisk;
   run;
   data boxfile;
       merge quartile whisk;
   proc print data=boxfile;
    /*----------------------------------------------*
     |  Annotate data set to draw boxes & whiskers    |
     *----------------------------------------------*/
   %let bx = &oaxis;
   %let by = &baxis;
   %let bz = &paxis;
   data &out;
       set boxfile;
       drop n lo_whisk hi_whisk q1 q3 iqr median mean
            center halfwid;
       length function $8 text $8;
       halfwid= &boxwidth / 2;
   %if ( &pos > 50 ) %then %do;
       center= &pos - halfwid; %end;
   %else %do;
       center= &pos + halfwid; %end;
       &bx.sys = '1';          /* data percentage coordinates for 'other' */
       &by.sys = '2';          /* data value coordinates for box axis      */
   %if ( &paxis ¬= %str() ) %then %do;
       &bz.sys = '1';          /* data percentage coordinates for 3rd axis*/
       &bz     = 1  ;
   %end;
   &bx =center-halfwid      ; &by = q1;       dot=1 ; link out; * box    ;
   &bx =center+halfwid      ; &by = q1;       dot=21; link out;
   &bx =center+halfwid      ; &by = q3;       dot=22; link out;
   &bx =center-halfwid      ; &by = q3;       dot=23; link out;
   &bx =center-halfwid      ; &by = q1;       dot=24; link out; * box    ;

   &bx =center-halfwid      ; &by = median  ; dot=3 ; link out; * median;
   &bx =center+halfwid      ; &by = median  ; dot=4 ; link out;

   &bx =center              ; &by = q1      ; dot=5 ; link out; * lo      ;
   &bx =center              ; &by = lo_whisk; dot=6 ; link out; * whisker;
   &bx =center              ; &by = q3      ; dot=7 ; link out; * hi      ;
   &bx =center              ; &by = hi_whisk; dot=8 ; link out; * whisker;
   &bx =center-halfwid/2    ; &by = lo_whisk; dot=9 ; link out;
```

```
&bx =center+halfwid/2      ; &by = lo_whisk; dot=10; link out;
&bx =center-halfwid/2      ; &by = hi_whisk; dot=11; link out;
&bx =center+halfwid/2      ; &by = hi_whisk; dot=12; link out;
&bx =center                ; &by = mean     ; dot=13; link out;
    return;

out:
   select;
      when (dot=1 | dot=3 | dot=5 | dot=7 | dot=9 | dot=11) do;
         line = .;
         function = 'MOVE';           output;
      end;
      when (dot=4 | dot=6 | dot=8 | dot=10 | dot=12
           | dot=21| dot=22| dot=23| dot=24 ) do;
         if dot=6 | dot=8
            then line = 3;
            else line = 1;
         function = 'DRAW';           output;
      end;
      when (dot = 13) do;
         text = 'STAR';
         function = 'SYMBOL';         output;
      end;
      otherwise;
   end;
   return;
run;
%mend boxaxis;

   /*------------------------------------------------------------*
    | BOXANNO macro - creates Annotate dataset for both X & Y |
    *------------------------------------------------------------*/
%macro boxanno(
     data=_last_,          /* Data set to be plotted  */
     xvar=,                /* Horizontal variable     */
     yvar=,                /* Vertical variable       */
     out=boxanno           /* Output annotate dataset */
     );

%boxaxis(
     data=&data, var=&xvar,
     baxis=x,    oaxis=y,   out=xanno);

%boxaxis(
     data=&data, var=&yvar,
     baxis=y,    oaxis=x,   out=yanno);
   /*----------------------------------------*
    | Concatenate the two Annotate datasets |
    *----------------------------------------*/
data &out;
     set xanno yanno;
%mend boxanno;
```

A1.4 The BOXPLOT Macro

The BOXPLOT macro draws side-by-side boxplots for the groups defined by one or more grouping (CLASS) variables in a data set.

A1.4.1 Parameters

DATA=_LAST_	name of the input data set.
CLASS=	grouping variables. The CLASS= variables can be character or numeric.
VAR=	name of the variable to be plotted on the ordinate.
ID=	character variable to identify each observation. If an ID= variable is specified, outside variables are labeled on the graph, using the first eight characters of the value of the ID variable (to reduce overplotting). Otherwise, outside points are not labeled.
WIDTH=.5	box width as proportion of the maximum. The default, WIDTH=.5, means that the maximum box width is half the spacing between boxes.
NOTCH=0	parameter that determines whether or not notched boxes are drawn. NOTCH=1 means draw notched boxes; NOTCH=0 means do not.
CONNECT=0	specification of the line style used to connect medians of adjacent groups. If CONNECT=0, the medians of adjacent groups are not to be connected.
F=0.5	for a notched boxplot, the parameter that determines the notch depth, from the center of the box, as a fraction of the halfwidth of each box. F must be between 0 and 1; the larger the value, the less deep is the notch.
FN=1	box width proportionality factor. The default, FN=1, means all boxes are the same width. If `FN=sqrt(n)` is specified, the boxes width will be proportional to the square root of the sample size of each group. Other functions of *n* are possible as well.
VARFMT=	name of a format for the ordinate variable.
CLASSFMT=	name of a format for the class variable. If the CLASS variable is a character variable, or there are two or more CLASS variables, the program maps the sorted values of the class variables into the integers 1, 2, . . . , *levels*, where *levels* is the number of distinct values of the class variables. A format provided for CLASSFMT should therefore provide labels corresponding to the numbers 1, 2, . . . , *levels*.

VARLAB=	label for the ordinate variable. If the label is not specified, the ordinate is labeled with the variable name.
CLASSLAB=	label for the class variables used to label the horizontal axis.
YORDER=	tick marks, and range for ordinate, in the form YORDER = *low* TO *high* BY *tick*.
ANNO=	name of an (optional) additional Annotate data set to be used in drawing the plot.
OUT=BOXSTAT	name of the output data set containing statistics used in drawing the boxplot. There is one observation for each group. The variables are N, MEAN, MEDIAN, Q1, Q3, IQR, LO_NOTCH, HI_NOTCH, LO_WHISK, and HI_WHISK.
NAME=BOXPLOT	name assigned to the graph in the graphic catalog.

A1.4.2   GOPTIONS Required

If there are many groups or the formatted labels of group names are long, you may need to increase the HPOS= option to allow a sufficient number of character positions for the labels.

A1.4.3   Program Listing

```
                              /* Description of Parameters:       */
%macro BOXPLOT(               /* ------------------------         */
        data=_LAST_,          /* Input dataset                    */
        class=,               /* Grouping variable(s)             */
        var=,                 /* Ordinate variable                */
        id=,                  /* Observation ID variable          */
        width=.5,             /* Box width as proportion of maximum */
        notch=0,              /* =0|1, 1=draw notched boxes       */
        connect=0,            /* =0 or line style to connect medians*/
        f=0.5,                /* Notch depth, fraction of halfwidth */
        fn=1,                 /* Box width proportional to &FN    */
        varfmt=,              /* Format for ordinate variable     */
        classfmt=,            /* Format for class variable(s)     */
        varlab=,              /* Label for ordinate variable      */
        classlab=,            /* Label for class variable(s)      */
        yorder=,              /* Tick marks, range for ordinate   */
        anno=,                /* Addition to ANNOTATE set         */
        out=boxstat,          /* Output data set: quartiles, etc. */
        name=BOXPLOT          /* Name for graphic catalog entry   */
        );

options nonotes;
%let _DSN_ = %upcase(&DATA);
%if &classlab = %str() %then %let classlab = &class;
%let CLASS = %upcase(&CLASS);
```

```
proc sort data=&DATA;
   by &CLASS;
run;
%let clvars = %nvar(&class);


 /*---------------------------------------*
  | Determine if &CLASS is char or numeric |
  *---------------------------------------*/
%let cltype=;
proc contents data=&DATA out=work noprint;
data _NULL_;
   length label2 $40;
   set work;
   if name="&CLASS"
      then if type=1 then call symput('CLTYPE', 'NUM');
                     else call symput('CLTYPE', 'CHAR');

   *-- find length of variable label and set y label angle --;
   %if &varlab ¬= %str() %then
      %str( label2 = "&varlab"; );
   %else
      %str( if name="&VAR" then label2=label; );
   if length(label2) <=8
      then call symput('YANGLE','');
      else call symput('YANGLE','a=90 r=0');
run;             /* Run required here */


 /*----------------------------------------------------------------*
  | If there is more than one class variables or class variable    |
  | is CHAR, create a numeric class variable, XCLASS. XCLASS        |
  | numbers the groups from 1,...number-of-groups. It is up to      |
  | the user to supply a format to associate proper group labels   |
  | with the XCLASS value.                                          |
  *----------------------------------------------------------------*/
%if ( &cltype=CHAR or &clvars > 1 ) %then %do;
   %let lclass = %scan( &CLASS, &clvars );
   data work;
      set &DATA;
      by &CLASS;
      if (first.&LCLASS) then xclass + 1;
      %if &cltype=CHAR and &clvars=1 and &classfmt=%str() %then
         %do;
         call symput('val'||left(put( xclass, 2. )), trim(&class) );
         %end;

   run;
   %let KLASS = xclass;
   %let data  = work;
   run;
%end;
%else %let KLASS = &CLASS;
```

```
   /*-------------------------------------------------*
   | Determine number of groups & quartiles of each |
   *-------------------------------------------------*/
proc means noprint data=&data;
    var &KLASS;
    output out=_grsum_ min=grmin max=grmax ;
    run;
proc univariate data=&data noprint;
    by &KLASS;
    var &VAR;
    output out=_qtile_
           n=n q1=q1  q3=q3  median=median qrange=iqr mean=mean;
data _qtile_;
    set _qtile_;
    By  &KLASS;
    Lo_Notch = Median - 1.58*IQR / sqrt(N);
    Hi_Notch = Median + 1.58*IQR / sqrt(N);
run;
data merged;
    merge &DATA _qtile_;
    by &KLASS;
  /*-------------------------------------------------*
   | Find outside & farout points                   |
   *-------------------------------------------------*/
data plotdat;
    set merged;
    keep &KLASS &VAR &ID outside;
    if &VAR ¬= .;
    outside=1;
    if &VAR < (Q1 -1.5*IQR) or &VAR > (Q3 +1.5*IQR)
       then outside=2;
    if &VAR < (Q1 -3.0*IQR) or &VAR > (Q3 +3.0*IQR)
       then outside=3;
run;
data _out_;
   set plotdat;
   if outside > 1 ;
proc sort data=_out_;
   by &KLASS &VAR ;
proc print data=_out_;
   id &ID &KLASS;
   title3 "Outside Observations in Data Set &_DSN_ ";
run;
  /*-------------------------------------------------*
   | If connnecting group medians, find them and append |
   *-------------------------------------------------*/
%if ( &connect ) %then %do;
   data connect;
      set _qtile_(keep=&KLASS Median
                  rename=(Median=&VAR));
      outside=0;
   proc append base=plotdat
              data=connect;
```

```
      run;
%end;

   /*------------------------------------------------------*
   |  Whiskers go from quartiles to most extreme values |
   |  which are *NOT* outside.                           |
   *------------------------------------------------------*/
data _in_;
   set plotdat;
   if outside = 1;               /* select inside points */
proc univariate data=_in_ noprint;
   by &KLASS;
   var &VAR;                     /* find min and max      */
   output out=_whisk_ min=lo_whisk max=hi_whisk;
run;
data &out;
   merge _qtile_ _whisk_ end=lastobs;
   by &KLASS;
   retain halfmax 1e23 fnmax -1e23;
   drop span halfmax fnmax offset grps;

   span = dif ( &KLASS );           /* x(k+1) - x(k) */
   if (_n_ > 1 )
      then halfmax = min( halfmax, span/2);
   fnmax = max( fnmax, &FN );

   if ( lastobs ) then do;
      if _n_=1 then halfmax=.5;
      call symput ('HALFMAX', left(put(halfmax,best.)) );
      put ' Maximum possible halfwidth is: ' halfmax /;
      call symput ('FNMAX',  left(put(fnmax,best.)) );
      grps=_n_;
      offset=max(5, 35-5*grps);
      call symput('OFFSET',left(put(offset,2.)) );
      put ' Number of groups: ' grps  'offset=' offset ;
   end;

proc print ;
   id &KLASS;
   title3 'BOXPLOT: Quartiles, notches and whisker values';
run;
   /*------------------------------------------------*
   | Annotate data set to draw boxes & whiskers   |
   *------------------------------------------------*/
data _dots_;
   set &out;
   retain halfmax &HALFMAX  k ;
   drop k halfmax halfwid hi_notch lo_notch iqr median mean q1  q3 ;
   drop grmin grmax ;
   if ( _n_ = 1) then do;
      set _grsum_;
      K = &WIDTH * HalfMax;
   end;
```

```
 halfwid = K * &FN / &FNMax ;
 length function text $8;
 XSYS = '2'; YSYS = '2';


 /*    Produce connect-the-dots X, Y pairs */
 X = &KLASS                       ; Y= Lo_Whisk ; dot =  1; link out;
 X = &KLASS                       ; Y= Q1       ; dot =  2; link out;
 X = &KLASS - halfwid             ; Y= Q1       ; dot =  3; link out;

%if ( &notch ) %then %do;
 X = &KLASS - halfwid             ; Y= Lo_Notch ; dot =  4; link out;
 X = &KLASS - (1-&F)*halfwid      ; Y= Median   ; dot =  5; link out;
 X = &KLASS - halfwid             ; Y= Hi_Notch ; dot =  6; link out;
%end;
 X = &KLASS - halfwid             ; Y= Q3       ; dot =  7; link out;
 X = &KLASS                       ; Y= Q3       ; dot =  8; link out;
 X = &KLASS                       ; Y= Hi_Whisk ; dot =  9; link out;
 X = &KLASS                       ; Y= Q3       ; dot = 10; link out;
 X = &KLASS + halfwid             ; Y= Q3       ; dot = 11; link out;
%if ( &notch ) %then %do;
 X = &KLASS + halfwid             ; Y= Hi_Notch ; dot = 12; link out;
%end;
 X = &KLASS+(1-&NOTCH*&F)*halfwid; Y= Median   ; dot = 13; link out;
 X = &KLASS-(1-&NOTCH*&F)*halfwid; Y= Median   ; dot = 14; link out;
 X = &KLASS+(1-&NOTCH*&F)*halfwid; Y= Median   ; dot = 15; link out;
%if ( &notch ) %then %do;
 X = &KLASS + halfwid             ; Y= Lo_Notch ; dot = 16; link out;
%end;
 X = &KLASS + halfwid             ; Y= Q1       ; dot = 17; link out;
 X = &KLASS                       ; Y= Q1       ; dot = 18; link out;

 X = &KLASS - halfwid/3           ; Y= Lo_Whisk ; dot = 19; link out;
 X = &KLASS - halfwid/3           ; Y= Hi_Whisk ; dot = 19; link out;
 X = &KLASS                       ; Y= Mean     ; dot = 20; link out;
 return;

out:
    Select;
       when ( dot=1 ) do;
          FUNCTION = 'MOVE';                    output;
          FUNCTION = 'POLY';                    output;
          End;
       when ( 1< dot <=18) do;
          FUNCTION = 'POLYCONT';                output;
          End;
       when ( dot=19) do;
          FUNCTION = 'MOVE';                    output;
          X = X + 2*halfwid/3 ;
          FUNCTION = 'DRAW';                    output;
          End;
```

```
          when ( dot=20) do;
             FUNCTION = 'MOVE';                          output;
             FUNCTION = 'SYMBOL'; TEXT='STAR';    output;
             End;
             Otherwise ;
       End;
       Return;
  run;


   /*------------------------------------------------------*
    |  Annotate data set to plot and label outside points |
    *------------------------------------------------------*/
  data _label_;
     set _out_;                              /* contains outliers only */
     by &KLASS;
     keep xsys ysys x y function text style position;
     length text function style $8;
     xsys = '2'; ysys = '2';
     y = &VAR;
     x = &KLASS ;
     function = 'SYMBOL';                        /* draw the point    */
     style = ' ';
     position = ' ';
     if OUTSIDE=2
        then do;  text='DIAMOND'; size=1.7; end;
        else do;  text='SQUARE '; size=2.3; end;
     output;
     %if &ID ¬= %str() %then %do;               /* if ID variable,   */
        if first.&KLASS then out=0;
        out+1;
        function = 'LABEL';                       /*  .. then label it */
        text = &ID;
        size=.9;
        style='SIMPLEX';
        x = &KLASS;
        if mod(out,2)=1                         /* on alternating sides*/
           then do; x=x -.05; position='4';  end;
           else do; x=x +.05; position='6';  end;
        output;
     %end;
  data _dots_;
     set _dots_ _label_ &anno ;
   /*----------------------------------------*
    | Clean up datasets no longer needed |
    *----------------------------------------*/
  proc datasets nofs nolist library=work memtype=(data);
     delete work _grsum_ merged _in_ _whisk_ _qtile_ _label_;
  options notes;
```

```
 /*-------------------------------------*
  | Symbols for connecting group medians |
  *-------------------------------------*/
%if &connect ¬= 0  %then %do;
   symbol1 C=BLACK V=NONE I=JOIN L=&connect r=1; /* connected medians  */
   symbol2 C=BLACK V=NONE R=3;                   /* rest done by annotate */
%end;
%else %do;
   symbol1 C=BLACK V=NONE R=3 i=none;            /* all done by annotate  */
%end;
title3;

proc gplot data=plotdat ;
   plot &VAR * &KLASS = outside
        / frame nolegend name="&name"
          vaxis=axis1 haxis=axis2 hminor=0
          annotate=_dots_;
   %if %length(&yorder) > 0 %then
      %let yorder = order=(&yorder);
   axis1
        &yorder
        value=(h=1.2) label =(&yangle h=1.5);
   axis2 value=(h=1.2) label =(h=1.5) offset=(&offset pct);
   %if &varfmt ¬= %str() %then %do; format &var    &varfmt ;    %end;
   %if &classfmt¬= %str() %then %do; format &KLASS &classfmt ;   %end;
   %if &varlab ¬= %str() %then %do; label  &var  = "&varlab";    %end;
   %if &classlab¬= %str() %then %do; label  &KLASS = "&classlab"; %end;
%mend boxplot;


  /*--------------------------------*
   | Count number of &CLASS variables |
   *--------------------------------*/
%macro nvar(varlist);
   %local wvar result;
   %let result = 1;
   %let wvar = %nrbquote(%scan( &varlist, &result));
   %do %until ( &wvar= );
       %let result = %eval( &result + 1);
       %let wvar = %nrbquote(%scan( &varlist, &result));
   %end;
   %eval( &result - 1)
%mend nvar;
```

A1.5 The CONTOUR Macro

The CONTOUR macro plots a bivariate scatterplot with a bivariate data ellipse for one or more groups.

A1.5.1 Parameters

DATA=_LAST_	name of the input data set.
X=	name of the X variable.
Y=	name of the Y variable.
GROUP=	group variable. If a GROUP= variable is specified, one ellipse is produced for each value of this variable in the data set. If no GROUP= variable is specified, a single ellipse is drawn for the entire sample. The GROUP= variable may be character or numeric.
PVALUE=.5	confidence coefficient(s) $(1-\alpha)$. This is the proportion of data from a bivariate normal distribution contained within the ellipse. Several values may be specified in a list (PVALUE=.5 .9, for example), in which case one ellipse is generated for each value.
STD=STDERR	error bar metric. STD=STDERR gives error bars equal to each mean plus or minus one standard error $(s/\sqrt{n}\,)$ for both variables. STD=STD gives error bars whose length is one standard deviation for both variables.
POINTS=40	number of points on each contour.
ALL=NO	parameter indicating whether the contour for the total sample should be drawn in addition to those for each group. If there is no GROUP= variable, ALL=YES just draws the ellipse twice.
OUT=CONTOUR	name of the output Annotate data set used to draw the ellipses, error bars, and group labels.
PLOT=YES	parameter that specifies whether the plot should be drawn. If YES, the macro plots the data together with the generated ellipses. Otherwise, only the output Annotate data set is generated.
I=NONE	SYMBOL statement interpolate option for drawing points. For example, I=RL includes the regression line as well.
NAME=CONTOUR	name assigned to the graphs in the graphic catalog.
COLORS=RED GREEN BLUE BLACK PURPLE YELLOW BROWN ORANGE	list of colors to use for each of the groups. If there are *g* groups, *g* colors should be specified if ALL=NO, and *g*+1 colors if ALL=YES. COLORS(*i*) is used for group *i*.

SYMBOLS=+
SQUARE
STAR−PLUS : $
=

list of symbols, separated by spaces, to use for plotting
points in each of the groups. SYMBOLS(*i*) is used for
group *i*.

A1.5.2   Program Listing

```
%macro CONTOUR(
        data=_LAST_,            /* input data set                */
        x=,                     /* X variable                    */
        y=,                     /* Y variable                    */
        group=,                 /* Group variable (optional)     */
        pvalue= .5,             /* Confidence coefficient (1-alpha) */
        std=STDERR,             /* error bar metric: STD or STDERR */
        points=40,              /* points on each contour        */
        all=NO,                 /* include contour for total sample?*/
        out=CONTOUR,            /* output data set               */
        plot=YES,               /* plot the results?             */
        i=none,                 /* SYMBOL statement interpolate opt */
        name=CONTOUR,           /* Name for graphic catalog entry */
        colors=RED GREEN BLUE BLACK PURPLE YELLOW BROWN ORANGE,
        symbols=+ square star -    plus   :     $    = );

%let all = %upcase(&all);
%if &x=%str() or &y=%str() %then %do;
   %put CONTOUR: X= and Y= variables must be specified;
   %goto DONE;
   %end;

proc iml;
start ellipse(c, x, y, npoints, pvalues, formean);
   /*-----------------------------------------------------------------*
    |  Computes elliptical contours for a scatterplot                 |
    |  C        returns the contours as consecutive pairs of columns  |
    |  X,Y      coordinates of the points                             |
    |  NPOINTS  scalar giving number of points around a contour       |
    |  PVALUES  column vector of confidence coefficients              |
    |  FORMEAN  0=contours for observations, 1=contours for means     |
    *-----------------------------------------------------------------*/

   xx = x||y;
   n  = nrow(x);
   *-- Correct for the mean --;
   mean = xx[+,]/n;
   xx = xx - mean @ j(n,1,1);

   *-- Find principal axes of ellipses --;
   xx = xx` * xx / (n-1);
   print 'Variance-Covariance Matrix',xx;
   call eigen(v, e, xx);
```

```
         *-- Set contour levels --;
         c =  2*finv(pvalues,2,n-1,0);
         if formean=1 then c = c / (n-1) ;
         print 'Contour values',pvalues c;
         a = sqrt(c*v[ 1 ] );
         b = sqrt(c*v[ 2 ] );

         *-- Parameterize the ellipse by angles around unit circle --;
         t = ( (1:npoints) - {1}) # atan(1)#8/(npoints-1);
         s = sin(t);
         t = cos(t);
         s = s` * a;
         t = t` * b;

         *-- Form contour points --;
         s = ( ( e*(shape(s,1)//shape(t,1) )) +
              mean` @ j(1,npoints*ncol(c),1) )` ;
         c = shape( s, npoints);
         *-- C returned as NCOL pairs of columns for contours--;
     finish;

     start dogroups(x, y, gp, pvalue);
        d  = design(gp);
        %if &all=YES %then %do;
           d = d || j(nrow(x),1,1);
        %end;
        do group = 1 to ncol(d);
           Print group;
           *-- select observations in each group;
           col = d[, group ];
           xg  = x[ loc(col), ];
           yg  = y[ loc(col), ];
           *-- Find ellipse boundary ;
           run ellipse(xyg,xg,yg,&points, pvalue, 0 );
           nr = nrow(xyg);

           *-- Output contour data for this group;
           cnames = { X Y PVALUE GP };
           do c=1 to ncol(pvalue);
              col=(2*c)-1 : 2*c ;
              xygp = xyg[,col] || j(nr,1,pvalue[c]) || j(nr,1,group);
              if group=1 & c=1
                 then create contour from xygp [colname=cnames];
              append from xygp;
           end;
        end;
     finish;

     *-- Get input data: X, Y, GP;
        use &data;
        read all var {&x}     into x [colname=lx];
        read all var {&y}     into y [colname=ly];
```

```
%if &group ¬= %str() %then %do;
   read all var {&group} into gp [colname=lg] ;
   %end;
%else %do;
   gp = j(nrow(x),1,1);
   %end;
   close &data;

*-- Find contours for each group;
   run dogroups(x, y, gp, { &pvalue} );
quit;

   /*---------------------------------*
    | Plot the contours using ANNOTATE |
    *---------------------------------*/
data contour;
   set contour ;
   by gp pvalue notsorted;
   length function color $8;
   xsys='2'; ysys='2';
   if first.pvalue then function='POLY';
                   else function='POLYCONT';
   color=scan("&colors",gp);
   line = 5;
run;
   /*---------------------------*
    | Crosses at Mean +- StdErr |
    *---------------------------*/
proc summary data=&data nway;
   class &group;
   var &x &y;
   output out=sumry mean=mx my &std=sx sy;
proc print;
data bars;
   set sumry end=eof;
   %if &group ¬= %str() %then %str(by &group;);
   length function color $8;
   retain g 0;
   drop _freq_ _type_ mx my sx sy g;
   xsys='2'; ysys='2';

   %if &group ¬= %str() %then %do;
      if first.&group then g+1;
   %end;
   color=scan("&colors",g);
   line=3;
   x = mx-sx; y=my;     function='MOVE'; output;
   x = mx+sx;           function='DRAW'; output;
   x = mx    ; y=my-sy; function='MOVE'; output;
              y=my+sy;  function='DRAW'; output;
```

```
*-- Write group label (convert numeric &group to character);
   %if &group ¬= %str() %then %do;
      length text $16;
      text = left(&group);
      position='3';
      size = 1.4;
      x = mx+.2*sx    ; y=my+.2*sy;
      function='LABEL'; output;
   %end;
   if eof then call symput('NGROUP',put(g,best.));
run;

data &out;
   set contour bars;

%if &group = %str() %then %let group=1;
%if %upcase(&plot)=YES %then %do;
   %gensym(n=&ngroup, h=1.2, i=&i, colors=&colors, symbols=&symbols );
   proc gplot data=&data  ;
      plot &y * &x = &group /
            annotate=&out
            nolegend frame
            vaxis=axis1 vminor=0
            haxis=axis2 hminor=0
            name="&name";
      axis1 offset=(3) value=(h=1.5) label=(h=1.5 a=90 r=0);
      axis2 offset=(3) value=(h=1.5) label=(h=1.5);
%end;
%done: ;
%mend contour;

   /*----------------------------------------------------*
    | Macro to generate SYMBOL statement for each GROUP |
    *----------------------------------------------------*/
%macro gensym(n=1, h=1.5, i=none,
               symbols=%str(- + : $ = X _ Y),
               colors=BLACK RED GREEN BLUE BROWN YELLOW ORANGE PURPLE);
   %*-- note: only 8 symbols & colors are defined;
    %*--    revise if more than 8 groups (recycle);
   %local chr col k;
   %do k=1 %to &n ;
      %let chr =%scan(&symbols, &k,' ');
      %let col =%scan(&colors, &k, ' ');
      symbol&k h=&h v=&chr c=&col i=&i;
   %end;
%mend gensym;
```

A1.6 The CORRESP Macro

The CORRESP macro performs correspondence analysis on a table of frequencies in a two-way (or higher-way) classification. The VAR= variables list specifies one of the classification variables. The observations in the input data set form the other classification variables.

The coordinates of the row and column points are output to the data set specified by the OUT= parameter. The labels for the points are output to the data set specified by the ANNO= parameter. See Section 10.3.2, "The CORRESP Macro," for details about plotting the results and equating axes.

A1.6.1   Parameters

DATA=_LAST_   name of the input data set

VAR=   column variables

ID=   ID variable: row labels

OUT=COORD   output data set for coordinates

ANNO=LABEL   name of the Annotate data set for row and column labels

ROWHT=1   height (in character cells) for the row labels

COLHT=1   height (in character cells) for the column labels

A1.6.2   The OUT= Data Set

The results from the analysis are saved in the OUT= data set. This data set contains two character variables (_TYPE_ and _NAME_) that identify the observations and two numeric variables (DIM1 and DIM2) that give the locations of each point in two dimensions.

The value of the _TYPE_ variable is 'OBS' for the observations that contain the coordinates for the rows of the table and 'VAR' for the observations that contain the coordinates for the columns. The _NAME_ variable contains the value of the variable specified by the ID= parameter for the row observations and the variable name for the column observations in the output data set.

A1.6.3 Program Listing

```
/*--------------------------------------------------------------*
 | CORRESP SAS - Macro for correspondence analysis using IML |
 *--------------------------------------------------------------*/
%macro CORRESP(
     data=_LAST_,        /* Name of input data set                 */
     var=,               /* Column variables                       */
     id=,                /* ID variable: row labels                */
     out=COORD,          /* output data set for coordinates        */
     anno=LABEL,         /* name of annotate data set for labels */
     rowht=1,            /* height for row labels                  */
     colht=1             /* height for col labels                  */
     );
  /*-----------------------------------------*
   | IML routine for Correspondence Analysis |
   *-----------------------------------------*/
Proc IML;
Start CORRESP(F,RowId,Vars);
  I = nrow(F);
  J = ncol(F);
  R = F[,+];                    * Row totals;
  C = F[+,];                    * Col totals;
  N = F[+];                     * Grand total;

  E  = R * C / N;              * Expected frequencies;
  D = (F - E) / sqrt(E);       * Standardized deviates;

  D = D / sqrt( N );
  DPD = D` * D;
  Inertia = trace(DPD);
  Chisq = N * Inertia;         * Total chi-square;
  DF = (I-1)*(J-1);
  reset noname;
  Print 'Overall Association', CHISQ[colname={'ChiSq'}]
                               DF[colname={DF} format=6.0];

  call eigen(values, vectors, dpd);
  k = min(I,J)-1;                   * number of non-zero eigenvalues;
  values  = values[1:k];
  cancorr = sqrt(values);          * singular values = Can R;
  chisq = n * values ;             * contribution to chi-square;
  percent = 100* values / inertia;
     *-- Cumulate by multiplying by lower triangular matrix of 1s;
  tri= (1:k)`*repeat(1,1,k)  >= repeat(1,k,1)*(1:k) ;
  cum = tri*percent;
  print 'Singular values, Inertia, and Chi-Square Decomposition',,
        cancorr [colname={' Singular   Values'} format=9.4]
        values  [colname={'Principal Inertias'} format=9.4]
        chisq   [colname={' Chi-    Squares'} format=9.3]
        percent [colname={'Percent'} format=8.2]
        cum     [colname={'Cum %  '} format=8.2];
```

```
      L = values[1:2];
      U = vectors[,1:2];
      Y = diag(1/sqrt(C/N)) * U * diag(sqrt(L));
      X = diag(N/R) * (F / N) * Y * diag(sqrt(1/L));
      Print 'Row Coordinates'    , X [Rowname=RowId Colname={DIM1 DIM2}];
      Print 'Column Coordinates', Y [Rowname=Vars  Colname={DIM1 DIM2}];

      OUT = X // Y;
      ID  = RowId // Vars`;
*     Call PGRAF(OUT,ID,'Dimension 1', 'Dimension 2', 'Row/Col Association');
      TYPE = repeat({"OBS "},I,1) // repeat({"VAR "},J,1);
      ID  = concat(TYPE, ID);
      Create &out  from OUT[rowname=ID colname={"Dim1" "Dim2"}];
      Append from OUT[rowname=ID];
    Finish;

    Use &data;
    Read all VAR {&var} into F [Rowname=&id Colname=Vars];

    Run CORRESP(F,&id,Vars);
    quit;
     /*--------------------------------*
      | Split ID into _TYPE_ and _NAME_ |
      *--------------------------------*/
    data &out;
       set &out;
       drop id;
       length _type_ $3 _name_ $16;
       _type_ = scan(id,1);
       _name_ = scan(id,2);
    proc print data=&out;
       id _type_ _name_;
     /*-----------------------------------------------*
      | Annotate row and column labels                |
      *-----------------------------------------------*/
    data &anno;
       set &out;
       length function text $8;
       xsys='2'; ysys='2';
       text = _name_;
       style='DUPLEX';
       x = dim1; y = dim2;

       if _type_ = 'OBS' then do;
          size= &rowht ;
          color='BLACK';
          position='5';
          function='LABEL   '; output;
          end;
```

```
        if _type_ = 'VAR' then do;
           color='RED  ';
           size= &colht;
           if dim1 >=0
              then position='6';              /* left justify    */
              else position='4';              /* right justify   */
           function='LABEL   '; output;
           end;
     %mend CORRESP;
```

A1.7 The DENSITY Macro

The DENSITY macro calculates a nonparametric density estimate of a data distribution as described in Section 3.4, "Histogram Smoothing and Density Estimation." The macro produces the output data set specified by the OUT= parameter, but leaves it to the user to call PROC GPLOT, so that the plot can be properly labeled. The output data set contains the variables DENSITY and WINDOW in addition to the variable specified by the VAR= parameter.

A typical plotting step, using the defaults, OUT=DENSPLOT and VAR=X, is shown below:

```
proc gplot data=densplot;
   plot density * X ;
   symbol1 i=join v=none;
```

A1.7.1 Parameters

DATA=_LAST_   name of the input data set.

OUT=DENSPLOT   name of the output data set.

VAR=X   name of the input variable (numeric).

WINDOW=   bandwidth (H) for kernel density estimate.

XFIRST=.   smallest X value at which density estimate is computed. If XFIRST = ., the minimum value of the VAR= variable is used.

XLAST=.   largest X value at which density estimate is computed. If XLAST = ., the maximum value of the VAR= variable is used.

XINC=.   step-size (increment) for computing density estimates. If XINC = ., the increment is calculated as values XINC = (XLAST-XFIRST)/60.

A1.7.2 Program Listing

```
%macro DENSITY(
     data=_LAST_,       /* Name of input data set          */
     out=DENSPLOT,      /* Name of output data set         */
     var=X,             /* Input variable (numeric)        */
     window=,           /* Bandwidth (H)                   */
     xfirst=.,          /* . or any real; smallest X value */
     xlast=.,           /* . or any real; largest X value  */
     xinc=. );          /* . or value>0; X-value increment */
                        /* Default: (XLAST-XFIRST)/60      */

data _in_;
   set &data;
   keep &var;
   if &var ¬= .;
proc sort data=_in_;
   by &var;

proc iml;

start WINDOW;    *-- Calculate default window width;
   mean = xa[+,]/n;
   css = ssq(xa - mean);
   stddev = sqrt(css/(n-1));
   q1 = floor(((n+3)/4) || ((n+6)/4));
   q1 = (xa[q1,]) [+,]/2;
   q3 = ceil(((3*n+1)/4) || ((3*n-2)/4));
   q3 = (xa[q3,]) [+,]/2;
   quartsig = (q3 - q1)/1.34;
   h  = .9*min(stddev,quartsig) * n##(-.2);   * Silvermans formula;
   finish;

start INITIAL;   *-- Translate parameter options;
   if xf=. then xf=xa[1,1];
   if xl=. then xl=xa[n,1];
   if xl <= xf then do;
      print 'Either largest X value chosen is too small';
      print 'or all data values are the same';
      stop;
      end;
   if dx=. | dx <= 0 then do;
      inc = (xl-xf)/60;
      rinc = 10 ## (floor(log10(inc))-1);
      dx = round(inc,rinc);
      end;
   if xf=xa[1,1] then xf=xf-dx;
   nx = int((xl-xf)/dx) +  3;
   finish;
```

```
*-- calculate density at specified x values;
start DENSITY;
   fnx = j(nx,3,0);
   vars = {"DENSITY"  "&VAR" "WINDOW"};
   create &out from fnx [colname=vars];
   sigmasqr = .32653;                     * scale constant for kernel  ;
   gconst = sqrt(2*3.14159*sigmasqr);
   nuh = n*h;
   x = xf - dx;
   do i = 1 to nx;
      x = x + dx;
      y = (j(n,1,x) - xa)/h;
      ky = exp(-.5*y#y / sigmasqr) / gconst;     * Gaussian kernel;
      fnx[i,1] = sum(ky)/(nuh);
      fnx[i,2] = x;
      end;
   fnx[,3] = round(h,.001);
   append from fnx;
   finish;

*-- Main routine ;
   use _in_;
   read all var "&var" into xa [colname=invar];
   n = nrow(xa);
   %if &window=%str() %then %do;
      run window;
      %end;
   %else %do;
      h = &window ;
      %end;

   xf    = &xfirst;
   xl    = &xlast;
   dx    = &xinc;
   run initial;
   run density;
   close &out;
   quit;
%mend DENSITY;
```

A1.8 The DOTPLOT Macro

The DOTPLOT macro produces grouped and ungrouped dot charts, as described in Section 2.5, "Dot Charts."

A1.8.1 Parameters

DATA=_LAST_	name of the input data set.
XVAR=	horizontal (response) variable.
XORDER=	plotting range for response. XORDER should be specified in the form XORDER = *low* TO *high* BY *step*.
XREF=	parameter that specifies the horizontal values at which reference lines are drawn for the response variable. If values are not specified, no reference lines are drawn.
YVAR=	vertical variable (observation label) for the dot chart. This should specify a character variable. At most, 16 characters of the value are used for the label.
YSORTBY=&XVAR	parameter that indicates how to sort the observations. The default, YSORTBY=&XVAR, indicates that observations should be sorted in ascending order of the response variable.
YLABEL=	label for *y* variable. If the label is not specified, the vertical axis is labeled with the name of the YVAR= variable.
GROUP=	vertical grouping variable. If it is specified, a grouped dot chart is produced with a separate panel for each value of the GROUP= variable.
GPFMT=	format for printing group variable value (include the period at the end of the format name).
CONNECT=DOT	parameter that specifies how to draw horizontal lines for each observation. Valid values are ZERO, DOT, AXIS, or NONE. The default, CONNECT=DOT, draws a dotted line from the *y* axis to the point. CONNECT=ZERO draws a line from an *x* value of 0 to the point. CONNECT=AXIS draws a line from the *y* axis to the plot frame at the maximum *x* value. CONNECT=NONE does not draw a line for the observation.
DLINE=2	line style for horizontal lines for each observation.
DCOLOR=BLACK	color of horizontal lines.
ERRBAR=	name of an input variable giving length of error bar for each observation. If the name is not specified, no error bars are drawn.
NAME=DOTPLOT	name assigned to the graph in the graphic catalog.

A1.8.2 GOPTIONS Required

The DOTPLOT macro plots each observation in a row of the graphics output area. Therefore, the VPOS= graphics option should specify a sufficient number of vertical character cells. The value for VPOS= should be specified in the following manner:

VPOS ≥ number of observations + number of groups + 10

A1.8.3 Program Listing

```
%macro dotplot(
        data=_LAST_,            /* input data set                         */
        xvar=,                  /* horizontal variable (response)         */
        xorder=,                /* plotting range of response             */
        xref=,                  /* reference lines for response variable  */
        yvar=,                  /* vertical variable (observation label)  */
        ysortby=&xvar,          /* how to sort observations               */
        ylabel=,                /* label for y variable                   */
        group=,                 /* vertical grouping variable             */
        gpfmt=,                 /* format for printing group variable     */
                                /* value (include the . at the end)       */
        connect=DOT,            /* draw lines to ZERO, DOT, AXIS, or NONE */
        dline=2,                /* style of horizontal lines              */
        dcolor=BLACK,           /* color of horizontal lines              */
        errbar=,                /* variable giving length of error bar    */
                                /* for each observation                   */
        name=DOTPLOT);          /* Name for graphic catalog entry         */

%if &yvar= %str() %then %do;
   %put DOTPLOT: Must specify y variable;
   %goto ENDDOT;
   %end;
%let connect=%upcase(&connect);
%if &ylabel = %str() %then %let ylabel=%upcase(&yvar);
%global nobs vref;
 /*----------------------------------------------------*
  | Sort observations in the desired order on Y axis |
  *----------------------------------------------------*/
%if &group ¬= %str() OR &ysortby ¬= %str() %then %do;
proc sort data=&data;
   by &group &ysortby;
%end;
```

```
   /*-------------------------------------------------------*
   | Add Sort_Key variable and construct macro variables |
   *-------------------------------------------------------*/
data _dot_dat;
  set &data;
  %if &group = %str() %then %do;
      %let group= _GROUP_;
      _group_ = 1;
  %end;
run;

data _dot_dat;
  set _dot_dat end=eof;
  retain vref ; drop vref;
  length vref $60;
      by &group;
  sort_key + 1;
  call symput( 'val' || left(put( sort_key, 3. )), trim(&yvar) );
  output;     /* output here so sort_key is in sync */

  if _n_=1 then vref='';
  if last.&group & ¬eof then do;
     sort_key+1;
     vref = trim(vref) || put(sort_key, 5.);
     call symput('val'|| left(put(sort_key, 3.)), '  ' );
     end;
  if eof then do;
     call symput('nobs', put(sort_key, 4.));
     call symput('vref', trim(vref));
     end;
run;

%if &nobs=0 %then %do;
   %put DOTPLOT: Data set &data has no observations;
   %goto ENDDOT;
   %end;
%makefmt(&nobs);

   /*-------------------------------------------------------*
   | Annotate data set to draw horizontal dotted lines |
   *-------------------------------------------------------*/
data _dots_;
   set _dot_dat;
      by &group;
   length function $ 8 text $ 20;
   text = ' ';
   %if &connect = ZERO
      %then %str(xsys = '2';) ;
      %else %str(xsys = '1';) ;
   ysys = '2';
   line = &dline;
   color = "&dcolor";
   y  = sort_key;
```

```
      x = 0;
      function ='MOVE'; output;

      function ='DRAW';
      %if &connect = DOT | &connect = ZERO
         %then %do;
            xsys = '2';
            x = &xvar; output;
         %end;
         %else %if &connect = AXIS
            %then %do;
            function='POINT';
            do x = 0 to 100 by 2;
               output;
                end;
            %end;

      %if &group ¬= _GROUP_ %then %do;
         if first.&group then do;
            xsys = '1';
            x = 98; size=1.5;
            function = 'LABEL';
            color='BLACK';
            position = 'A';
            %if &gpfmt ¬= %str()
               %then %str(text = put(&group, &gpfmt ) ;) ;
               %else %str(text = &group ;) ;
            output;
         end;
      %end;
%if &errbar ¬= %str() %then %do;
data _err_;
   set _dot_dat;
   xsys = '2'; ysys = '2';
   y = sort_key;
   x = &xvar - &errbar ;
   function = 'MOVE ';   output;
   text = '|';
   function = 'LABEL';   output;
   x = &xvar + &errbar ;
   function = 'DRAW ';   output;
   function = 'LABEL';   output;
data _dots_;
   set _dots_ _err_;
%end;
```

```
/*----------------------------------------------*
| Draw the dot plot, plotting formatted Y vs. X |
*----------------------------------------------*/
proc gplot data= _dot_dat  ;
   plot sort_key * &xvar
        /vaxis=axis1 vminor=0
         haxis=axis2 frame
         name="&name"
      %if &vref ¬= %str()
      %then    vref=&vref ;
      %if &xref ¬= %str()
      %then    href=&xref lhref=21 chref=red ;
         annotate=_dots_;
   label   sort_key="&ylabel";
   format  sort_key _yname_.;
   symbol1 v='-' h=1.4 c=black;
   axis1   order=(1 to &nobs by 1)    label=(f=duplex)
           major=none value=(j=r f=simplex);
   axis2   %if &xorder ¬= %str() %then order=(&xorder) ;
           label=(f=duplex) offset=(1);
%enddot:
%mend dotplot;


    /*-----------------------------------------*
    |  Macro to generate a format of the form  |
    |    1 ="&val1"  2="&val2" ...             |
    |  for observation labels on the y axis.   |
    *-----------------------------------------*/
%macro makefmt(nval);
   %if &sysver < 6 & "&sysscp"="CMS"
       %then %do;
          x set cmstype ht;           /* For SAS 5.18 on CMS, must   */
          x erase _yname_ text *;      /* erase format so that dotplot */
          x set cmstype rt;           /* can be used more than once   */
       %end;                          /* in a single SAS session      */
   %local i ;

   proc format;
       value _yname_
     %do i=1 %to &nval ;
        &i = "&&val&i"
        %end;
        ;
%mend makefmt;
```

A1.9 The LOWESS Macro

The LOWESS macro performs robust, locally weighted scatterplot smoothing as described in Section 4.4.2, "Lowess Smoothing." The data and the smoothed curve are plotted if PLOT=YES is specified. The smoothed response variable is returned in the output data set named by the OUT= parameter.

A1.9.1   Parameters

DATA=_LAST_	name of the input data set.
OUT=SMOOTH	name of the output data set. The output data set contains the X=, Y=, and ID= variables plus the variables _YHAT_, _RESID_, and _WEIGHT_. _YHAT_ is the smoothed value of the Y= variable, _RESID_ is the residual, and _WEIGHT_ is the combined weight for that observation in the final iteration.
X = X	name of the independent (X) variable.
Y = Y	name of the dependent (Y) variable to be smoothed.
ID=	name of an optional character variable to identify observations.
F = .50	lowess window width, the fraction of the observations used in each locally weighted regression.
ITER=2	total number of iterations.
PLOT=NO	parameter that specifies whether the plot is to be drawn. If PLOT=YES is specified, a high-resolution plot is drawn by the macro.
NAME=LOWESS	name assigned to the graph in the graphic catalog.

A1.9.2   Program Listing

```
%macro LOWESS(
        data=_LAST_,        /* name of input data set           */
        out=SMOOTH,         /* name of output data set          */
        x = X,              /* name of independent variable     */
        y = Y,              /* name of Y variable to be smoothed */
        id=,                /* optional row ID variable         */
        f = .50,            /* lowess window width              */
        iter=2,             /* total number of iterations       */
        plot=NO,            /* draw the plot?                   */
        name=LOWESS);       /* name for graphic catalog entry   */
proc sort data=&data;
   by &x;

proc iml;
start WLS( X, Y, W, B, I );     *-- Weighted least squares;
   x = j(nrow(x), 1, 1) || x;
   xpx = x` * diag( w ) * x;
   xpy = x` * diag( w ) * y;
```

```
      if abs(det(xpx)) > .00001
         then b    = inv(xpx) * xpy;
         else do;
            b = (y[loc(w¬=0)])[:] // { 0 } ;
            print 'Singular matrix for observation', I;
         end;
   finish;

   start MEDIAN( W, M);         * calculate median ;
      n = nrow( W );
      R = rank( W );
      i = int((n+1)/2);
      i =  i || n-i+1;
      M = W[ R[i] ];
      M = .5 # M[+];
   finish;

   start ROBUST( R, WTS);       * calculate robustness weights;
      run median(abs(R), M);
      W = R / (6 # M);          * bisquare function;
      WTS = (abs(W) < 1) # (1 - W##2) ## 2;
   finish;

   start LOWESS( X, Y, F, STEPS, YHAT, RES, DELTA);
      n = nrow(X);
      if n < 2 then do;
         yhat = y;
         return;
         end;
      q = round( f * n);       * # nearest neighbors;
      res  = y;
      yhat = J(n,1,0);
      delta= J(n,1,1);          * robustness weights;
      if steps <= 0 then steps=1;
      do it = 1 to steps;
         do i = 1 to n;
            dist = abs( x - x[i] );     * distance to each other pt;
            r = rank( dist );
            s = r; s[r]=1:n;
            near =  s[1:q] ;            * find the q nearest;
            nx = x [ near ];
            ny = y [ near ];
            d  = dist[ near[q] ];       * distance to q-th nearest;
            if d > 0 then do;
               u = abs( nx - x[i] ) / d ;
               wts = (u < 1) # (1 - u##3) ## 3; * neighborhood wts;
               wts = delta[ near ] # wts;
               if sum(wts[2:q]) > .0001 then do;
                  run wls( nx, ny, wts, b, i );
                  yhat[i] = (1 || x[i]) * b;     * smoothed value;
                  end;
               else yhat[i] = y[i];
            end;
```

```
              else do;
                  yhat[i] = ny [+] /q;
              end;
          end;
          res = y - yhat;
          run robust(res,delta);
       end;
   finish;

   *-- Main routine;
     use &data;
     %if &id.NULL=NULL %then %let rowid=;
     %else %let rowid=rowname=&id;
     read all var{&x &y} into xy[ colname=vars &rowid ];
     close &data;
     x = xy[,1];
     y = xy[,2];
     run lowess(x, y, &f, &iter, yhat, res, weight);
     xyres =x || y || yhat || res || weight;
     cname = vars || {"_YHAT_" "_RESID_" "_WEIGHT_" };
     print "Data, smoothed fit, residuals and weights",
           xyres[ colname=cname &rowid ];

   *-- Output results to data set &out ;
     xys =    yhat || res || weight;
     cname = {"_YHAT_" "_RESID_" "_WEIGHT_" };

     create &out from xys [ colname=cname ];
         append from xys;
   quit;
     /*-------------------------------------------*
     | Merge data with smoothed results.         |
     | (In a data step to retain variable labels) |
     *-------------------------------------------*/
   data &out;
     merge &data(keep=&x &y &id)
           &out ;
     label _yhat_ = "Smoothed &y"
           _weight_='Lowess weight';
   %if %upcase(&PLOT)=YES %then %do;
   proc gplot data=&out ;
     plot &y      * &x = 1
          _yhat_ * &x = 2
          / overlay frame
            vaxis=axis1 haxis=axis2
            name="&name" ;
     symbol1 v=+ h=1.5  i=none c=black;
     symbol2 v=none i=join c=red;
     axis1   label=(h=1.5 f=duplex a=90 r=0) value=(h=1.3);
     axis2   label=(h=1.5 f=duplex)          value=(h=1.3);
   %end;
   %mend LOWESS;
```

A1.10   The NQPLOT Macro

The NQPLOT macro produces normal Q-Q plots for single variables. The parameters MU= and SIGMA= determine how the comparison line, representing a perfect fit to a normal distribution, is estimated.

A1.10.1   Parameters

DATA=_LAST_	name of the input data set.
VAR=X	name of the variable to be plotted.
OUT=NQPLOT	name of the output data set.
MU=MEDIAN	estimate of the mean of the reference normal distribution. Specify MU=MEAN, MU=MEDIAN, or MU=*numeric value*.
SIGMA=HSPR	estimate of the standard deviation of the reference normal distribution. Specify SIGMA=STD, SIGMA=HSPR, or SIGMA=*numeric value*.
STDERR=YES	parameter that specifies whether standard errors should be plotted around curves.
DETREND=YES	parameter that specifies whether the detrended version should be plotted. If DETREND=YES is specified, the detrended version is plotted, too.
LH=1.5	height, in character cells, for the axis labels.
ANNO=	name of an optional input Annotate data set. This can be used to add labels or other information to the plot.
NAME=NQPLOT	name assigned to the graphs in the graphic catalog.
GOUT=	name of the graphic catalog used to store the graphs for later replay.

A1.10.2 Program Listing

```
*-----------------------------------------------------------------*
* SAS Macro for Normal Quantile-Comparison Plot                   *
*                                                                 *
* minimal syntax: %nqplot (data=dataset,var=variable);            *
*-----------------------------------------------------------------*;
%macro nqplot (
        data=_LAST_,     /* input data set                        */
        var=x,           /* variable to be plotted                */
        out=nqplot,      /* output data set                       */
        mu=MEDIAN,       /* est of mean of normal distribution:   */
                         /*   MEAN, MEDIAN or literal value       */
        sigma=HSPR,      /* est of std deviation of normal:       */
                         /*   STD, HSPR, or literal value         */
        stderr=YES,      /* plot std errors around curves?        */
        detrend=YES,     /* plot detrended version?               */
        lh=1.5,          /* height for axis labels                */
        anno=,           /* name of input annotate data set       */
        name=NQPLOT,     /* name of graphic catalog entries       */
        gout=);          /* name of graphic catalog               */

%let stderr=%UPCASE(&stderr);
%let sigma=%UPCASE(&sigma);
%let detrend=%UPCASE(&detrend);
%if &sigma=HSPR    %then %let sigma=HSPR/1.349;
%if &anno¬=%str()  %then %let anno=ANNOTATE=&anno;
%if &gout¬=%str()  %then %let gout=GOUT=&gout;


data pass;
  set &data;
  _match_=1;
  if &var ne . ;                   * get rid of missing data;

proc univariate noprint;          * find n, median and hinge-spread;
   var &var;
   output out=n1 n=nobs median=median qrange=hspr mean=mean std=std;
data n2; set n1;
   _match_=1;

data nqplot;
   merge pass n2;
   drop _match_;
   by _match_;

proc sort data=nqplot;
   by &var;
run;
```

```
data &out;
   set nqplot;
   drop sigma hspr nobs median std mean ;
   sigma = &sigma;
   _p_=(_n_ - .5)/nobs;                      * cumulative prob.;
   _z_=probit(_p_);                          * unit-normal Quantile;
   _se_=(sigma/((1/sqrt(2*3.1415926))*exp(-(_z_**2)/2)))
       *sqrt(_p_*(1-_p_)/nobs);              * std. error for normal quantile;
  _normal_= sigma * _z_ + &mu ;             * corresponding normal quantile;
   _resid_ = &var - _normal_;                * deviation from normal;
   _lower_ = _normal_ - 2*_se_;              * +/- 2 SEs around fitted line;
   _upper_ = _normal_ + 2*_se_;
   _reslo_  = -2*_se_;                       * +/- 2 SEs ;
   _reshi_   = 2*_se_;
  label _z_='Normal Quantile'
        _resid_='Deviation From Normal';
  run;
proc gplot data=&out &anno &gout ;
  plot &var    * _z_= 1
        _normal_ * _z_= 2
  %if &stderr=YES %then %do;
        _lower_ * _z_= 3
        _upper_ * _z_= 3 %end;
       / overlay frame
         vaxis=axis1 haxis=axis2
         hminor=1 vminor=1
         name="&name" ;
%if &detrend=YES %then %do;
  plot _resid_ * _z_= 1
  %if &stderr=YES %then %do;
        _reslo_ * _z_= 3
        _reshi_ * _z_= 3 %end;
       / overlay
         vaxis=axis1 haxis=axis2
         vref=0 frame
         hminor=1 vminor=1
         name="&name" ;
%end;
%let vh=1;             *-- value height;
%if &lh >= 1.5 %then %let vh=1.5;
%if &lh >= 2.0 %then %let vh=1.8;
  symbol1 v=+ h=1.1 i=none c=black l=1;
  symbol2 v=none   i=join  c=blue  l=3 w=2;
  symbol3 v=none   i=join  c=green l=20;
  axis1  label=(f=duplex a=90 r=0 h=&lh) value=(h=&vh);
  axis2  label=(f=duplex h=&lh) value=(h=&vh);
run;
%mend;
```

A1.11 The OUTLIER Macro

The OUTLIER macro calculates robust Mahalanobis distances for each observation in a data set. The results are robust in that potential outliers do not contribute to the distance of any other observations. A high-resolution plot can be constructed from the output data set; see the examples in Section 9.3, "Detecting Multivariate Outliers."

The macro makes one or more passes through the data. Each pass assigns 0 weight to observations whose DSQ values have Prob (χ^2)<PVALUE. The number of passes should be determined empirically so that no new observations are trimmed on the last step.

A1.11.1 Parameters

DATA=_LAST_   name of the data set to analyze.

VAR=_NUMERIC_   list of input variables.

ID=   name of an optional ID variable to identify observations.

OUT=CHIPLOT   name of the output data set for plotting. The robust squared distances are named DSQ. The corresponding theoretical quantiles are named EXPECTED. The variable _WEIGHT_ has the value 0 for observations identified as possible outliers.

PVALUE=.1   probability value of χ^2 statistic used to trim observations.

PASSES=2   number of passes of the iterative trimming procedure.

PRINT=YES   parameter that specifies whether the OUT= data set should be printed.

A1.11.2 Program Listing

```
%macro OUTLIER(
        data=_LAST_,      /* Data set to analyze       */
        var=_NUMERIC_,    /* input variables           */
        id=,              /* ID variable for observations */
        out=CHIPLOT,      /* Output dataset for plotting */
        pvalue=.1,        /* Prob < pvalue -> weight=0 */
        passes=2,         /* Number of passes          */
        print=YES);       /* Print OUT= data set?      */
```

```
/*----------------------------------------------------------*
 | Add WEIGHT variable. Determine number of observations |
 | and variables, and create macro variables.           |
 *----------------------------------------------------------*/
data in;
   set &data end=lastobs;
   array invar{*} &var;
   _weight_ = 1;                  /* Add weight variable */
   if ( lastobs ) then do;
      call symput('NOBS', _n_);
      call symput('NVAR', left(put(dim(invar),3.)) );
      end;

%do pass = 1 %to &PASSES;
   %if &pass=1 %then %let in=in;
             %else %let in=trimmed;
   /*----------------------------------------------------------------*
    | Transform variables to scores on principal components.        |
    | Observations with _WEIGHT_=0 are not used in the calculation, |
    | but get component scores based on the remaining observations. |
    *----------------------------------------------------------------*/
   proc princomp std noprint data=&in out=prin;
      var &var;
      freq _weight_;


   /*----------------------------------------------------------*
    | Calculate Mahalanobis D**2 and its probability value. For |
    | standardized principal components, D**2 is just the sum   |
    | of squares. Output potential outliers to separate dataset.|
    *----------------------------------------------------------*/
   data out1     (keep=pass case &id dsq prob)
        trimmed (drop=pass case );
      set prin ;
      pass = &pass;
      case = _n_;

      dsq = uss(of prin1-prin&nvar);    /* Mahalanobis D**2 */
      prob = 1 - probchi(dsq, &nvar);
      _weight_ = (prob > &pvalue);
      output trimmed;
      if _weight_ = 0 then do;
         output out1   ;
         end;
   run;
   proc append base=outlier data=out1;
%end;
   proc print data=outlier;
   title2 'Observations trimmed in calculating Mahalanobis distance';
```

```
/*-----------------------------------------*
 | Prepare for Chi-Square probability plot. |
 *-----------------------------------------*/
proc sort data=trimmed;
   by dsq;
data &out;
   set trimmed;
   drop prin1 - prin&nvar;
   _weight_ = prob > &pvalue;
   expected = 2 * gaminv(_n_/(&nobs+1), (&nvar/2));

%if &print=yes %then %do;
proc print data=&out;
   %if &id ¬=%str() %then
   %str(id &id;);
   title2 'Possible multivariate outliers have _WEIGHT_=0';
%end;

%if &ID = %str() %then %let SYMBOL='*';
               %else %let SYMBOL=&ID;
proc plot data=&out;
   plot dsq      * expected = &symbol
        expected * expected = '.'   /overlay hzero vzero;
title2 'Chi-Squared probability plot for multivariate outliers';
run;
%done:
proc datasets nofs nolist;
   delete outlier out1;
%mend outlier;
```

A1.12   The PARTIAL Macro

The PARTIAL macro draws partial regression residual plots as described in Section 5.5, "Partial Regression Plots." This implementation of the program is specific to Version 5 of the SAS System.

A1.12.1   Parameters

DATA = _LAST_	name of the input data set.
YVAR =	name of the dependent variable.
XVAR =	list of independent variables. The list of variables must be given explicitly; the range notation X1-Xn cannot be used.
ID =	name of an optional character variable used to label observations. If ID= is not specified, the observations are identified by the numbers 1, 2, ...
LABEL=INFL	parameter that specifies which points on the plot should be labeled with the value of the ID= variable. If LABEL=NONE, no points are labeled; if LABEL=ALL, all

points are labeled; otherwise (LABEL=INFL), only
potentially influential observations (those with large
leverage values or large studentized residuals) are labeled.

OUT =   name of the output data set containing partial residuals.
This data set contains $(p+1)$ pairs of variables, where p is
the number of XVAR= variables. The partial residuals
for the intercept are named UINTCEPT and VINTCEPT. If
XVAR=X1 X2 X3, the partial residuals for X1 are
named UX1 and VX1, and so on. In each pair, the U
variable contains the partial residuals for the independent
(X) variable, and the V variable contains the partial
residuals for the dependent (Y) variable.

GOUT=GSEG   name of graphic catalog used to store the graphs for later
replay.

NAME=PARTIAL   name assigned to the graphs in the graphic catalog.

A1.12.2   Computing Note

In order to follow the description in the text, the program computes one
regression analysis for each regressor variable (including the intercept). Velleman
and Welsch (1981) show how the partial regression residuals and other regression
diagnostics can be computed more efficiently—from the results of a single
regression using all predictors. They give an outline of the computations in the
MATRIX procedure language.

A1.12.3   Program Listing

```
/*--------------------------------------------------------------------*
 * PARTIAL SAS - IML macro program for partial regression residual    *
 *               plots.                                               *
 *--------------------------------------------------------------------*/
%macro PARTIAL(
        data = _LAST_,   /* name of input data set                 */
        yvar =,          /* name of dependent variable             */
        xvar =,          /* list of independent variables          */
        id =,            /* ID variable                            */
        label=INFL,      /* label ALL, NONE, or INFLuential obs    */
        out =,           /* output data set: partial residuals     */
        gout=gseg,       /* name of graphic catalog                */
        name=PARTIAL);   /* name of graphic catalog entries        */

%let label = %UPCASE(&label);
Proc IML;
start axes (xa, ya, origin, len, labels);
   col= 'BLACK';
   ox = origin[1,1];
   oy = origin[1,2];
   call gxaxis(origin,len,xa,7) color=col format='7.1';
   call gyaxis(origin,len,ya,7) color=col format='7.1';
```

```
        xo = ox + len/2 - length(labels[1])/2;
        yo = oy - 8;
        call gscript(xo,yo,labels[1]) color=col;
        xo = ox - 12;
        yo = oy + len;
        if nrow(labels)>1 | ncol(labels)>1
            then call gscript(xo,yo,labels[2]) angle=270 rotate=90 color=col;
    finish;

    *-----Find partial residuals for each variable-----;
    start partial(x, y, names, obs, uv, uvname );
        k = ncol(x);
        n = nrow(x);
        yname = names[,k+1];
        k1= k + 1;                      *-- number of predictors;
        x = j( n , 1 , 1) || x;    *-- add column of 1s;
        name1 = { 'INTCEPT' };
        names = name1 || names[,1:k];

        *----- module to fit one regression ----------;
        start reg (y, x, b, res, yhat, h, rstudent);
            n = nrow(x);
            p = ncol(x);
            xpx = x` * x;
            xpy = x` * y;
            xpxi= inv(xpx);
            b   = xpxi * xpy;
            yhat= x * b;
            res = y - yhat;
            h   = vecdiag(x * xpxi * x`);
            sse = ssq(res);
            sisq= j(n,1,sse) - (res##2) / (1-h);
            sisq= sisq / (n-p-1);
            rstudent = res / sqrt( sisq # (1-h) );
        finish;

    run reg( y,  x, b, res, yhat, hat, rstudent );
    print "Full regression";
    print "Regression weights" , b[ rowname=names ];
    lev = hat > 2*k1/n;
    if any( lev ) then do;
        l = loc(lev)`;
        xl=  x[l ,];
        Print "High leverage points", L XL [colname=names ];
    end;
    flag = lev | (rstudent > 2);
```

```
do i = 1 to k1;
   name = names[,i];
   reset noname;
   free others;
   do j = 1 to k1;
      if j ¬=i then others = others || j;
   end;
   run reg( y,     x[, others], by, ry, fy, hy, sry );
   run reg( x[,i],x[, others], bx, rx, fx, hx, srx );
   uv = uv || ry ||rx;
   uvname = uvname || concat({'U'},name)
                   || concat({'V'},name);

   if i>1 then do;
   /**-------------------------------**
    | Start IML graphics             |
    **-------------------------------**/
   %if &sysver < 6 %then %do;
      %let lib=%scan(&gout,1,'.');
      %let cat=%scan(&gout,2,'.');
      %if &cat=%str() %then %do;
          %let cat=&lib;
          %let lib=work;
          %end;
      call gstart gout={&lib &cat}
           name="&name"
           descript="Partial regression plot for &data";
   %end;
   %else %do;      /* Version 6 */
      call gstart("&gout");
      call gopen("&name",1,"Partial regression plot for &data");
   %end;

      labels = concat( {"Partial "}, name )  ||
             concat( {"Partial "}, yname ) ;
      run axes(rx,ry,{15 15}, 75, labels) ;
      call gpoint(rx,ry) color = 'BLACK';

      *-- Draw regression line from slope;
      xs = rx[<>] // rx[><];
      ys = b[i] * xs;
      call gdraw(xs, ys, 3, 'BLUE');

      *-- Mark influential points and large residuals;
      %if &label ¬= NONE %then %do;
         outy = ry[ loc(flag) ];
         outx = rx[ loc(flag) ];
         outl = obs[ loc(flag) ];
         call gpoint(outx, outy,19) color ='RED';
         call gtext(outx,outy,outl) color ='RED';
      %end;
```

```
          %if &label = ALL %then %do;
              outy = ry[ loc(¬flag) ];
              outx = rx[ loc(¬flag) ];
              outl = obs[ loc(¬flag) ];
              call gtext(outx,outy,outl) color ='BLACK';
          %end;
          call gshow;
        end;
    end;
    print "Partial Residuals", uv[ colname=uvname];
finish;  /* end of partial */

*-----read the data and prepare partial regression plots----;
    use &data;
    %if &id ¬= %str() %then %do;
        read all var{&xvar} into  x[ rowname=&id colname=xname ];
    %end;
    %else %do;
        read all var{&xvar} into  x[ colname=xname ];
        %let id = obs;
        obs = char(1:nrow(x),3,0);
    %end;
    read all var{&yvar } into  y[ colname=yname ];
    names = xname || yname;
    run partial(x, y, names, &id, uv, uvname);
    %if &out ¬= %str() %then %do;
        create &out from uv;
        append from uv;
    %end;
quit;
%mend PARTIAL;
```

A1.13 The SCATMAT Macro

The SCATMAT macro draws a scatterplot matrix for all pairs of variables specified in the VAR= parameter. The default variable list is all numeric variables (VAR=_NUMERIC_) in the data set, but the program will not do more than ten variables. This limit could easily be extended, but the plots would most likely be too small to see.

If a classification variable is specified with the GROUP= parameter, the value of that variable determines the shape and color of the plotting symbol. The macro GENSYM defines the SYMBOL statements for the different groups, which are assigned according to the sorted value of the grouping variable. The default values for the SYMBOLS= and COLORS= parameters allow for up to eight different plotting symbols and colors. If no GROUP= variable is specified, all observations are plotted using the first symbol and color.

A1.13.1 Parameters

DATA=_LAST_	name of the data set to be plotted.
VAR=_NUMERIC_	list of variables to be plotted. The macro allows the variables to be specified in any of the forms allowed in the VAR statement, that is, as a variable list (VAR=X1 X2 X3) or range (VAR=X1-X3 or VAR=PRICE--REPAIR).
GROUP=	name of an optional grouping variable used to define the plot symbols and colors.
SYMBOLS=%str(− + : $ = X _ Y)	list of symbols, separated by spaces, to use for plotting points in each of the groups. The *i*th element of SYMBOLS is used for group *i*.
COLORS=BLACK RED GREEN BLUE BROWN YELLOW ORANGE PURPLE	list of colors to use for each of the groups. If there are *g* groups, *g* colors should be specified. The *i*th element of COLORS is used for group *i*.
GOUT=GSEG	name of the graphics catalog used to store the final scatterplot matrix constructed by the GREPLAY procedure. The individual plots are stored in WORK.GSEG.

A1.13.2 Program Listing

```
%macro SCATMAT(
        data =_LAST_,           /* data set to be plotted        */
        var  =_NUMERIC_,        /* variables to be plotted - can be */
                                /* a list or X1-X4 or VARA--VARB   */
        group=,                 /* grouping variable (plot symbol) */
        symbols=%str(- + : $ = X _ Y),
        colors=BLACK RED GREEN BLUE BROWN YELLOW ORANGE PURPLE,
        gout=GSEG);             /* graphic catalog for plot matrix */

 options nonotes dquote;
*-- Parse variables list;
 %let var = %upcase(&var);
 data _null_;
 set &data (obs=1);
    %if %index(&var,-) > 0 or "&var"="_NUMERIC_" %then %do;

        * find the number of variables in the list and
           convert shorthand variable list to long form;
    length _vname_ $ 8 _vlist_ $ 200;
    array _xx_ &var;
    _vname_ = ' ';
```

```
        do over _xx_;
           call vname(_xx_,_vname_);
           if _vname_ ne "&group" then do;
              nvar + 1;
              if nvar = 1 then startpt = 1;
                           else startpt = length(_vlist_) + 2;
              endpt = length(_vname_);
              substr(_vlist_,startpt,endpt) = _vname_;
           end;
        end;
        call symput( 'VAR', _vlist_ );
     %end;
     %else %do;
        * find the number of variables in the list;
        nvar = n(of &var);
     %end;
     call symput('NVAR',trim(left(put(nvar,2.))));
   RUN;
%if &nvar < 2 or &nvar > 10 %then %do;
   %put Cannot do a scatterplot matrix for &nvar variables ;
   %goto DONE;
   %end;


   /*-------------------------------------------------------*
    | Determine grouping variable and plotting symbol(s) |
    *-------------------------------------------------------*/
%if &group = %str() %then %do;
   %let NGROUPS=1;
   %let plotsym=1;       /* SYMBOL for data panels  */
   %let plotnam=2;       /* for variable name panel */
   %end;
%else %do;
   %let plotsym=&group;
   *-- How many levels of group variable? --;
   proc freq data = &data;
      tables &group / noprint out=_DATA_;
   data _null_;
      set end=eof;
      ngroups+1;
      if eof then do;
         call symput( 'NGROUPS', put(ngroups,3.) );
      end;
    run;
    %let plotnam=%eval(&ngroups+1);
%end;

%gensym(n=&ngroups, ht=&nvar, symbols=&symbols, colors=&colors);
goptions NODISPLAY;          * device dependent;

title h=0.1 ' ';
%let plotnum=0;    * number of plots made;
%let replay = ;    * replay list;
```

```
%do i = 1 %to &nvar;                         /* rows */
   %let vi = %scan(&var , &i );
   proc means noprint data=&data;
       var &vi;
       output out=minmax min=min max=max;

   %do j = 1 %to &nvar;                       /* cols */
      %let vj = %scan(&var , &j );
      %let plotnum = %eval(&plotnum+1);
      %let replay  = &replay &plotnum:&plotnum ;
      %*put plotting &vi vs. &vj ;

      %if &i = &j %then %do;                  /* diagonal panel */
         data title;
            length text $8;
            set minmax;
            xsys = '1'; ysys = '1';
            x = 50; y = 50;
            text = "&vi";
            size = 2 * &nvar;
            function = 'LABEL';  output;

            x = 6; y = 6; position = '6';
            text = left(put(min, best6.));
            size = &nvar;
            output;

            x = 95; y = 95; position = '4';
            text = trim(put(max, best6.));
            size = &nvar;
            output;

         proc gplot data = &data;
            plot &vi * &vi = &plotnam
            / frame anno=title vaxis=axis1 haxis=axis1;
         axis1 label=none value=none major=(h=-%eval(&nvar-1))
               minor=none offset=(2);
         run;
      %end;

      %else %do;                             /* off-diagonal panel */
         proc gplot data = &data;
            plot &vi * &vj = &plotsym
            / frame nolegend vaxis=axis1 haxis=axis1;
         axis1 label=none value=none major=none minor=none offset=(2);
         run;
      %end;

   %end; /* cols */
%end;    /* rows */

goptions DISPLAY;          * device dependent;
```

```
%macro TDEF(nv, size, shift );
%* ------------------------------------------------------------;
%* Generate a TDEF statement for a scatterplot matrix          ;
%* Start with (1,1) panel in upper left, and copy it across & down;
%* ------------------------------------------------------------;
%local i j panl panl1 lx ly;

    TDEF scat&nv DES="scatterplot matrix &nv x &nv"
    %let panl=0;
    %let lx = &size;
    %let ly = %eval(100-&size);
    %do i = 1 %to &nv;
    %do j = 1 %to &nv;
        %let panl  = %eval(&panl + 1);
        %if &j=1 %then
          %do;
              %if &i=1 %then %do;      %* (1,1) panel;
                &panl/
                  ULX=0   ULY=100   URX=&lx URY=100
                  LLX=0   LLY=&ly    LRX=&lx LRY=&ly
                  %end;
              %else
                %do;                      %* (i,1) panel;
                    %let panl1 = %eval(&panl - &nv );
                  &panl/ copy= &panl1 xlatey= -&shift
                    %end;
          %end;
        %else
          %do;
                %let panl1 = %eval(&panl - 1);
                &panl/ copy= &panl1 xlatex= &shift
          %end;
    %end;
    %end;
      %str(;);      %* end the TDEF statement;
%mend TDEF;


proc greplay igout=gseg
             gout=&gout  nofs
            template=scat&nvar
            tc=templt ;
%if &nvar = 2 %then %do;
  TDEF scat2 DES="scatterplot matrix 2x2"
          1/ ULX=0  ULY=100   URX=52  URY=100
             LLX=0  LLY=52    LRX=52  LRY=52
          2/ copy=1 XLATEX= 48      /* Panels are numbered: */
          3/ copy=1 XLATEY=-48              /*    1   2   */
          4/ copy=3 XLATEX= 48;             /*    3   4   */
%end;

%if &nvar = 3 %then   %TDEF(&nvar,34,33);
%if &nvar = 4 %then   %TDEF(&nvar,25,25);
%if &nvar = 5 %then   %TDEF(&nvar,20,20);
```

```
%if &nvar = 6 %then   %TDEF(&nvar,17,16);
%if &nvar = 7 %then   %TDEF(&nvar,15,14);
%if &nvar = 8 %then   %TDEF(&nvar,13,12);
%if &nvar = 9 %then   %TDEF(&nvar,12,11);
%if &nvar =10 %then   %TDEF(&nvar,10,10);

   TREPLAY &replay;
%DONE:
  options notes;
%mend SCATMAT;



   /*-------------------------------------------------*
   |  Macro to generate SYMBOL statement for each GROUP |
   *-------------------------------------------------*/
%macro gensym(n=1, ht=1.5,
              symbols=%str(- + : $ = X _ Y),
              colors=BLACK RED GREEN BLUE BROWN YELLOW ORANGE PURPLE);
   %*-- note: only 8 symbols & colors are defined;
   %*--     revise if more than 8 groups (recycle);
   %local chr col k;
   %do k=1 %to &n ;
      %let chr =%scan(&symbols, &k,' ');
      %let col =%scan(&colors, &k, ' ');
      SYMBOL&k H=&HT V=&chr C=&col;
   %end;
   %let k=%eval(&n+1);
      SYMBOL&k v=none;
%mend gensym;
```

A1.14 The STARS Macro

The STARS macro draws a star plot of the multivariate observations in a data set, as described in Section 8.4, "Star Plots." Each observation is depicted by a star-shaped figure with one ray for each variable, whose length is proportional to the size of that variable.

A1.14.1 Missing Data

The scaling of the data in the PROC IML step makes no allowance for missing values.

A1.14.2 Parameters

DATA=_LAST_   name of the data set to be displayed.

VAR=_NUMERIC_   list of variables, in the order to be placed around the star, starting from angle=0 (horizontal) and proceeding counterclockwise.

ID=	character observation identifier variable (required).
MINRAY=.1	minimum ray length, 0<=MINRAY<1.
ACROSS=5	number of stars across a page.
DOWN=6	number of stars down a page. If the product of ACROSS and DOWN is less than the number of observations, multiple graphs are produced.

A1.14.3 Program Listing

```
%macro STARS(
        data=_LAST_,        /* Data set to be displayed            */
        var=_NUMERIC_,      /* Variables, as ordered around the    */
                            /* star from angle=0 (horizontal)      */
        id=,                /* Observation identifier  (char)      */
        minray=.1,          /* Minimum ray length, 0<=MINRAY<1     */
        across=5,           /* Number of stars across a page       */
        down=6              /* Number of stars down a page         */
        );
   /*-----------------------------------------------------*
    |  Scale each variable to range from MINRAY to 1.0    |
    *-----------------------------------------------------*/

PROC IML;
  reset;
  use &DATA;
  read all var{&VAR} into  X[ rowname=&ID colname=VARS ];
  n = nrow( x);
  min = J( n , 1 ) * X[>< ,];
  max = J( n , 1 ) * X[<> ,];
  c = &MINRAY ;
  X = c + ( 1 - c ) * ( X - min ) / ( max - min );
  create SCALED from X[ rowname=&ID colname=VARS ];
  append from X[ rowname= &ID ];
  quit;
run;
  %put &DATA dataset variables scaled to range &MINRAY to 1;


  /*----------------------------------------*
   |  Find out how many variables and obs.  |
   *----------------------------------------*/
data _null_;
  file print;
  array p(k) &var ;
  point=1;
  set scaled point=point nobs=nobs;
  do over p;           /* Loop to count variables*/
  end;
  k = k-1;
  put @10 "STARS plots for data set &DATA" /;
  put @10 'Number of variables   = ' k /;
```

```
      put @10 'Number of observations = ' nobs /;
      call symput('NV'  , put(k,    2.));
      call symput('NOBS', put(nobs,5.));
      stop;     /* Don't forget this ! */
   run;
    /*----------------------------------------------------*
     |  Text positions corresponding to rays of varying   |
     |  angle around the star                             |
     *----------------------------------------------------*/
   proc format;
      value posn    0-22.5  = '6'  /* left, centered */
                 22.6-67.5  = 'C'  /* left, above    */
                 67.6-112.5 = 'B'  /* centered, above */
                 112.6-157.5= 'A'  /* right, above    */
                 157.6-202.5= '4'  /* right, centered */
                 202.6-247.5= '7'  /* right, below    */
                 247.6-292.5= 'E'  /* centered, below */
                 other='F';        /* left, below     */
   run;


    /*------------------------------------------*
     |  Construct Annotate data set to draw and |
     |  label the star for each observation.    |
     *------------------------------------------*/
   data stars;
      length function varname $8;
      array p(k) &var ;

      retain s1-s&nv c1-c&nv;
      retain cols   &across      /* number of observations per row */
             rows   &down        /* number of rows per page        */
             xsys   '1'          /* use data percentage coordinates */
             ysys   '1'          /* for both X and Y               */
             lx ly page 0        /* cell X,Y and page  counters    */
             rx ry r;            /* cell radii                     */
      array s(k)   s1-s&nv;             /* sines of angle     */
      array c(k)   c1-c&nv;             /* cosines of angle   */

      drop cols rows rx ry cx cy s1-s&nv c1-c&nv &var;
      drop varname showvar;
      *--- precompute ray angles;
      if page=0 then do;
         do k= 1 to &nv;
            ang = 2 * 3.1415926 * (k-1)/&nv;
            s = sin( ang );
            c = cos( ang );
            p = 1.0;                      /* For variable key */
         end;
         x0 = 50; y0 = 50;
         r  = 30;
         size=2;
         text = 'Variable Assignment Key';
         x  = x0; y = 10;
```

```
               function = 'LABEL';   output;
               showvar=1;
               link DrawStar;                /* Do variable key */
               page+1;
               lx = 0;
               ly = 0;
      end;

      set scaled end=lastobs;
      label  =&id;
      showvar=0;

      *--- set size of one cell;
      if _n_=1 then do;
         rx= (100/cols)/2;
         ry= (100/rows)/2;
         r = .95 * min(rx,ry);
      end;

      /* (CX,CY) specify location of lower left corner */
      /*  as percent of data area                     */
      cx  = 100 * (lx) / cols;
      cy  = 100 * ((rows-1)-ly) / rows;

      function = 'LABEL';         /* Label the observation centered */
      size = round(r/12,.1);      /* at bottom of the cell          */
      size = min(max(.8,size),2); /* .8 <= SIZE <= 2                */
      text = &id;
      position='5';
      x =rx+cx; y=2+cy;
      output;

      x0 = cx + rx;                    /* Origin for this star */
      y0 = cy + ry;

      link drawstar;
      if ( lastobs ) then do;
         call symput('PAGES',trim(left(page)));
         put 'STARS plot will produce ' page 'page(s).';
      end;
      lx + 1;                 /* next column */
      if lx = cols then do;
         lx = 0;
         ly + 1; end;         /* next row    */
      if ly = rows then do;
         lx = 0;
         ly = 0;
         page + 1; end;       /* next page   */
      return;
```

```
DrawStar:
    *-- Draw star outline;
    do k = 1 to &nv;
        x = x0 + p * r * c;
        y = y0 + p * r * s;
        if k=1 then function = 'POLY';
                else function = 'POLYCONT';
        output;
    end;

    *-- draw rays from center to each point;
    *-- label with the variable name if showvar=1;
    do k = 1 to &nv;
        x=x0; y=y0;
        function='MOVE';   output;
        x = x0 + p * r * c;
        y = y0 + p * r * s;
        function = 'DRAW'; output;
        if showvar = 1 then do;
            ang = 2 * 3.1415926 * (k-1)/&nv;
            varname= ' ';
            call vname(p,varname) ;
            text = trim(left(varname));
            position = left(put(180*ang/3.14159,posn.));
            function = 'LABEL';  output;
        end;
    end;
  return;
run;                              /* Force SAS to do it (DONT REMOVE) */


  /*-----------------------------------------*
  |  Plot each page with GSLIDE:             |
  |   - Copy observations for current page   |
  |   - Draw plot                            |
  |   - Delete page data set                 |
  *-----------------------------------------*/
%do pg = 0 %to &pages;
   data slide&pg;                       /* Select current page to plot */
      set stars;
      if page = &pg;

   proc gslide annotate=slide&pg ;  /* Plot current page          */
   title;
   run;
   proc delete data=slide&pg;              /* Delete temporary data set  */
%end; /* end of page */

%mend STARS;
```

A1.15 The SYMPLOT Macro

The SYMPLOT macro produces any of the plots for diagnosing symmetry of a distribution described in Section 3.6, "Plots for Assessing Symmetry."

A1.15.1 Parameters

DATA=_LAST_   name of the input data to be analyzed.

VAR=   name of the variable to be plotted. Only one variable may be specified.

PLOT=MIDSPR   type of plots: NONE, or one or more of UPLO, MIDSPR, MIDZSQ, or POWER. One plot is produced for each keyword included in the PLOT= parameter.

TRIM=0   number or percent of extreme observations to be trimmed. If TRIM=*number* is specified, the highest and lowest *number* observations are not plotted. If TRIM=*percent* PCT is specified, the highest and lowest *percent*% of the observations are not plotted. The TRIM= option is most useful in the POWER plot.

OUT=SYMPLOT   name of the output data set.

NAME=SYMPLOT   name assigned to the graphs in the graphic catalog.

A1.15.2 Program Listing

```
%macro SYMPLOT(
        data=_LAST_,    /* data to be analyzed                 */
        var=,           /* variable to be plotted              */
        plot=MIDSPR,    /* Type of plot(s): NONE, or any of    */
                        /* UPLO, MIDSPR, MIDZSQ, or POWER      */
        trim=0,         /* # or % of extreme obs. to be trimmed */
        out=symplot,    /* output data set                     */
        name=SYMPLOT);  /* name for graphic catalog entry      */

%let plot = %upcase(&plot);

data analyze;
   set &data;
   keep &var;
   if &var =. then delete;

proc univariate data=analyze noprint;
   var &var;
   output out=stats n=nobs median=median;

%let pct  = %upcase(%scan(&trim,2));
```

```
data stats;
   set stats;
   trim = %scan(&trim,1) ;
   %if &pct = PCT %then %do;
   trim = floor( trim * nobs / 100 );
   %end;
   put 'SYMPLOT:' trim 'Observations trimmed at each extreme' ;

proc sort data=analyze out=sortup;
   by &var;
proc sort data=analyze out=sortdn;
   by descending &var;
   /* merge x(i) and x(n+1-i)    */
data &out;
   merge sortup(rename=(&var=frombot))         /* frombot = x(i)     */
         sortdn(rename=(&var=fromtop));        /* fromtop = x(n+1-i) */
   if _n_=1 then set stats;                     /* get nobs, median   */
   depth = _n_ ;
   if depth > trim ;                            /* trim extremes      */
   zsq = ( probit((depth-.5)/nobs) )**2;
   mid = (fromtop + frombot) / 2;
   spread = fromtop - frombot;
   lower = median - frombot;
   upper = fromtop - median;
   mid2  = mid - median;
   spread2 = (lower**2 + upper**2 ) / (4*median) ;
   if _n_ > (nobs+1)/2 then stop;
   label mid =  "Mid value of &var"
         lower= 'Lower distance to median'
         upper= 'Upper distance to median'
         zsq  = 'Squared Normal Quantile'
         mid2 = "Centered Mid Value of &var"
         spread2 = 'Squared Spread'
         ;
run;

%if %index(&PLOT,POWER) > 0 %then %do;
   *-- Annotate POWER plot with slope and power;
proc reg data=&out outest=parms noprint ;
   model mid2 = spread2;
data label;
   set parms(keep=spread2);
   xsys='1'; ysys='1';
   length text $12 function $8;
   x = 10;   y=90;
   function = 'LABEL';
   size = 1.4;
   style = 'DUPLEX';
   power = round(1-spread2, .5);
   position='6'; text = 'Slope: ' || put(spread2,f5.2);  output;
   position='9'; text = 'Power: ' || put(power,  f5.2);  output;
   %if &trim ¬= 0 %then %do;
   %if &pct=PCT %then %let pct=%str( %%);
```

```
            position='3'; text = 'Trim : ' || put(%scan(&trim,1),  f3. )||"&pct";
               output;
            %end;
         %end;

         %if %length(&PLOT) > 0 &
            &PLOT ¬= NONE %then %do;        /* Something to plot? */
      proc gplot data=&out  ;
         *-- Upper vs. Lower plot;
         %if %index(&PLOT,UPLO) > 0 %then %do;
         plot upper * lower = 1
              upper * upper = 2
              / overlay
                vaxis=axis1 haxis=axis2 vm=1 hm=1 name="&name";
         symbol1 v=+ c=black;
         symbol2 v=none i=join c=black l=20;
         %end;
         axis1 label=(h=1.5 a=90 r=0) value=(h=1.2) offset=(2);
         axis2 label=(h=1.5) value=(h=1.5);

         *-- Mid vs. Spread plot;
         %if %index(&PLOT,MIDSPR) > 0 %then %do;
         plot mid    * spread = 1
              median* spread = 2
              / overlay
                vaxis=axis1 haxis=axis2 vm=1 hm=1 name="&name";
         symbol1 v=+     i=rl   c=black;
         symbol2 v=none i=join c=red l=20;
         %end;
         *-- Mid vs. ZSQ     plot;
         %if %index(&PLOT,MIDZSQ) > 0 %then %do;
         plot mid    * zsq   = 1
              median* zsq   = 2
              / overlay
                vaxis=axis1 haxis=axis2 vm=1 hm=1 name="&name";
         symbol1 v=+     i=rl   c=black;
         symbol2 v=none i=join c=red l=20;
         %end;
         *-- Mid2 vs. Spread2    plot;
         %if %index(&PLOT,POWER) > 0 %then %do;
         plot mid2  * spread2= 1
                / overlay vref=0 lvref=20 cvref=red anno=label
                  vaxis=axis1 haxis=axis2 vm=1 hm=1 name="&name";
         symbol1 v=+     i=rl   c=black;
         symbol2 v=none i=join c=red l=20;
         %end;
      run;
      %end;
      %mend SYMPLOT;
```

A1.16   The TWOWAY Macro

The TWOWAY macro carries out analysis of two-way experimental design data with one observation per cell, including Tukey's 1 degree of freedom test for non-additivity as described in Section 7.6, "Displaying Two-Way Tables for $n=1$ Designs." Two plots may be produced: a graphical display of the fit and residuals for the additive model and a diagnostic plot for removable non-additivity.

A1.16.1   Parameters

DATA=_LAST_	name of the data set to be analyzed. One factor in the design is specified by the list of variables in the VAR= parameter. The other factor is defined by the observations in the data set.
VAR=	list of variables (columns of the table) to identify the levels of the first factor.
ID=	row identifier, a character variable to identify the levels of the second factor.
RESPONSE=Response	label for the response variable on the vertical axis of the two-way FIT plot.
PLOT=FIT DIAGNOSE	parameter that specifies the kinds of plots to be drawn. The PLOT parameter can contain one or more of the keywords FIT, DIAGNOSE, and PRINT. FIT requests a high-resolution plot of fitted values and residuals for the additive model. DIAGNOSE requests a high-resolution diagnostic plot for removable non-additivity. PRINT produces both of these plots in printed form.
NAME=TWOWAY	name assigned to the graphs in the graphic catalog.
GOUT=GSEG	name of the graphic catalog used to save the output for later replay. The default is WORK.GSEG, which is erased at the end of the session. To save graphs in a permanent catalog, a two-part name must be used.

A1.16.2   GOPTIONS Required

The HSIZE= and VSIZE= values in the GOPTIONS statement should be adjusted to equate the data units on the horizontal and vertical axes of the FIT plot so that the corners are square.

A1.16.3 Program Listing

```
%macro TWOWAY(
        data=_LAST_,            /* Data set to be analyzed      */
        var=,                   /* list of variables: cols of table*/
        id=,                    /* row identifier: char variable  */
        response=Response,      /* Label for response on 2way plot */
        plot=FIT DIAGNOSE,      /* What plots to do?            */
        name=TWOWAY,            /* Name for graphic catalog plots */
        gout=GSEG);             /* Name for graphic catalog     */

%if &var = %str() %then %do;
    %put ERROR: You must supply a VAR= variable list for columns;
    %goto DONE;
%end;
%if &id = %str() %then %do;
    %put ERROR: You must supply an ID= character variable;
    %goto DONE;
%end;
%let plot = %upcase(&plot);

proc iml;
   reset;
     use &data;
     read all into  y[colname=clabel rowname=&id] var { &var };
     r = nrow( y);
     c = ncol( y);
     rowmean = y[ , :];
     colmean = y[ : ,];
     allmean = y[ :  ];                          * grand mean ;

     roweff = rowmean - allmean;                 * row effects;
     coleff = colmean - allmean;                 * col effects;
     data = ( y         || rowmean || roweff ) //
            ( colmean || allmean || 0 ) //
            ( coleff  || 0       || 0 );

     rl   = &id    // {'COLMEAN','COLEFF'};
     cl   = clabel || {'ROWMEAN' 'ROWEFF'};
     print , data [ rowname = rl colname = cl ];

     jc = j( r , 1);
     jr = j( 1 , c);
     e = y - (rowmean * jr) - (jc * colmean) + allmean;
     print 'Interaction Residuals ',
           e [ rowname = rl colname = cl ];

     sse = e[ ## ];
     dfe = ( r - 1 ) # ( c - 1 );
     ssrow = roweff[## ,];  ssa = c * ssrow;
```

```
   sscol = coleff[ ,##];  ssb = r * sscol;
   product = ( roweff * coleff ) # y;
   d = product[ + ]  / ( ssrow # sscol );

   ssnon = ( ( product[ + ] ) ## 2 ) / ( ssrow # sscol );
   sspe = sse - ssnon;
   ss    = ssa   // ssb // sse // ssnon // sspe ;
   df    = (r-1) //(c-1)// dfe //  1     // dfe-1;

   ms = ss / df ; mspe=sspe/(dfe-1);
   f  = ms / (ms[{3 3 3 5 5},]);

   source= { "Rows","Cols","Error","Non-Add","Pure Err"};
   srt = "SOURCE   ";
   sst = "   SS    ";
   dft = "   DF    ";
   mst = "   MS    ";
   ft  = "   F     ";
   reset noname;
   print 'ANALYSIS OF VARIANCE SUMMARY TABLE ',
         'with Tukey 1 df test for Non - Additivity ',,
         source[ colname=srt ]
         ss[ colname=sst format=9.3]  df[ colname=dft format=best8. ]
         ms[ colname=mst format=9.3]  f [ colname=ft];

   re = ( roweff * jr );
   cf = ( jc * coleff )  + allmean;

   compare = ( roweff * coleff ) / allmean;
   compare = shape( e ,0 , 1) ||
             shape( compare ,0 , 1) ||
             shape( re ,0 ,1) ||
             shape( cf ,0, 1);
   vl = { 'RESIDUAL' 'COMPARE' 'ROWEFF' 'COLFIT'};
   create compare from compare[ colname=vl ];
   append from compare;

   /* Calculate slope of Residuals on Comparison values */
   /* for possible power transformation                 */
   xy = compare[,{2 1}];
   slope = sum(xy[,1] # xy[,2]) / xy[##,1];
   slope = d || slope || (1-slope);
   print 'D = Coefficient of alpha ( i ) * beta ( j ) ',
         'Slope of regression of Residuals on Comparison values',
         '1 - slope = power for transformation',,
         slope[ colname={D Slope Power}];
  ;
start twoway;
```

```
/*-----------------------------------------------------------*
| Calculate points for lines in two-way display of fitted    |
| value. Each point is (COLFIT+ROWEFF, COLFIT-ROWEFF).       |
*-----------------------------------------------------------*/
  do i=1 to r;
      clo  = coleff[><]+allmean;
      from = from // (clo-roweff[i] || clo+roweff[i]);
      chi  = coleff[<>]+allmean;
      to   = to   // (chi-roweff[i] || chi+roweff[i]);
      labl = labl || rl[i];
      end;
  do j=1 to c;
      rlo  = roweff[><];
      to   = to   // (coleff[j]+allmean-rlo || coleff[j]+allmean+rlo);
      rhi  = roweff[<>];
      from = from // (coleff[j]+allmean-rhi || coleff[j]+allmean+rhi);
      labl = labl || cl[j];
      end;


  /*---------------------*
  | Find large residuals |
  *---------------------*/
  do i=1 to r;
  do j=1 to c;
     if abs(e[i, j]) > sqrt(mspe) then do;
         from = from // ((cf[i,j]-re[i,j])||(cf[i,j]+re[i,j]));
         to   = to   // ((cf[i,j]-re[i,j])||(cf[i,j]+re[i,j]+e[i,j]));
         end;
       end; end;


  /*------------------------------*
  | Start IML graphics            |
  *------------------------------*/
  %if &sysver < 6 %then %do;
     %let lib=%scan(&gout,1,'.');
     %let cat=%scan(&gout,2,'.');
     %if &cat=%str() %then %do;
         %let cat=&lib;
         %let lib=work;
         %end;
     call gstart gout={&lib &cat}
         name="&name" descript="Two-way plot for dataset &data";
  %end;
  %else %do;      /* Version 6 */
     call gstart("&gout");
     call gopen("&name",1,"Two-way plot for dataset &data");
  %end;
```

```
      /**-------------------------------**
      | Find scales for the two-way plot |
      **-------------------------------**/
      call gport(({10 10, 90 90});
      call gyaxis( {10 10}, 80, from[,2]//to[,2], 5, 0, '5.0') ;
      call gscale( scale2, from[,2]//to[,2], 5);
      call gscript(3, 40, "&Response",'DUPLEX',3) angle=90;

      call gscale( scale1, from[,1]//to[,1], 5);
      window = scale1[1:2] || scale2[1:2];
      call gwindow(window);

      /*-------------------------------*
      | Draw lines for fit and residuals |
      *-------------------------------*/
      l = nrow(from);
      call gdrawl( from[1:r+c,],   to[1:r+c,])  style=1 color={"BLACK"};
      call gdrawl( from[r+c+1:l,], to[r+c+1:l,]) style=3 color={"RED"};

      /*---------------------------------------*
      | Plot row and column labels at margins; |
      *---------------------------------------*/
      xoffset=.04 * (to[<>,1]-to[><,1]);
      yoffset=0;
      do i=1 to r+c;
         if i>r then do;
            yoffset=-.04 * (to[<>,2]-to[><,2]);
            end;
         call gtext(xoffset+to[i,1],yoffset+to[i,2],labl[i]);
      end;
      call gshow;
      call gstop;
finish;

      %if %index(&plot,FIT) > 0 %then %do;
      run twoway;
      %end;
quit;

data compare;
   set compare;
   fit = colfit + roweff;
   data= fit + residual;
   diff= colfit - roweff;

   /* Print values for fit and diagnostic plots */
proc print data=compare;
   var data roweff colfit fit diff residual compare;
%if %index(&plot,PRINT) > 0 %then %do;
```

```
proc plot;
    plot data* diff = '+'
         fit * diff = '*'    / overlay;
proc plot;
    plot residual * compare / vpos=45;
%end;

%if %index(&plot,DIAGNOSE) > 0 %then %do;
proc gplot data=compare gout=&gout;
    plot residual * compare
       / vaxis=axis1 haxis=axis2
         vminor=1 hminor=1 name="&name" ;
    symbol1 v=- h=1.4 c=black i=rl;
    axis1 label=(a=90 r=0 h=1.5 f=duplex) value=(h=1.3);
    axis2 label=(h=1.5 f=duplex) value=(h=1.3);
    label residual = 'INTERACTION RESIDUAL'
          compare  = 'COMPARISON VALUE';
%end;

%DONE:
%mend TWOWAY;
```

Appendix **2** Data Sets

A2.1   Introduction

This appendix lists the major data sets used in this book. Program examples in the text that use these data sets are signaled by the %INCLUDE statement, with the name of the SAS file in uppercase. See the entries under *data sets*, *examples* in the index for examples using each set of data.

A2.2   The AUTO Data Set: Automobiles Data

The AUTO data set contains the following variables for 74 automobile models from the 1979 model year:

MODEL   make and model

ORIGIN   region of origin (America, Europe, or Japan)

PRICE   price in dollars

MPG   gas mileage in miles per gallon

REP77   repair records for 1977, rated on a five-point scale (5=best, 1=worst)

REP78   repair records for 1978, rated on a five-point scale (5＝best, 1＝worst)

HROOM   headroom in inches

RSEAT   rear seat clearance (distance from front seat back to rear seat back) in inches

TRUNK   trunk space in cubic feet

WEIGHT   weight in pounds

LENGTH   length in inches

TURN   turning diameter (clearance required to make a U-turn) in feet

DISPLA   engine displacement in cubic inches

GRATIO   gear ratio for high gear

Source: The data set is listed in Chambers et al. (1983, pp. 352—355). The original data come from various sources, primarily *Consumer Reports* (April, 1979) and the U. S. Environmental Protection Agency statistics on fuel consumption.*

```
data auto;
    input model $ 1-17 origin $ 20  a24 price
          mpg rep78 rep77 hroom rseat trunk weight length
          turn displa gratio;
    label model  = 'MAKE & MODEL'
          price  = 'PRICE'
          mpg    = 'MILEAGE'
          rep78  = 'REPAIR RECORD 1978'
          rep77  = 'REPAIR RECORD 1977'
          hroom  = 'HEADROOM (IN.)'
          rseat  = 'REAR SEAT (IN.)'
          trunk  = 'TRUNK SPACE (CU FT)'
          weight = 'WEIGHT (LBS)'
          length = 'LENGTH (IN.)'
          turn   = 'TURN CIRCLE (FT)'
          displa = 'DISPLACEMENT (CU IN)'
          gratio = 'GEAR RATIO';
    cards;
AMC CONCORD       A    4099 22 3 2 2.5 27.5 11 2930 186 40 121 3.58
AMC PACER         A    4749 17 3 1 3.0 25.5 11 3350 173 40 258 2.53
AMC SPIRIT        A    3799 22 . . 3.0 18.5 12 2640 168 35 121 3.08
AUDI 5000         E    9690 17 5 2 3.0 27.0 15 2830 189 37 131 3.20
AUDI FOX          E    6295 23 3 3 2.5 28.0 11 2070 174 36  97 3.70
BMW 320I          E    9735 25 4 4 2.5 26.0 12 2650 177 34 121 3.64
BUICK CENTURY     A    4816 20 3 3 4.5 29.0 16 3250 196 40 196 2.93
BUICK ELECTRA     A    7827 15 4 4 4.0 31.5 20 4080 222 43 350 2.41
BUICK LE SABRE    A    5788 18 3 4 4.0 30.5 21 3670 218 43 231 2.73
BUICK OPEL        A    4453 26 . . 3.0 24.0 10 2230 170 34 304 2.87
```

```
BUICK REGAL          A    5189 20 3 3 2.0 28.5 16 3280 200 42 196 2.93
BUICK RIVIERA        A   10372 16 3 4 3.5 30.0 17 3880 207 43 231 2.93
BUICK SKYLARK        A    4082 19 3 3 3.5 27.0 13 3400 200 42 231 3.08
CAD. DEVILLE         A   11385 14 3 3 4.0 31.5 20 4330 221 44 425 2.28
CAD. ELDORADO        A   14500 14 2 2 3.5 30.0 16 3900 204 43 350 2.19
CAD. SEVILLE         A   15906 21 3 3 3.0 30.0 13 4290 204 45 350 2.24
CHEV. CHEVETTE       A    3299 29 3 3 2.5 26.0  9 2110 163 34 231 2.93
CHEV. IMPALA         A    5705 16 4 4 4.0 29.5 20 3690 212 43 250 2.56
CHEV. MALIBU         A    4504 22 3 3 3.5 28.5 17 3180 193 41 200 2.73
CHEV. MONTE CARLO    A    5104 22 2 3 2.0 28.5 16 3220 200 41 200 2.73
CHEV. MONZA          A    3667 24 2 2 2.0 25.0  7 2750 179 40 151 2.73
CHEV. NOVA           A    3955 19 3 3 3.5 27.0 13 3430 197 43 250 2.56
DATSUN 200-SX        J    6229 23 4 3 1.5 21.0  6 2370 170 35 119 3.89
DATSUN 210           J    4589 35 5 5 2.0 23.5  8 2020 165 32  85 3.70
DATSUN 510           J    5079 24 4 4 2.5 22.0  8 2280 170 34 119 3.54
DATSUN 810           J    8129 21 4 4 2.5 27.0  8 2750 184 38 146 3.55
DODGE COLT           A    3984 30 5 4 2.0 24.0  8 2120 163 35  98 3.54
DODGE DIPLOMAT       A    5010 18 2 2 4.0 29.0 17 3600 206 46 318 2.47
DODGE MAGNUM XE      A    5886 16 2 2 3.5 26.0 16 3870 216 48 318 2.71
DODGE ST. REGIS      A    6342 17 2 2 4.5 28.0 21 3740 220 46 225 2.94
FIAT STRADA          E    4296 21 3 1 2.5 26.5 16 2130 161 36 105 3.37
FORD FIESTA          A    4389 28 4 . 1.5 26.0  9 1800 147 33  98 3.15
FORD MUSTANG         A    4187 21 3 3 2.0 23.0 10 2650 179 42 140 3.08
HONDA ACCORD         J    5799 25 5 5 3.0 25.5 10 2240 172 36 107 3.05
HONDA CIVIC          J    4499 28 4 4 2.5 23.5  5 1760 149 34  91 3.30
LINC. CONTINENTAL    A   11497 12 3 4 3.5 30.5 22 4840 233 51 400 2.47
LINC. CONT MARK V    A   13594 12 3 4 2.5 28.5 18 4720 230 48 400 2.47
LINC. VERSAILLES     A   13466 14 3 3 3.5 27.0 15 3830 201 41 302 2.47
MAZDA GLC            J    3995 30 4 4 3.5 25.5 11 1980 154 33  86 3.73
MERC. BOBCAT         A    3829 22 4 3 3.0 25.5  9 2580 169 39 140 2.73
MERC. COUGAR         A    5379 14 4 3 3.5 29.5 16 4060 221 48 302 2.75
MERC. COUGAR XR-7    A    6303 14 4 4 3.0 25.0 16 4130 217 45 302 2.75
MERC. MARQUIS        A    6165 15 3 2 3.5 30.5 23 3720 212 44 302 2.26
MERC. MONARCH        A    4516 18 3 . 3.0 27.0 15 3370 198 41 250 2.43
MERC. ZEPHYR         A    3291 20 3 3 3.5 29.0 17 2830 195 43 140 3.08
OLDS. 98             A    8814 21 4 4 4.0 31.5 20 4060 220 43 350 2.41
OLDS. CUTLASS        A    4733 19 3 3 4.5 28.0 16 3300 198 42 231 2.93
OLDS. CUTL SUPR      A    5172 19 3 4 2.0 28.0 16 3310 198 42 231 2.93
OLDS. DELTA 88       A    5890 18 4 4 4.0 29.0 20 3690 218 42 231 2.73
OLDS. OMEGA          A    4181 19 3 3 4.5 27.0 14 3370 200 43 231 3.08
OLDS. STARFIRE       A    4195 24 1 1 2.0 25.5 10 2720 180 40 151 2.73
OLDS. TORONADO       A   10371 16 3 3 3.5 30.0 17 4030 206 43 350 2.41
PEUGEOT 604 SL       E   12990 14 . . 3.5 30.5 14 3420 192 38 163 3.58
PLYM. ARROW          A    4647 28 3 3 2.0 21.5 11 2360 170 37 156 3.05
PLYM. CHAMP          A    4425 34 5 4 2.5 23.0 11 1800 157 37  86 2.97
PLYM. HORIZON        A    4482 25 3 . 4.0 25.0 17 2200 165 36 105 3.37
PLYM. SAPPORO        A    6486 26 . . 1.5 22.0  8 2520 182 38 119 3.54
PLYM. VOLARE         A    4060 18 2 2 5.0 31.0 16 3330 201 44 225 3.23
PONT. CATALINA       A    5798 18 4 4 4.0 29.0 20 3700 214 42 231 2.73
PONT. FIREBIRD       A    4934 18 1 2 1.5 23.5  7 3470 198 42 231 3.08
PONT. GRAND PRIX     A    5222 19 3 3 2.0 28.5 16 3210 201 45 231 2.93
PONT. LE MANS        A    4723 19 3 3 3.5 28.0 17 3200 199 40 231 2.93
PONT. PHOENIX        A    4424 19 . . 3.5 27.0 13 3420 203 43 231 3.08
```

```
PONT. SUNBIRD      A    4172 24 2 2 2.0 25.0  7 2690 179 41 151 2.73
RENAULT LE CAR     E    3895 26 3 3 3.0 23.0 10 1830 142 34  79 3.72
SUBARU             J    3798 35 5 4 2.5 25.5 11 2050 164 36  97 3.81
TOYOTA CELICA      J    5899 18 5 5 2.5 22.0 14 2410 174 36 134 3.06
TOYOTA COROLLA     J    3748 31 5 5 3.0 24.5  9 2200 165 35  97 3.21
TOYOTA CORONA      J    5719 18 5 5 2.0 23.0 11 2670 175 36 134 3.05
VW RABBIT          E    4697 25 4 3 3.0 25.5 15 1930 155 35  89 3.78
VW RABBIT DIESEL   E    5397 41 5 4 3.0 25.5 15 2040 155 35  90 3.78
VW SCIROCCO        E    6850 25 4 3 2.0 23.5 16 1990 156 36  97 3.78
VW DASHER          E    7140 23 4 3 2.5 37.5 12 2160 172 36  97 3.74
VOLVO 260          E   11995 17 5 3 2.5 29.5 14 3170 193 37 163 2.98
;
```

A2.3 The BASEBALL Data Set: Baseball Data

The BASEBALL data set contains variables that measure batting and fielding performance for 322 regular and substitute hitters in the 1986 year, their career performance statistics, and their salary at the start of the 1987 season.

NAME	hitter's name
ATBAT	times at bat
HITS	hits
HOMER	home runs
RUNS	runs
RBI	runs batted in
WALKS	walks
YEARS	years in the major leagues
ATBATC	career times at bat
HITSC	career hits
HOMERC	career home runs
RUNSC	career runs scored
RBIC	career runs batted in
POSITION	player's position
PUTOUTS	put outs
ASSISTS	assists
ERRORS	errors
SALARY	annual salary, expressed in units of $1,000
BATAVG	batting average, calculated as 1,000*(HITS/ATBAT)
BATAVGC	career batting average, calculated as 1,000*(HITSC/ATBATC)

Player's position: If a substitute played 70 percent of his games at one position, that is the only position listed for him in the data set. If he did not play 70 percent of his games at one position, but played 90 percent of his games at two positions, he is listed with a combination position, such as 'S2' for shortstop and second base or 'CO' for catcher and outfield. If a player failed to meet either the 70 percent or 90 percent requirement listed above, he is listed as a utility player ('UT').

Two character formats are provided for the POSITION variable. $POSFMT describes POSITION as coded in the data set. $POS can be used to recode the POSITION variable into a smaller number of possible values.

Source: The 1986 and career statistics are taken from *The 1987 Baseball Encyclopedia Update* published by Collier Books, Macmillan Publishing Company, New York.* The salary data are taken from *Sports Illustrated* (April 20, 1987).** The salary of any player not included in that article is listed as a missing value and shown as a period. This data set and additional data sets on pitchers and teams were analyzed by several authors at a special poster session at the 1988 meetings of the Statistical Graphics Section of the American Statistical Association, titled "Why They Make What They Make - An Analysis of Major League Baseball Salaries."

```
Title 'Baseball Hitters Data';

 /* Formats to specify the coding of some of the variables */
proc format;
   value $league
     'N' ='National'
     'A' ='American';
   value $team
     'ATL'='Atlanta      '
     'BAL'='Baltimore    '
     'BOS'='Boston       '
     'CAL'='California   '
     'CHA'='Chicago A    '
     'CHN'='Chicago N    '
     'CIN'='Cincinnati   '
     'CLE'='Cleveland    '
     'DET'='Detroit      '
     'HOU'='Houston      '
     'KC '='Kansas City  '
     'LA '='Los Angeles  '
     'MIL'='Milwaukee    '
     'MIN'='Minnesota    '
     'MON'='Montreal     '
     'NYA'='New York A   '
     'NYN'='New York N   '
     'OAK'='Oakland      '
```

```
                      'PHI'='Philadelphia '
                      'PIT'='Pittsburgh   '
                      'SD '='San Diego    '
                      'SEA'='Seattle      '
                      'SF '='San Francisco'
                      'STL'='St. Louis    '
                      'TEX'='Texas        '
                      'TOR'='Toronto      ';
                  value $posfmt
                    '1B' = 'First Base'
                    '2B' = 'Second Base'
                    'SS' = 'Short Stop'
                    '3B' = 'Third Base'
                    'RF' = 'Right Field'
                    'CF' = 'Center Field'
                    'LF' = 'Left Field'
                    'C ' = 'Catcher'
                    'DH' = 'Designated Hitter'
                    'OF' = 'Outfield'
                    'UT' = 'Utility'
                    'OS' = 'Outfield & Short Stop'
                    '3S' = 'Third Base & Short Stop'
                    '13' = 'First & Third Base'
                    '30' = 'Third Base & Outfield'
                    '01' = 'Outfield & First Base'
                    'S3' = 'Short Stop & Third Base'
                    '32' = 'Third & Second Base'
                    'DO' = 'Designated Hitter & Outfield'
                    'OD' = 'Outfield & Designated Hitter'
                    'CD' = 'Catcher & Designated Hitter'
                    'CS' = 'Catcher & Short Stop'
                    '23' = 'Second & Third Base'
                    '10' = 'First Base and Outfield'
                    '2S' = 'Second Base and Short Stop';
                  /* Recode position to short list */
                  value $pos
                    'CS','CD'        ='C '
                    'OS','01','OD'   ='OF'
                    'CF','RF','LF'   ='OF'
                    '10','13'        ='1B'
                    '2S','23'        ='2B'
                    'DO'             ='DH'
                    'S3'             ='SS'
                    '32','3S','30'   ='3B' ;

              data baseball;
                input name $1-14
                      league $15 team $16-18 position $19-20
                      atbat 3. hits 3. homer 3. runs 3. rbi 3. walks 3. years 3.
                      atbatc 5. hitsc 4. homerc 4. runsc 4. rbic 4. walksc 4.
                      putouts 4. assists 3. errors 3. salary 4.;
```

```
batavg = round(1000 * (hits / atbat));
batavgc= round(1000 * (hitsc/ atbatc));
label
    name    = "Hitter's name"
    atbat   = 'Times at Bat'
    hits    = 'Hits'
    homer   = 'Home Runs'
    runs    = 'Runs'
    rbi     = 'Runs Batted In'
    walks   = 'Walks'
    years   = 'Years in the Major Leagues'
    atbatc  = 'Career Times at Bat'
    hitsc   = 'Career Hits'
    homerc  = 'Career Home Runs'
    runsc   = 'Career Runs Scored'
    rbic    = 'Career Runs Batted In'
    position= 'Position(s)'
    putouts = 'Put Outs'
    assists = 'Assists'
    errors  = 'Errors'
    salary  = 'Salary (in 1000$)'
    batavg  = 'Batting Average'
    batavgc = 'Career Batting Average';
cards;
Andy Allanson  ACLEC 293 66  1 30 29 14  1  293   66   1  30   29   14 446 33 20    .
Alan Ashby     NHOUC 315 81  7 24 38 39 14 3449  835  69 321  414  375 632 43 10  475
Alvin Davis    ASEA1B479130 18 66 72 76  3 1624  457  63 224  266  263 880 82 14  480
Andre Dawson   NMONRF496141 20 65 78 37 11 56281575 225 828  838  354 200 11  3  500
A Galarraga    NMON1B321 87 10 39 42 30  2  396  101  12  48   46   33 805 40  4   92
A Griffin      AOAKSS594169  4 74 51 35 11 44081133  19 501  336  194 282421 25  750
Al Newman      NMON2B185 37  1 23  8 21  2  214   42   1  30    9   24  76127  7   70
A Salazar      AKC SS298 73  0 24 24  7  3  509  108   0  41   37   12 121283  9  100
Andres Thomas  NATLSS323 81  6 26 32  8  2  341   86   6  32   34    8 143290 19   75
A Thornton     ACLEDH401 92 17 49 66 65 13 52061332 253 784  890  866   0  0  01100
Alan Trammell  ADETSS574159 21107 75 59 10 46311300  90 702  504  488 238445 22  517
Alex Trevino   NLA C 202 53  4 31 26 27  9 1876  467  15 192  186  161 304 45 11  513
A Van.Slyke    NSTLRF418113 13 48 61 47  4 1512  392  41 205  204  203 211 11  7  550
Alan Wiggins   ABAL2B239 60  0 30 11 22  6 1941  510   4 309  103  207 121151  6  700
Bill Almon     NPITUT196 43  7 29 27 30 13 3231  825  36 376  290  238  80 45  8  240
Billy Beane    AMINOF183 39  3 20 15 11  3  201   42   3  20   16   11 118  0  0    .
Buddy Bell     NCIN3B568158 20 89 75 73 15 80682273 1771045  993  732 105290 10  775
B Biancalana   AKC SS190 46  2 24  8 15  5  479  102   5  65   23   39 102177 16  175
Bruce Bochte   AOAK1B407104  6 57 43 65 12 52331478 100 643  658  653 912 88  9    .
Bruce Bochy    NSD C 127 32  8 16 22 14  8  727  180  24  67   82   56 202 22  2  135
Barry Bonds    NPITCF413 92 16 72 48 65  1  413   92  16  72   48   65 280  9  5  100
Bobby Bonilla  ACHAO1426109  3 55 43 62  1  426  109   3  55   43   62 361 22  2  115
Bob Boone      ACALC  22 10  1  4  2  1  6   84   26   2   9    9    3 812 84 11    .
Bob Brenly     NSF C 472116 16 60 62 74  6 1924  489  67 242  251  240 518 55  3  600
Bill Buckner   ABOS1B629168 18 73102 40 18 84242464 16410081072  4021067157 14  777
Brett Butler   ACLECF587163  4 92 51 70  6 2695  747  17 442  198  317 434  9  3  765
Bob Dernier    NCHNCF324 73  4 32 18 22  7 1931  491  13 291  108  180 222  3  3  708
Bo Diaz        NCINC 474129 10 50 56 40 10 2331  604  61 246  327  166 732 83 13  750
Bill Doran     NHOU2B550152  6 92 37 81  5 2308  633  32 349  182  308 262329 16  625
```

```
Brian Downing   ACALLF513137 20 90 95 90 14 52011382 166 763 734 784 267  5  3 900
Bobby Grich     ACAL2B313 84  9 42 30 39 17 68901833 2241033 8641087 127221  7   .
Billy Hatcher   NHOUCF419108  6 55 36 22  3  591 149   8  80  46  31 226  7  4 110
Bob Horner      NATL1B517141 27 70 87 52  9 3571 994 215 545 652 3371378102  8   .
Brook Jacoby    ACLE3B583168 17 83 80 56  5 1646 452  44 219 208 136 109292 25 613
Bob Kearney     ASEAC 204 49  6 23 25 12  7 1309 308  27 126 132  66 419 46  5 300
Bill Madlock    NLA 3B379106 10 38 60 30 14 62071906 146 859 803 571  72170 24 850
Bobby Meacham   ANYASS161 36  0 19 10 17  4 1053 244   3 156  86 107  70149 12   .
Bob Melvin      NSF C 268 60  5 24 25 15  2  350  78   5  34  29  18 442 59  6  90
Ben Oglivie     AMILDH346 98  5 31 53 30 16 59131615 235 784 901 560   0  0  0   .
Bip Roberts     NSD 2B241 61  1 34 12 14  1  241  61   1  34  12  14 166172 10   .
B Robidoux      AMIL1B181 41  1 15 21 33  2  232  50   4  20  29  45 326 29  5  68
Bill Russell    NLA UT216 54  0 21 18 15 18 73181926  46 796 627 483 103 84  5   .
Billy Sample    NATLOF200 57  6 23 14 14  9 2516 684  46 371 230 195  69  1  1   .
B Schroeder     AMILUT217 46  7 32 19  9  4  694 160  32  86  76  32 307 25  1 180
Butch Wynegar   ANYAC 194 40  7 19 29 30 11 41831069  64 486 493 608 325 22  2   .
Chris Bando     ACLEC 254 68  2 28 26 22  6  999 236  21 108 117 118 359 30  4 305
Chris Brown     NSF 3B416132  7 57 49 33  3  932 273  24 113 121  80  73177 18 215
C Castillo      ACLEOD205 57  8 34 32  9  5  756 192  32 117 107  51  58  4  4 248
Cecil Cooper    AMIL1B542140 12 46 75 41 16 70992130 235 9871089 431 697 61  9   .
Chili Davis     NSF RF526146 13 71 70 84  6 2648 715  77 352 342 289 303  9  9 815
Carlton Fisk    ACHAC 457101 14 42 63 22 17 65211767 2811003 977 619 389 39  4 875
Curt Ford       NSTLOF214 53  2 30 29 23  2  226  59   2  32  32  27 109  7  3  70
Cliff Johnson   ATORDH 19  7  0  1  2  1  4   41  13   1   3   4   4   0  0  0   .
C Lansford      AOAK3B591168 19 80 72 39  9 44781307 113 634 563 319  67147  41200
Chet Lemon      ADETCF403101 12 45 53 39 12 51501429 166 747 666 526 316  6  5 675
C Maldonado     NSF OF405102 18 49 85 20  6  950 231  29  99 138  64 161 10  3 415
C Martinez      NSD O1244 58  9 28 25 35  4 1335 333  49 164 179 194 142 14  2 340
Charlie Moore   AMILC 235 61  3 24 39 21 14 39261029  35 441 401 333 425 43  4   .
C Reynolds      NHOUSS313 78  6 32 41 12 12 3742 968  35 409 321 170 106206  7 417
Cal Ripken      ABALSS627177 25 98 81 70  6 3210 927 133 529 472 313 240482 131350
Cory Snyder     ACLEOS416113 24 58 69 16  1  416 113  24  58  69  16 203 70 10  90
Chris Speier    NCHN3S155 44  6 21 23 15 16 66311634  98 698 661 777  53 88  3 275
C Wilkerson     ATEX2S236 56  0 27 15 11  4 1115 270   1 116  64  57 125199 13 230
Dave Anderson   NLA 3S216 53  1 31 15 22  4  926 210   9 118  69 114  73152 11 225
Doug Baker      AOAKOF 24  3  0  1  0  2  3  159  28   0  20  12   9  80  4  0   .
Don Baylor      ABOSDH585139 31 93 94 62 17 75461982 31511411179 727   0  0  0 950
D Bilardello    NMONC 191 37  4 12 17 14  4  773 163  16  61  74  52 391 38  8   .
Daryl Boston    ACHACF199 53  5 29 22 21  3  514 120   8  57  40  39 152  3  5  75
Darnell Coles   ADET3B521142 20 67 86 45  4  815 205  22  99 103  78 107242 23 105
Dave Collins    ADETLF419113  1 44 27 44 12 44841231  32 612 344 422 211  2  1   .
D Concepcion    NCINUT311 81  3 42 30 26 17 82472198 100 950 909 690 153223 10 320
D Daulton       NPHIC 138 31  8 18 21 38  3  244  53  12  33  32  55 244 21  4   .
Doug DeCinces   ACAL3B512131 26 69 96 52 14 53471397 221 712 815 548 119216 12 850
Darrell Evans   ADET1B507122 29 78 85 91 18 77611947 347117511521380 808108  2 535
Dwight Evans    ABOSRF529137 26 86 97 97 15 66611785 2911082 949 989 280 10  5 933
Damaso Garcia   ATOR2B424119  6 57 46 13  9 36511046  32 461 301 112 224286  8 850
Dan Gladden     NSF CF351 97  4 55 29 39  4 1258 353  16 196 110 117 226  7  3 210
Danny Heep      NNYNOF195 55  5 24 33 30  8 1313 338  25 144 149 153  83  2  1   .
D Henderson     ASEAOF388103 15 59 47 39  6 2174 555  80 285 274 186 182  9  4 325
Donnie Hill     AOAK23339 96  4 37 29 23  4 1064 290  11 123 108  55 104213  9 275
Dave Kingman    AOAKDH561118 35 70 94 33 16 66771575 442 9011210 608 463 32  8   .
Davey Lopes     NCHN3O255 70  7 49 35 43 15 63111661 1541019 608 820  51 54  8 450
```

```
Don Mattingly  ANYA1B677238 31117113 53  5 2223 737  93 349 401 1711377100  61975
Darryl Motley  AKC RF227 46  7 23 20 12  5 1325 324  44 156 158  67  92  2  2    .
Dale Murphy    NATLCF614163 29 89 83 75 11 50171388 266 813 822 617 303  6  61900
Dwayne Murphy  AOAKCF329 83  9 50 39 56  9 3828 948 145 575 528 635 276  6  2 600
Dave Parker    NCINRF637174 31 89116 56 14 67272024 247 9781093 495 278  9  91042
Dan Pasqua     ANYALF280 82 16 44 45 47  2  428 113  25  61  70  63 148  4  2 110
D Porter       ATEXCD155 41 12 21 29 22 16 54091338 181 746 805 875 165  9  1 260
D Schofield    ACALSS458114 13 67 57 48  4 1350 298  28 160 123 122 246389 18 475
Don Slaught    ATEXC 314 83 13 39 46 16  5 1457 405  28 156 159  76 533 40  4 432
D Strawberry   NNYNRF475123 27 76 93 72  4 1810 471 108 292 343 267 226 10  61220
Dale Sveum     AMIL3B317 78  7 35 35 32  1  317  78   7  35  35  32  45122 26  70
D Tartabull    ASEARF511138 25 76 96 61  3  592 164  28  87 110  71 157  7  8 145
Dickie Thon    NHOUSS278 69  3 24 21 29  8 2079 565  32 258 192 162 142210 10    .
Denny Walling  NHOU3B382119 13 54 58 36 12 2133 594  41 287 294 227  59156  9 595
Dave Winfield  ANYARF565148 24 90104 77 14 72872083 30511351234 791 292  9  51861
Enos Cabell    NLA 1B277 71  2 27 29 14 15 59521647  60 753 596 259 360 32  5    .
Eric Davis     NCINLF415115 27 97 71 68  3  711 184  45 156 119  99 274  2  7 300
Eddie Milner   NCINCF424110 15 70 47 36  7 2130 544  38 335 174 258 292  6  3 490
Eddie Murray   ABAL1B495151 17 61 84 78 10 56241679 275 8841015 7091045 88 132460
Ernest Riles   AMILSS524132  9 69 47 54  2  972 260  14 123  92  90 212327 20    .
Ed Romero      ABOSSS233 49  2 41 23 18  8 1350 336   7 166 122 106 102132 10 375
Ernie Whitt    ATORC 395106 16 48 56 35 10 2303 571  86 266 323 248 709 41  7    .
Fred Lynn      ABALCF397114 23 67 67 53 13 55891632 241 906 926 716 244  2  4    .
Floyd Rayford  ABAL3B210 37  8 15 19 15  6  994 244  36 107 114  53  40115 15    .
F Stubbs       NLA LF420 95 23 55 58 37  3  646 139  31  77  77  61 206 10  7    .
Frank White    AKC 2B566154 22 76 84 43 14 61001583 131 743 693 300 316439 10 750
George Bell    ATORLF641198 31101108 41  5 2129 610  92 297 319 117 269 17 101175
Glenn Braggs   AMILLF215 51  4 19 18 11  1  215  51   4  19  18  11 116  5 12  70
George Brett   AKC 3B441128 16 70 73 80 14 66752095 20910721050 695  97218 161500
Greg Brock     NLA 1B325 76 16 33 52 37  5 1506 351  71 195 219 214 726 87  3 385
Gary Carter    NNYNC 490125 24 81105 62 13 60631646 271 847 999 680 869 62  81926
Glenn Davis    NHOU1B574152 31 91101 64  3  985 260  53 148 173  951253111 11 215
George Foster  NNYNLF284 64 14 30 42 24 18 70231925 348 9861239 666  96  4  4    .
Gary Gaetti    AMIN3B596171 34 91108 52  6 2862 728 107 361 401 224 118334 21 900
Greg Gagne     AMINSS472118 12 63 54 30  4  793 187  14 102  80  50 228377 26 155
G Hendrick     ACALOF283 77 14 45 47 26 16 68401910 259 9151067 546 144  6  5 700
Glenn Hubbard  NATL2B408 94  4 42 36 66  9 3573 866  59 429 365 410 282487 19 535
Garth Iorg     ATOR32327 85  3 30 44 20  8 2140 568  16 216 208  93  91185 12 363
Gary Matthews  NCHNLF370 96 21 49 46 60 15 69861972 2311070 955 921 137  5  9 733
Graig Nettles  NSD 3B354 77 16 36 55 41 20 87162172 384117212671057  83174 16 200
Gary Pettis    ACALCF539139  5 93 58 69  5 1469 369  12 247 126 198 462  9  7 400
Gary Redus     NPHILF340 84 11 62 33 47  5 1516 376  42 284 141 219 185  8  4 400
G Templeton    NSD SS510126  2 42 44 35 11 55621578  44 703 519 256 207358 20 738
Gorman Thomas  ASEADH315 59 16 45 36 58 13 46771051 268 681 782 697   0  0  0    .
Greg Walker    ACHA1B282 78 13 37 51 29  5 1649 453  73 211 280 138 670 57  5 500
Gary Ward      ATEXLF380120  5 54 51 31  8 3118 900  92 444 419 240 237  8  1 600
Glenn Wilson   NPHIRF584158 15 70 84 42  5 2358 636  58 265 316 134 331 20  4 663
Harold Baines  ACHARF570169 21 72 88 38  7 37541077 140 492 589 263 295 15  5 950
Hubie Brooks   NMONSS306104 14 50 58 25  7 2954 822  55 313 377 187 116222 15 750
H Johnson      NNYN3S220 54 10 30 39 31  5 1185 299  40 145 154 128  50136 20 298
Hal McRae      AKC DH278 70  7 22 37 18 18 71862081 190 9351088 643   0  0  0 325
H Reynolds     ASEA2B445 99  1 46 24 29  4  618 129   1  72  31  48 278415 16  88
Harry Spilman  NSF 1B143 39  5 18 30 15  9  639 151  16  80  97  61 138 15  1 175
```

```
H Winningham    NMONOF185 40   4 23 11 18   3  524 125    7  58  37  47  97  2   2  90
J Barfield      ATORRF589170 40107108 69   6 2325 634  128 371 376 238 368 20  31238
Juan Beniquez   ABALUT343103   6 48 36 40  15 43381193   70 581 421 325 211 56  13 430
Juan Bonilla    ABAL2B284 69   1 33 18 25   5 1407 361    6 139  98 111 122140   5   .
J Cangelosi     ACHALF438103   2 65 32 71   2  440 103    2  67  32  71 276  7   9 100
Jose Canseco    AOAKLF600144  33 85117 65   2  696 173   38 101 130  69 319  4  14 165
Joe Carter      ACLERF663200  29108121 32   4 1447 404   57 210 222  68 241  8   6 250
Jack Clark      NSTL1B232 55   9 34 23 45  12 44051213  194 702 705 625 623 35  31300
Jose Cruz       NHOULF479133  10 48 72 55  17 74722147  153 9801032 854 237  5   4 773
Julio Cruz      ACHA2B209 45   0 38 19 42  10 3859 916   23 557 279 478 132205   5   .
Jody Davis      NCHNC 528132  21 61 74 41   6 2641 671   97 273 383 226 885105   81008
Jim Dwyer       ABALDO160 39   8 18 31 22  14 2128 543   56 304 268 298  33  3   0 275
Julio Franco    ACLESS599183  10 80 74 32   5 2482 715   27 330 326 158 231374  18 775
Jim Gantner     AMIL2B497136   7 58 38 26  11 38711066   40 450 367 241 304347  10 850
Johnny Grubb    ADETDH210 70  13 32 51 28  15 40401130   97 544 462 551   0  0   0 365
J Hairston      ACHAUT225 61   5 32 26 26  11 1568 408   25 202 185 257 132  9   0   .
Jack Howell     ACAL3B151 41   4 26 21 19   2  288  68    9  45  39  35  28 56   2  95
John Kruk       NSD LF278 86   4 33 38 45   1  278  86    4  33  38  45 102  4   2 110
J Leonard       NSF LF341 95   6 48 42 20  10 2964 808   81 379 428 221 158  4   5 100
Jim Morrison    NPIT3B537147  23 58 88 47  10 2744 730   97 302 351 174  92257  20 278
John Moses      ASEACF399102   3 56 34 34   5  670 167    4  89  48  54 211  9   3  80
J Mumphrey      NCHNOF309 94   5 37 32 26  13 46181330   57 616 522 436 161  3   3 600
Joe Orsulak     NPITRF401100   2 60 19 28   4  876 238    2 126  44  55 193 11   4   .
Jorge Orta      AKC DH336 93   9 35 46 23  15 57791610  128 730 741 497   0  0   0   .
Jim Presley     ASEA3B616163  27 83107 32   3 1437 377   65 181 227  82 110308  15 200
Jamie Quirk     AKC CS219 47   8 24 26 17  12 1188 286   23 100 125  63 260 58   4   .
Johnny Ray      NPIT2B579174   7 67 78 58   6 3053 880   32 366 337 218 280479   5 657
Jeff Reed       AMINC 165 39   2 13  9 16   3  196  44    2  18  10  18 332 19   2  75
Jim Rice        ABOSLF618200  20 98110 62  13 71272163  351 11041289 564 330 16   82413
Jerry Royster   NSD UT257 66   5 31 26 32  14 3910 979   33 518 324 382  87166  14 250
John Russell    NPHIC 315 76  13 35 60 25   3  630 151   24  68  94  55 498 39  13 155
Juan Samuel     NPHI2B591157  16 90 78 26   4 2020 541   52 310 226  91 290440  25 640
John Shelby     ABALOF404 92  11 54 49 18   6 1354 325   30 188 135  63 222  5   5 300
Joel Skinner    ACHAC 315 73   5 23 37 16   4  450 108    6  38  46  28 227 15   3 110
Jeff Stone      NPHIOF249 69   6 32 19 20   4  702 209   10  97  48  44 1,03  8   2   .
Jim Sundberg    AKC C 429 91  12 41 42 57  13 55901397   83 578 579 644 686 46   4 825
Jim Traber      ABALUT212 54  13 28 44 18   2  233  59   13  31  46  20 243 23   5   .
Jose Uribe      NSF SS453101   3 46 43 61   3  948 218    6  96  72  91 249444  16 195
Jerry Willard   AOAKC 161 43   4 17 26 22   3  707 179   21  77  99  76 300 12   2   .
J Youngblood    NSF OF184 47   5 20 28 18  11 3327 890   74 419 382 304  49  2   0 450
Kevin Bass      NHOURF591184  20 83 79 38   5 1689 462   40 219 195  82 303 12   5 630
Kal Daniels     NCINOF181 58   6 34 23 22   1  181  58    6  34  23  22  88  0   3  87
Kirk Gibson     ADETRF441118  28 84 86 68   8 2723 750  126 433 420 309 190  2  21300
Ken Griffey     ANYAOF490150  21 69 58 35  14 61261839  121 983 707 600  96  5  31000
K Hernandez     NNYN1B551171  13 94 83 94  13 60901840  128 969 900 9171199149   51800
Kent Hrbek      AMIN1B550147  29 85 91 71   6 2816 815  117 405 474 3191218104  101310
Ken Landreaux   NLA OF283 74   4 34 29 22  10 39191062   85 505 456 283 145  5   7 738
K McReynolds    NSD CF560161  26 89 96 66   4 1789 470   65 233 260 155 332  9   8 625
K Mitchell      NNYNOS328 91  12 51 43 33   2  342  94   12  51  44  33 145 59   8 125
K Moreland      NCHNRF586159  12 72 79 53   9 3082 880   83 363 477 295 181 13  41043
Ken Oberkfell   NATL3B503136   5 62 48 83  10 3423 970   20 408 303 414  65258   8 725
Ken Phelps      ASEADH344 85  24 69 64 88   7  911 214   64 150 156 187   0  0   0 300
Kirby Puckett   AMINCF680223  31119 96 34   3 1928 587   35 262 201  91 429  8   6 365
```

```
K Stillwell    NCINSS279 64  0 31 26 30  1  279   64   0  31  26  30 107205 16  75
Leon Durham    NCHN1B484127 20 66 65 67  7 3006  844 116 436 458 3771231 80  71183
Len Dykstra    NNYNCF431127  8 77 45 58  2  667  187   9 117  64  88 283  8   3 203
Larry Herndon  ADETOF283 70  8 33 37 27 12 44791222  94 557 483 307 156  2   2 225
Lee Lacy       ABALRF491141 11 77 47 37 15 42911240  84 615 430 340 239  8   2 525
Len Matuszek   NLA O1199 52  9 26 28 21  6  805  191  30 113 119  87 235 22   5 265
Lloyd Moseby   ATORCF589149 21 89 86 64  7 3558  928 102 513 471 351 371  6   6 788
Lance Parrish  ADETC 327 84 22 53 62 38 10 42731123 212 577 700 334 483 48   6 800
Larry Parrish  ATEXDH464128 28 67 94 52 13 58291552 210 740 840 452   0  0   0 588
Luis Rivera    NMONSS166 34  0 20 13 17  1  166   34   0  20  13  17  64119  9   .
Larry Sheets   ABALDH338 92 18 42 60 21  3  682  185  36  88 112  50   0  0   0 145
Lonnie Smith   AKC LF508146  8 80 44 46  9 3148  915  41 571 289 326 245  5   9   .
Lou Whitaker   ADET2B584157 20 95 73 63 10 47041320  93 724 522 576 276421 11 420
Mike Aldrete   NSF 1O216 54  2 27 25 33  1  216   54   2  27  25  33 317 36   1  75
Marty Barrett  ABOS2B625179  4 94 60 65  5 1696  476  12 216 163 166 303450 14 575
Mike Brown     NPITOF243 53  4 18 26 27  4  853  228  23 101 110  76 107  3   3   .
Mike Davis     AOAKRF489131 19 77 55 34  7 2051  549  62 300 263 153 310  9   9 780
Mike Diaz      NPITO1209 56 12 22 36 19  2  216   58  12  24  37  19 201  6   3  90
M Duncan       NLA SS407 93  8 47 30 30  2  969  230  14 121  69  68 172317 25 150
Mike Easler    ANYADH490148 14 64 78 49 13 34001000 113 445 491 301   0  0   0 700
M Fitzgerald   NMONC 209 59  6 20 37 27  4  884  209  14  66 106  92 415 35   3   .
Mel Hall       ACLELF442131 18 68 77 33  6 1416  398  47 210 203 136 233  7   7 550
M Hatcher      AMINUT317 88  3 40 32 19  8 2543  715  28 269 270 118 220 16   4   .
Mike Heath     NSTLC 288 65  8 30 36 27  9 2815  698  55 315 325 189 259 30  10 650
Mike Kingery   AKC OF209 54  3 25 14 12  1  209   54   3  25  14  12 102  6   3  68
M LaValliere   NSTLC 303 71  3 18 30 36  3  344   76   3  20  36  45 468 47   6 100
Mike Marshall  NLA RF330 77 19 47 53 27  6 1928  516  90 247 288 161 149  8   6 670
M Pagliarulo   ANYA3B504120 28 71 71 54  3 1085  259  54 150 167 114 103283 19 175
Mark Salas     AMINC 258 60  8 28 33 18  3  638  170  17  80  75  36 358 32   8 137
Mike Schmidt   NPHI3B 20  1  0  0  0  0  2   41    9   2   6   7   4  78220 62127
Mike Scioscia  NLA C 374 94  5 36 26 62  7 1968  519  26 181 199 288 756 64  15 875
M Tettleton    AOAKC 211 43 10 26 35 39  3  498  116  14  59  55  78 463 32   8 120
Milt Thompson  NPHICF299 75  6 38 23 26  3  580  160   8  71  33  44 212  1   2 140
Mitch Webster  NMONCF576167  8 89 49 57  4  822  232  19 132  83  79 325 12   8 210
Mookie Wilson  NNYNOF381110  9 61 45 32  7 3015  834  40 451 249 168 228  7   5 800
Marvell Wynne  NSD OF288 76  7 34 37 15  4 1644  408  16 198 120 113 203  3   3 240
Mike Young     ABALLF369 93  9 43 42 49  5 1258  323  54 181 177 157 149  1   6 350
Nick Esasky    NCIN1B330 76 12 35 41 47  4 1367  326  55 167 198 167 512 30   5   .
Ozzie Guillen  ACHASS547137  2 58 47 12  2 1038  271   3 129  80  24 261459 22 175
O McDowell     ATEXCF572152 18105 49 65  2  978  249  36 168  91 101 325 13   3 200
Omar Moreno    NATLRF359 84  4 46 27 21 12 49921257  37 699 386 387 151  8   5   .
Ozzie Smith    NSTLSS514144  0 67 54 79  9 47391169  13 583 374 528 229453 151940
Ozzie Virgil   NATLC 359 80 15 45 48 63  7 1493  359  61 176 202 175 682 93  13 700
Phil Bradley   ASEALF526163 12 88 50 77  4 1556  470  38 245 167 174 250 11   1 750
Phil Garner    NHOU3B313 83  9 43 41 30 14 58851543 104 751 714 535  58141 23 450
P Incaviglia   ATEXRF540135 30 82 88 55  1  540  135  30  82  88  55 157  6  14 172
Paul Molitor   AMIL3B437123  9 62 55 40  9 41391203  79 676 390 364  82170 151260
Pete O'Brien   ATEX1B551160 23 86 90 87  5 2235  602  75 278 328 2731224115 11   .
Pete Rose      NCIN1B237 52  0 15 25 30 24140534256 160216513141566 523 43   6 750
Pat Sheridan   ADETOF236 56  6 41 19 21  5 1257  329  24 166 125 105 172  1   4 190
Pat Tabler     ACLE1B473154  6 61 48 29  6 1966  566  29 250 252 178 846 84   9 580
R Belliard     NPITSS309 72  0 33 31 26  5  354   82   0  41  32  26 117269 12 130
Rick Burleson  ACALUT271 77  5 35 29 33 12 49331358  48 630 435 403  62 90   3 450
```

```
Randy Bush      AMINLF357 96   7 50 45 39  5 1394 344   43 178 192 136 167  2  4 300
Rick Cerone     AMILC 216 56   4 22 18 15 12 2796 665   43 266 304 198 391 44  4 250
Ron Cey         NCHN3B256 70  13 42 36 44 16 70581845  312 9651128 990  41118 81050
Rob Deer        AMILRF466108  33 75 86 72  3  652 142   44 102 109 102 286  8  8 215
Rick Dempsey    ABALC 327 68  13 42 29 45 18 3949 939   78 438 380 466 659 53  7 400
Rich Gedman     ABOSC 462119  16 49 65 37  7 2131 583   69 244 288 150 866 65  6   .
Ron Hassey      ANYAC 341110   9 45 49 46  9 2331 658   50 249 322 274 251  9  4 560
R Henderson     ANYACF608160  28130 74 89  8 40711182  103 862 417 708 426  4  61670
R Jackson       ACALDH419101  18 65 58 92 20 95282510  548150916591342  0  0  0 488
Ricky Jones     ACALRF 33  6   0  2  4  7  1   33   6    0   2   4   7 205  5  4   .
Ron Kittle      ACHADH376 82  21 42 60 35  5 1770 408  115 238 299 157  0  0  0 425
Ray Knight      NNYN3B486145  11 51 76 40 11 39671102   67 410 497 284  88204 16 500
Randy Kutcher   NSF OF186 44   7 28 16 11  1  186  44    7  28  16  11  99  3  1   .
Rudy Law        AKC OF307 80   1 42 36 29  7 2421 656   18 379 198 184 145  2  2   .
Rick Leach      ATORDO246 76   5 35 39 13  6· 912 234   12 102  96  80  44  0  1 250
Rick Manning    AMILOF205 52   8 31 27 17 12 51341323   56 643 445 459 155  3  2 400
R Mulliniks     ATOR3B348 90  11 50 45 43 10 2288 614   43 295 273 269  60176  6 450
Ron Oester      NCIN2B523135   8 52 44 52  9 3368 895   39 377 284 296 367475 19 750
Rey Quinones    ABOSSS312 68   2 32 22 24  1  312  68    2  32  22  24  86150 15  70
R Ramirez       NATLS3496119   8 57 33 21  7 3358 882   36 365 280 165 155371 29 875
Ronn Reynolds   NPITLF126 27   3  8 10  5  4  239  49    3  16  13  14 190  2  9 190
Ron Roenicke    NPHIOF275 68   5 42 42 61  6  961 238   16 128 104 172 181  3  2 191
Ryne Sandberg   NCHN2B627178  14 68 76 46  6 3146 902   74 494 345 242 309492  5 740
R Santana       NNYNSS394 86   1 38 28 36  4 1089 267    3  94  71  76 203369 16 250
Rick Schu       NPHI3B208 57   8 32 25 18  3  653 170   17  98  54  62  42 94 13 140
Ruben Sierra    ATEXOF382101  16 50 55 22  1  382 101   16  50  55  22 200  7  6  98
Roy Smalley     AMINDH459113  20 59 57 68 12 53481369  155 713 660 735  0  0  0 740
R Thompson      NSF 2B549149   7 73 47 42  1  549 149    7  73  47  42 255450 17 140
Rob Wilfong     ACAL2B288 63   3 25 33 16 10 2682 667   38 315 259 204 135257  7 342
R Williams      NLA CF303 84   4 35 32 23  2  312  87    4  39  32  23 179  5  3   .
Robin Yount     AMILCF522163   9 82 46 62 13 70372019 1531043 827 535 352  9  11000
Steve Balboni   AKC 1B512117  29 54 88 43  6 1750 412  100 204 276 1551236 98 18 100
Scott Bradley   ASEAC 220 66   5 20 28 13  3  290  80    5  27  31  15 281 21  3  90
Sid Bream       NPIT1B522140  16 73 77 60  4  730 185   22  93 106  861320166 17 200
S Buechele      ATEX3B461112  18 54 54 35  2  680 160   24  76  75  49 111226 11 135
S Dunston       NCHNSS581145  17 66 68 21  2  831 210   21 106  86  40 320465 32 155
S Fletcher      ATEXSS530159   3 82 50 47  6 1619 426   11 218 149 163 196354 15 475
Steve Garvey    NSD 1B557142  21 58 81 23 18 87592583 27111381299 4781160 53  71450
Steve Jeltz     NPHISS439 96   0 44 36 65  4  711 148    1  68  56  99 229406 22 150
S Lombardozzi   AMIN2B453103   8 53 33 52  2  507 123    8  63  39  58 289407  6 105
Spike Owen      ASEASS528122   1 67 45 51  4 1716 403   12 211 146 155 209372 17 350
Steve Sax       NLA 2B633210   6 91 56 59  6 3070 872   19 420 230 274 367432 16  90
Tony Armas      ABOSCF 16  2   0  1  0  0  2   28   4    0   1   0   0 247  4  8   .
T Bernazard     ACLE2B562169  17 88 73 53  8 3181 841   61 450 342 373 351442 17 530
Tom Brookens    ADETUT281 76   3 42 25 20  8 2658 657   48 324 300 179 106144  7 342
Tom Brunansky   AMINRF593152  23 69 75 53  6 2765 686  133 369 384 321 315 10  6 940
T Fernandez     ATORSS687213  10 91 65 27  4 1518 448   15 196 137  89 294445 13 350
Tim Flannery    NSD 2B368103   3 48 28 54  8 1897 493    9 207 162 198 209246  3 327
Tom Foley       NMONUT263 70   1 26 23 30  4  888 220    9  83  82  86  81147  4 250
Tony Gwynn      NSD RF642211  14107 59 52  5 2364 770   27 352 230 193 337 19  4 740
Terry Harper    NATLOF265 68   8 26 30 29  7 1337 339   32 135 163 128  92  5  3 425
Toby Harrah     ATEX2B289 63   7 36 41 44 17 74021954 19511115 9191153 166211  7   .
Tommy Herr      NSTL2B559141   2 48 61 73  8 3162 874   16 421 349 359 352414  9 925
```

```
Tim Hulett      ACHA3B520120 17 53 44 21  4  927 227  22 106  80  52  70144 11 185
Terry Kennedy   NSD C  19  4  1  2  3  1  1   19   4   1   2   3   1 692 70  8 920
Tito Landrum    NSTLOF205 43  2 24 17 20  7  854 219  12 105  99  71 131  6  1 287
Tim Laudner     AMINC 193 47 10 21 29 24  6 1136 256  42 129 139 106 299 13  5 245
Tom O'Malley    ABAL3B181 46  1 19 18 17  5  937 238   9  88  95 104  37 98  9  .
Tom Paciorek    ATEXUT213 61  4 17 22  3 17 40611145  83 488 491 244 178 45  4 235
Tony Pena       NPITC 510147 10 56 52 53  7 2872 821  63 307 340 174 810 99 181150
T Pendleton     NSTL3B578138  1 56 59 34  3 1399 357   7 149 161  87 133371 20 160
Tony Perez      NCIN1B200 51  2 14 29 25 23 97782732 37912721652 925 398 29  7  .
Tony Phillips   AOAK2B441113  5 76 52 76  5 1546 397  17 226 149 191 160290 11 425
Terry Puhl      NHOUOF172 42  3 17 14 15 10 40861150  57 579 363 406  65  0  0 900
Tim Raines      NMONLF580194  9 91 62 78  8 33721028  48 604 314 469 270 13  6  .
Ted Simmons     NATLUT127 32  4 14 25 12 19 83962402 24210481348 819 167 18  6 500
Tim Teufel      NNYN2B279 69  4 35 31 32  4 1359 355  31 180 148 158 133173  9 278
Tim Wallach     NMON3B480112 18 50 71 44  7 3031 771 110 338 406 239  94270 16 750
Vince Coleman   NSTLLF600139  0 94 29 60  2 1236 309   1 201  69 110 300 12  9 160
Von Hayes       NPHI1B610186 19107 98 74  6 2728 753  69 399 366 2861182 96 131300
Vance Law       NMON2B360 81  5 37 44 37  7 2268 566  41 279 257 246 170284  3 525
Wally Backman   NNYN2B387124  1 67 27 36  7 1775 506   6 272 125 194 186290 17 550
Wade Boggs      ABOS3B580207  8107 71105  5 2778 978  32 474 322 417 121267 191600
Will Clark      NSF 1B408117 11 66 41 34  1  408 117  11  66  41  34 942 72 11 120
Wally Joyner    ACAL1B593172 22 82100 57  1  593 172  22  82 100  571222139 15 165
W Krenchicki    NMON13221 53  2 21 23 22  8 1063 283  15 107 124 106 325 58  6  .
Willie McGee    NSTLCF497127  7 65 48 37  5 2703 806  32 379 311 138 325  9  3 700
W Randolph      ANYA2B492136  5 76 50 94 12 55111511  39 897 451 875 313381 20 875
W Tolleson      ACHA3B475126  3 61 43 52  6 1700 433   7 217  93 146  37113  7 385
Willie Upshaw   ATOR1B573144  9 85 60 78  8 3198 857  97 470 420 3321314131 12 960
Willie Wilson   AKC CF631170  9 77 44 31 11 49081457  30 775 357 249 408  4  31000
;
```

A2.4   The CITYTEMP Data Set: City Temperatures Data

The data set CITYTEMP contains the mean monthly temperature in January and July in 64 selected North American cities. The city names are listed in full in the variable CITY and abbreviated to the first three letters in the variable CTY.

Source: The data come from the *SAS User's Guide: Statistics, Version 5 Edition.* They appear in the files PRINCOMP SAS (Version 5) and PRINCOEX SAS (Version 6) in the SAS Sample Library.

```
title 'Mean temperature in January and July for selected cities';
data citytemp;
   input cty $1-3 city $1-15 january july;
cards;
MOBILE          51.2 81.6
PHOENIX         51.2 91.2
LITTLE ROCK     39.5 81.4
SACRAMENTO      45.1 75.2
DENVER          29.9 73.0
HARTFORD        24.8 72.7
WILMINGTON      32.0 75.8
WASHINGTON DC   35.6 78.7
```

```
JACKSONVILLE      54.6 81.0
MIAMI             67.2 82.3
ATLANTA           42.4 78.0
BOISE             29.0 74.5
CHICAGO           22.9 71.9
PEORIA            23.8 75.1
INDIANAPOLIS      27.9 75.0
DES MOINES        19.4 75.1
WICHITA           31.3 80.7
LOUISVILLE        33.3 76.9
NEW ORLEANS       52.9 81.9
PORTLAND, MAINE   21.5 68.0
BALTIMORE         33.4 76.6
BOSTON            29.2 73.3
DETROIT           25.5 73.3
SAULT STE MARIE   14.2 63.8
DULUTH             8.5 65.6
MINNEAPOLIS       12.2 71.9
JACKSON           47.1 81.7
KANSAS CITY       27.8 78.8
ST LOUIS          31.3 78.6
GREAT FALLS       20.5 69.3
OMAHA             22.6 77.2
RENO              31.9 69.3
CONCORD           20.6 69.7
ATLANTIC CITY     32.7 75.1
ALBUQUERQUE       35.2 78.7
ALBANY            21.5 72.0
BUFFALO           23.7 70.1
NEW YORK          32.2 76.6
CHARLOTTE         42.1 78.5
RALEIGH           40.5 77.5
BISMARCK           8.2 70.8
CINCINNATI        31.1 75.6
CLEVELAND         26.9 71.4
COLUMBUS          28.4 73.6
OKLAHOMA CITY     36.8 81.5
PORTLAND, OREG    38.1 67.1
PHILADELPHIA      32.3 76.8
PITTSBURGH        28.1 71.9
PROVIDENCE        28.4 72.1
COLUMBIA          45.4 81.2
SIOUX FALLS       14.2 73.3
MEMPHIS           40.5 79.6
NASHVILLE         38.3 79.6
DALLAS            44.8 84.8
EL PASO           43.6 82.3
HOUSTON           52.1 83.3
SALT LAKE CITY    28.0 76.7
BURLINGTON        16.8 69.8
NORFOLK           40.5 78.3
RICHMOND          37.5 77.9
SPOKANE           25.4 69.7
```

```
CHARLESTON, WV  34.5 75.0
MILWAUKEE       19.4 69.9
CHEYENNE        26.6 69.1
;
```

A2.5 The CRIME Data Set: State Crime Data

The data set CRIME contains the rates of occurrence (per 100,000 population) of seven types of crime in each of the 50 U.S. states. The state names are listed in full in the variable STATE and abbreviated to standard two-letter codes in the variable ST.

Source: The data come from the *SAS User's Guide: Statistics*. They appear in the files PRINCOMP SAS (Version 5) and PRINCOEX SAS (Version 6) in the SAS Sample Library.

```
data crime;
    input state $1-15 murder rape robbery assault burglary larceny
          auto st $;
    cards;
ALABAMA         14.2 25.2  96.8 278.3 1135.5 1881.9 280.7  AL
ALASKA          10.8 51.6  96.8 284.0 1331.7 3369.8 753.3  AK
ARIZONA          9.5 34.2 138.2 312.3 2346.1 4467.4 439.5  AZ
ARKANSAS         8.8 27.6  83.2 203.4  972.6 1862.1 183.4  AR
CALIFORNIA      11.5 49.4 287.0 358.0 2139.4 3499.8 663.5  CA
COLORADO         6.3 42.0 170.7 292.9 1935.2 3903.2 477.1  CO
CONNECTICUT      4.2 16.8 129.5 131.8 1346.0 2620.7 593.2  CT
DELAWARE         6.0 24.9 157.0 194.2 1682.6 3678.4 467.0  DE
FLORIDA         10.2 39.6 187.9 449.1 1859.9 3840.5 351.4  FL
GEORGIA         11.7 31.1 140.5 256.5 1351.1 2170.2 297.9  GA
HAWAII           7.2 25.5 128.0  64.1 1911.5 3920.4 489.4  HI
IDAHO            5.5 19.4  39.6 172.5 1050.8 2599.6 237.6  ID
ILLINOIS         9.9 21.8 211.3 209.0 1085.0 2828.5 528.6  IL
INDIANA          7.4 26.5 123.2 153.5 1086.2 2498.7 377.4  IN
IOWA             2.3 10.6  41.2  89.8  812.5 2685.1 219.9  IA
KANSAS           6.6 22.0 100.7 180.5 1270.4 2739.3 244.3  KS
KENTUCKY        10.1 19.1  81.1 123.3  872.2 1662.1 245.4  KY
LOUISIANA       15.5 30.9 142.9 335.5 1165.5 2469.9 337.7  LA
MAINE            2.4 13.5  38.7 170.0 1253.1 2350.7 246.9  ME
MARYLAND         8.0 34.8 292.1 358.9 1400.0 3177.7 428.5  MD
MASSACHUSETTS    3.1 20.8 169.1 231.6 1532.2 2311.3 1140.1 MA
MICHIGAN         9.3 38.9 261.9 274.6 1522.7 3159.0 545.5  MI
MINNESOTA        2.7 19.5  85.9  85.8 1134.7 2559.3 343.1  MN
MISSISSIPPI     14.3 19.6  65.7 189.1  915.6 1239.9 144.4  MS
MISSOURI         9.6 28.3 189.0 233.5 1318.3 2424.2 378.4  MO
MONTANA          5.4 16.7  39.2 156.8  804.9 2773.2 309.2  MT
NEBRASKA         3.9 18.1  64.7 112.7  760.0 2316.1 249.1  NE
NEVADA          15.8 49.1 323.1 355.0 2453.1 4212.6 559.2  NV
NEW HAMPSHIRE    3.2 10.7  23.2  76.0 1041.7 2343.9 293.4  NH
NEW JERSEY       5.6 21.0 180.4 185.1 1435.8 2774.5 511.5  NJ
NEW MEXICO       8.8 39.1 109.6 343.4 1418.7 3008.6 259.5  NM
NEW YORK        10.7 29.4 472.6 319.1 1728.0 2782.0 745.8  NY
NORTH CAROLINA  10.6 17.0  61.3 318.3 1154.1 2037.8 192.1  NC
```

NORTH DAKOTA	0.9	9.0	13.3	43.8	446.1	1843.0	144.7	ND
OHIO	7.8	27.3	190.5	181.1	1216.0	2696.8	400.4	OH
OKLAHOMA	8.6	29.2	73.8	205.0	1288.2	2228.1	326.8	OK
OREGON	4.9	39.9	124.1	286.9	1636.4	3506.1	388.9	OR
PENNSYLVANIA	5.6	19.0	130.3	128.0	877.5	1624.1	333.2	PA
RHODE ISLAND	3.6	10.5	86.5	201.0	1489.5	2844.1	791.4	RI
SOUTH CAROLINA	11.9	33.0	105.9	485.3	1613.6	2342.4	245.1	SC
SOUTH DAKOTA	2.0	13.5	17.9	155.7	570.5	1704.4	147.5	SD
TENNESSEE	10.1	29.7	145.8	203.9	1259.7	1776.5	314.0	TN
TEXAS	13.3	33.8	152.4	208.2	1603.1	2988.7	397.6	TX
UTAH	3.5	20.3	68.8	147.3	1171.6	3004.6	334.5	UT
VERMONT	1.4	15.9	30.8	101.2	1348.2	2201.0	265.2	VT
VIRGINIA	9.0	23.3	92.1	165.7	986.2	2521.2	226.7	VA
WASHINGTON	4.3	39.6	106.2	224.8	1605.6	3386.9	360.3	WA
WEST VIRGINIA	6.0	13.2	42.2	90.9	597.4	1341.7	163.3	WV
WISCONSIN	2.8	12.9	52.2	63.7	846.9	2614.2	220.7	WI
WYOMING	5.4	21.9	39.7	173.9	811.6	2772.2	282.0	WY

;

A2.6 The DIABETES Data Set: Diabetes Data

Reaven and Miller (1979) examined the relationship among blood chemistry measures of glucose tolerance and insulin in 145 nonobese adults classified as subclinical (chemical) diabetics, overt diabetics, and normals. The data set DIABETES contains the following variables:

PATIENT	patient number
RELWT	relative weight, expressed as a ratio of actual weight to expected weight, given the person's height
GLUFAST	fasting plasma glucose
GLUTEST	test plasma glucose, a measure of glucose intolerance
SSPG	steady state plasma glucose, a measure of insulin resistance
INSTEST	plasma insulin during test, a measure of insulin response to oral glucose
GROUP	clinical group (1=overt diabetic, 2=chemical diabetic, 3=normal)

Source: The data set is listed in Andrews and Herzberg (1985, chap. 36).*

---

* From *Data: A Collection of Problems from Many Fields for the Student and Research Worker* by D. F. Andrews and A. M. Herzberg. Copyright © 1985 by Springer-Verlag New York Inc. Reprinted with permission.

```
title 'Diabetes Data';
proc format;
    value gp    1='Overt Diabetic ' 2='Chem. Diabetic' 3='Normal';
data diabetes;
    input patient relwt glufast glutest instest sspg group;
    label relwt   = 'Relative weight'
          glufast = 'Fasting Plasma Glucose'
          glutest = 'Test Plasma Glucose'
          sspg    = 'Steady State Plasma Glucose'
          instest = 'Plasma Insulin during Test'
          group   = 'Clinical Group';
cards;
     1  0.81  80  356 124   55 3
     2  0.95  97  289 117   76 3
     3  0.94 105  319 143  105 3
     4  1.04  90  356 199  108 3
     5  1.00  90  323 240  143 3
     6  0.76  86  381 157  165 3
     7  0.91 100  350 221  119 3
     8  1.10  85  301 186  105 3
     9  0.99  97  379 142   98 3
    10  0.78  97  296 131   94 3
    11  0.90  91  353 221   53 3
    12  0.73  87  306 178   66 3
    13  0.96  78  290 136  142 3
    14  0.84  90  371 200   93 3
    15  0.74  86  312 208   68 3
    16  0.98  80  393 202  102 3
    17  1.10  90  364 152   76 3
    18  0.85  99  359 185   37 3
    19  0.83  85  296 116   60 3
    20  0.93  90  345 123   50 3
    21  0.95  90  378 136   47 3
    22  0.74  88  304 134   50 3
    23  0.95  95  347 184   91 3
    24  0.97  90  327 192  124 3
    25  0.72  92  386 279   74 3
    26  1.11  74  365 228  235 3
    27  1.20  98  365 145  158 3
    28  1.13 100  352 172  140 3
    29  1.00  86  325 179  145 3
    30  0.78  98  321 222   99 3
    31  1.00  70  360 134   90 3
    32  1.00  99  336 143  105 3
    33  0.71  75  352 169   32 3
    34  0.76  90  353 263  165 3
    35  0.89  85  373 174   78 3
    36  0.88  99  376 134   80 3
    37  1.17 100  367 182   54 3
    38  0.85  78  335 241  175 3
    39  0.97 106  396 128   80 3
    40  1.00  98  277 222  186 3
    41  1.00 102  378 165  117 3
```

42	0.89	90	360	282	160	3
43	0.98	94	291	94	71	3
44	0.78	80	269	121	29	3
45	0.74	93	318	73	42	3
46	0.91	86	328	106	56	3
47	0.95	85	334	118	122	3
48	0.95	96	356	112	73	3
49	1.03	88	291	157	122	3
50	0.87	87	360	292	128	3
51	0.87	94	313	200	233	3
52	1.17	93	306	220	132	3
53	0.83	86	319	144	138	3
54	0.82	86	349	109	83	3
55	0.86	96	332	151	109	3
56	1.01	86	323	158	96	3
57	0.88	89	323	73	52	3
58	0.75	83	351	81	42	3
59	0.99	98	478	151	122	2
60	1.12	100	398	122	176	3
61	1.09	110	426	117	118	3
62	1.02	88	439	208	244	2
63	1.19	100	429	201	194	2
64	1.06	80	333	131	136	3
65	1.20	89	472	162	257	2
66	1.05	91	436	148	167	2
67	1.18	96	418	130	153	3
68	1.01	95	391	137	248	3
69	0.91	82	390	375	273	3
70	0.81	84	416	146	80	3
71	1.10	90	413	344	270	2
72	1.03	100	385	192	180	3
73	0.97	86	393	115	85	3
74	0.96	93	376	195	106	3
75	1.10	107	403	267	254	3
76	1.07	112	414	281	119	3
77	1.08	94	426	213	177	2
78	0.95	93	364	156	159	3
79	0.74	93	391	221	103	3
80	0.84	90	356	199	59	3
81	0.89	99	398	76	108	3
82	1.11	93	393	490	259	3
83	1.19	85	425	143	204	2
84	1.18	89	318	73	220	3
85	1.06	96	465	237	111	2
86	0.95	111	558	748	122	2
87	1.06	107	503	320	253	2
88	0.98	114	540	188	211	2
89	1.16	101	469	607	271	2
90	1.18	108	486	297	220	2
91	1.20	112	568	232	276	2
92	1.08	105	527	480	233	2
93	0.91	103	537	622	264	2
94	1.03	99	466	287	231	2

```
 95  1.09 102   599 266  268  2
 96  1.05 110   477 124   60  2
 97  1.20 102   472 297  272  2
 98  1.05  96   456 326  235  2
 99  1.10  95   517 564  206  2
100  1.12 112   503 408  300  2
101  0.96 110   522 325  286  2
102  1.13  92   476 433  226  2
103  1.07 104   472 180  239  2
104  1.10  75   455 392  242  2
105  0.94  92   442 109  157  2
106  1.12  92   541 313  267  2
107  0.88  92   580 132  155  2
108  0.93  93   472 285  194  2
109  1.16 112   562 139  198  2
110  0.94  88   423 212  156  2
111  0.91 114   643 155  100  2
112  0.83 103   533 120  135  2
113  0.92 300  1468  28  455  1
114  0.86 303  1487  23  327  1
115  0.85 125   714 232  279  1
116  0.83 280  1470  54  382  1
117  0.85 216  1113  81  378  1
118  1.06 190   972  87  374  1
119  1.06 151   854  76  260  1
120  0.92 303  1364  42  346  1
121  1.20 173   832 102  319  1
122  1.04 203   967 138  351  1
123  1.16 195   920 160  357  1
124  1.08 140   613 131  248  1
125  0.95 151   857 145  324  1
126  0.86 275  1373  45  300  1
127  0.90 260  1133 118  300  1
128  0.97 149   849 159  310  1
129  1.16 233  1183  73  458  1
130  1.12 146   847 103  339  1
131  1.07 124   538 460  320  1
132  0.93 213  1001  42  297  1
133  0.85 330  1520  13  303  1
134  0.81 123   557 130  152  1
135  0.98 130   670  44  167  1
136  1.01 120   636 314  220  1
137  1.19 138   741 219  209  1
138  1.04 188   958 100  351  1
139  1.06 339  1354  10  450  1
140  1.03 265  1263  83  413  1
141  1.05 353  1428  41  480  1
142  0.91 180   923  77  150  1
143  0.90 213  1025  29  209  1
144  1.11 328  1246 124  442  1
145  0.74 346  1568  15  253  1
;
```

A2.7 The DRAFTUSA Data Set: Draft Lottery Data

The DRAFTUSA data set contains a rank ordering of the days of the year from the draft lottery conducted by the U.S. Selective Service in December of 1969. The priority number assigned to each day of the year is the order in which draft-eligible men born on that day would have been drafted into the armed forces in 1970.

The data set DRAFTUSA contains the following variables:

DAY   day of the month (1—31)

MONTH   month of the year (1—12)

PRIORITY   draft priority number (1—366)

Source: The data appear in Fienberg (1971).*

```
title 'USA Draft Lottery Data';
proc format;
     value mon    1='Jan'  2='Feb'  3='Mar'  4='Apr'  5='May'  6='Jun'
                  7='Jul'  8='Aug'  9='Sep' 10='Oct' 11='Nov' 12='Dec';
data draftusa;
     input day mon1-mon12;
     drop  i mon1-mon12;
     array mon{12} mon1-mon12;
     do i = 1 to 12;
        month=i;
        priority = mon{i};
        if priority ¬=. then output;
        end;
```

* Date cards;	Jan	Feb	Mar	Apr	May	Jun	Jul	Aug	Sep	Oct	Nov	Dec ;
1	305	086	108	032	330	249	093	111	225	359	019	129
2	159	144	029	271	298	228	350	045	161	125	034	328
3	251	297	267	083	040	301	115	261	049	244	348	157
4	215	210	275	081	276	020	279	145	232	202	266	165
5	101	214	293	269	364	028	188	054	082	024	310	056
6	224	347	139	253	155	110	327	114	006	087	076	010
7	306	091	122	147	035	085	050	168	008	234	051	012
8	199	181	213	312	321	366	013	048	184	283	097	105
9	194	338	317	219	197	335	277	106	263	342	080	043
10	325	216	323	218	065	206	284	021	071	220	282	041
11	329	150	136	014	037	134	248	324	158	237	046	039
12	221	068	300	346	133	272	015	142	242	072	066	314
13	318	152	259	124	295	069	042	307	175	138	126	163
14	238	004	354	231	178	356	331	198	001	294	127	026

---

* From "Randomization and Social Affairs: The 1970 Draft Lottery" by Stephen E. Fienberg, SCIENCE, Volume 171. Copyright © 1971 by the American Association for the Advancement of Science.

```
15  017  089  169  273  130  180  322  102  113  171  131  320
16  121  212  166  148  055  274  120  044  207  254  107  096
17  235  189  033  260  112  073  098  154  255  288  143  304
18  140  292  332  090  278  341  190  141  246  005  146  128
19  058  025  200  336  075  104  227  311  177  241  203  240
20  280  302  239  345  183  360  187  344  063  192  185  135
21  186  363  334  062  250  060  027  291  204  243  156  070
22  337  290  265  316  326  247  153  339  160  117  009  053
23  118  057  256  252  319  109  172  116  119  201  182  162
24  059  236  258  002  031  358  023  036  195  196  230  095
25  052  179  343  351  361  137  067  286  149  176  132  084
26  092  365  170  340  357  022  303  245  018  007  309  173
27  355  205  268  074  296  064  289  352  233  264  047  078
28  077  299  223  262  308  222  088  167  257  094  281  123
29  349  285  362  191  226  353  270  061  151  229  099  016
30  164   .   217  208  103  209  287  333  315  038  174  003
31  211   .   030   .   313   .   193  011   .   079   .   100
;
proc sort;
   by month;
```

A2.8 The DUNCAN Data Set: Duncan Occupational Prestige Data

The DUNCAN data set gives measures of income, education, and occupational prestige for 45 occupational titles for which income and education data were available in the 1950 U.S. Census. The variables are defined as follows:

JOB abbreviated job title.

TITLE census occupational category.

INCOME proportion of males in a given occupational category reporting income of $3,500 or more in the 1950 U.S. Census.

EDUC proportion of males in each occupation with at least a high school education in that census.

PRESTIGE percent of people rating the general standing of someone engaged in each occupation as good or excellent, using a five-point scale. The survey of almost 3,000 people was conducted by the National Opinion Research Center.

Source: The data come from Duncan (1961).*

```
title 'Duncan Occupational Prestige Data';
data duncan;
   input job $ 1-15 title $ 16-50 income educ prestige;
   case=_n_;
   index=mod(case, 10);
   label income='Income'        /* % males >= $3500    */
         educ='Education'       /* % males h.s. grad.  */
         prestige='Prestige';   /* % good or excellent */
cards;
Accountant      accountant for a large business   62  86  82
Pilot           airline pilot                     72  76  83
Architect       architect                         75  92  90
Author          author of novels                  55  90  76
Chemist         chemist                           64  86  90
Minister        minister                          21  84  87
Professor       college professor                 64  93  93
Dentist         dentist                           80 100  90
Reporter        reporter on a daily newspaper     67  87  52
Civil Eng.      civil engineer                    72  86  88
Undertaker      undertaker                        42  74  57
Lawyer          lawyer                            76  98  89
Physician       physician                         76  97  97
Welfare Wrkr.   welfare worker for city government 41 84  59
PS Teacher      instructor in the public schools  48  91  73
RR Conductor    railroad conductor                76  34  38
Contractor      building contractor               53  45  76
Factory Owner   owner of a factory employing 100  60  56  81
Store Manager   manager of a small store in a city 42 44  45
Banker          banker                            78  82  92
Bookkeeper      bookkeeper                        29  72  39
Mail carrier    mail carrier                      48  55  34
Insur. Agent    insurance agent                   55  71  41
Store clerk     clerk in a store                  29  50  16
Carpenter       carpenter                         21  23  33
Electrician     electrician                       47  39  53
RR Engineer     railroad engineer                 81  28  67
Machinist       trained machinist                 36  32  57
Auto repair     automobile repairman              22  22  26
Plumber         plumber                           44  25  29
Gas stn attn    filling-station attendant         15  29  10
Coal miner      coal miner                         7   7  15
Motorman        streetcar motorman                42  26  19
Taxi driver     taxi-driver                        9  19  10
Truck driver    truck-driver                      21  15  13
Machine opr.    machine-operator in a factory     21  20  24
Barber          barber                            16  26  20
```

```
Bartender     bartender              16  28   7
Shoe-shiner   shoe-shiner             9  17   3
Cook          restaurant cook        14  22  16
Soda clerk    soda fountain clerk    12  30   6
Watchman      night watchman         17  25  11
Janitor       janitor                 7  20   8
Policeman     policeman              34  47  41
Waiter        restaurant waiter       8  32  10
;
```

A2.9 The FUEL Data Set: Fuel Consumption Data

The FUEL data set gives the following variables for each of the 48 contiguous U.S. states:

AREA	state area (square miles)
POP	1971 state population (thousands)
TAX	1972 motor fuel tax (cents per gallon)
NLIC	1971 number licensed drivers (thousands)
DRIVERS	1971 proportion of licensed drivers
INC	1972 per capita personal income
ROAD	1971 length of federal highways (miles)
FUEL	1972 per capita fuel consumption

Note that one variable, NLIC, is a measure that varies directly with state population. INC, on the other hand, is scaled on a per-person basis. To make the variables comparable, NLIC is also represented in scaled form,

DRIVERS = NLIC / POP = proportion of population with driver's license

Source: The data are from Weisberg (1985).* The original source is the *American Almanac* for 1974. Fuel consumption data come from the 1974 *World Almanac and Book of Facts*. State area, which comes from *The Times Atlas of the World*, was added to the original data set.

```
title 'Fuel Consumption across the US';
data fuel;
   input state $ area pop tax nlic inc road drivers fuel;
   label area  = 'Area (sq. mi.)'
         pop   = 'Population (1000s)'
         tax   = 'Motor fuel tax (cents/gal.)'
         nlic  = 'Number licensed drivers (1000s)'
         drivers= 'Proportion licensed drivers'
```

```
                    inc   = 'Per Capita Personal income ($)'
                    road  = 'Length Federal Highways (mi.)'
                    fuel  = 'Fuel consumption (/person)';
          *STATE   AREA     POP    TAX    NLIC    INC   ROAD   DRIVERS   FUEL ;
          cards;
           AL     50767    3510    7.00   1801   3333   6594    0.513    554
           AR     52078    1978    7.50   1081   3357   4121    0.547    628
           AZ    113508    1945    7.00   1173   4300   3635    0.603    632
           CA    156299   20468    7.00  12130   5002   9794    0.593    524
           CO    103595    2357    7.00   1475   4449   4639    0.626    587
           CT      4872    3082   10.00   1760   5342   1333    0.571    457
           DE      1932     565    8.00    340   4983    602    0.602    540
           FL     54153    7259    8.00   4084   4188   5975    0.563    574
           GA     58056    4720    7.50   2731   3846   9061    0.579    631
           IA     55965    2883    7.00   1689   4318  10340    0.586    635
           ID     82412     756    8.50    501   3635   3274    0.663    648
           IL     55645   11251    7.50   5903   5126  14186    0.525    471
           IN     35932    5291    8.00   2804   4391   5939    0.530    580
           KS     81778    2258    7.00   1496   4593   7834    0.663    649
           KY     39669    3299    9.00   1626   3601   4650    0.493    534
           LA     44521    3720    8.00   1813   3528   3495    0.487    487
           MA      7824    5787    7.50   3060   4870   2351    0.529    414
           MD      9837    4056    9.00   2073   4897   2449    0.511    464
           ME     30995    1029    9.00    540   3571   1976    0.525    541
           MI     56954    9082    7.00   5213   4817   6930    0.574    525
           MN     79548    3896    7.00   2368   4332   8159    0.608    566
           MO     68945    4753    7.00   2719   4206   8508    0.572    603
           MS     47233    2263    8.00   1309   3063   6524    0.578    577
           MT    145388     719    7.00    421   3897   6385    0.586    704
           NC     48843    5214    9.00   2835   3721   4746    0.544    566
           ND     69300     632    7.00    341   3718   4725    0.540    714
           NE     76644    1525    8.50   1033   4341   6010    0.677    640
           NH      8993     771    9.00    441   4092   1250    0.572    524
           NJ      7468    7367    8.00   4074   5126   2138    0.553    467
           NM    121335    1065    7.00    600   3656   3985    0.563    699
           NV    109894     527    6.00    354   5215   2302    0.672    782
           NY     47377   18366    8.00   8278   5319  11868    0.451    344
           OH     41004   10783    7.00   5948   4512   8507    0.552    498
           OK     68655    2634    6.58   1657   3802   7834    0.629    644
           OR     96184    2182    7.00   1360   4296   4083    0.623    610
           PA     44888   11926    8.00   6312   4447   8577    0.529    464
           RI      1055     968    8.00    527   4399    431    0.544    410
           SC     30203    2665    8.00   1460   3448   5399    0.548    577
           SD     75952     579    7.00    419   4716   5915    0.724    865
           TN     41155    4031    7.00   2088   3640   6905    0.518    571
           TX    262017   11649    5.00   6595   4045  17782    0.566    640
           UT     82073    1126    7.00    572   3745   2611    0.508    591
           VA     39704    4764    9.00   2463   4258   4686    0.517    547
           VT      9273     462    9.00    268   3865   1586    0.580    561
```

WA	66511	3443	9.00	1966	4476	3942	0.571	510
WI	54426	4520	7.00	2465	4207	6580	0.545	508
WV	24119	1781	8.50	982	4574	2619	0.551	460
WY	96989	345	7.00	232	4345	3905	0.672	968

```
;
```

A2.10   The IRIS Data Set: Iris Data

The IRIS data set gives measurements on 50 flowers from each of three species of iris.

SPEC_NO	species number (1=*Setosa*, 2=*Versicolor*, 3=*Virginica*)
SPECIES	species name
SEPALLEN	sepal length in millimeters (mm.)
SEPALWID	sepal width in mm.
PETALLEN	petal length in mm.
PETALWID	petal width in mm.

Source: The data were collected by Edgar Anderson (1935) and made popular by Fisher (1936). The data are contained in several files in the SAS Sample Library, CANDISC SAS (Version 5) or CANDIEX SAS (Version 6), for example.

```
title 'Fisher (1936) Iris Data';
data iris;
   input sepallen sepalwid petallen petalwid spec_no @@;
   select(spec_no);
      when (1)  species='Setosa    ';
      when (2)  species='Versicolor';
      when (3)  species='Virginica ';
      otherwise ;
   end;
   label sepallen='Sepal length in mm.'
         sepalwid='Sepal width  in mm.'
         petallen='Petal length in mm.'
         petalwid='Petal width  in mm.';
   cards;
50 33 14 02 1  64 28 56 22 3  65 28 46 15 2  67 31 56 24 3
63 28 51 15 3  46 34 14 03 1  69 31 51 23 3  62 22 45 15 2
59 32 48 18 2  46 36 10 02 1  61 30 46 14 2  60 27 51 16 2
65 30 52 20 3  56 25 39 11 2  65 30 55 18 3  58 27 51 19 3
68 32 59 23 3  51 33 17 05 1  57 28 45 13 2  62 34 54 23 3
77 38 67 22 3  63 33 47 16 2  67 33 57 25 3  76 30 66 21 3
49 25 45 17 3  55 35 13 02 1  67 30 52 23 3  70 32 47 14 2
64 32 45 15 2  61 28 40 13 2  48 31 16 02 1  59 30 51 18 3
55 24 38 11 2  63 25 50 19 3  64 32 53 23 3  52 34 14 02 1
49 36 14 01 1  54 30 45 15 2  79 38 64 20 3  44 32 13 02 1
67 33 57 21 3  50 35 16 06 1  58 26 40 12 2  44 30 13 02 1
77 28 67 20 3  63 27 49 18 3  47 32 16 02 1  55 26 44 12 2
```

```
50 23 33 10 2    72 32 60 18 3    48 30 14 03 1    51 38 16 02 1
61 30 49 18 3    48 34 19 02 1    50 30 16 02 1    50 32 12 02 1
61 26 56 14 3    64 28 56 21 3    43 30 11 01 1    58 40 12 02 1
51 38 19 04 1    67 31 44 14 2    62 28 48 18 3    49 30 14 02 1
51 35 14 02 1    56 30 45 15 2    58 27 41 10 2    50 34 16 04 1
46 32 14 02 1    60 29 45 15 2    57 26 35 10 2    57 44 15 04 1
50 36 14 02 1    77 30 61 23 3    63 34 56 24 3    58 27 51 19 3
57 29 42 13 2    72 30 58 16 3    54 34 15 04 1    52 41 15 01 1
71 30 59 21 3    64 31 55 18 3    60 30 48 18 3    63 29 56 18 3
49 24 33 10 2    56 27 42 13 2    57 30 42 12 2    55 42 14 02 1
49 31 15 02 1    77 26 69 23 3    60 22 50 15 3    54 39 17 04 1
66 29 46 13 2    52 27 39 14 2    60 34 45 16 2    50 34 15 02 1
44 29 14 02 1    50 20 35 10 2    55 24 37 10 2    58 27 39 12 2
47 32 13 02 1    46 31 15 02 1    69 32 57 23 3    62 29 43 13 2
74 28 61 19 3    59 30 42 15 2    51 34 15 02 1    50 35 13 03 1
56 28 49 20 3    60 22 40 10 2    73 29 63 18 3    67 25 58 18 3
49 31 15 01 1    67 31 47 15 2    63 23 44 13 2    54 37 15 02 1
56 30 41 13 2    63 25 49 15 2    61 28 47 12 2    64 29 43 13 2
51 25 30 11 2    57 28 41 13 2    65 30 58 22 3    69 31 54 21 3
54 39 13 04 1    51 35 14 03 1    72 36 61 25 3    65 32 51 20 3
61 29 47 14 2    56 29 36 13 2    69 31 49 15 2    64 27 53 19 3
68 30 55 21 3    55 25 40 13 2    48 34 16 02 1    48 30 14 01 1
45 23 13 03 1    57 25 50 20 3    57 38 17 03 1    51 38 15 03 1
55 23 40 13 2    66 30 44 14 2    68 28 48 14 2    54 34 17 02 1
51 37 15 04 1    52 35 15 02 1    58 28 51 24 3    67 30 50 17 2
63 33 60 25 3    53 37 15 02 1
;
```

A2.11 The NATIONS Data Set: Infant Mortality Data

The NATIONS data set gives the following information on 105 nations in 1970:

NATION	name of the nation
REGION	region of the world
INCOME	per capita income
IMR	infant mortality rate (per 1,000 live births)
OILEXPRT	oil exporting country (0=no, 1=yes)
IMR80	infant mortality rate, 1980
GNP80	GNP per capital, 1980

Source: The 1970 data are from Leinhardt and Wasserman (1979).* Infant mortality rate is missing for four nations, and the 1970 rates for several nations

---

* Schuessler, K. F. (ed.) *Sociological Methodology 1979*. San Francisco: Jossey-Bass, 1978, pp. 320—321 and 343 (tables).

appear excessively high. The data for 1980 come from *The New Book of World Rankings* (Kurian 1984).*

```
proc format;
  value region 1='Americas'    2='Africa'
               3='Europe'      4='Asia/Oceania';
  value oil    1='Yes'         0='No';
data nations;
  input nation $ 1-21 income imr region oilexprt imr80 gnp80;
  label income=  'Per Capita Income'
        imr=     'Infant Mortality Rate'
        oilexprt='Oil Exporting Country'
        imr80 =  'Infant Mortality Rate, 1980'
        gnp80 =  'Per Capita GNP, 1980' ;
  format region region. oilexprt oil.;
cards;
Afghanistan             75   400.0   4   0   185.0     .
Algeria                400    86.3   2   1    20.5   1920
Argentina             1191    59.6   1   0    40.8   2390
Australia             3426    26.7   4   0    12.5   9820
Austria               3350    23.7   3   0    14.8  10230
Bangladesh             100   124.3   4   0   139.0    120
Belgium               3346    17.0   3   0    11.2  12180
Benin                   81   109.6   2   0   109.6    300
Bolivia                200    60.4   1   0    77.3    570
Brazil                 425   170.0   1   0    84.0   2020
Britain               2503    17.5   3   0    12.6   7920
Burma                   73   200.0   4   0   195.0    180
Burundi                 68   150.0   2   0   150.0    200
Cambodia               123   100.0   4   0     .        .
Cameroon               100   137.0   2   0   157.0    670
Canada                4751    16.8   1   0    12.0  10130
Central Afr. Republic  122   190.0   2   0   190.0    300
Chad                    70   160.0   2   0   160.0    120
Chile                  590    78.0   1   0    40.1   2160
Colombia               426    62.8   1   0    46.6   1180
Congo                  281   180.0   2   0   180.0    730
Costa Rica             725    54.4   1   0    22.3   1730
Denmark               5029    13.5   3   0     9.1  12950
Dominican Republic     406    48.8   1   0     .        .
Ecuador                250    78.5   1   1    72.1   1220
Egypt                  210   114.0   2   0     .        .
El Salvador            319    58.2   1   0    50.8    590
Ethiopia                79    84.2   2   0    84.2    140
Finland               3312    10.1   3   0     .        .
France                3403    12.9   3   0     9.6  11730
Ghana                  217    63.7   2   0   156.0    420
Greece                1760    27.8   3   0    18.7   4520
```

Guatemala	302	79.1	1	0	69.2	1110
Guinea	79	216.0	2	0	216.0	290
Haiti	100	.	1	0	130.0	270
Honduras	284	39.3	1	0	31.4	560
India	93	60.6	4	0	122.0	240
Indonesia	110	125.0	4	1	125.0	420
Iran	1280	.	4	1	108.1	.
Iraq	560	28.1	4	1	29.9	3020
Ireland	2009	17.8	3	0	14.9	4880
Israel	2526	22.1	4	0	16.0	4500
Italy	2298	25.7	3	0	15.3	6480
Ivory Coast	387	138.0	2	0	138.0	1150
Jamaica	727	26.2	1	0	16.2	1030
Japan	3292	11.7	3	0	8.0	9890
Jordan	334	21.3	4	0	14.9	1620
Kenya	169	55.0	2	0	54.1	420
Laos	71	.	4	0	175.0	.
Lebanon	631	13.6	4	0	13.6	.
Liberia	197	159.2	2	0	159.2	520
Libya	3010	300.0	2	1	130.0	8640
Madagascar	120	102.0	2	0	102.0	350
Malawi	130	148.3	2	0	142.1	230
Malaysia	295	32.0	4	0	31.8	1670
Mali	50	120.0	2	0	120.0	190
Mauritania	174	187.0	2	0	187.0	320
Mexico	684	60.9	1	0	60.2	2050
Morocco	279	149.0	2	0	149.0	860
Nepal	90	.	4	0	133.0	140
Netherlands	4103	11.6	3	0	8.5	11470
New Zealand	3723	16.2	4	0	13.8	7090
Nicaragua	507	46.0	1	0	42.9	720
Niger	70	200.0	2	0	200.0	330
Nigeria	220	58.0	2	1	157.0	1010
Norway	4102	11.3	3	0	8.6	12650
Pakistan	102	124.3	4	0	124.0	350
Panama	754	34.1	1	0	22.0	1730
Papua New Guinea	477	10.2	4	0	128.0	780
Paraguay	347	38.6	1	0	38.6	1340
Peru	335	65.1	1	0	70.3	950
Philippines	230	67.9	4	0	47.6	720
Portugal	956	44.8	3	0	38.9	2350
Rwanda	61	132.9	2	0	127.0	200
Saudi Arabia	1530	650.0	4	1	118.0	11260
Sierra Leone	148	170.0	2	0	136.0	270
Singapore	1268	20.4	4	0	13.2	4480
Somalia	85	158.0	2	0	177.0	.
South Africa	1000	71.5	2	0	50.0	2290
South Korea	344	58.0	4	0	37.0	1520
South Yemen	96	80.0	4	0	170.0	420
Spain	1256	15.1	3	0	15.1	5350
Sri Lanka	162	45.1	4	0	42.4	270
Sudan	125	129.4	2	0	93.6	470
Sweden	5596	9.6	3	0	7.3	13520

Switzerland	2963	12.8	3	0	8.6	16440
Syria	334	21.7	4	0	13.0	1340
Taiwan	261	19.1	4	0	.	.
Tanzania	120	162.5	2	0	165.0	265
Thailand	210	27.0	4	0	25.5	670
Togo	160	127.0	2	0	127.0	410
Trinidad & Tobago	732	26.2	1	0	24.4	4370
Tunisia	434	76.3	2	0	125.0	1310
Turkey	435	153.0	4	0	153.0	1460
Uganda	134	160.0	2	0	160.0	280
United States	5523	17.6	1	0	13.0	11360
Upper Volta	82	180.0	2	0	182.0	190
Uruguay	799	40.4	1	0	48.5	2820
Venezuela	1240	51.7	1	1	33.7	3630
Vietnam	130	100.0	4	0	115.0	.
West Germany	5040	20.4	3	0	14.7	13590
Yemen	77	50.0	4	0	160.0	460
Yugoslavia	406	43.3	3	0	32.2	2620
Zaire	118	104.0	2	0	104.0	220
Zambia	310	259.0	2	0	259.0	560

;

A2.12 The SALARY Data Set: Salary Survey Data

The data set SALARY contains data from a salary survey (fictitious) of 46 computer professionals in a large corporation designed to investigate the roles of experience, education, and management responsibility as determinants of salary (Chatterjee and Price 1977). The data set SALARY contains the following variables:

EXPRNC   experience (years)

EDUC   education (1=high school, 2=B.S. degree, 3=advanced degree)

MGT   management responsibility (0=no, 1=yes)

GROUP   code for education—management group (1—6)

SALARY   salary, expressed in increments of $1,000

Source: The data are taken from Chatterjee and Price (1977, Table 4.1).*

```
Title 'Salary survey data';
* Formats for group codes;
proc format;
     value glfmt    1='HS' 2='BS' 3='AD' 4='HSM' 5='BSM' 6='ADM';
     value edfmt    1='High School' 2='B.S. Degree' 3='Advanced Degree';
     value mgfmt    0='Non-management' 1='Management';
```

```
data salary;
   input case exprnc educ mgt salary;
   label exprnc = 'Experience (years)'
         educ   = 'Education'
         mgt    = 'Management responsibility'
         salary = 'Salary (in $1000s)';

   salary = salary / 1000;
   group = 3*(mgt=1)+educ;
   format group glfmt.;
cards;
 1    1   1   1   13876
 2    1   3   0   11608
 3    1   3   1   18701
 4    1   2   0   11283
 5    1   3   0   11767
 6    2   2   1   20872
 7    2   2   0   11772
 8    2   1   0   10535
 9    2   3   0   12195
10    3   2   0   12313
11    3   1   1   14975
12    3   2   1   21371
13    3   3   1   19800
14    4   1   0   11417
15    4   3   1   20263
16    4   3   0   13231
17    4   2   0   12884
18    5   2   0   13245
19    5   3   0   13677
20    5   1   1   15965
21    6   1   0   12336
22    6   3   1   21352
23    6   2   0   13839
24    6   2   1   22884
25    7   1   1   16978
26    8   2   0   14803
27    8   1   1   17404
28    8   3   1   22184
29    8   1   0   13548
30   10   1   0   14467
31   10   2   0   15942
32   10   3   1   23174
33   10   2   1   23780
34   11   2   1   25410
35   11   1   0   14861
36   12   2   0   16882
37   12   3   1   24170
38   13   1   0   15990
39   13   2   1   26330
40   14   2   0   17949
41   15   3   1   25685
42   16   2   1   27837
```

```
43  16  2  0  18838
44  16  1  0  17483
45  17  2  0  19207
46  20  1  0  19346
;
```

A2.13 The SPENDING Data Set: School Spending Data

The SPENDING data set lists the estimated expenditure on public school education in each of the 50 U.S. states plus the District of Columbia in 1970, and several related predictor variables (per capita income, proportion of young people, and the degree of urbanization in each state).

ST	state two-letter postal abbreviation.
STATE	state two-digit FIPS code.
REGION	geographic region.
GROUP	geographic subregion.
SPENDING	public school expenditures per capita (not per student) in 1970. Schools include elementary and secondary schools, as well as other programs under the jurisdiction of local school boards, but not state universities.
INCOME	personal income per capita, 1968.
YOUTH	proportion of persons below the age of 18 in 1969, per 1,000 state population.
URBAN	proportion of persons classified as urban in the 1970 census, per 1,000 state population.

Source: The data are taken from Anscombe (1981).* His source is the *Statistical Abstract of the United States, 1970.*

```
Proc Format;
   Value $REGION
          'NE' = 'North East'    'NC' = 'North Central'
          'SO' = 'South Region' 'WE' = 'West Region';
   Value $GROUP
          'NE' = 'New England'       'MA' = 'Mid Atlantic'
          'ENC'= 'East North Central' 'WNC'= 'West North Central'
          'SA' = 'South Atlantic'     'ESC'= 'East South Central'
          'WSC'= 'West South Central' 'MT' = 'Mountain States'
          'PA' = 'Pacific States';
```

---

* From *Computing in Statistical Science through APL* by Francis John Anscombe. Copyright © 1981 by Springer-Verlag New York Inc.

```
Data Schools;
   Input ST $  SPENDING INCOME YOUTH URBAN REGION $ GROUP $;
   STATE=STFIPS(ST);
   LABEL ST = 'State'
         SPENDING='School Expenditures 1970'
         INCOME  ='Personal Income 1968'
         YOUTH   ='Young persons 1969'
         URBAN   ='Proportion Urban';
cards;
ME  189  2824  350.7   508  NE  NE
NH  169  3259  345.9   564  NE  NE
VT  230  3072  348.5   322  NE  NE
MA  168  3835  335.3   846  NE  NE
RI  180  3549  327.1   871  NE  NE
CT  193  4256  341.0   774  NE  NE
NY  261  4151  326.2   856  NE  MA
NJ  214  3954  333.5   889  NE  MA
PA  201  3419  326.2   715  NE  MA
OH  172  3509  354.5   753  NC  ENC
IN  194  3412  359.3   649  NC  ENC
IL  189  3981  348.9   830  NC  ENC
MI  233  3675  369.2   738  NC  ENC
WI  209  3363  360.7   659  NC  ENC
MN  262  3341  365.4   664  NC  WNC
IA  234  3265  343.8   572  NC  WNC
MO  177  3257  336.1   701  NC  WNC
ND  177  2730  369.1   443  NC  WNC
SD  187  2876  368.7   446  NC  WNC
NE  148  3239  349.9   615  NC  WNC
KS  196  3303  339.9   661  NC  WNC
DE  248  3795  375.9   722  SO  SA
MD  247  3742  364.1   766  SO  SA
DC  246  4425  352.1  1000  SO  SA
VA  180  3068  353.0   631  SO  SA
WV  149  2470  328.8   390  SO  SA
NC  155  2664  354.1   450  SO  SA
SC  149  2380  376.7   476  SO  SA
GA  156  2781  370.6   603  SO  SA
FL  191  3191  336.0   805  SO  SA
KY  140  2645  349.3   523  SO  ESC
TN  137  2579  342.8   588  SO  ESC
AL  112  2337  362.2   584  SO  ESC
MS  130  2081  385.2   445  SO  ESC
AR  134  2322  351.9   500  SO  WSC
LA  162  2634  389.6   661  SO  WSC
OK  135  2880  329.8   680  SO  WSC
TX  155  3029  369.4   797  SO  WSC
MT  238  2942  368.9   534  WE  MT
ID  170  2668  367.7   541  WE  MT
WY  238  3190  365.6   605  WE  MT
CO  192  3340  358.1   785  WE  MT
NM  227  2651  421.5   698  WE  MT
AZ  207  3027  387.5   796  WE  MT
```

```
UT  201  2790  412.4  804  WE  MT
NV  225  3957  385.1  809  WE  MT
WA  215  3688  341.3  726  WE  PA
OR  233  3317  332.7  671  WE  PA
CA  273  3968  348.4  909  WE  PA
AK  372  4146  439.7  484  WE  PA
HI  212  3513  382.9  831  WE  PA
;
```

A2.14 The TEETH Data Set: Mammals' Teeth Data

The data set TEETH lists the number of each of eight types of teeth found in 32 species of mammals. The data set contains the following variables:

MAMMAL	name of mammal
ID	observation number, as a two-digit character string
V1	number of top incisors
V2	number of bottom incisors
V3	number of top canines
V4	number of bottom canines
V5	number of top premolars
V6	number of bottom premolars
V7	number of top molars
V8	number of bottom molars

Source: The data come from the *SAS User's Guide: Statistics*. The data are contained in several files in the SAS Sample Library, CLUSTER2 SAS (Version 5) or CLUSTER SAS (Version 6), for example.

```
data teeth;
   title "Mammals' Teeth Data";
   input mammal $ 1-16 @21 (v1-v8) (1.);
   length id $2;
   id=put(_n_,z2.);
   format v1-v8 1.;
   label v1='Top incisors'
         v2='Bottom incisors'
         v3='Top canines'
         v4='Bottom canines'
         v5='Top premolars'
         v6='Bottom premolars'
         v7='Top molars'
         v8='Bottom molars';
   cards;
BROWN BAT          23113333
MOLE               32103333
```

```
        SILVER HAIR BAT    23112333
        PIGMY BAT          23112233
        HOUSE BAT          23111233
        RED BAT            13112233
        PIKA               21002233
        RABBIT             21003233
        BEAVER             11002133
        GROUNDHOG          11002133
        GRAY SQUIRREL      11001133
        HOUSE MOUSE        11000033
        PORCUPINE          11001133
        WOLF               33114423
        BEAR               33114423
        RACCOON            33114432
        MARTEN             33114412
        WEASEL             33113312
        WOLVERINE          33114412
        BADGER             33113312
        RIVER OTTER        33114312
        SEA OTTER          32113312
        JAGUAR             33113211
        COUGAR             33113211
        FUR SEAL           32114411
        SEA LION           32114411
        GREY SEAL          32113322
        ELEPHANT SEAL      21114411
        REINDEER           04103333
        ELK                04103333
        DEER               04003333
        MOOSE              04003333
        ;
```

A2.15 The WHEAT Data Set: Broadbalk Wheat Data

The data set WHEAT contains the yields, in bushels of dressed grain per acre, from two plots, labeled 9a and 7b, in the Broadbalk wheat experiments over the 30 years from 1855 to 1884. Anscombe (1981) reports that the plots had been treated identically with the same type and amount of fertilizers, except that plot 9a had received nitrogen in the form of nitrate of soda, whereas 7b had received ammonium salts. The data set includes the following variables:

YEAR   year of experiment, 1855—1884

YIELDIFF   yield difference, plots 9a and plot 7b

RAIN   winter rainfall, November to February (inches)

PLOT9A   yield for plot 9a (bushels per acre)

PLOT7B   yield for plot 7b (bushels per acre)

Source: The data are taken from Fisher (1925, Table 29).* See also Anscombe (1981, p. 121). Andrews and Herzberg (1985, chap. 5) list more extensive data from the Broadbalk wheat experiments, from 1855 to 1925. These data, however, are given in units of tons per hectare and do not list the a,b subplots separately.

```
*-------------------------------------------------------------------*
* Yield difference of two plots in Broadbalk wheat field (bushels/acre)
* Source: Fisher, Statistical Methods for Research Workers, table 29
* See also: Anscombe, Statistical computing with APL, p121.
*-------------------------------------------------------------------*
;
title 'Broadbalk wheat experiment';

data wheat;
      retain year 1855;
      input yieldiff rain plot9a plot7b;
      output;
      year+1;
      label yieldiff ='Yield difference, plots 9a & 7b'
            rain     ='Rainfall, Nov to Feb (inches)';
cards;
 -3.38   5.1   29.62   33.00
 -4.53   8.1   32.38   36.91
 -1.09   7.9   43.75   44.84
 -1.38   5.2   37.56   38.94
 -4.66   6.2   30.00   34.66
  4.90   9.7   32.62   27.72
 -1.19   7.2   33.75   34.94
  7.56   7.9   43.44   35.88
  1.90   7.9   55.56   53.66
  5.28   6.0   51.06   45.78
  3.84   8.9   44.06   40.22
  2.59  11.3   32.50   29.91
  6.97   9.4   29.13   22.16
  8.62   7.8   47.81   39.19
 10.75  10.9   39.00   28.25
  4.13   9.5   45.50   41.37
 12.13   7.1   34.44   22.31
 11.63   8.2   40.69   29.06
 13.06  13.3   35.81   22.75
 -1.37   6.4   38.19   39.56
  3.87   9.3   30.50   26.63
  7.81  10.5   33.31   25.50
 21.00  17.3   40.12   19.12
  5.00  11.0   37.19   32.19
  4.69  12.8   21.94   17.25
 -0.25   5.1   34.06   34.31
```

```
 9.31   11.2   35.44   26.13
-2.94   11.4   31.81   34.75
 7.07   14.4   43.38   36.31
 2.69    8.7   40.44   37.75
;
```

Appendix 3 Color Output

Output A3.1   *Scatterplot Matrix for Baseball Data (American League: Black +; National League: Red ×.)*

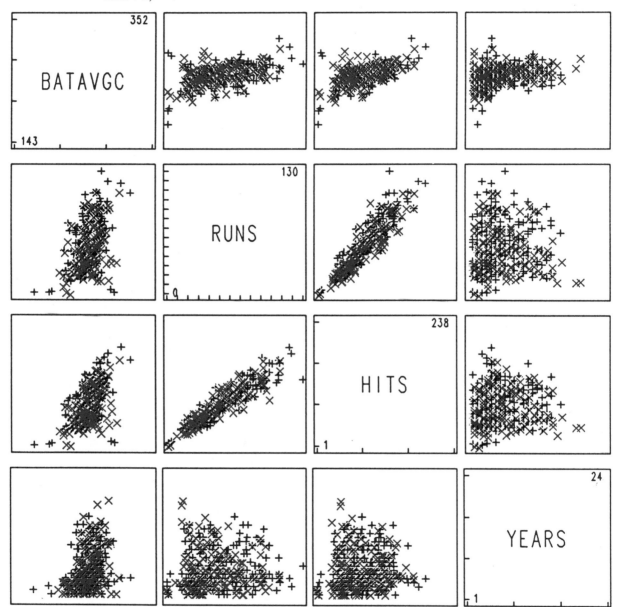

(A black-and-white version of this output, with explanatory text, appears as Output 1.9 in Chapter 1, "Introduction to Statistical Graphics.")

Output A3.2
C$_P$ Plot for
Baseball Data

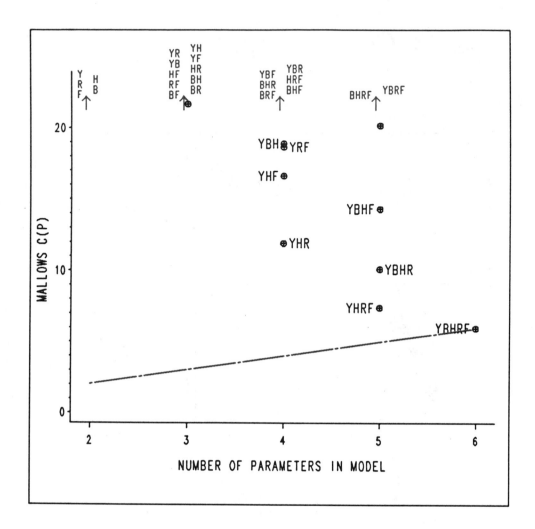

(A black-and-white version of this output, with explanatory text, appears as Output 1.12 in Chapter 1.)

Output A3.3   *Matching Graphic Form to the Question*

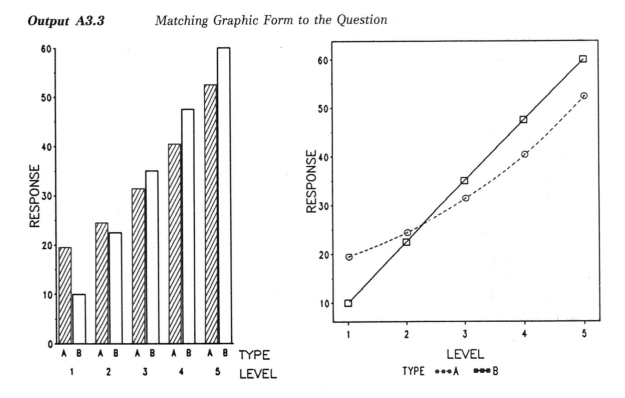

(A black-and-white version of this output, with explanatory text, appears as Output 1.15 in Chapter 1.)

Output A3.4   *Visual Grouping: Comparing Bars in a Grouped Bar Chart*

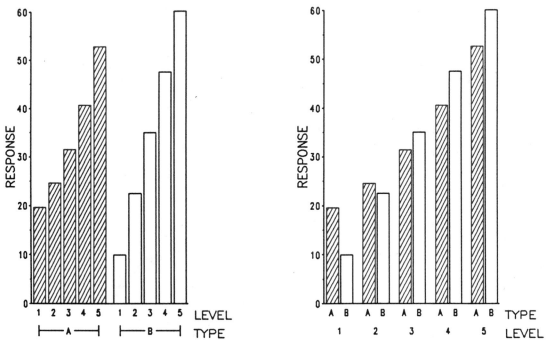

(A black-and-white version of this output, with explanatory text, appears as Output 1.16 in Chapter 1.)

Output A3.5   *Visual Grouping: Comparing Connected Points in Line Graphs*

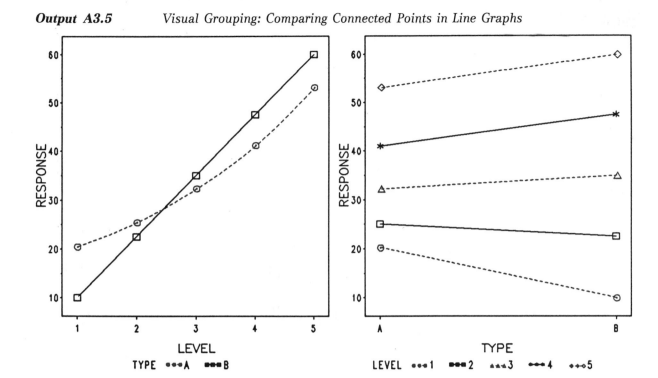

(A black-and-white version of this output, with explanatory text, appears as Output 1.17 in Chapter 1.)

Output A3.6   *Bar Chart of Mean Hits with Standard Error Bars*

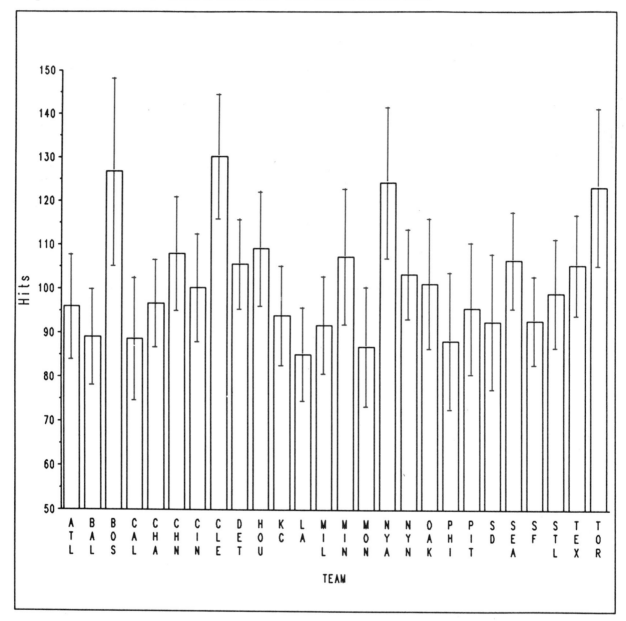

(A black-and-white version of this output, with explanatory text, appears as Output 2.6 in Chapter 2, "Graphical Methods for Univariate Data.")

Output A3.7
Parametric Plot of Chi Square Density Function

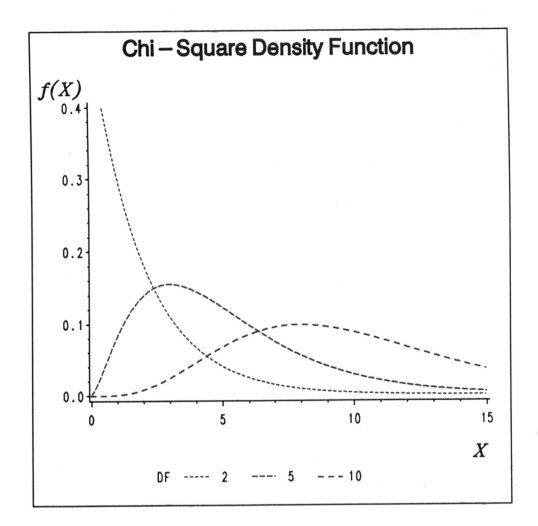

(A black-and-white version of this output, with explanatory text, appears as Output 3.4 in Chapter 3, "Plotting Theoretical and Empirical Distributions.")

Output A3.8
*Plot of the F
Density Function*

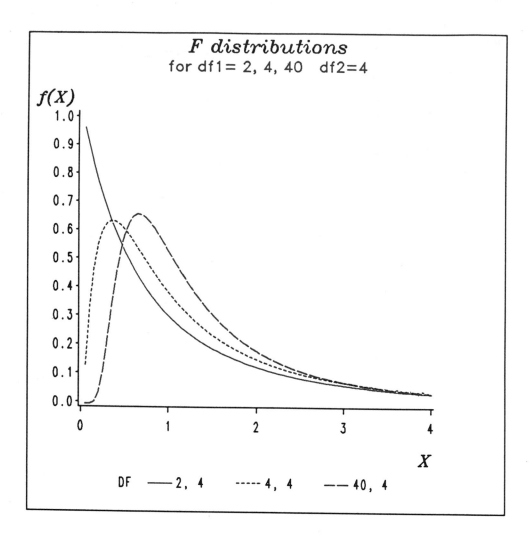

(A black-and-white version of this output, with explanatory text, appears as Output 3.5 in Chapter 3.)

Output A3.9
*Percentiles of
Student's t
Distribution*

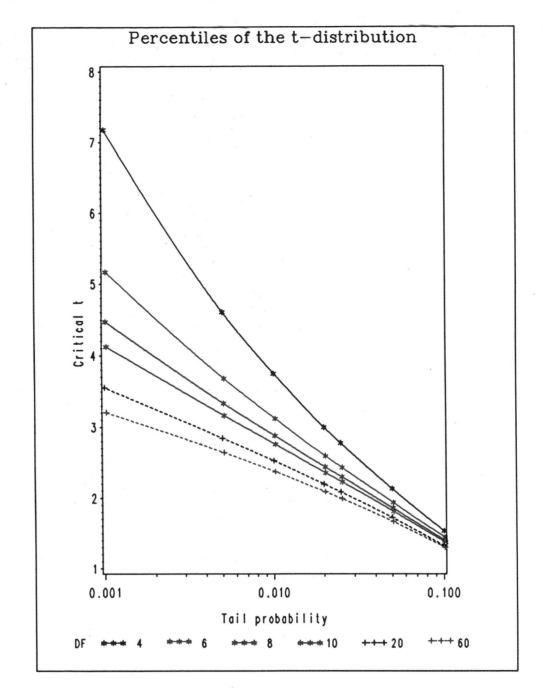

(A black-and-white version of this output, with explanatory text, appears as Output 3.6 in Chapter 3.)

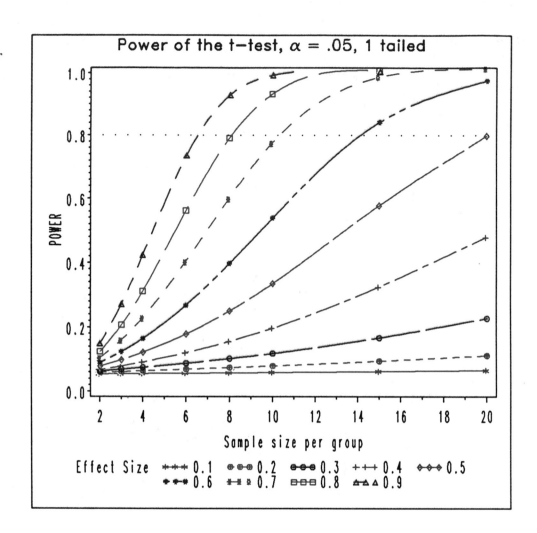

(A black-and-white version of this output, with explanatory text, appears as Output 3.8 in Chapter 3.)

Output A3.11
*Contour Plot of
Bivariate Normal
Distribution*

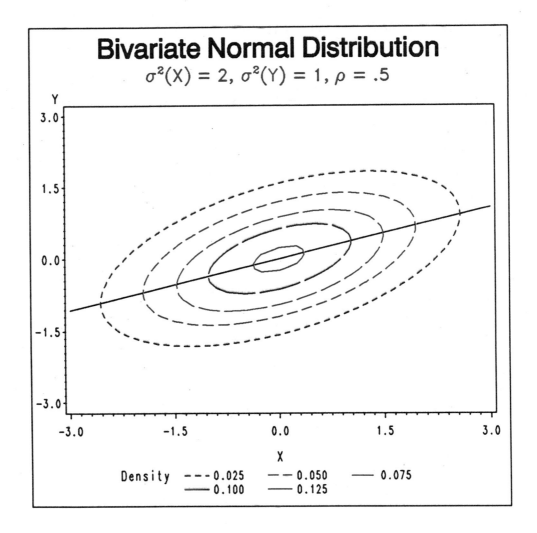

(A black-and-white version of this output, with explanatory text, appears as Output 3.10 in Chapter 3.)

Output A3.12   *Annotated Plot of PRESTIGE by INCOME*

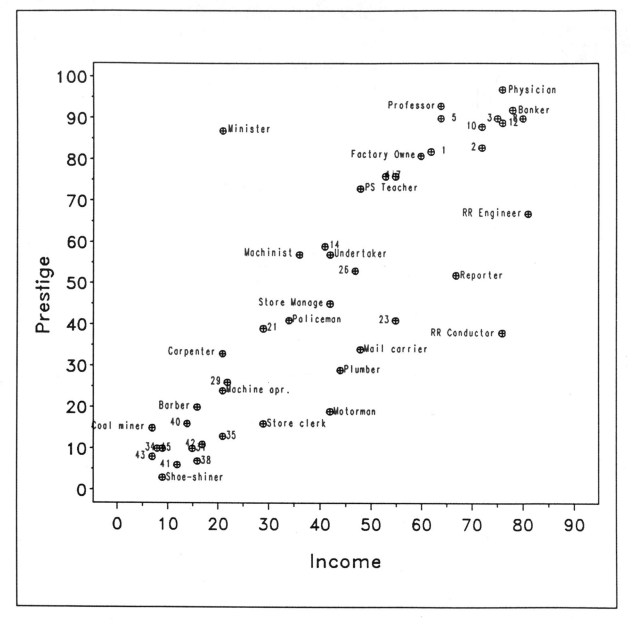

(A black-and-white version of this output, with explanatory text, appears as Output 4.12 in Chapter 4, "Scatterplots.")

Output A3.13   *Plot of Auto Data with Boxplots for All Axes from the G3D Procedure*

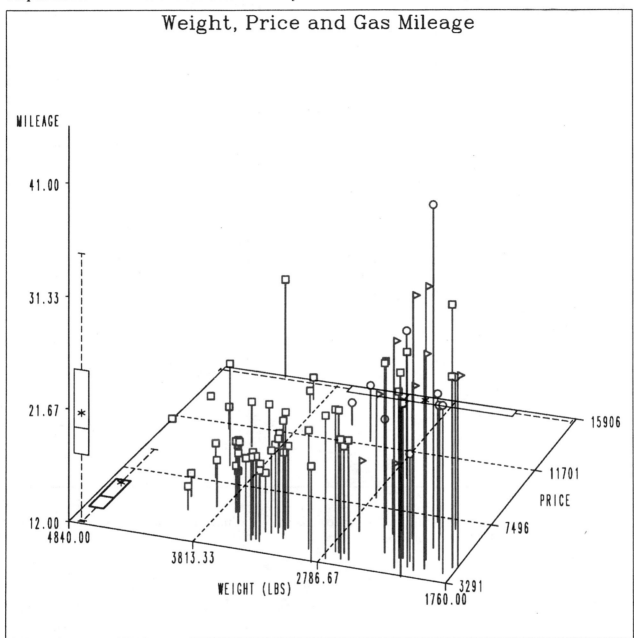

(A black-and-white version of this output, with explanatory text, appears as Output 4.17 in Chapter 4.)

Output A3.14
*WEIGHT Plotted
against PRICE of
Automobiles, with
Data Ellipse for
Each Region of
Origin*

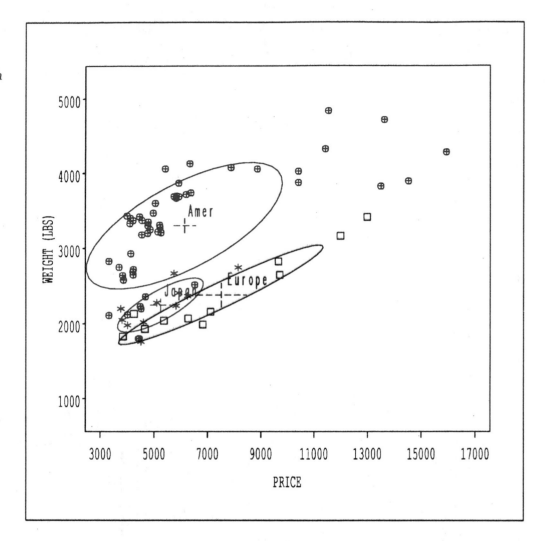

(A black-and-white version of this output, with explanatory text, appears as Output 4.22 in Chapter 4.)

Output A3.15
Index Plot of Yield
Difference
Accompanied by
Rainfall

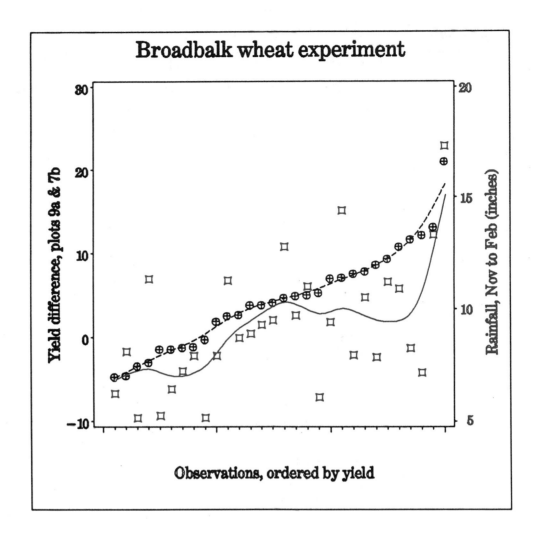

(A black-and-white version of this output, with explanatory text, appears as Output 4.29 in Chapter 4.)

Output A3.16   *Bubble Plot of Weight Plotted against Price for Automobiles Data*

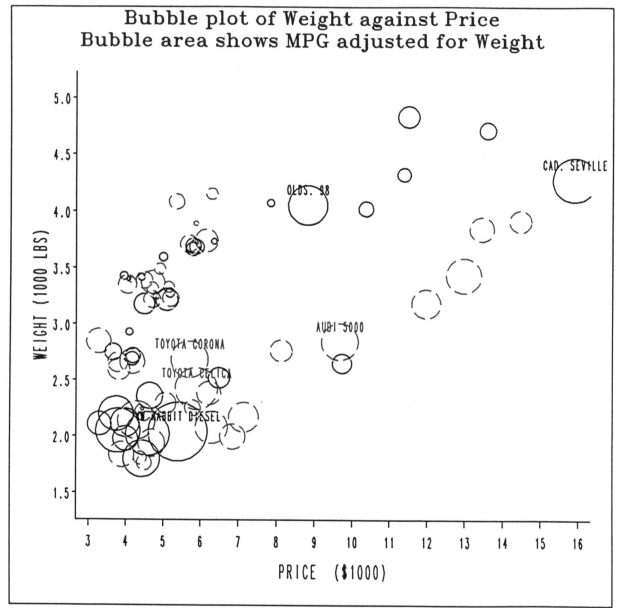

(A black-and-white version of this output, with explanatory text, appears as Output 4.30 in Chapter 4.)

Output A3.17
Three-Dimensional Scatterplot of Iris Data

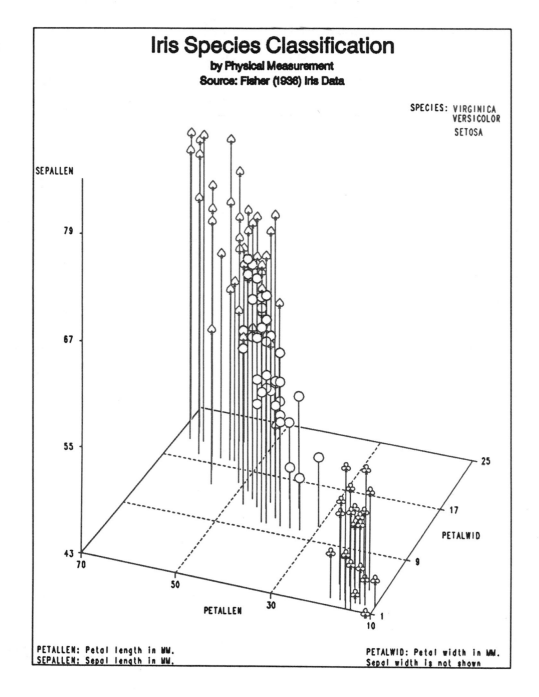

(A black-and-white version of this output, with explanatory text, appears as Output 4.31 in Chapter 4.)

Output A3.18   *Plot of Salary Survey Data with Confidence Bands*

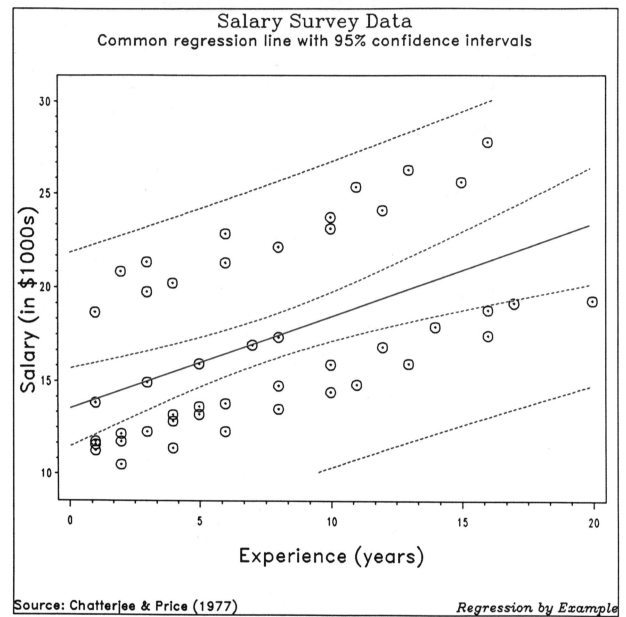

(A black-and-white version of this output, with explanatory text, appears as Output 5.4 in Chapter 5, "Plotting Regression Data.")

Output A3.19 *Salary Survey Data, with a Separate Regression Line for Each Group*

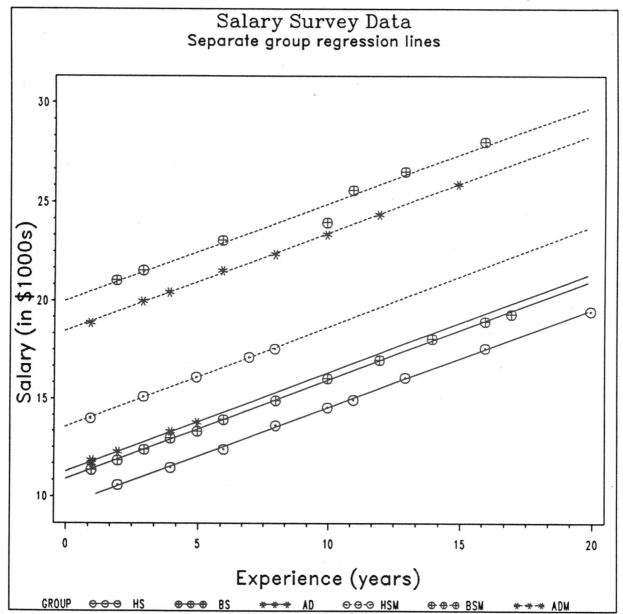

(A black-and-white version of this output, with explanatory text, appears as Output 5.5 in Chapter 5.)

Output A3.20 *Residuals from Model Including Education and Management Dummy Variables*

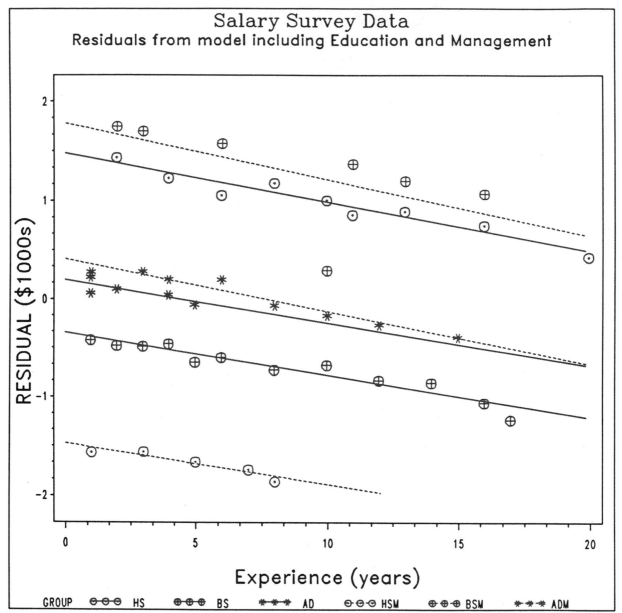

(A black-and-white version of this output, with explanatory text, appears as Output 5.7 in Chapter 5.)

Output A3.21   *Residuals from Model Including Education and Management Dummy Variables*

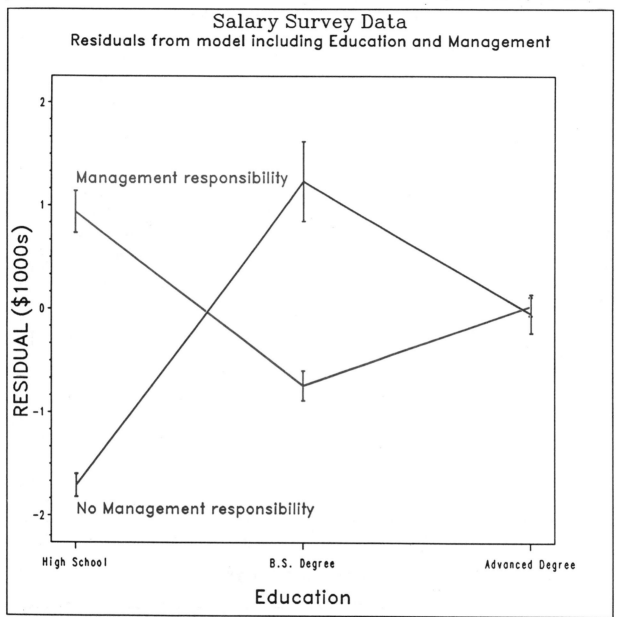

(A black-and-white version of this output, with explanatory text, appears as Output 5.8 in Chapter 5.)

Output A3.22   *Residuals from Model with Education—Management Interaction*

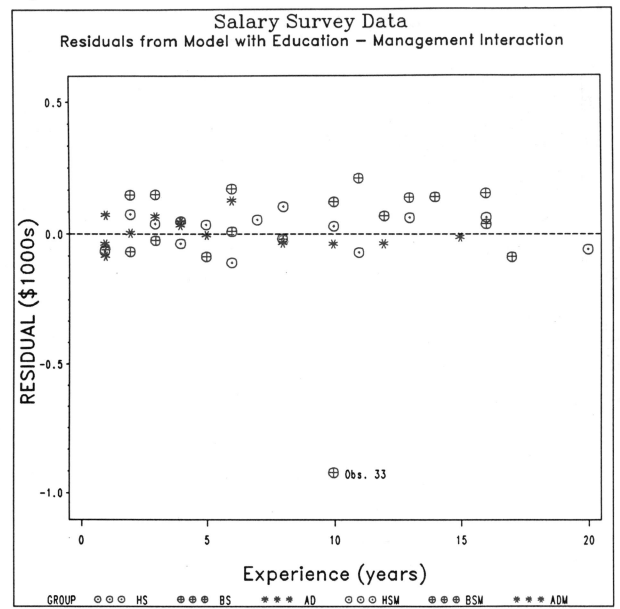

(A black-and-white version of this output, with explanatory text, appears as Output 5.11 in Chapter 5.)

Output A3.23
Illustration of
Leverage and
Influence

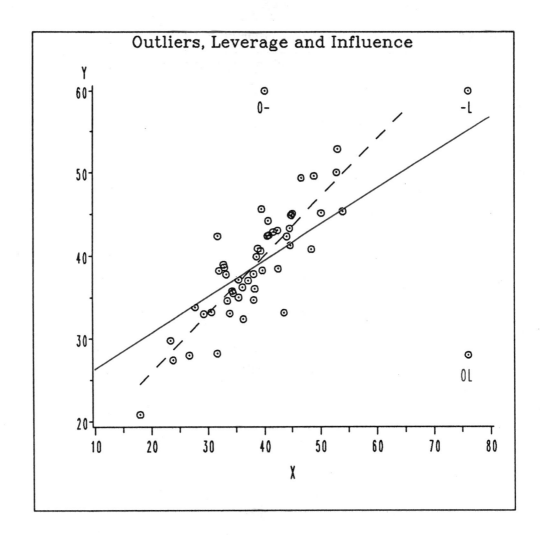

(A black-and-white version of this output, with explanatory text, appears as Output 5.18 in Chapter 5.)

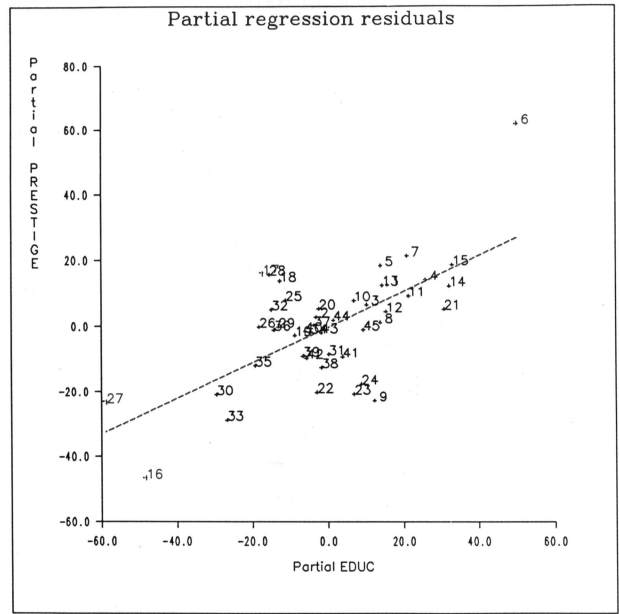

(A black-and-white version of this output, with explanatory text, appears as Output 5.28 in Chapter 5.)

Output A3.25   *Response Surface Plot of Occupational Prestige Data*

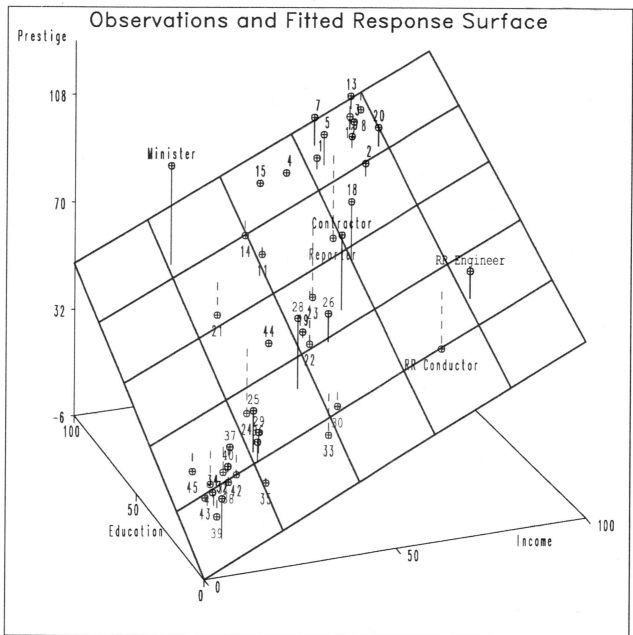

(A black-and-white version of this output, with explanatory text, appears as Output 5.29 in Chapter 5.)

Output A3.26   *Quadratic Response Surface for Occupational Prestige Data*

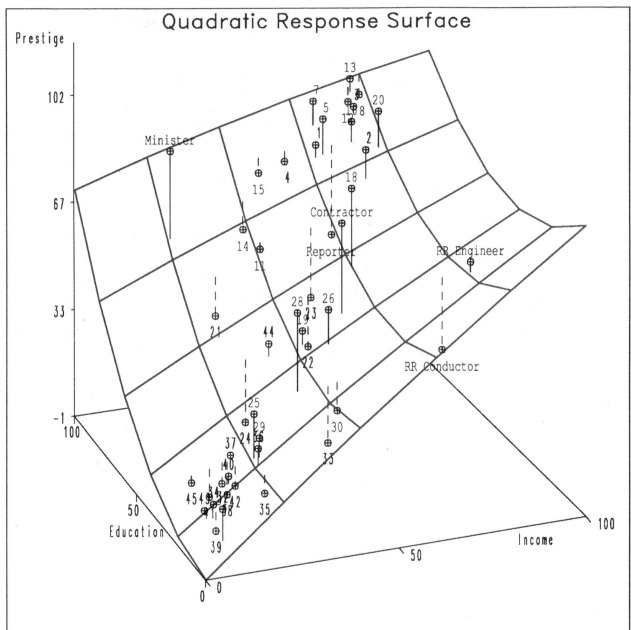

(A black-and-white version of this output, with explanatory text, appears as Output 5.31 in Chapter 5.)

Output A3.27
C_P *Plot for Fuel*
Consumption Data

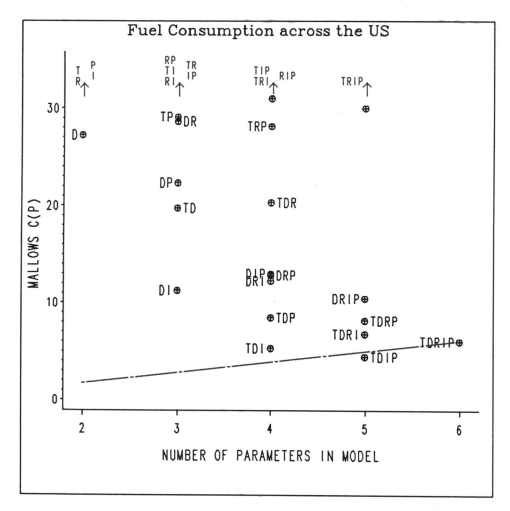

(A black-and-white version of this output, with explanatory text, appears as Output 5.34 in Chapter 5.)

Output A3.28   *Map of U.S. Public School Expenditures in 1970*

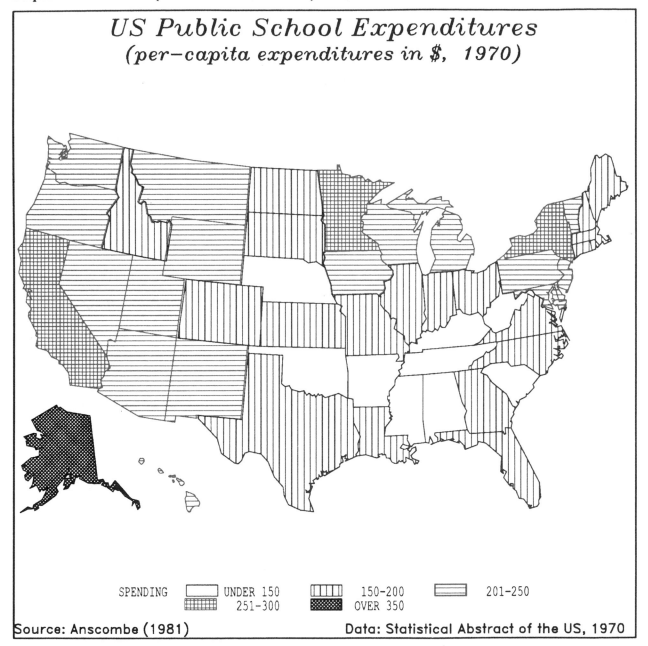

(A black-and-white version of this output, with explanatory text, appears as Output 5.35 in Chapter 5.)

Output A3.29 *Residual Map for Public School Expenditures*

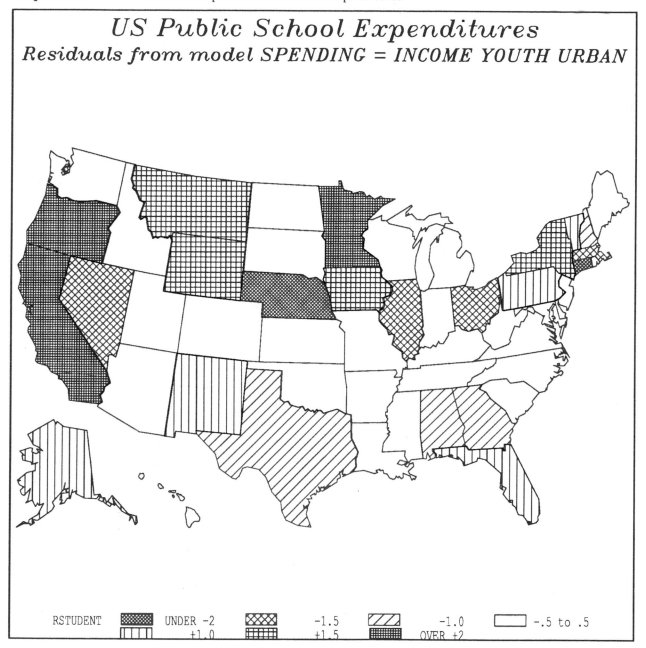

(A black-and-white version of this output, with explanatory text, appears as Output 5.36 in Chapter 5.)

Output A3.30   *Bubble Map of School Spending Residuals*

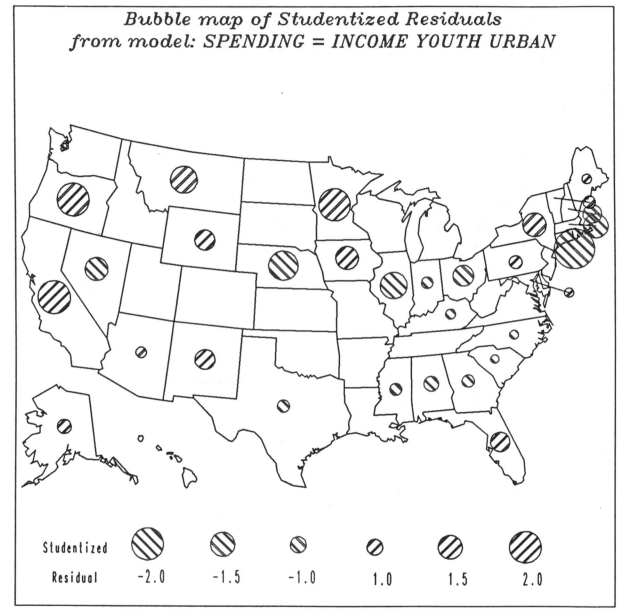

(A black-and-white version of this output, with explanatory text, appears as Output 5.37 in Chapter 5.)

Output A3.31
*Plot of Means and
Standard Errors
for Stern Data*

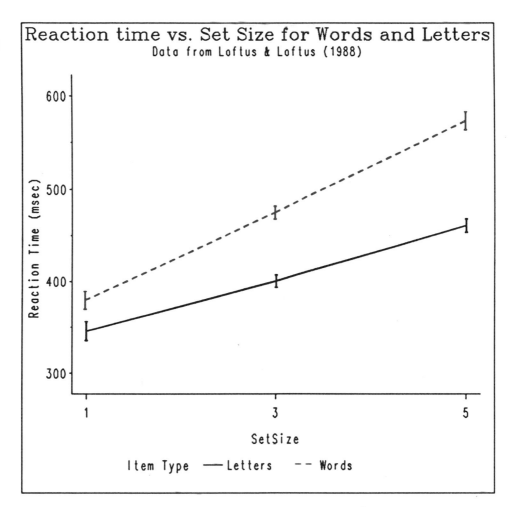

(A black-and-white version of this output, with explanatory text, appears as Output
7.12 in Chapter 7, "Plotting ANOVA Data.")

Output A3.32
Plot of Stern Data,
Showing Means on
a Second Variable

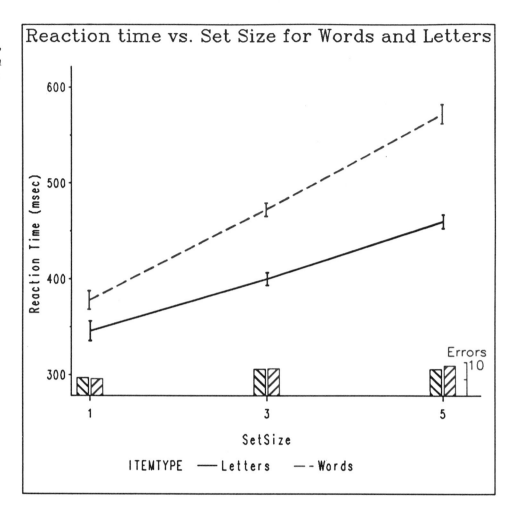

(A black-and-white version of this output, with explanatory text, appears as Output 7.14 in Chapter 7.)

Output A3.33   *Three-Way Plot of A*B Means for Each Level of Factor C*

(A black-and-white version of this output, with explanatory text, appears as Output 7.17 in Chapter 7.)

Output A3.34   *Side-By-Side Display of A*C and B*C Interactions*

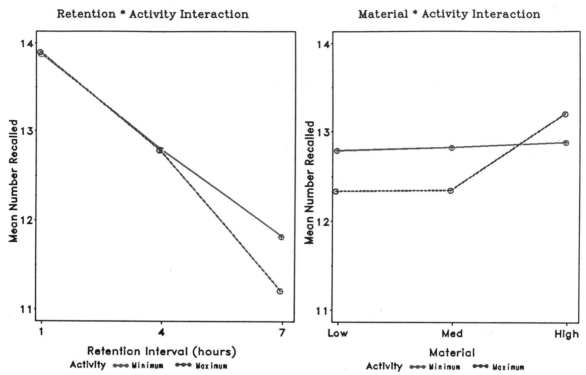

(A black-and-white version of this output, with explanatory text, appears as Output 7.18 in Chapter 7.)

Output A3.35 *Mean Learning Curves for D8 Data*

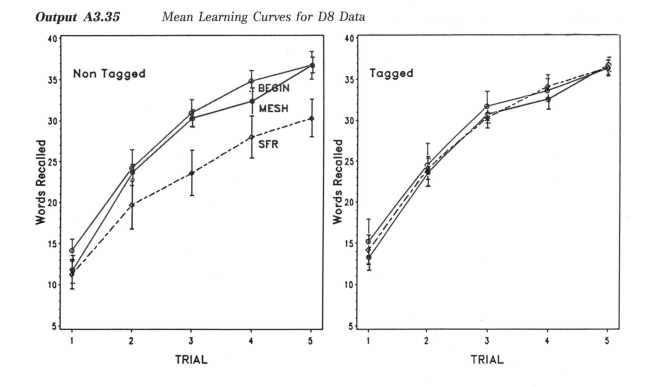

(A black-and-white version of this output, with explanatory text, appears as Output 7.21 in Chapter 7.)

Output A3.36   *Biplot of Sentence Data to Diagnose Model for Non-Additivity*

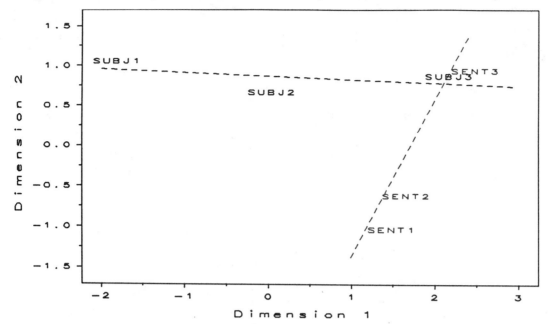

(A black-and-white version of this output, with explanatory text, appears as Output 7.30 in Chapter 7.)

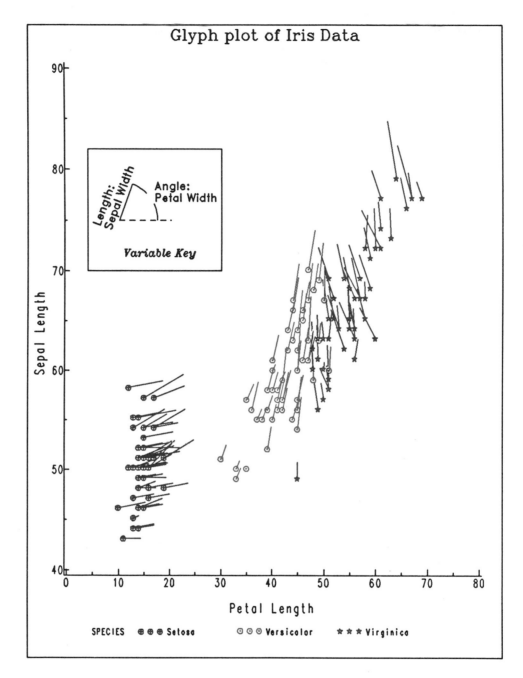

(A black-and-white version of this output, with explanatory text, appears as Output 8.1 in Chapter 8, "Displaying Multivariate Data.")

Output A3.38   *Draftsman's Display of Price, Weight, and Repair Record for Automobiles Data*

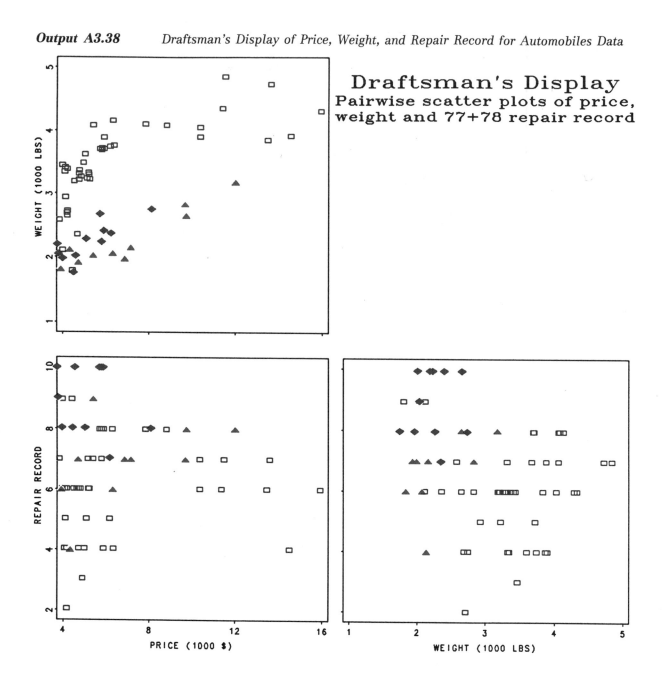

(A black-and-white version of this output, with explanatory text, appears as Output 8.3 in Chapter 8.)

Output A3.39   *Scatterplot Matrix for Automobiles Data*

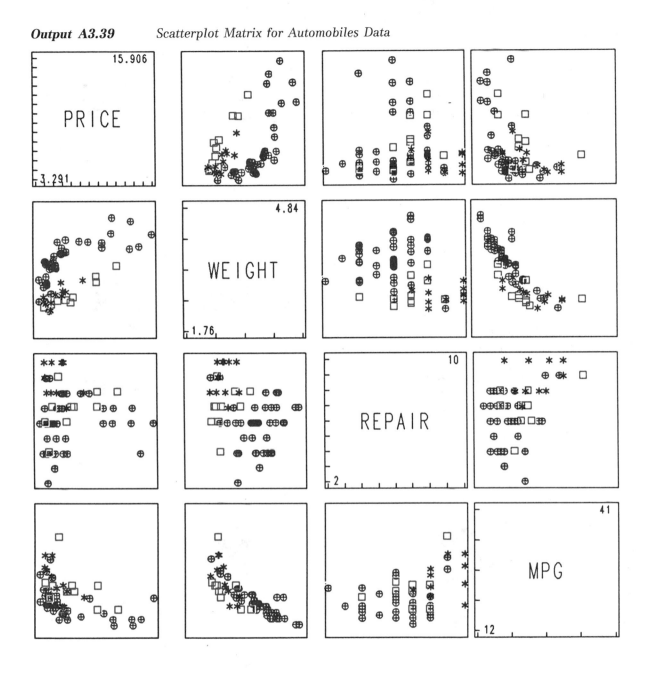

(A black-and-white version of this output, with explanatory text, appears as Output 8.4 in Chapter 8.)

Output A3.40   *Scatterplot Matrix for Iris Data*

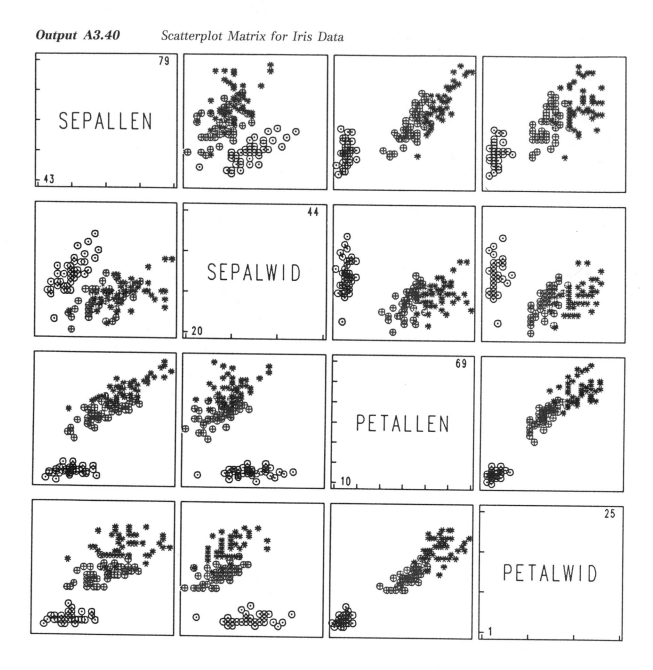

(A black-and-white version of this output, with explanatory text, appears as Output 8.5 in Chapter 8.)

Output A3.41   *Naive Profile Plot of Raw Crime Data*

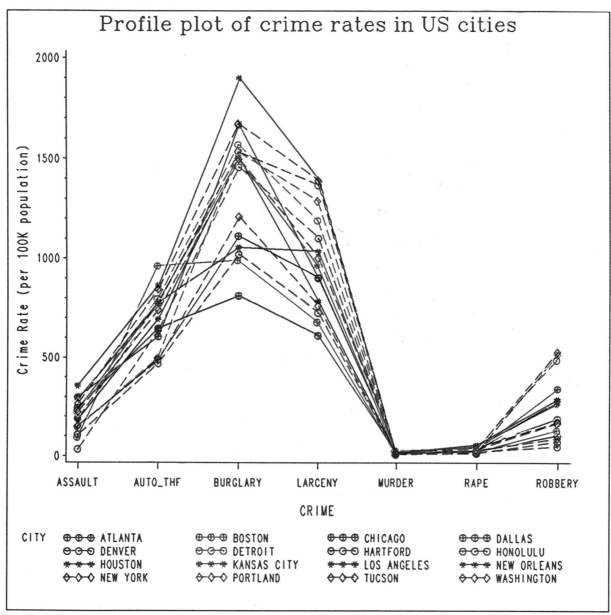

(A black-and-white version of this output, with explanatory text, appears as Output 8.9 in Chapter 8.)

Output A3.42   *Profile Plot of Standardized Crime Data*

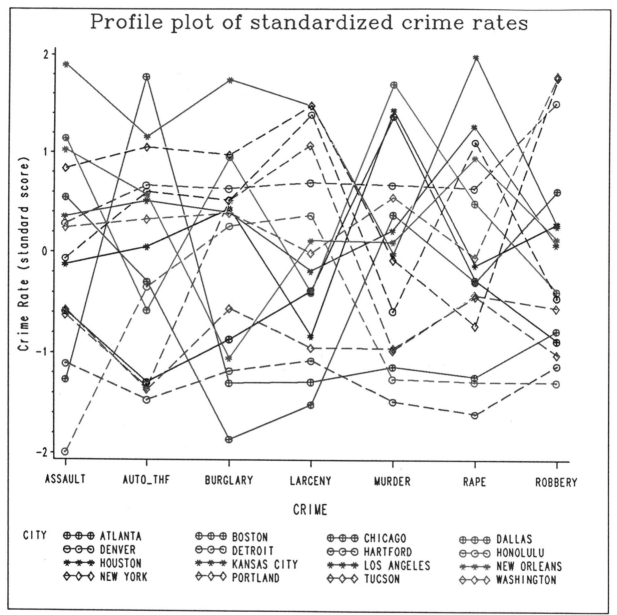

(A black-and-white version of this output, with explanatory text, appears as Output 8.10 in Chapter 8.)

(Output A3.43 and Output A3.44 appear on the following two pages. Black-and-white versions, with explanatory text, appear as Output 8.11 and Output 8.12 in Chapter 8.)

Output A3.43   *Optimal Linear Profiles for Crime Data*

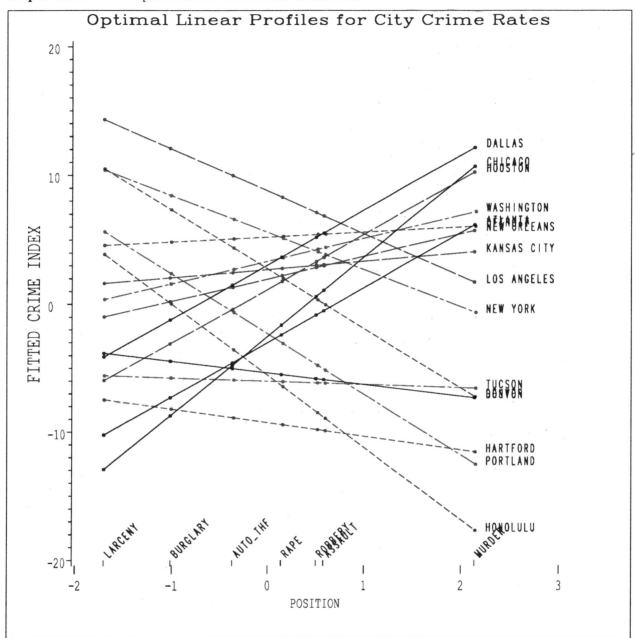

Output A3.44   *Andrews Function Plot for Multivariate Data*

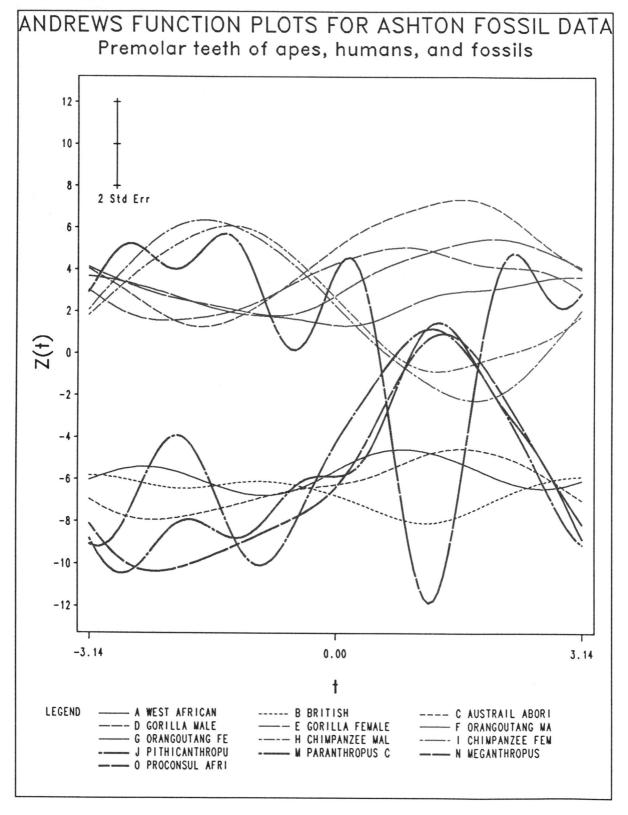

Output A3.45   *Plot of Principal Component Scores on PRIN1 and PRIN2*

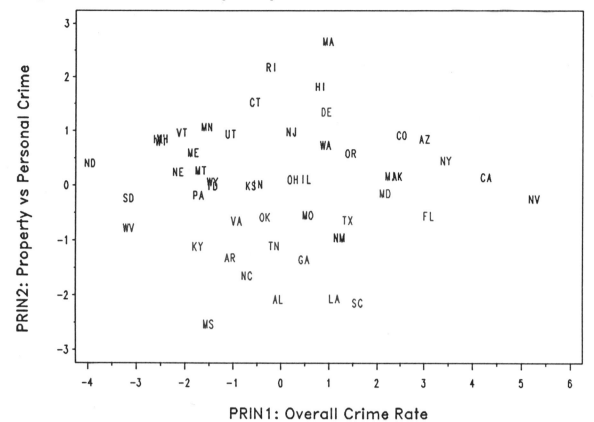

(A black-and-white version of this output, with explanatory text, appears as Output 9.17 in Chapter 9, "Multivariate Statistical Methods.")

Output A3.46 *Plot of Principal Component Scores on PRIN1 and PRIN3*

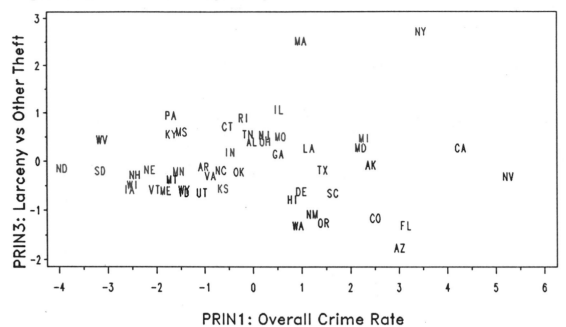

(A black-and-white version of this output, with explanatory text, appears as Output 9.18 in Chapter 9.)

Output A3.47   *Plot of CAN1 and CAN2 for Iris Data*

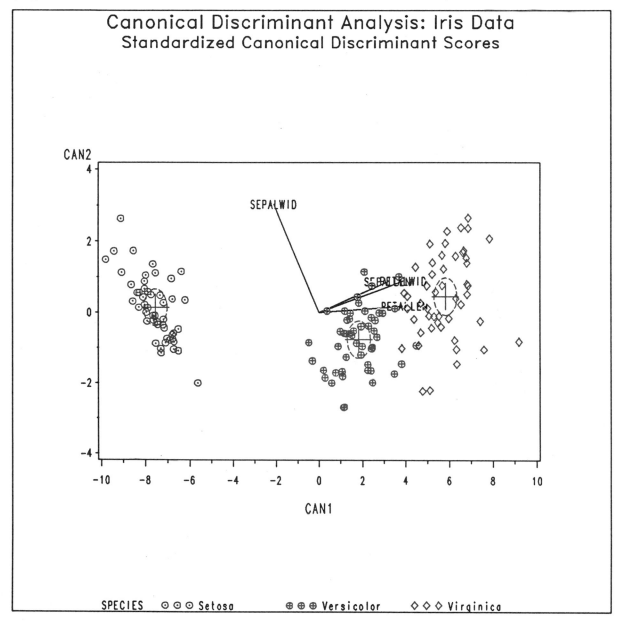

(A black-and-white version of this output, with explanatory text, appears as Output 9.30 in Chapter 9.)

Output A3.48   *Plot of CAN1 and CAN2 for Diabetes Data*

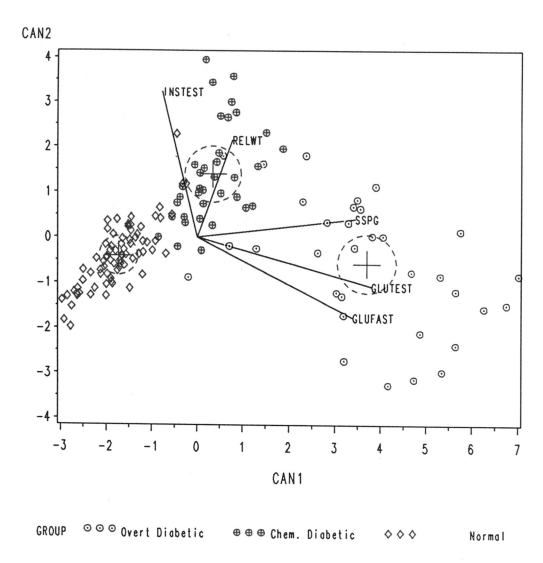

(A black-and-white version of this output, with explanatory text, appears as Output 9.31 in Chapter 9.)

Output A3.49
*Association Plot
for Hair and Eye
Color Data*

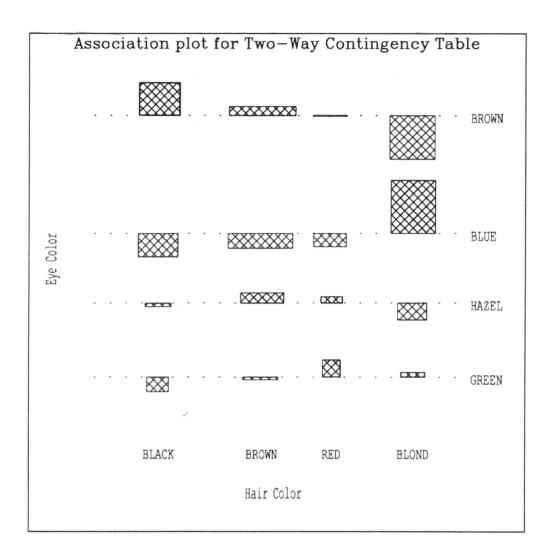

(A black-and-white version of this output, with explanatory text, appears as Output 10.3 in Chapter 10, "Displaying Categorical Data.")

Output A3.50
*Condensed Column
Proportion Mosaic,
Constant Row and
Column Spacing*

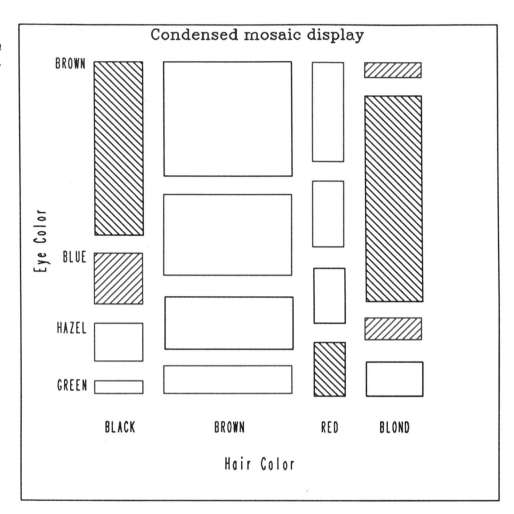

(A black-and-white version of this output, with explanatory text, appears as Output 10.8 in Chapter 10.)

Output A3.51   *Two-Dimensional Correspondence Analysis of Suicide Data*

Suicide Rates by Age, Sex and Method

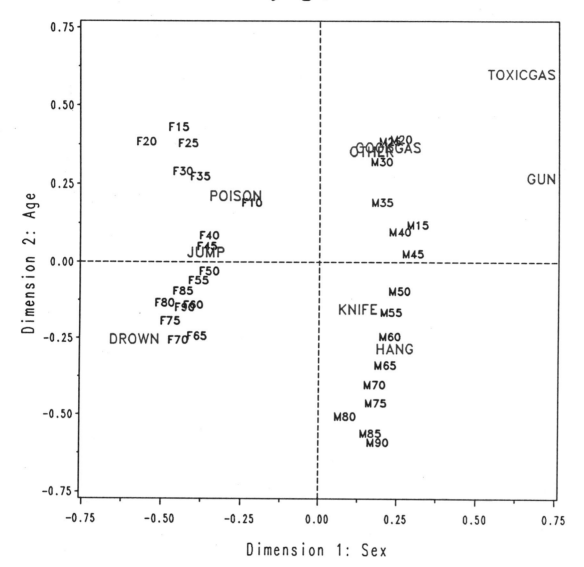

(A black-and-white version of this output, with explanatory text, appears as Output 10.14 in Chapter 10.)

682

References

Anderson, E. (1935), "The Irises of the Gaspé Peninsula," *Bulletin of the American Iris Society*, 35, 2—5.

Andrews, D. F. (1972), "Plots of High Dimensional Data," *Biometrics*, 28, 125—136.

Andrews, D. F. and Herzberg, A. M. (1985), *Data: A Collection of Problems from Many Fields for the Student and Research Worker*, New York: Springer-Verlag.

Andrews, H. P., Snee, R. D., and Sarner, M. H. (1980), "Graphical Display of Means," *The American Statistician*, 34, 195—199.

Anscombe, F. J. (1973), "Graphs in Statistical Analysis," *The American Statistician*, 27, 17—21.

Anscombe, F. J. (1981), *Computing in Statistical Science through APL*, New York: Springer-Verlag.

Ashton, E. H., Healy M. J. R., and Lipton, S. (1975), "The Descriptive Use of Discriminant Functions in Physical Anthropology," *Journal of the Royal Statistical Society*, Series B, 146, 552—572.

Atkinson, A. C. (1981), "Two Graphical Displays for Outlying and Influential Observations in Regression," *Biometrika*, 68, 13—20.

Atkinson, A. C. (1987), *Plots, Transformations and Regression: An Introduction to Graphical Methods of Diagnostic Regression Analysis*, New York: Oxford University Press Inc.

Becker, R. A. and Cleveland, W. S. (1987), "Brushing Scatterplots," *Technometrics*, 29, 127—142.

Becker, R. A., Chambers, J. M., and Wilks, A. R. (1988), *The New S Language*, Pacific Grove, CA: Brooks/Cole Publishing Company.

Belsley, D. A., Kuh, E., and Welsch, R. E. (1980), *Regression Diagnostics: Identifying Influential Data and Sources of Collinearity*, New York: John Wiley & Sons, Inc.

Benoit, P. (1985a), "Box-and-Whisker Plots Using the ANNOTATE Facility," *SAS Communications*, X(3), 34—35.

Benoit, P. (1985b), "Statistical Graphics Made Possible by the SAS/GRAPH ANNOTATE Facility under VMS," *Proceedings of the SAS User's Group International Conference*, 10, 228—234.

Bertin, J. (1983), *Semiology of Graphics*, trans. W. Berg, Madison, WI: University of Wisconsin Press.

Blettner, V. (1988), "Creating Custom Symbols with the GFONT Procedure," *Proceedings of the SAS User's Group International Conference*, 13, 705—709.

Blom, G. (1958), *Statistical Estimates and Transformed Beta Variables*, New York: John Wiley & Sons, Inc.

Bolorforoush, M. and Wegman, E. J. (1988), "On Some Graphical Representations of Multivariate Data," *Computing Science and Statistics: Proceedings of the 20th Symposium on the Interface*, 121—126.

Bowerman, B. L. and O'Connell, R. T. (1990), *Linear Statistical Models*, Second Edition, Boston, MA: PWS/Kent Publishing Company.

Bradu, D. and Gabriel, R. K. (1978), "The Biplot as a Diagnostic Tool for Models of Two-Way Tables," *Technometrics*, 20, 47—68.

Brown, R. L., Durban, J., and Evans, J. M. (1975), "Techniques for Testing the Constancy of Regression Relationships," *Journal of the Royal Statistical Society*, Series B, 37, 149—163.

Bruntz, S. M., Cleveland, W. S., Kleiner, B., and Warner, J. L. (1974), "The Dependence of Ambient Ozone on Solar Radiation, Wind, Temperature, and Mixing Height," in *Symposium on Atmospheric Diffusion and Air Pollution*, Boston: American Meteorological Society, 125—128.

Cary, A. J. L. (1983), "SAS Macros for *F*-test Power Computations in Balanced Experimental Designs," *Proceedings of the SAS User's Group International Conference*, 8, 671—676.

Chambers, J. M., Cleveland, W. S., Kleiner, B., and Tukey, P. A. (1983), *Graphical Methods for Data Analysis*, Belmont, CA: Wadsworth Publishing Company.

Chatterjee, S. and Price, B. (1977), *Regression Analysis by Example*, New York: John Wiley & Sons, Inc.

Cleveland, W. S. (1979), "Robust Locally Weighted Regression and Smoothing Scatterplots," *Journal of the American Statistical Association*, 74, 829—836.

Cleveland, W. S. (1984a), "Graphs in Scientific Publications," *The American Statistician*, 38, 261—269.

Cleveland, W. S. (1984b), "Graphical Methods for Data Presentation: Full Scale Breaks, Dot Charts, and Multibased Logging," *The American Statistician*, 38, 270—280.

Cleveland, W. S. (1985), *The Elements of Graphing Data*, Monterey, CA: Wadsworth Advanced Books and Software.

Cleveland, W. S. and Devlin, S. J. (1988), "Locally Weighted Regression: An Approach to Regression Analysis by Local Fitting," *Journal of the American Statistical Association*, 83, 596—610.

Cleveland, W. S. and McGill, R. (1984a), "Graphical Perception: Theory, Experimentation and Application to the Development of Graphical Methods," *Journal of the American Statistical Association*, 79, 531—554.

Cleveland, W. S. and McGill, R. (1984b), "The Many Faces of a Scatterplot," *Journal of the American Statistical Association*, 79, 807—822.

Cleveland, W. S. and McGill, R. (1985), "Graphical Perception and Graphical Methods for Analyzing Scientific Data," *Science*, 229, 828—833.

Cleveland, W. S. and McGill, R. (1986), "An Experiment in Graphical Perception," *International Journal of Man-Machine Studies*, 25, 491—500.

Cleveland, W. S. and McGill, R. (1987), "Graphical Perception: The Visual Decoding of Quantitative Information on Graphical Displays of Data (with discussion), *Journal of the Royal Statistical Society*, Series A, 150, 192—229.

Cleveland, W. S., Devlin, S. J., and Grosse, E. (1988), "Regression by Local Fitting: Methods, Properties, and Computational Algorithms," *Journal of Econometrics*, 37, 87—114.

Cleveland, W. S., Kettenring, J. R., and McGill, R. (1976), "Graphical Methods in Multivariate Analysis and Regression," paper presented at the meeting of the American Public Health Association.

Cohen, J. (1977), *Statistical Power Analysis for the Behavioral Sciences*, New York: Academic Press, Inc.

Cook, R. D. and Weisberg, S. (1982), *Residuals and Influence in Regression*, New York: Chapman and Hall.

Daniel, C. (1959), "Use of Half-Normal Plots in Interpreting Two-Level Factorial Experiments," *Technometrics*, 1, 311—341.

Duncan, O. D. (1961), "A Socioeconomic Index for All Occupations," in *Occupations and Social Status*, eds. A. J. Reiss, Jr., O. D. Duncan, P. K. Hatt, and C. C North, New York: The Free Press, 109—138.

Dunn, R. (1987), "Variable-Width Framed Rectangle Charts for Statistical Mapping," *The American Statistician*, 41, 153—156.

Emerson, J. D. (1983), "Mathematical Aspects of Transformation," in *Understanding Robust and Exploratory Data Analysis*, eds. D. C. Hoaglin, F. Mosteller, and J. W. Tukey, New York: John Wiley & Sons, Inc., 247–282.

Emerson, J. D. and Hoaglin, D. C. (1983), "Stem-and-Leaf Displays," in *Understanding Robust and Exploratory Data Analysis*, eds. D. C. Hoaglin, F. Mosteller, and J. W. Tukey, New York: John Wiley & Sons, Inc., 7–32.

Emerson, J. D. and Stoto, M. A. (1982), "Exploratory Methods for Choosing Power Transformations," *Journal of the American Statistical Association*, 77, 103–108.

Fienberg, S. E. (1971), "Randomization and Social Affairs: The 1970 Draft Lottery," *Science*, 171, 255–261.

Fisher, R. A. (1925), *Statistical Methods for Research Workers*, London: Oliver & Boyd.

Fisher, R. A. (1936), "The Use of Multiple Measurements in Taxonomic Problems," *Annals of Eugenics*, 8, 379–388.

Flores, F. and Flack, V. F. (1990), "Program to Generate Atkinson's and Resistant Envelopes for Normal Probability Plots of Regression Residuals," *Proceedings of the SAS User's Group International Conference*, 15, 1345–1352.

Fox, J. (1984), *Linear Statistical Models and Related Methods*, New York: John Wiley & Sons, Inc.

Fox, J. (1987), "Effect Displays for Generalized Linear Models," in *Sociological Methodology*, ed C. C. Clogg, San Francisco: Jossey-Bass. 347–361.

Fox, J. (1988), "Graphical and Exploratory Methods of Data Analysis," course notes, Institute for Survey Research, York University.

Fox, J. (1990), "Describing Univariate Distributions," in *Modern Methods of Data Analysis*, eds. J. Fox and S. Long, Beverly Hills, CA: Sage Publications, 58–125.

Fox, J. (1991), *Regression Diagnostics*, Beverly Hills, CA: Sage Publications.

Freund, R. J. and Littell, R. C. (1991), *SAS System for Regression, Second Edition*, Cary, NC: SAS Institute Inc.

Freund, R. J., Littell, R. C., and Spector, P. C. (1991), *SAS System for Linear Models, Third Edition*, Cary, NC: SAS Institute Inc.

Friedman, H. P., Farrell, E. S., Goldwyn, R. M., Miller, M., and Sigel, J. (1972), "A Graphic Way of Describing Changing Multivariate Patterns," *Proceedings of the Sixth Interface Symposium on Computer Science and Statistics*, Berkeley, CA: University of California Press, 56–59.

Friendly, M. L. and Franklin, P. (1980), "Interactive Presentation in Multitrial Free Recall," *Memory and Cognition*, 8, 265–270.

Gabriel, K. R. (1971), "The Biplot Graphic Display of Matrices with Application to Principal Components Analysis," *Biometrics*, 58(3), 453–467.

Gabriel, K. R. (1978), "A Simple Method of Multiple Comparisons of Means," *Journal of the American Statistical Association*, 73, 724–729.

Gabriel, K. R. (1980), "Biplot," in *Encyclopedia of Statistical Sciences*, Volume 1, eds. N. L. Johnson and S. Kotz, New York: John Wiley & Sons, Inc., 263–271.

Gabriel, K. R. (1981), "Biplot Display of Multivariate Matrices for Inspection of Data and Diagnosis," in *Interpreting Multivariate Data*, ed. V. Barnett, London: John Wiley & Sons, Inc., 147–173.

Gnanadesikan, R. (1977), *Methods for Statistical Data Analysis of Multivariate Observations*, New York: John Wiley & Sons, Inc.

Gnanadesikan, R. and Kettenring, J. R. (1972), "Robust Estimates, Residuals, and Outlier Detection with Multiresponse Data," *Biometrics*, 28, 81–124.

Goodall, C. (1990), "A Survey of Smoothing Techniques," in *Modern Methods of Data Analysis*, eds. J. Fox and J. S. Long, Beverly Hills, CA: Sage Publications, 126—176.

Greenacre, M. (1984), *Theory and Applications of Correspondence Analysis*, London: Academic Press, Inc.

Greenacre, M. and Hastie, T. (1987), "The Geometric Interpretation of Correspondence Analysis," *Journal of the American Statistical Association*, 82, 437—447.

Hamming, R. W. (1977), *Digital Filters*, Englewood Cliffs, NJ: Prentice Hall, Inc.

Harris, A. H. (1972), "Time-Distance Records and the Power Function," *Perception & Psychophysics*, 12, 289—290.

Hartigan, J. A. (1975a), *Clustering Algorithms*, New York: John Wiley & Sons, Inc.

Hartigan, J. A. (1975b), "Printer Graphics for Clustering," *Journal of Statistical Computing and Simulation*, 4, 187—213.

Hartigan, J. A. and Kleiner, B. (1981), "Mosaics for Contingency Tables," *Computer Science and Statistics: Proceedings of the 13th Symposium on the Interface*, New York: Springer-Verlag, 268—273.

Hartigan, J. A. and Kleiner, B. (1984), "A Mosaic of Television Ratings," *The American Statistician*, 38, 32—35.

Heijden, P. G. M. van der and de Leeuw, J. (1985), "Correspondence Analysis Used Complementary to Loglinear Analysis," *Psychometrika*, 50, 429—447.

Heuer, J. (1979), *Selbstmord bei Kinder und Jugendlichen* (Suicide by children and youth), Stuttgard: Ernst Klett Verlag.

Hoaglin, D. C. (1985a), "Using Quantiles to Study Shape," in *Exploring Data Tables, Trends and Shapes*, eds. D. C. Hoaglin, F. Mosteller, and J. W. Tukey, New York: John Wiley & Sons, Inc., 417—460.

Hoaglin, D. C. (1985b), "Summarizing Shape Numerically: The *g*-and-*h* Distributions," in *Exploring Data Tables, Trends and Shapes*, eds. D. C. Hoaglin, F. Mosteller, and J. W. Tukey, New York: John Wiley & Sons, Inc., 461—513.

Hoaglin, D. C., Mosteller, F., and Tukey, J. W., eds. (1983), *Understanding Robust and Exploratory Data Analysis*, New York: John Wiley & Sons, Inc.

Hoaglin, D. C., Mosteller, F., and Tukey, J. W., eds. (1985), *Exploring Data Tables, Trends and Shapes*, New York: John Wiley & Sons, Inc.

Hochberg, Y., Weiss, G., and Hart, S. (1982), "On Graphical Procedures for Multiple Comparisons," *Journal of the American Statistical Association*, 77, 767—772.

Householder, A. S. and Young, G. (1938), "Matrix Approximation and Latent Roots," *American Mathematical Monthly*, 45, 165—171.

Inselberg, A. (1985), "The Plane with Parallel Coordinates," *The Visual Computer*, 1, 69—91.

Inselberg, A. and Dinsmore, B. (1988), "Visualizing Multi-Dimensional Geometry with Parallel Coordinates," *Computing Science and Statistics: Proceedings of the 20th Symposium on the Interface*, 115—120.

Johnson, E. G. and Tukey, J. W. (1987), "Graphical Exploratory Analysis of Variance Illustrated on a Splitting of the Johnson and Tsao Data," in *Design, Data, and Analysis*, ed. C. L. Mallows, New York: John Wiley & Sons, Inc., 171—244.

Johnson, R. A. and Wichern, D. W. (1982), *Applied Multivariate Statistical Analysis*, Englewood Cliffs, NJ: Prentice Hall, Inc.

Keppel, G. (1973), *Design and Analysis: A Researcher's Handbook*, Englewood Cliffs, NJ: Prentice Hall, Inc.

Kosslyn, S. M. (1985), "Graphics and Human Information Processing: A Review of Five Books," *Journal of the American Statistical Association*, 80, 499—512.

Kosslyn, S. M. (1989), "Understanding Charts and Graphs," *Applied Cognitive Psychology*, 3, 185—226.

Kramer, C. Y. (1956), "Extension of Multiple Range Tests to Group Means with Unequal Numbers of Replications," *Biometrics*, 12, 307—310.

Kuhfeld, W. F. (1986), "Metric and Nonmetric Plotting Models," *Psychometrika*, 51, 155—161.

Kurian, G. T. (1984), *The New Book of World Rankings*, New York: Facts on File, Inc.

Larsen, W. A. and McCleary, S. J. (1972), "The Use of Partial Residual Plots in Regression Analysis," *Technometrics*, 14, 781—790.

Lebart, L., Morineau, A., and Tabard, N. (1977), *Techniques de la Description Statistique*, Paris: Dunod.

Leinhardt, S. and Wasserman, S. S., (1979), "Exploratory Data Analysis: An Introduction to Selected Methods," in *Sociological Methodology*, ed. K. F. Schuessler, San Francisco: Jossey-Bass, 311—365.

Lewandowsky, S. and Spence, I. (1989), "The Perception of Statistical Graphs," in *Sociological Methods & Research*, 18, 200—242.

Loftus, G. R. and Loftus, E. F. (1988), *Essence of Statistics*, New York: Alfred A. Knopf, Inc.

Lohr, V. I. and O'Brien, R. G. (1984), "Power Analysis for Univariate Linear Models: The SAS System Makes it Easy," *Proceedings of the SAS User's Group International Conference*, 9, 847—852.

Longbotham, C. R. (1987), "Nonparametric Density Estimation," *Proceedings of the SAS User's Group International Conference*, 12, 907—909.

Mallows, C. L. (1973), "Some Comments on C_p," *Technometrics*, 15, 661—675.

Mandel, J. (1961), "Non-Additivity in Two-Way Analysis of Variance," *Journal of the American Statistical Association*, 56, 878—888.

Mandel, J. (1969), "A Method for Fitting Empirical Surfaces to Physical or Chemical Data," *Technometrics*, 11, 411—429.

Mariam, S. G. and Griffin, D. W. (1986), "One Degree of Freedom Test for Transformable Nonadditivity - A Generalized SAS Macro," *Proceedings of the SAS User's Group International Conference*, 11, 675—682.

McGill, R., Tukey, J. W., and Larsen, W. (1978), "Variations of Box Plots," *The American Statistician*, 32, 12—16.

McWhirter, N. and McWhirter, R. (1973), *Guinness Book of World Records*, New York: Bantam Books.

Monette, G. (1990), "Geometry of Multiple Regression and Interactive 3-D Graphics," in *Modern Methods of Data Analysis*, eds. J. Fox and S. Long, Beverly Hills, CA: Sage Publications, 209—256.

Monlezun, C. J. (1979), "Two-Dimensional Plots for Interpreting Interactions in the Three-Factor Analysis of Variance Model," *The American Statistician*, 33, 63—69.

Morrison, D. F. (1976), *Multivariate Statistical Methods* (2nd ed.; 3rd ed. is 1990), New York: McGraw-Hill, Inc.

Muhlbaier, L. H. (1987), "JITTERing and Other Graphics Macros for Exploratory Data Analysis," *Proceedings of the SAS User's Group International Conference*, 12, 916—919.

Neter, J., Wasserman, W., and Kutner, M. H. (1985), *Applied Linear Statistical Models*, Second Edition, Homewood, IL: Richard D. Irwin, Inc.

Nicholson, W. L. and Littlefield, R. J. (1983), "Interactive Color Graphics for Multivariate Data," *Computer Science and Statistics: Proceedings of the 14th Symposium on the Interface*, eds. K. W. Heiner, R. S. Sacher, and J. W. Wilkinson, New York: Springer-Verlag, 211—219.

Nishisato, S. (1980), *Analysis of Categorical Data: Dual Scaling and its Applications*, Toronto: University of Toronto Press.

Noma, E. (1987), "Heuristic Method for Label Placement in Scatterplots," *Psychometrika*, 52, 463—468.

O'Brien, R. G. and Lohr, V. I. (1984), "Power Analysis for Linear Models: The Time Has Come," *Proceedings of the SAS User's Group International Conference*, 9, 840—846.

Olmstead, A. (1985), "Box Plots Using SAS/Graph Software," *Proceedings of the SAS User's Group International Conference*, 10, 888—894.

Rawlings, J. O. (1988), *Applied Regression Analysis: A Research Tool*, Pacific Grove, CA: Brooks/Cole Publishing Company.

Reaven, G. M. and Miller, R. G. (1979), "An Attempt to Define the Nature of Chemical Diabetes Using a Multidimensional Analysis," *Diabetologia*, 16, 17—24.

Rosenblatt, M. (1956), "Remarks on Some Non-Parametric Estimates of a Density Function," *Annals of Mathematical Statistics*, 25, 166—172.

SAS Institute Inc. (1983), *SUGI Supplemental Library User's Guide, 1983 Edition*, Cary, NC: SAS Institute Inc.

SAS Institute Inc. (1985), *SAS/GRAPH User's Guide, Version 5 Edition*, Cary, NC: SAS Institute Inc.

SAS Institute Inc. (1985), *SAS/IML User's Guide, Version 5 Edition*, Cary, NC: SAS Institute Inc.

SAS Institute Inc. (1985), *SAS User's Guide: Basics, Version 5 Edition*, Cary, NC: SAS Institute Inc.

SAS Institute Inc. (1985), *SAS User's Guide: Statistics, Version 5 Edition*, Cary, NC: SAS Institute Inc.

SAS Institute Inc. (1986), *SAS/QC User's Guide, Version 5 Edition*, Cary, NC: SAS Institute Inc.

SAS Institute Inc. (1986), *SUGI Supplemental Library User's Guide, Version 5 Edition*, Cary, NC: SAS Institute Inc.

SAS Institute Inc. (1988), *SAS/GRAPH User's Guide, Release 6.03 Edition*, Cary, NC: SAS Institute Inc.

SAS Institute Inc. (1989), *Multivariate Statistical Methods: Practical Applications Course Notes*, Cary, NC: SAS Institute Inc.

SAS Institute Inc. (1989), *SAS/QC Software: Reference, Version 6, First Edition*, Cary, NC: SAS Institute Inc.

SAS Institute Inc. (1989), SAS Technical Report P-188, *SAS/QC Software Examples, Version 6*, Cary, NC: SAS Institute Inc.

SAS Institute Inc. (1990), *SAS/GRAPH Software: Reference, Version 6, Volume 1* and *Volume 2*, Cary, NC: SAS Institute Inc.

SAS Institute Inc. (1990), *SAS Procedures Guide, Version 6, Third Edition*, Cary, NC: SAS Institute Inc.

SAS Institute Inc. (1990), *SAS/STAT User's Guide, Version 6, Fourth Edition*, Cary, NC: SAS Institute Inc.

Schmid, C. F. (1983), *Statistical Graphics: Design Principles and Practices*, New York: John Wiley & Sons, Inc.

Schmid, C. F. and Schmid, S. E. (1979), *Handbook of Graphic Presentation*, New York: John Wiley & Sons, Inc.

Seber, G. A. F. (1984), *Multivariate Observations*, New York: John Wiley & Sons, Inc.

Silverman, B. W. (1982), "Kernel Density Estimation Using the Fast Fourier Transform," *Applied Statistics*, 31, 93—99.

Silverman, B. W. (1986), *Density Estimation for Statistics and Data Analysis*, New York: Chapman & Hall.

Simkin, D. and Hastie, R. (1987), "An Information-Processing Analysis of Graph Perception," *Journal of the American Statistical Association*, 82, 454—465.

Snee, R. D. (1974), "Graphical Display of Two-Way Contingency Tables," *The American Statistician*, 28, 9—12.

Spjøtvoll, E. (1977), "Alternatives to Plotting C_p in Multiple Regression," *Biometrika*, 64, 1—8.

Stevens, J. (1986), *Applied Multivariate Statistics for the Social Sciences*, Hillsdale, NJ: Lawrence Erlbaum Associates, Inc.

Strunk, W., Jr. and White, E. B. (1979), *The Elements of Style*, Third Edition, New York: Macmillan Publishing Company, Inc.

Tabachnick, B. G. and Fidell, L. S. (1989), *Using Multivariate Statistics*, Second Edition, New York: Harper & Row Pubs., Inc.

Terrell, G. R. and Scott, D. W. (1985), "Oversmoothed Nonparametric Density Estimates," *Journal of the American Statistical Association*, 80, 209—214.

Timm, N. H. (1975), *Multivariate Analysis with Applications in Education and Psychology*, Belmont, CA: Brooks/Cole Publishing Company.

Tufte, E. R. (1983), *The Visual Display of Quantitative Information*, Cheshire, CT: Graphics Press.

Tukey, J. W. (1949), "One Degree of Freedom for Nonadditivity," *Biometrics*, 5, 232—242.

Tukey, J. W. (1953), *The Problem of Multiple Comparisons*, unpublished manuscript, Princeton University.

Tukey, J. W. (1977), *Exploratory Data Analysis*, Reading, MA: Addison Wesley Publishing Co., Inc.

Tukey, J. W. and Tukey, P. A. (1983), "Some Graphics for Studying Four-Dimensional Data," *Computer Science and Statistics: Proceedings of the 14th Symposium on the Interface*, eds. K. W. Heiner, R. S. Sacher, and J. W. Wilkinson, New York: Springer-Verlag, 60—66.

Tukey, P. A. and Tukey, J. W. (1981), "Graphical Display of Data Sets in 3 or More Dimensions," in *Interpreting Multivariate Data*, ed. V. Barnett, Chichester, U.K.: John Wiley & Sons, Inc., 189—275.

Velleman, P. F. and Hoaglin, D. C. (1981), *Applications, Basics, and Computing of Exploratory Data Analysis*, North Situate, MA: Duxbury Press.

Velleman, P. F. and Welsch, R. E. (1981), "Efficient Computing of Regression Diagnostics," *The American Statistician*, 35, 234—242.

Wainer, H. (1984), "How to Display Data Badly," *The American Statistician*, 38, 137—147.

Wainer, H. and Thissen, D. (1981), "Graphical Data Analysis," *Annual Review of Psychology*, 32, 191—241.

Wakimoto, K. and Taguri, M. (1978), "Constellation Graphical Method for Representing Multidimensional Data," *Annals of the Institute for Statistical Mathematics*, 30 (Part A), 97—104.

Wang, C. M. (1985), "Applications and Computing of Mosaics," *Computational Statistics & Data Analysis*, 3, 89—97.

Weisberg, S. (1985), *Applied Linear Regression*, Second Edition, New York: John Wiley & Sons, Inc.

Winer, B. J. (1971), *Statistical Principles in Experimental Design*, Second Edition, New York: McGraw-Hill Book Co.

Index

Your Turn

If you have comments or suggestions about *SAS System for Statistical Graphics, First Edition*, please send them to us on a photocopy of this page or send us electronic mail.

For comments about this book, please return the photocopy to

> SAS Institute Inc.
> Publications Division
> SAS Campus Drive
> Cary, NC 27513
> **e-mail:** yourturn@unx.sas.com

For suggestions about the software, please return the photocopy to

> SAS Institute Inc.
> Technical Support Division
> SAS Campus Drive
> Cary, NC 27513
> **e-mail:** suggest@unx.sas.com